THE FUTURE OF SOLAR SYSTEM EXPLORATION, 2003-2013
Community Contributions to the
NRC Solar System Exploration Decadal Survey

COVER ILLUSTRATION:

Astronauts Exploring a Near-Earth Asteroid
by William K. Hartmann.

The high public interest in solar system exploration is inspired in part by the idea that we are destined one day to live and work there. It is a reason that many of the authors of this book were inspired to become planetary scientists. This art depicts our first steps beyond the Earth-Moon system (upper right). It is a scenario that could very well occur in the next decade, according to Jones et al. (this volume, pg. 141).

A SERIES OF BOOKS ON RECENT DEVELOPMENTS IN ASTRONOMY AND ASTROPHYSICS

Publisher

THE ASTRONOMICAL SOCIETY OF THE PACIFIC
390 Ashton Avenue, San Francisco, California, USA 94112-1722
Phone: (415) 337-1100 **E-Mail: catalog@astrosociety.org**
Fax: (415) 337-5205 **Web Site: www.astrosociety.org**

A listing of all other ASP Conference Series Volumes and IAU Volumes published by the ASP is cited at the back of this volume

ASTRONOMICAL SOCIETY OF THE PACIFIC
CONFERENCE SERIES

Volume 272

THE FUTURE OF SOLAR EXPLORATION, 2003-2013
Community Contributions to the NRC Solar System Exploration Decadal Survey

Between June 2001 and January 2002, planetary scientists from around the United States and other nations created 24 self-selected panels to discuss issues facing solar system exploration in 2003-2013 and how those issues should be addressed. Each panel generated a white paper, containing their conclusions and recommendations. These were forwarded to the National Research Council as community input in support of the Solar System Exploration Decadal Survey and comprise the chapters of this book.

Edited by

Mark V. Sykes
Steward Observatory, University of Arizona, Tucson, USA

Library of Congress Cataloging in Publication Data
Main entry under title

Card Number: 2002106589
ISBN: 1-58381-113-3

ASP Conference Series - First Edition

Printed in United States of America by Sheridan Books, Chelsea, Michigan

CONTENTS

Foreword

In early 2001, Dr. Edward Weiler, Associate Administrator for Space Science at NASA, asked the National Research Council to conduct a survey to determine the nature and status of contemporary Solar System Exploration (SSE), find out why it remains compelling today, and to construct a prioritized list of future flight missions and supporting ground-based activities. The survey was conceived on the pattern of a highly successful series of decadal surveys of Astronomy and Astrophysics sponsored by various government agencies done over the past forty years. The hallmark of these surveys was wide community involvement and the development of community consensus regarding prioritization of major initiatives.

This was a novel idea. Previous strategic planning documents for SSE had been done through the auspices of the NRC's Space Studies Board and its Committee on Lunar and Planetary Exploration with but modest community involvement. As a result of this lack of involvement, new starts and delays for planetary missions often became highly controversial and community action polarized and vociferous. The challenge for this initial Solar System Exploration Decadal Survey was to bring the knowledge and ideas of the exceptionally diverse SSE community together for the first time in a convincing way, generate a system of priorities that would represent a practical and harmonious consensus, and do all of this in less than half the time typically allotted to surveys of this kind.

This challenge has, I believe, been met. Not only as the result of the work of the survey Steering Committee and its five panels, but also as a result of the hard and timely work of the ad hoc community groups who are the authors of the chapters in this book. All of the chapters in this book were submitted to the Survey by the end of 2001 in time for major impact on the deliberations of the survey panels and the Steering Committee. In a real sense the recommendations of the survey and its system of priorities, still embargoed at the time of this writing by the NRC pending detailed peer review, are built, not just on the work of the panels but, through them, on the materials to be found in this book.

It is not often possible to record a sense of appreciation and gratitude for the work done by an entire community, however, this foreword gives me such an opportunity. And, on behalf of the SSE Decadal survey and its panels I would like to express my thanks to the authors and editor of this book for producing in these chapters part of the foundation for the survey that should determine much of the exploration of the solar system that will be accomplished over the next decade.

Michael Belton
Chair,
NRC Solar System Exploration Decadal Survey
Tucson, Arizona
23 April 2002

Preface

Decadal surveys identify and prioritize scientific issues and the means of addressing them. For the past forty years they have shaped the activities and planning of our national agencies in the funding of astronomy and astrophysics. They have carried substantial weight with Congress, because they have been held up as the consensus recommendations of the American astronomical community. In that time, modern planetary sciences was born and grew with the Space Age. Now we have concluded our first community-wide assessment of where we should be going in the future.

The Solar System Exploration Decadal Survey has been a uniquely democratic effort. The planetary professional societies have worked to support the National Research Council decadal panels by giving every planetary scientist an opportunity to provide substantive input, on the premise that each person, from graduate student to senior researcher, has had something of importance to contribute. Whether that contribution was small or large, unique or widely held, was secondary to the act of participating. Planetary scientists responded with energy and imagination.

The Division for Planetary Sciences of the American Astronomical Society funded the creation of an interactive website for community-initiated input to the decadal survey. This site was co-sponsored by the Meteoritical Society, the Planetary Geology Division of the Geological Society of America, and the Planetary Sciences Section of the American Geophysical Union. The site was maintained on computers at the headquarters of the American Astronomical Society in Washington, DC.

The community decadal website hosted real-time forums and facilitated the creation of community-based panels who discussed and wrote up white papers on issues and priorities within and across subdisciplines. These panels were self-selected both in topic, membership, and leadership. Their members were distributed around the United States with participants in a number of other countries as distant as Europe and Asia. They conducted their business by email, telecons, and on the real-time forums. Ultimately, 373 scientists formed 24 panels and submitted their white papers to the NRC decadal panels. This book is the product of those scientists.

The professional societies also cooperated on the electronic dissemination of announcements regarding NRC-sponsored forums held around the country, requests for input from NRC discipline panels, as well as updates on community-initiated activities.

The value of this effort transends the decadal survey. These chapters provide a snapshot of what planetary scientists consider to be the important issues in all areas of solar system exploration, and how they think those issues can best be addressed. The essential elements motivating scientific investigations in those areas have been distilled. These documents will be a significant resource for future proposals for research grants, telescope time, and spacecraft missions. They will be of great value when the next decadal survey is initiated.

I would like to thank every scientist who participated in the decadal survey and worked to generate these papers. I would like to thank Michael Belton and his NRC decadal panels, for their encouragement and support of the community-based decadal activities, the open forums they sponsored around the country, and for the great effort they personally undertook to digest the more than 400 pages of detailed input contained within this volume. As a community, we can take pride in the conduct and product of this survey.

Mark V. Sykes
Chair, 2000-2001
Division for Planetary Sciences
American Astronomical Society
Tucson, Arizona
April 29, 2002

The Future of Solar System Exploration, 2003-2013
ASP Conference Series, Vol. 272, 2002
Mark V. Sykes, ed.

Mercury

A.E. Potter[1], R.M. Killen[2], & B. Hapke[3]

[1] *National Solar Observatory, Tucson, AZ*
[2] *Southwest Research Institute, San Antonio, TX*
[3] *Department of Geology and Planetary Science, University of Pittsburgh, Pittsburgh, PA*

Abstract. Mercury is the least-explored planet except for Pluto. The Mariner 10 mission yielded much information about the planet, but raised new questions. Groundbased observations of the polar craters and atmosphere have also raised new questions. The Messenger mission will address many of these, but in order to maximize the return from this mission, a program of laboratory, observational, and theoretical research on Mercury should be maintained. Some thought should also be given to follow-on Mercury missions, in particular, a lander.

REPORT

1. Current State of Knowledge

Mercury is the least-explored planet except for Pluto. It is the closest planet to the sun, and has the largest diurnal range in surface temperature. It has the largest uncompressed density of any planet, and hence the largest mass fraction of nickel-iron.

Earth-based radar observations have yielded much information, finding that the planet is locked into a 3:2 spin-orbit resonance, and mapping low-resolution surface topography over much of the planet, including the poles. There appear to be deposits of ice in permanently shadowed polar craters that are visible to Earth radars.

The most detailed information that we have for Mercury came from three flybys of the Mariner 10 spacecraft in 1974 and 1975 (Figure 1). These mapped part of the surface, the magnetosphere, atmosphere, and ionosphere. However, the mapping was incomplete, and many questions remain.

The surface mapping provided a first-order understanding of the geology and history of the planet. Mercury's surface shows craters, intercrater plains, large basins, smooth plains, and tectonic features. The Mercurian craters have been age-dated using stratigraphic relationships and crater counts (assuming a lunar cratering history) to give a time history for the ancient evolution of the planet. These data suggest that Mercury may have experienced early global melting, similar to the lunar "magma ocean". This process would have concentrated low-density plagioclase into the uppermost part of the crust. Intercrater plains that may be partly volcanic in origin have obliterated the earliest craters

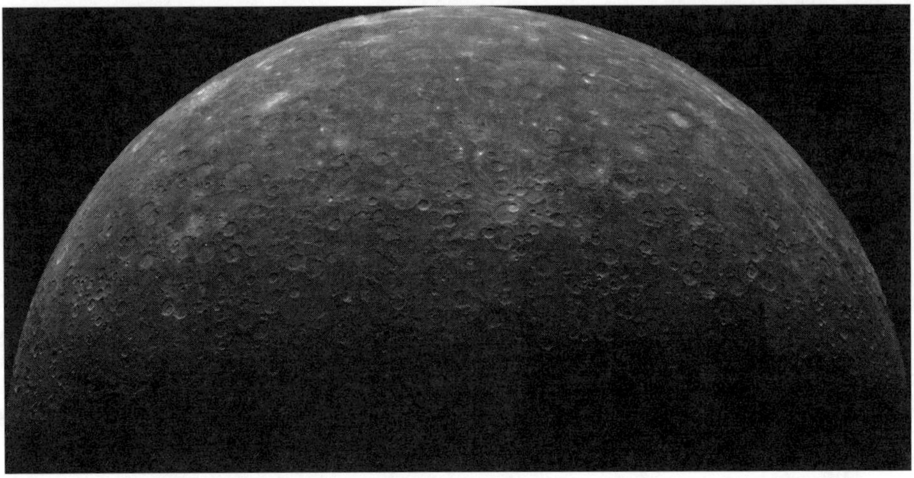

Figure 1. Mariner 10 image of Mercury (NASA).

on Mercury. The large Mercury basins mark large impacts that occurred during a time of high impact rates. Following some of these major impacts, massive extrusions of flood lavas formed the smooth plains. Since this time, the cratering rate has been low. Lobate scarps are widely distributed over the planet. They are interpreted as thrust faults resulting from global compression following the end of the major impact period, and might possibly be the result of solidification of the planetary core.

The composition of the Mercury surface is poorly understood. The crust appears to be low in iron and titanium, with only 2-6% FeO, and is probably feldspathic. Ground-based mid-infrared spectra of the surface suggest that a variety of different minerals may be present, but the data are difficult to interpret unambiguously. The photometric and polarimetric properties are similar to the Moon's, implying a lunar-like regolith. However, the regolith is more mature than that of the Moon, consisting of more than 50% impact generated vapor deposits.

One surprising observation of Mariner 10 was the discovery that Mercury has a small external magnetic field, suggesting the existence of an internal magnetic field, and possibly an internal hydromagnetic dynamo. The surface maps show that volcanic activity stopped early in planetary history, so that the existence of a fluid dynamo at the present time is hard to understand. The magnetosphere that arises from the magnetic field is similar in some respects to the Earth's magnetosphere, but there are important differences because of the weakness of the field and the closeness of Mercury to the Sun. Recent theoretical analyses suggest that the direction of the interplanetary magnetic field (IMF) is important, in that a south IMF will interact with the magnetic field so as to open Mercury to the solar wind. In extreme conditions, the magnetosphere could be open to the solar wind all the way to the equator.

The ultraviolet photometers on Mariner 10 detected a tenuous atmosphere - but only atomic hydrogen, helium, and possibly atomic oxygen were found. The altitude distribution of atomic hydrogen was peculiar, in that there appeared to be both a low-temperature component and a high-temperature component. Although expected to be present, argon was not detected. Recent theoretical considerations suggest that argon may be lost rapidly from the planet. A surprising feature of the Mercury atmosphere is the presence of sodium, potassium, and calcium, all observed by means of ground-based spectroscopy. These species are highly variable - their source and loss processes are still under debate.

2. Key Science Questions

- What is the geological history of Mercury? We have a first approximation from the Mariner 10 data, but there are many questions that can be answered only by more data of better spatial resolution that cover the entire planet.
- What processes led to the formation of Mercury? Why is the metal/silicate ratio the highest among all other planets and satellites?
- What is the composition of the Mercury crust?
- What is the structure and state of Mercury's core?
- What is the nature and origin of the Mercury magnetic field? Is the magnetic field a dipole?
- Is there a dynamo that produces the magnetic field? If so, where is it located?
- How does the magnetosphere respond to interplanetary conditions? That is, how does the direction of the IMF, the properties of the solar wind, etc. affect the magnetosphere?
- What is the composition of the magnetosphere? Do heavy ions play a role? How are currents closed in the absence of an ionosphere?
- How conducting is the Mercury surface?
- What are the radar-reflective materials in the polar craters? Are they really water ice?
- What are the major constituents of the atmosphere? What are their sources and sinks?

The questions above are listed in approximate order of priority.

3. Recommendations

A. Support the Messenger mission to Mercury. This spacecraft will be in orbit around Mercury for about a year in 2009. It includes instruments to image the surface in stereo at visible and near-IR wavelengths, measure its composition, measure the magnetic field, perform quantitative altimetry to measure libration, and identify neutral and charged species in the atmosphere. The payload addresses most of the scientific questions outlined above. Details of the funding for this Discovery mission are available from NASA budget documents.

B. Support the ESA Bepi-Colombo mission to Mercury as appropriate for U.S. interests. This mission addresses much the same list of questions as does the Messenger mission, and may include the addition of a lander to perform in-situ analyses. U.S. participation in the experiment teams would be a very cost effective way to obtain support for Mercury research. Modest funding for travel, publication, and some data processing expenses would allow U.S. researchers to obtain unique Mercury data. A level of $150 K/year could support 10-15 U.S. investigators.

C. A lander mission to follow the Messenger orbiter mission should be considered. This would allow all kinds of measurements impossible for an orbiter, including general relativity investigations. These could include full element analysis of surface composition, in-situ radiometric dating, seismic and heat flow measurements, trenching and coring of the regolith, etc. These measurements would answer fundamental questions about the origin of the planet. It would be worthwhile to support some small studies as to the feasibility and probable scientific return from a lander mission. A total funding of about $300K should be sufficient.

D. The flight missions should be supported by a program of theoretical, astronomical, and laboratory studies related to Mercury. Examples include theoretical studies of the space weather-magnetosphere interaction, astronomical observations of the atmospheric constituents Na, K, and Ca, radar mapping of the planet and its polar regions, mid-IR investigations of surface composition, and laboratory and theoretical studies of the interaction of solar particles and photons with the Mercury regolith. The annual funding level should be of the order of $300K/year.

4. Maintaining Capabilities for the Future

It is essential to maintain a research program related to Mercury during the time that the Messenger mission is built and launched. It is certain that new insights will be gained, and these can (or at least should) influence the detailed design of the observations to be done when Messenger finally arrives at Mercury.

The Future of Solar System Exploration, 2003-2013
ASP Conference Series, Vol. 272, 2002
Mark V. Sykes, ed.

Divergent Evolution Among Earth-like Planets: The Case for Venus Exploration

D. Crisp[1], M.A. Allen[1], V.G. Anicich[1], R.E. Arvidson[2], S.K. Atreya[3], K.H. Baines[1], W.B. Banerdt[1], G.L. Bjoraker[4], S.W. Bougher[5], B.A. Campbell[6], R.W. Carlson[1], G. Chin[4], A. Chutjian[1], R.T. Clancy[7], B.C. Clark[8], T.E. Cravens[9], A.D. Del Genio[10], L.W. Esposito[11], B. Fegley[2], M. Flasar[4], J.L. Fox[12], P.J. Gierasch[13], R.M. Goody[14], D.H. Grinspoon[15], S. Gulkis[1], V.L. Hansen[16], R.R. Herrick[17], D.L. Huestis[18], D.M. Hunten[5], M.A. Janssen[1], J. Jenkins[19], C.L. Johnson[20], G.M. Keating[21], A.J. Kliore[1], S.S. Limaye[22], J.G. Luhmann[23], J.I. Lunine[5], P. Mahaffy[4], P.J. McGovern[17], V.S. Meadows[1], F.P. Mills[1], H.B. Niemann[4], T.C. Owen[24], K.I. Oyama[25], R.O. Pepin[26], J.J. Plaut[1], D.C. Reuter[4], M.I. Richardson[27], C.T. Russell[28], R.S. Saunders[1], J.T. Schofield[1], G. Schubert[28], D.A. Senske[29], M.K. Shepard[30], T.G. Slanger[18], S.E. Smrekar[1], D.J. Stevenson[27], D.V. Titov[31], E.A. Ustinov[1], R.E. Young[32], & Y.L. Yung[27]

[1] *Jet Propulsion Laboratory, California Institute of Technology, Pasadena, CA*
[2] *Department of Earth and Planetary Science, Washington University, St Louis, MO*
[3] *Department of Atmospheric, Ocean, and Space Sciences, University of Michigan, Ann Arbor, MI*
[4] *NASA Goddard Space Flight Center, Greenbelt, MD*
[5] *Lunar and Planetary Laboratory, University of Arizona, Tucson, AZ*
[6] *National Air and Space Museum, Washington, DC*
[7] *Space Science Institute, Boulder, CO*
[8] *Lockheed Martin Astronautics, Denver, CO*
[9] *Department of Physics and Astronomy, University of Kansas, Lawrence, KS*
[10] *NASA Goddard Institute for Space Studies, New York, NY*
[11] *Atmospheric and Space Physics Laboratory, University of Colorado, Boulder, CO*
[12] *Department of Physics, Wright State University, Dayton, OH*
[13] *Department of Astronomy, Cornell University, Ithaca, NY*
[14] *Division of Applied Sciences, Harvard University, Cambridge, MA*
[15] *Southwest Research Institute, Boulder, CO*
[16] *Department of Geological Sciences, Southern Methodist University, Dallas, TX*
[17] *Lunar and Planetary Institute, Houston, TX*
[18] *Molecular Physics Laboratory, SRI International, Menlo Park, CA*
[19] *SETI Institute, Mountain View, CA*
[20] *Scripps Institute for Oceanography, University of California, San Diego, CA*

[21] *George Washington University at NASA Langley Research Center, Hampton VA*
[22] *Space Science and Engineering Center, University of Wisconsin-Madison, Madison, WI*
[23] *Space Science Laboratory, University of California, Berkeley, CA*
[24] *Institute for Astronomy, University of Hawaii, Honolulu, HI*
[25] *Institute of Space and Aeronautical Science, Japan*
[26] *School of Physics and Astronomy, University of Minnesota, Minneapolis, MN*
[27] *Geological and Planetary Sciences, California Institute of Technology, Pasadena, CA*
[28] *Dept Earth and Space Sciences, University of California, Los Angeles, CA*
[29] *NASA Headquarters, Washington, DC*
[30] *Department of Geography and Earth Science, Bloomsburg University, Bloomsburg, PA*
[31] *Max-Planck-Institut für Aeronomie, Katlenburg-Lindau, Germany*
[32] *NASA Ames Research Center, Moffett Field, CA*

Abstract. The planet Venus is our most Earth-like neighbor in size, mass, and distance from the sun. In spite of these similarities, and the intense scrutiny that it received early in the space age, the Venus surface and atmosphere are characterized by some of the most enigmatic features seen anywhere in the solar system. A reinvigorated Venus exploration program is essential to the development of a comprehensive understanding of the origin and evolution of Earth-like terrestrial planets. The present NASA inner planets strategy, which focuses exclusively on Mars, will provide an incomplete, and possibly misleading description of processes that produce these objects. If Venus-like terrestrial planets are common, this approach will also impede efforts to interpret observations of extrasolar terrestrial planets, which are expected to become available by the end of the decade. Here, we propose a Venus exploration program that has been designed to explain the origin and divergent evolution of the interiors, surfaces, and atmospheres of the terrestrial planets in our solar system, and provide greater insight into the conditions that may affect the habitability of terrestrial planets in other solar systems.

EXECUTIVE SUMMARY

Venus was once thought to be Earth's twin because of its similar size, mass, and solar distance. These factors, combined with its relative proximity made it the target of intense scrutiny during the first three decades of the space age. The first planetary spacecraft, Mariner 2, flew past Venus 40 years ago on 14 December 1962. That mission and subsequent observational and theoretical work indicate that although Venus and Earth were apparently formed nearby in the solar nebula, sharing common inventories of refractory and volatile constituents, they followed dramatically different evolutionary paths. While the Earth evolved into the only known oasis for life, Venus developed an almost unimaginably hostile environment, characterized by a massive (90-bar) CO_2 atmosphere, a hellish (730 K, 854 °F) surface temperature, and a global cloud deck composed of sulfuric acid (H_2SO_4) particles. Some of the processes that contributed to the divergent evolution of these two planets can be attributed to the fact that Venus is closer to the sun (0.7 AU), and receives about twice as much solar radiation as Earth. The intense solar insolation may have helped to induce a runaway greenhouse in the early Venus atmosphere producing surface temperatures high enough to ensure that the water vapor, CO_2, and sulfur gases emitted by volcanic processes remained in the atmosphere.

Other aspects of the environment of Earth's sister planet are among the most enigmatic in the solar system. For example, although the volatile inventory of Venus is not as well characterized as that of Earth and Mars, existing measurements indicate that the relative abundances of the noble gases in its atmosphere are much more solar-like than those on the other two terrestrial planets, whose atmospheres appear to be strongly altered. Similarly, contemporary Venus is almost 100,000 times drier than Earth while the deuterium-to-hydrogen ratio is ~150 times larger, suggesting that Venus may have lost the equivalent of an ocean of water. The specific processes and timing of this large loss of water are currently not known. The atmospheric dynamics are equally perplexing. Although the surface of Venus rotates very slowly, with a ~242 day period, the cloud-level atmosphere (~70 km) appears to rotate almost as a solid body, with a period of 4 days, or about 60 times faster than the surface. The mechanism(s) driving this atmospheric superrotation have evaded explanation since its discovery more than three decades ago.

A planet-wide cloud deck completely hides Venus' surface at visible wavelengths, but radar observations by the Pioneer Venus, Venera 15, and Magellan orbiters show that it is also mysterious. In particular, the cratering record indicates that the average age of the surface is about 750 million years, but the crater mophology suggests that there has been very little geologic activity since that time. In particular, while the surface records a rich geologic history, it lacks plate boundary features and other evidence of plate tectonics. In addition, while there are numerous features that appear to be of volcanic origin, there is currently no direct evidence for active volcanism or other processes that would indicate whether the planet is geologically dead or alive.

The processes that produced the surface morphology and their implications for the evolution of terrestrial planet interiors remain controversial. Some have interpreted these features as evidence for an internal cooling mechanism characterized by episodic, catastrophic, global resurfacing events, the last one of which occurred ~750 million years ago. Others believe that these surface features simply indicate that the Venus lithosphere thickened significantly over the past ~1 billion years, such that the interior now loses heat primarily through sluggish or stagnant lid tectonics. The available data are not adequate to clearly differentiate between these two mechanisms, or to resolve the effects of other processes that may have affected the evolution of the Venus interior, such as the slow rotation rate, the low present-day water abundance, or the high surface temperatures.

As NASA embarks on its search for extrasolar terrestrial planets, with a focus on origins, it is interesting to note that Venus has been ignored for over a decade. A reinvigorated Venus exploration program could yield great dividends by dramatically improving our understanding of the origin, evolution, and habitability of terrestrial planets. Specifically, a comprehensive inventory of the noble gases and isotopes of carbon, hydrogen, oxygen, and nitrogen in the Venus atmosphere would help us to understand our past by clarifying the conditions in the primordial solar nebula that produced the unique character of our home planet. A detailed investigation of the processes controlling the current surface and atmospheric environment on Venus will help us to obtain a deeper understanding of the processes operating on Earth, including greenhouse effects, atmospheric photochemistry, heterogeneous chemistry, thermal chemistry, atmospheric dynamics, surface chemical weathering, and intraplate geology. A more comprehensive understanding of Venus could yield important clues about our future by providing constraints on atmospheric compositional stability, and perhaps the geophysical evolution of the Earth.

For the reasons described above, coordinated Venus and Mars programs are essential to the development of a comprehensive understanding of the origin and evolution of Earth-like terrestrial planets. The present NASA inner planets strategy, which focuses exclusively on Mars, will provide an incomplete, and possibly misleading description of processes that produced the current environments of Venus, Earth, and Mars. This could also impede efforts to interpret observations of extrasolar terrestrial planets, which may be available within a decade.

Although it is not possible to address all of these questions with a single dedicated Venus mission, we can dramatically improve our understanding of our sister planet within the constraints of a coordinated program of small (Discovery Class), medium (New Frontiers Class), and large (~$1 B FY01) missions conducted by NASA and our international partners over the next 10 to 20 years. We strongly urge NASA to expand the inner planets program to include a Venus mission line. This mission line need not be as large or ambitious as the Mars Exploration Program, but its objectives and schedule should be designed to maintain an appropriate balance with that program.

The NASA mission with the highest priority and greatest technological maturity is the Noble Gas/Trace Gas Explorer. The other missions listed below have similar priorities, but different levels of technological maturity, and should fly as their critical technologies become available.

- The Noble Gas and Trace Gas Explorer is the highest priority mission because its data are vital to our understanding of the origin of Venus. This small mission requires a single entry probe that will carry the state-of-the-art instruments needed to complete the noble gas and trace gas inventories between the cloud tops and the surface.

- The Global Geological Process Mapping Orbiter is a small to medium class mission. It will carry a C- and/or X-band radar designed for stereo or interferometric imaging, to provide global maps of the surface at horizontal resolutions of 25 to 50 meters. These data are needed to identify and characterize the geologic processes that have shaped the Venus surface.

- The Atmospheric Composition Orbiter is a small mission that will carry remote sensing instruments for characterizing spatial and temporal variations in the clouds and trace gases throughout the atmosphere. This mission will collect the data needed to characterize the radiative, chemical, and dynamical processes that are maintaining the thermal structure and composition of the present atmosphere.

- The Atmospheric Dynamics Explorer is a small to medium mission that will deploy 12 to 24 long-lived balloons over a range of latitudes and levels of the Venus atmosphere to identify the mechanisms responsible for maintaining the atmospheric superrotation.

- The Surface and Interior Explorer is a large mission that will deploy three or more long-lived landers on the Venus surface. Each lander will carry a seismometer for studies of the interior structure, as well as in-situ instruments for characterizing the surface mineralogy and elemental composition. This mission requires significant technology development.

- A sample return mission will eventually be needed to conduct investigations of the Venus surface and atmosphere that cannot be conducted by instruments on remote sensing platforms or on entry probes. This will probably require a large mission and significant technology development.

Two international Venus missions complement the missions described above. The first is the ESA Venus Express mission. This mission would use the spare Mars Express bus to conduct a remote sensing investigation of the trace gas abundances in the lower atmosphere and the cloud properties. It may also carry a surface-penetrating radar (with NASA contributions) for studies of the subsurface. This mission has been selected by the ESA Space Science Advisory Committee, but funding has not yet been allocated. The second is the Japanese (ISAS) Venus Climate Orbiter. This mission will carry ultraviolet, visible, and near infrared imaging systems for studies of the cloud level dynamics. It may also carry space physics instruments for studies of the exosphere. This mission has been approved by the ISAS Space Science Committee as the third Japanese deep space mission (Planet C), and is seeking approval as a new start.

REPORT

1. Unifying Scientific Themes

1.1. Past: Origin of terrestrial planets in our solar system

A detailed analysis of the noble gas inventories on Venus, Earth, and Mars may provide the best possible understanding of the conditions of the primordial solar nebula from which these planets were formed (Pepin 1991, 1997). However, a much more comprehensive, accurate inventory of noble gases and their isotopic ratios is needed for Venus to construct a coherent picture of these conditions. The volatile inventory of Venus is poorly constrained, but existing evidence suggests that the Venus noble gas inventory may be more representative of the primordial solar nebula than that on either Earth or Mars. In-situ atmospheric measurements of noble gases, their isotopes, and other key volatiles such as H_2O, CO, OCS, and SO_2 would contribute directly to our understanding of the origin and evolution of terrestrial planets in this solar system. This information will be essential for interpreting observations of terrestrial planets around other stars.

1.2. Present: What processes shape the terrestrial planets

Because the atmospheric, surface, and interior processes have evolved so differently on Venus, Earth and Mars, coordinated investigations of these planets provide opportunities to identify those factors most important to the environment of a terrestrial planet, and how processes currently acting on Earth might produce different outcomes under slightly different conditions. In particular, such a comparison could delineate the roles of the exogenic processes (e.g., solar insolation, impacts, tidal effects) that define its local environment in the solar system, from the endogenic physical, chemical, and thermodynamic processes that represent the planet's response to that environment.

1.3. Future: What does Venus tell us about the fate of the Earth's environment?

Is Venus a future Earth? Probably not. Nevertheless, processes occurring on Venus today are relevant to predicting how the Earth might evolve. For example, the massive CO_2 greenhouse in the Venus atmosphere is driven by radiative processes similar to those operating in the Earth's atmosphere as the atmospheric CO_2 concentration continues to increase. Venus' past may also be relevant to the Earth's future. Stellar evolution models predict that the solar luminosity will continue to increase, such that in about a billion years, the Earth could suffer a runaway greenhouse. Venus is thought to be a prime example of such an event. There is a similar analogy concerning interior processes. Will the Earth eventually evolve to an internal state where plate tectonics disappear, and heat loss from the interior evolves to the sluggish or stagnant lid conditions like those that are thought to now characterize Venus?

2. Current Knowledge and Future Directions

2.1. What led to Earth's unique character?

Because of their proximity, it is reasonable to assume that Venus, Earth, and Mars originated in the same part of the primordial solar nebula, with similar inventories of volatile and refractory elements. A comprehensive theory of the origin and evolution of terrestrial planets must therefore explain the observed properties of all three bodies. Precise measurements of noble gases in the atmospheres and surface materials of these three planets will provide the best clues to each planet's original supply of volatile elements. This information must be combined with measurements of its internal structure, surface age and record of geological processes, and the atmospheric composition and thermal structure to characterize the evolution of each planet's environment over time. A more detailed understanding of these processes is essential to interpret observations of extrasolar terrestrial planets, once they become available (Owen 2000). Each of these requirements is described in more detail below.

Noble gases: The isotopic constraints on models of atmospheric history are much weaker for Venus than for Earth and Mars because the Pioneer Venus and Venera missions obtained no measurements of the heaviest noble gases, krypton (Kr) and xenon (Xe). Interestingly, in spite of their large uncertainties, the available measurements of noble gases on Venus indicate high abundances and solar-like elemental ratios (Pepin 1991, Owen et al. 1992). This suggests that at least the heavier noble gases on Venus have not evolved substantially from their primordial states. Neon and argon isotope ratios also appear to be displaced toward solar values compared to their terrestrial counterparts (Figure 1). Venus consequently occupies a unique position among this triad of terrestrial planets in that its atmosphere may have experienced much less alteration from its initial composition by planet-specific fractionating loss mechanisms, while Earth and Mars appear to have been much more highly processed.

A more complete assessment and understanding of the composition of the atmosphere of Venus is therefore enormously important in the context of models for the origin and evolution of terrestrial planet volatiles. Measurement of the isotopic composition of Venusian Xe would immediately provide a critical test of two very different proposed mechanisms for generating the strong fractionation pattern, relative to solar, displayed by atmospheric Xe on Earth and Mars. Did this separation occur in the planetesimal stage prior to planetary formation, implying a similar composition on Venus, or afterward, when energy deposited in a giant impact on Earth and Mars blew off parts of their atmospheres while the Venusian Xe remained near its initial solar composition? An equally important question is whether the nonradiogenic heavy noble gases on Venus are relatively unaltered samples of the ancient gaseous nebula. Improved constraints on the nebular gas composition during planetary formation are needed to discriminate between models that propose direct (gravitational condensation) or indirect (cometary accretion) nebular sources of primordial volatiles on terrestrial planets (Owen and Bar-Nun 1995).

Trace gases and their isotopes: The concentrations and isotopic ratios of other volatile species are also poorly known for Venus. Pioneer Venus and Venera measurements indicate that the Venus atmosphere is 96.5% CO_2 and 3.5% N_2. Other trace gases include H_2O, SO_2, OCS, CO, HCl, and HF, but there are large uncertainties in their abundances, especially below the cloud tops (Esposito et al. 1997). Excited atomic and molecular oxygen have been detected via airglow emission from the upper atmosphere, but ground-state oxygen has not yet been detected in the bulk atmosphere. An improved understanding of the distribution of these trace gases is needed both for models of solar nebula formation and planetary accretion, and to understand their role in the greenhouse mechanism that maintains the planet's high surface temperature.

The Pioneer Venus and Venera probes returned measurements of the H_2O abundance that varied from less than 20 to more than 5000 parts per million by volume (Hoffman et al. 1980, Moroz 1983). Recent measurements from Earth-based near-infrared observations of the night side (Bézard et al. 1990, de Bergh et al. 1995, Meadows and Crisp 1996) indicate H_2O mixing ratios closer to 30 ± 10 ppm. This result is supported by a reanalysis of results from the Pioneer Venus Large Probe Neutral Mass Spectrometer (Donahue and Hodges 1993) and the Venera 13-14 Spectrophotometers. These values are 4 to 5 orders of magnitude

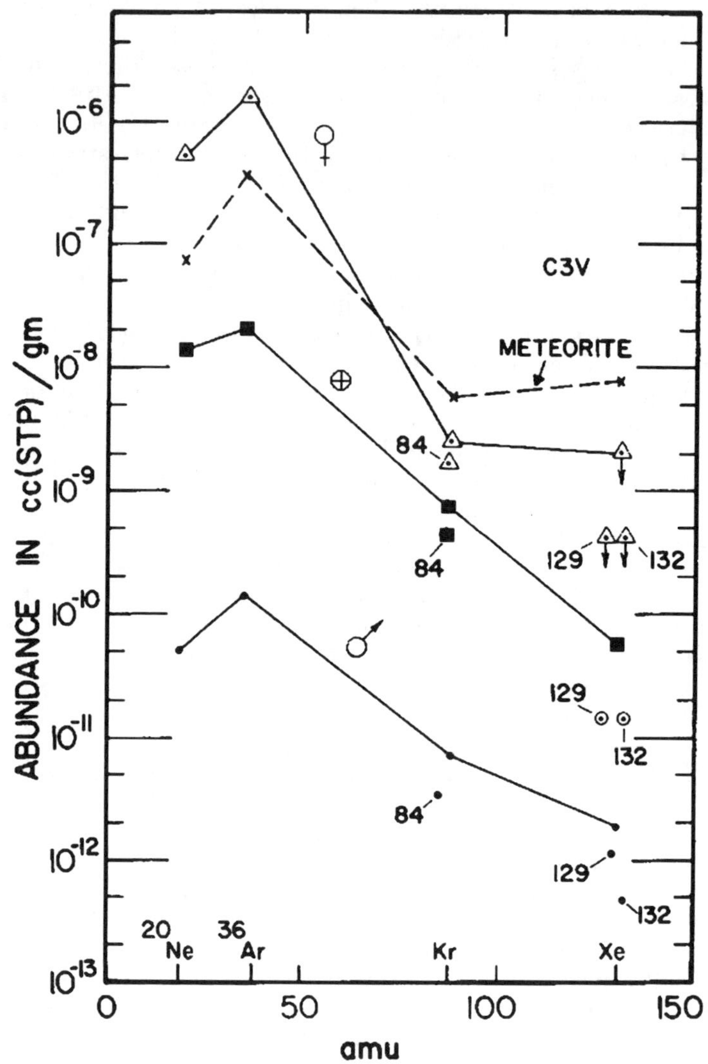

Figure 1. Comparison of noble gas abundances on Venus, Earth, and Mars with that of C3V carbonaceous chondrites (from Donahue and Pollock 1983).

smaller than terrestrial values. This indicates that either the primordial abundance of volatiles on Venus was significantly different than that of Earth, which is difficult to explain with current models of solar system formation (Wetherill 1981), or that Venus lost much of its primordial water inventory over time.

The latter explanation is supported by the anomalously high deuterium-to-hydrogen (D/H) ratio of Venusian water, which is 150 ± 30 times greater than that of standard mean ocean water on Earth (de Bergh et al. 1991, Donahue and Hodges 1993). Although models indicate that this high D/H ratio could have been produced by the photo-dissociation of water vapor at levels above the cloud tops, and the subsequent preferential loss of hydrogen over deuterium, the loss rates and dominant loss mechanisms are still too uncertain to determine when Venus lost the bulk of its water. This information is essential for understanding the early runaway greenhouse mechanism, and for studies of the role of water in the evolution of the Venus crust and volcanotectonic processes.

The abundance and distribution of the sulfur bearing gases, SO_2, OCS, H_2S, and H_2SO_4 are also poorly known at levels below the cloud tops (Bézard et al. 1993, Pollack et al. 1993). Improved measurements of these gases are needed both for studies of the processes that maintain the global H_2SO_4 clouds and to constrain the current rates of volcanism and surface chemical weathering (Fegley et al. 1997).

CRITICAL MEASUREMENTS NEEDED: A complete, accurate inventory of the noble gases and their isotopes is essential to our understanding of the origin and evolution of Venus. Measurements of the isotopes of Xe are of highest priority because their concentrations have not been measured on Venus, and this gas is expected to provide the best possible estimate of the initial noble gas inventory because it should be least modified by atmospheric escape. A more complete and accurate altitude-dependent description of the trace gas inventory is needed below the cloud tops. The noble gas measurements could be made by a single entry probe carrying a state-of-the-art mass spectrometer with a noble gas enrichment system. This probe should also carry instruments to record trace gas profiles between the cloud tops and the surface.

Surface properties: A comprehensive description of the processes that have shaped the Venus surface and controlled the exchange of volatiles between the interior and the atmosphere is also essential to our understanding of the origin and divergent evolution of the environments of Venus, Earth, and Mars. Images of the surface from the Venera landers (Florenskiy et al. 1983) show rocks intermixed with soil. X-ray fluorescence spectrometers aboard these Venera spacecraft indicate that the composition of these rocks is consistent with that of terrestrial basalts (e.g., Barsukov et al. 1986). Efforts to locate the Venera landing sites using Magellan data indicate that these properties characterize the lowland plains, which may have formed by extensive lava flows. The composition of the surface materials that cover highland regions, giving them anomalously low microwave emissivities, is not yet known (Fegley et al. 1997). However, laboratory studies indicate that the foundations of these regions are also likely to be basaltic.

Age of the Venus surface: Improved constraints on the age of the Venus surface would be of particular value in studies of the evolution of its interior and atmosphere. The cratering record offers the only available chronometer for timing absolute age of Venus' surface. Existing observations from the Venera 15 and Magellan radar mapping missions and Earth-based radars indicate that the number of craters is small (~950), and that their distribution is remarkably uniform across the planet. The cratering statistics derived from Magellan observations indicate that the average age of the surface is ~750 million year (Ma), which is comparable to the mean age of the Earth's surface. This young age was surprising because unlike Earth, Venus shows no evidence of plate tectonics.

The spatial uniformity of the crater distribution, combined with the relatively pristine condition of the existing crater population has been interpreted as evidence that the surface was subject to a global resurfacing event 350 to 1000 Ma ago, followed by a period of much slower geologic activity (Strom et al. 1994). If catastrophic resurfacing events occur episodically with a mean period of 300–500 Ma, this process could provide an interior heat release mechanism as efficient as plate tectonics (Parmentier and Hess 1992, Head et al. 1994). This interpretation is intriguing but it remains controversial for several reasons. First, the accuracy of the cratering chronometer has been questioned because it has never been adequately calibrated for a planet with a massive atmosphere. Second, the cratering densities are too low to constrain the ages of specific regions or geologic features on the Venus surface (Herrick and Phillips 1994, Hauck et al. 1998, Campbell 1999). Third, some geophysical models show that the cratering record and other surface properties can be explained adequately by a more gradual evolutionary transition from a thin, mobile lithosphere, to a thicker, stagnant lithosphere (Phillips and Hansen 1998), precluding the need for episodic catastrophic resurfacing events. In short, although the Venus surface apparently records a rich volcanic and tectonic history, uncertainties in the time scales of the resurfacing processes have continued to produce significant controversy. In fact, the existing data are not adequate to determine whether the planet is currently volcanically active.

Other aspects of the Venus geologic record are slowly being unraveled. If Venus had an internal heat budget similar to Earth, it apparently cools by mechanisms other than plate tectonics (Solomon et al. 1992). Earth's surface is transected by large scale linear structures marking divergent and convergent plate boundaries–regions where new crust is formed and recycled, respectively. Earth also displays two types of crust that have different compositions: thick low-density continental crust and thinner, higher density oceanic crust. Earth's crustal differentiation and plate tectonics together result in a surface comprised of large tracts of old crust (continents) and young crust (ocean basins).

In contrast, Venus displays a surface characterized by globally distributed circular, rather than linear, structures. The circular structures include the ~950 impact craters that pepper Venus' surface (Schaber et al. 1992, Phillips et al. 1992), recording an average surface age of ~750 Ma (McKinnon et al. 1997). The near random spatial distribution of the craters indicates that, unlike Earth, Venus lacks large tracts of very old and very young surfaces (Phillips 1993).

Venus hosts two distinct sizes of endogenic circular features. These include about 20 large (1600–2600 km diameter) quasi-circular volcanic rises and crustal plateaus, and 360–400 relatively small (100–1010 km diameter, 200 km median diameter) quasi-circular coronae. The volcanic rises and possibly the crustal

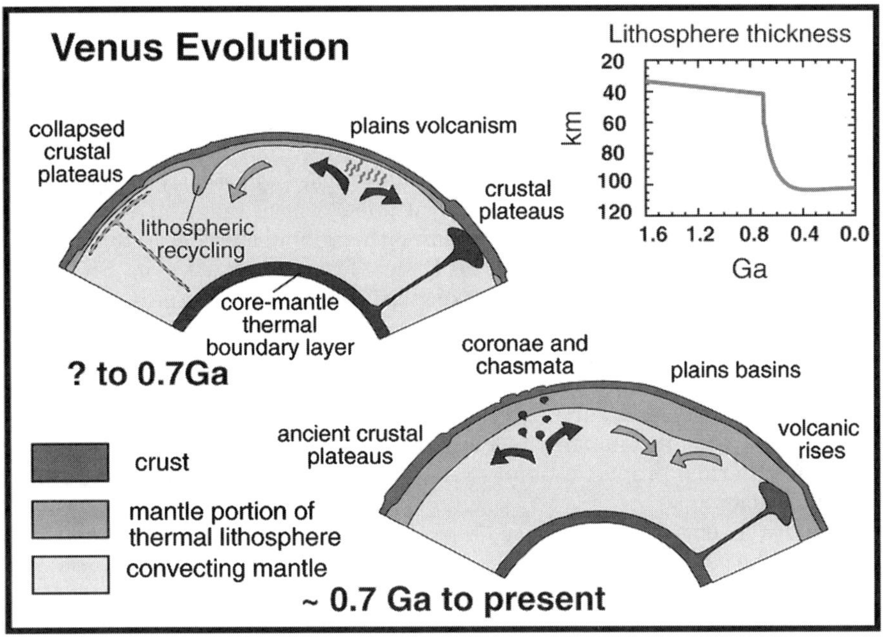

Figure 2. One model for the evolution of the Venus interior showing possible early and present processes (from Phillips and Hansen 1998)

plateaus are thought to be the surface signatures of deep mantle plumes on a contemporary thick lithosphere (McGill et al. 1981, Phillips and Malin 1984, Smrekar et al. 1997) and an ancient thin lithosphere (Hansen et al. 1997; Hansen and Willis 1998; Hansen et al. 1999, 2000; Phillips and Hansen 1994, 1998), respectively. Coronae are also widely accepted as the surface signature of mantle diapirs (Squyres et al. 1992; Janes et al. 1992; Stofan et al. 1991, 1992, 1997).

The physical mechanisms that form these structures and their implications for the evolution of the Venus interior are still controversial. For example, the bimodality of Venus' large (plateaus/rises) and small (coronae) diapiric structures, and their distribution on the surface, may reflect the different locations and mechanisms of diapir formation (Hansen 2001). In these models, thermally driven deep mantle plumes form along a warm core-mantle boundary layer and rise through the mantle to the lithosphere, forming plateaus and rises and transferring heat from the core (Figure 2). Compositional diapirs that form at relatively shallow levels in the upper mantle may rise to form coronae chains and transfer heat from the mantle (Phillips and Hansen 1994). Coronae clusters may result from compositional diapers spawned locally by plumes of deep mantle origin (Hansen 2002). Alternatively, they might be associated with heat transfer processes originating elsewhere in the mantle, but triggered by thermal anomalies from the core-mantle boundary. Improved constraints on the internal structure, heat flow, seismisity, mineralogy, and reology are needed to discriminate and validate these models.

CRITICAL MEASUREMENTS NEEDED: Venus clearly records detailed geologic histories across its surface. A series of carefully coordinated steps will be needed to unravel these histories and to infer their implications for the evolution of the planet. The best existing constraints on these surface properties have been derived from the moderate-resolution, global, Magellan data sets. The next logical step in this process would be to collect imagery and altimetry observations with horizontal and vertical resolutions similar to those currently being obtained for Mars. This could be accomplished by a polar orbiter carrying a state-of-the-art X or C-band radar designed to acquire either global stereo or interferometric measurements (e.g., the recent Shuttle Radar Topography Mission, SRTM).

Given this global context, the next step would be to obtain improved constraints on the chemical composition and mechanical properties of the selected surface regions to better constrain models of the cratering record and endogenic processes that have altered the surface. This could be accomplished by deploying in-situ instruments on a small network of surface landers (three to five). Each lander would carry instruments to measure long and short period seismic activity, surface thermal properties, and sensors to characterize the surface chemistry and mineralogy.

Finally, given a comprehensive understanding of the geological context, samples should be collected for age dating. This last step will require advanced technologies for precision landing, sample acquisition, processing, and may require the return of the sample to Earth. Advanced age dating techniques must also be developed and calibrated for this application, because the methods commonly used to date terrestrial geologic samples would not be reliable for application to samples that formed and evolved in the Venus surface environment.

The role of volatiles in interior evolution: Volatiles in the interior and atmosphere play a key role in determining the geologic style of a planet. The planetary volatile inventory is directly coupled to the interior evolution through its role in controlling the rheology of the lithosphere. Water may be the key element enabling plate tectonics on Earth. The role of water in weakening the terrestrial lower lithosphere may allow the initiation of subduction (Regnauer-Lieb et al. 2001). Weakening of the upper mantle by water may also be responsible for the low viscosity zone at the top of Earth's asthenosphere. This zone may lubricate the motion of tectonic plates and promote a convective pattern favorable to plate tectonics. In fact, water may be necessary to produce the chemically distinct, buoyant crust that forms the continents.

The extremely low abundance of water on Venus may have contributed to its entirely distinct style of lithospheric deformation. Despite the high surface temperatures on Venus, the strength of the crust is believed to be comparable to that of the mantle due to the low abundance of water in the interior (Mackwell et al. 1998). The very high ratios of gravity and topography indicate a strong coupling between mantle convection and deformation of the lithosphere. Thus no significant low viscosity zone exists on Venus. The strong Venusian lithosphere and the lack of an underlying low viscosity zone inhibit lateral motion of the lithosphere. The current low lithospheric recycling rate reduces the amount of heat loss to the exterior, possibly leading to a 'stagnant lid' mode of convection where the interior heats up due to the strong, insulating outer layer (Weinstein and Olson 1992, Moresi and Solomatov 1998). Recycling of the crust occurs through delamination of the lower crust, if at all (Parmentier and Hess 1992).

The evolution of volatiles and tectonics may be closely coupled on Venus by the feedback between geologic processes and climate change. Early in its history, when Venus apparently had a thin, presumably mobile lithosphere, widespread pressure release melting of the mantle may have resulted in globally widespread volcanism. This volcanism would have released water and other greenhouse gases into the Venus atmosphere, raising surface temperatures at least 100 K above present day values (Bullock and Grinspoon 1996, 2001) and perhaps as high as 1000 K (Phillips et al. 2001). As interior heat escaped and the planet cooled, the lithosphere thickened, changing from a thin, presumably mobile state to a thick stagnant lid. This change resulted in a significant decrease in pressure release melting and a decrease in global volcanic activity, which in turn led to cooling of the global surface temperature to today's value as volcanic greenhouse gases were no longer produced at such rapid rates. The change from a thin to a thick lithosphere is recorded by ancient crustal plateaus and contemporary volcanic rises which record the surface signature of deep mantle plumes on thin and thick lithosphere, respectively. The global cooling may have resulted in widespread low strain deformation of the surface (Anderson and Smrekar 1999, Solomon et al. 1999).

CRITICAL MEASUREMENTS NEEDED: Interpretations of Magellan data have produced significantly different, but plausible, competing global geological histories. The available imagery and altimetry are inadequate to resolve the regional and local details necessary to differentiate among these hypotheses. Imagery and altimetry data comparable in resolution to that provided by the Mars Global Surveyor mission would provide a similar advancement in our understanding of Venus. An orbiter with an X or C-band radar could be used to obtain global stereo imaging, or used as an interferometer with the addition of a long (tens of meters) boom, to yield altimetry data with a spatial resolution of 25 to 50 m. Higher resolution data could be acquired over more limited regions with optical and infrared imagers and radar altimeters on balloons or aircraft.

Measurements that constrain the present day composition of the atmosphere and the nature of chemical weathering at the surface are also needed to define the role of volatiles in the history of Venus. These measurements could be obtained by a lander that collects images, spectra, and uses in-situ instruments to collect other critical geochemical data at the surface. These data would be used to identify the surface mineral composition, soil types, and weathering products. A long term seismic network is needed to constrain present day rates of geologic activity. A seismic network is also needed to provide constraints on the crustal thickness and to determine whether the core is solid or liquid. Heat flow measurements would help constrain the present day geothermal gradient and the rate of heat loss from the interior, providing important constraints on the recent history of the lithosphere.

2.2. What common processes shape Earth-like planets?

The Venus greenhouse mechanism: Climate models show that greenhouse gases in the massive, 90-bar atmosphere, combined with the thermally-opaque planet wide sulfuric acid cloud deck, can maintain the anomalously high surface temperatures on Venus (Crisp and Titov 1997). The high infrared opacity of the major atmospheric constituent, CO_2, plays a key role in producing a strong greenhouse effect, while trace gases (SO_2 and H_2O) and the planet wide H_2SO_4 cloud deck fill spectral windows at all infrared wavelengths longer than 2.5 μm where CO_2 is transparent. Existing measurements show that SO_2, H_2O and other trace gases are spatially and temporally variable on Venus, but these measurements are not adequate to constrain their horizontal and vertical distributions at levels below the cloud tops. Furthermore, at the high temperatures and pressures of the lower atmosphere of Venus, the opacities of many gases are not well known, and the spectral distributions of solar and thermal radiation are poorly constrained. Nevertheless, radiative transfer models based on the best available estimates of the composition have successfully reproduced the atmospheric thermal structure. While these results support the greenhouse explanation for high surface temperature, they are not adequate to resolve the relative roles of specific radiative processes.

Several greenhouse processes that contribute to the Earth's climate are much more effective, and easier to study in the atmosphere of Venus. For example, gas absorption by weak absorption lines, the far wings of strong lines, and pressure-induced absorption are essential components of the Venus greenhouse. These factors also produce measurable changes in the Earth's atmosphere, but are much more difficult to study there. Aerosol scattering and absorption also produce a much larger radiative forcing of the climate within the main Venus cloud deck, where they play a critical role in the Venus greenhouse. Preliminary studies of these clouds provided the theoretical basis for studies of the effects of volcanic aerosols emitted by the Mount St. Helens, Chichon, and Pinatubo eruptions on the Earth's climate. Studies of the impacts of these radiative processes on Venus have also provided many of the tools currently being used to gain greater insight into the early climates of Earth and Mars.

CRITICAL MEASUREMENTS NEEDED: Improved, spectrally resolved measurements of the solar and thermal radiation fields are needed at altitudes between the cloud tops and the surface. These measurements must be combined with simultaneous constraints on the abundances and distributions of radiatively active trace gases and aerosols to clearly identify their effects on the atmospheric thermal balance. A single entry probe with a suite of state-of-the-art instruments for trace gas measurements (mass spectrometer, gas chromatograph, tunable diode lasers, grating spectrometers) could provide dramatic improvements in our understanding of the vertical profiles of all known radiatively active gases in the Venus atmosphere. These measurements should be combined with UV, near IR and microwave remote sensing observations from orbit to characterize the horizontal distributions of these species.

Improved, spatially resolved measurements of the Venus clouds are also needed to understand their effects on the solar and thermal radiation fields. Pioneer Venus and Venera measurements show that the clouds reflect ~80% of the incident solar radiation and absorb over half of the solar energy that is deposited on Venus. An unknown UV absorber in the upper cloud is responsible for about half of this absorption. In-situ observations in the upper cloud (58–60 km) may needed to identify the composition of this absorber. The effects of the clouds on the thermal radiation field are not well understood, in part because the middle and lower clouds are spatially inhomogeneous. Entry probe and near IR remote sensing observations document spatial variations in cloud opacity, but provide limited information about their vertical distribution, their radiative impacts or the mechanisms responsible for their production and maintenance. Long term in-situ observations of the the cloud particle properties (shape, size, number density, and composition) must be combined with simultaneous measurements of the solar and thermal radiation fields above, within, and below the clouds to resolve these uncertainties and quantify the effects of clouds on the greenhouse.

Atmospheric superrotation: Since its confirmation over 30 years ago, the superrotation of the Venus atmosphere has remained one of the most intriguing enigmas in planetary science (Gierasch et al. 1997). At cloud-top levels, the Venus atmosphere rotates almost 60 times faster than the solid surface of the planet. The wind direction is westward everywhere between the surface and ~70 km (Figure 3). At low latitudes, the wind speeds increase with altitude from values near zero at the surface, to values near 100 m/sec in the upper cloud (60 - 70 km). Several candidate mechanisms have been proposed for generating and maintaining these winds, but this anomalous circulation has not yet been simulated by even the most comprehensive global general circulation models. This raises concerns about our understanding of the dynamics of slowly rotating planetary atmospheres and the range of validity of the best available climate and weather prediction models.

Above the clouds, the amplitude of the superrotation decreases rapidly with altitude and the equator to pole temperature gradient reverses, such that the poles are warmer than the equator throughout much of the mesosphere (70 to 90 km; Taylor et al. 1980). This deceleration is thought to be a consequence of the interaction of atmospheric thermal tides with the westward superrotating zonal winds (Crisp 1986, Baker and Leovy 1987). It is interesting to note, however, that westward superrotating winds also dominate the circulation again at altitudes above 100 km in the thermosphere. Modeling studies (Zhang et al. 1996) suggest that these superrotating winds may be forced by the interaction of gravity waves and the subsolar to antisolar flow at these levels, but the observations are not adequate to verify this theory.

CRITICAL MEASUREMENTS NEEDED: A global description of both the steady and transient components of the wind field is needed at altitudes between the surface and the cloud tops. At levels within the clouds, these measurements can be obtained by tracking cloud feature motions from an orbiting spacecraft. The winds at levels near the cloud tops can be inferred by tracking motions of the UV features (Limaye 1988). These measurements can be combined with near-infrared observations of the night side, which reveal motions in the middle and lower cloud decks (Crisp et al. 1991) to yield a comprehensive, 3-dimensional description of the cloud dynamics. To be of use in diagnosing the atmospheric circulation, the altitude of the features being tracked must be known to within a few kilometers. This information can be acquired by deploying UV and IR imaging spectrometers rather than simple imagers.

Wind measurements at levels below the cloud base must be acquired by in-situ sensors, because there are no known tracers at those levels that can be tracked by remote sensing observations. High temporal resolution delta-delta-VLBI tracking of long-lived balloons would be best for this application because

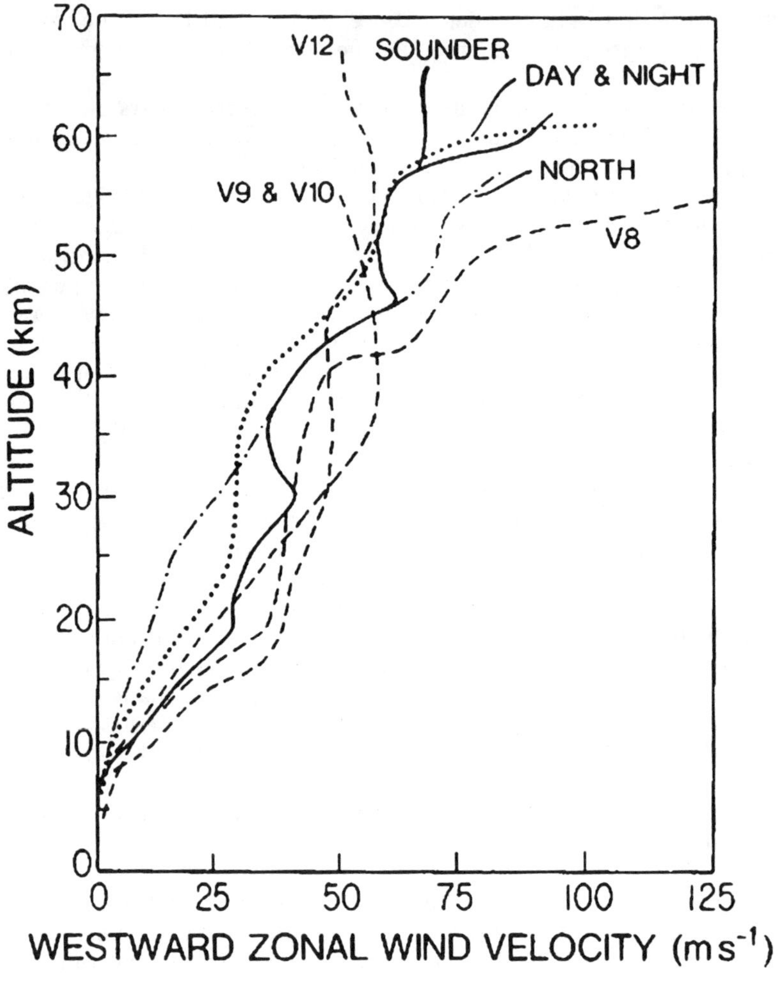

Figure 3. Wind velocity profiles measured by the Pioneer Venus Day, Night, North, and Sounder probes are compared to those measured by the Venera 8, 9, 10, and 12 (V8, V9, V10, and V12). The winds blow from east to west all altitudes and latitudes where measurements exist (from Schubert et al. 1980).

this would reveal both the mean and transient components of the flow (Crisp et al. 1990). One approach would deploy strings of three to five very simple, long-lived balloons at equatorial, mid-latitude, and high-latitude sites. At each site, the balloons could be deployed successively from a single entry probe as it falls through the atmosphere, and float at altitudes near 5, 20, 30, 40, and 50 km. A lifetime of about one week should be adequate to resolve most of the important components of the circulation. Besides transmitters, the balloons should carry pressure and temperature sensors to define the balloon altitude and atmospheric thermal structure. Relative wind sensors (and/or accelerometers) are needed to detect gusts and other transient phenomena. The atmospheric structure experiment should also be carried to the surface by the entry system to yield detailed profiles of pressures and temperature throughout the lower atmosphere at the balloon entry sites.

Lightning: Lightning plays an important role in the electrical and chemical compositions of the atmospheres of Earth and Jupiter. There is tentative evidence for its presence on Venus (Hunten 1995), but its role in the Venus environment is still controversial.

CRITICAL MEASUREMENTS NEEDED: Long-term simultaneous optical and electromagnetic observations, above, within, and below the clouds are needed to determine whether lightning plays a major role on Venus.

Middle atmosphere composition: The composition of the Venus middle atmosphere is the product of multiple local chemical processes and global circulation (Lellouch et al. 1997). The dominant constituent in the atmosphere, CO_2, is photodissociated to form carbon monoxide (CO) and atomic oxygen (O). Over time, CO and O_2 (from atomic O recombination) would become primary atmospheric constituents in the absence of catalytic processes that increase the rate at which CO_2 is reformed. However, CO and O_2 are transported to lower altitudes where their abundances decrease as a result of conversion to CO_2 through a series of homogeneous and/or heterogeneous reactions involving sulfur oxides (SO_x), hydrogen oxides (HO_x), or chlorine (Cl_x) radicals. Observations of intense $O_2(a^1\Delta_g)$ emission imply that atomic O recombination is highly efficient.

In addition to CO oxidation, the oxidation of SO_2 to form H_2SO_4 is a second loss process for O_2. H_2SO_4 condenses to produce aerosols that provide the sites for heterogeneous chemical reactions, leading to chemical processing not possible in pure gas phase reactions. This is balanced by transport to the troposphere and subsequent conversion back to SO_2. Mass exchange between the troposphere and the middle atmosphere must also result in transport of the hydrogen-, sulfur-, nitrogen-, and chlorine constituents that serve as sources for transient molecules involved in catalytic chemistry. Spatial and temporal changes in composition are forced by diurnal and latitudinal variations in insolation and transport, but very few measurements are available for even a rough characterization of these behaviors. Changes in the tropospheric composition resulting from volcanic eruptions may also lead to variability in middle atmosphere constituent abundances, but the evidence for this is still controversial.

CRITICAL MEASUREMENTS NEEDED: Spatially resolved global measurements of the vertical distributions of mesospheric densities and temperatures at vertical resolutions finer than 1 scale height (< 5 km) are needed to define the atmospheric thermal structure. Global measurements of the abundances of CO, O_2, O_3, ClO, H_2O, H_2O_2, NO, SO, SO_2, HCl and other important trace species are need at similar spatial resolution. These measurements should be performed over one to two Venus years to establish the amplitude of long-term secular variations.

Middle atmosphere dynamics: Thermally driven atmospheric tides are forced by the motion of the subsolar point across the upper cloud, where most of the sunlight is absorbed. The solar thermal tides assume the form of a solar-fixed wave disturbance that propagates eastward (upwind) and upward through the superrotating mean flow. The tides are readily apparent in the Pioneer Venus infrared temperature retrievals, but the height resolution is not sufficient to define the vertical structure through the middle atmosphere.

A key question about atmosphere dynamics is why the superrotation reaches a maximum near 70 km and then decays at higher altitudes (Lellouch et al. 1997). Two candidate mechanisms have been proposed. Atmospheric thermal tides induced by solar heating within and above the upper cloud are expected to produce downward momentum transports and poleward heat transport throughout the mesosphere. Vertically propagating, small scale gravity waves, discovered by Magellan radio occultation measurements, may also create sufficient wave drag to affect the mean zonal flow at some mesospheric levels. The relative roles of these two mechanisms are not yet understood.

The Venus middle atmosphere also serves as a dynamical transition from zonal circulation of the lower atmosphere to the extremely rapid and efficient subsolar-to-antisolar circulation within the thermosphere. This circulation transports the products of CO_2 photolysis (CO, O, and O_2) in the day side middle atmosphere to the nightside thermosphere, creating large night side enhancements in CO and (presumably) O_2 at high altitudes. Very little is known about the dynamical transition between zonal and the subsolar-to-antisolar flow within the middle atmosphere. However, existing measurements of the thermal structure and Doppler velocities indicate large amplitude, global scale changes in the upper mesospheric circulation on time scales of less than 1 year. The mechanisms responsible for these changes have not yet been identified.

CRITICAL MEASUREMENTS NEEDED: Direct Doppler measurements of global and vertical distributions of zonal and meridional winds are needed between the cloud tops and >100 km. These measurements need a vertical resolution of one scale height (~ 5 km) or less. Like the compositional measurements, these data should be collected over long enough time scale to establish the scale of long-term secular variations.

Thermospheric composition and dynamics: Venus has no appreciable intrinsic magnetic field. Hence, the solar wind interacts directly with its ionosphere forming a magnetic barrier and ionopause. Most of the survey-level upper atmospheric investigations were carried out by the Pioneer Venus Probes and Orbiter (Kasprzak et al. 1997). In particular, UV imaging and limb scanning measurements were made by the Pioneer Venus Orbiter (PVO) Ultraviolet Spectrometer (OUVS) throughout the solar cycle, illustrating the variable nature of the chemical, energetic and dynamical processes of the thermosphere, ionosphere, and exosphere over time. In addition, PVO orbital decay measurements of atmospheric densities and temperatures and neutral mass spectrometer (NMS) measurements of several species (CO_2, CO, N_2, O, N, He) were made near the equator spanning altitudes between 130 and 250 km over a fraction of the 11-year solar cycle. Atmospheric drag measurements by Magellan provided a more global description of the density structure, and extended the observations over the entire solar cycle (Keating and Hsu 1993). These measurements also revealed a 4-day oscillation in the thermospheric density structure that is thought to be related to the thermospheric superrotation.

CRITICAL MEASUREMENTS NEEDED: There is a compelling need to follow up on the survey-level orbital decay, NMS, and OUVS measurements with a monitoring program focused on the regular observation of atmospheric densities, neutral winds, and atmospheric species subject to volatile escape. Direct wind measurements at thermospheric levels could be obtained by a Fabry-Perot interferometer or a baffled neutral mass spectrometer on a Venus orbiter. Densities and temperatures can be inferred from accelerometer measurements of orbit decay. In addition, tracers of upper atmospheric winds, including strong IR (O_2 1.27-μm) and UV (NO 198.0-nm) emissions, can be monitored to reveal spatial and temporal variations of the circulation.

Ionospheric structure, composition, and dynamics: The processes contributing to the volatile escape of hydrogen and oxygen from Venus are not well constrained for the present epoch (Fox and Kliore 1997, Nagy and Cravens 1997). Hence, they are difficult to extrapolate into the past for studies of Venus water loss. Pioneer Venus measurements provided an initial sampling of the UV emissions and tailward ion flow needed to identify potential loss mechanisms, but these measurements were not adequate to quantify the loss processes.

CRITICAL MEASUREMENTS NEEDED: Both H and O coronas were identified and periodically measured by the OUVS. However, an ongoing monitoring program of key UV emissions (130.4-nm, 135.6-nm, 121.6-nm, etc.) and tailward ion density flows (H^+, O^+, etc.) is needed to properly constrain loss mechanisms and their rates over the solar cycle. In addition, O_2^+ densities (lower ionosphere) are needed simultaneously to constrain the dissociative recombination rates which control the exospheric hot-O seed population for ion-pickup (O^+) loss. Such data sets will enable model extrapolations of volatile escape rates to be made (into the past) with enhanced reliability.

3. Interconnections: The Volatile History of Venus

As noted above, the volatile inventory of Venus appears to be more representative of the primordial solar nebula than Earth and Mars. A coordinated investigation of the volatile inventories of Venus and Mars may be essential to understand the origin and evolution of terrestrial planets, a critical focus of the Astrophysical Origins Program. An investigation of Venus volatiles would also provide an obvious link to NASA's current Mars strategy, where water plays a central role.

For a terrestrial planet, an investigation of volatiles provides critical constraints on the atmospheric origin and compositional history, emergence of a coupled atmosphere-surface system and climate evolution. A coordinated analysis of water and other volatiles on Venus, Earth, and Mars will provide improved constraints on the effects of atmospheric cratering, thermal escape, sputtering, and other exogenic processes that have contributed to the loss of oxygen and water from their atmospheres. An understanding of the early loss of water by these planets is also needed to understand the evolution of their surfaces. In particular, on Venus, the water loss rate affects the rheology and differentiation rate of the interior. These factors affect the surface morphology and its ability to record past geological processing.

Improved constraints on the history of water and other volatiles on Venus and Mars will provide benefits that extend well beyond an improved understanding of their relationship to the initial state of a particular planet and the mechanisms that drove it down its specific evolutionary track. Attempts to decipher planetary histories in the broader context of the evolution of the solar system as a whole are focusing more and more attention on the sources and processing of volatiles in the primordial solar accretion disk, in primitive meteorites, and on the terrestrial planets as a class. The proposed measurement of Venus noble gases, trace gases, and surface chemical composition and weathering products directly address this need.

4. Required Associated Activities

4.1. Technology development

Little or no technology development would be needed to enable the orbiter and in-situ measurements of atmospheric noble gases and trace gas distributions. In fact, several technologically mature Discovery missions have been proposed to address these issues over the past decade. The radar technologies needed to enable improved high-resolution global mapping of the surface have been developed and validated in Earth orbit over the past decade. In contrast, a substantial amount of technology development is needed to enable long-lived in-situ measurements in the high-temperature, high-pressure environment of the deep atmosphere and the surface.

The Atmosphere Dynamics Explorer described above employs long-lived balloons deployed over a range of altitudes. This mission requires technologies for lightweight, low-power, high-temperature electronics and power systems that can operate over the full range of temperatures encountered between the 730 K surface and the 240 K cloud tops. This mission also requires the development of advanced balloon materials and deployment systems.

The Surface and Interior Explorer will deploy three or more landers at the surface to acquire seismological and geochemical data up to a year. This mission requires the development of a broad range of technologies that can survive and operate in this environment. The acquisition of seismic data is a challenge because seismometers require excellent coupling to the surface, and a minimum of mechanical noise from winds and the spacecraft. In-situ chemical analysis of surface samples requires technologies for acquiring samples (perhaps from a range of near-surface depths) and delivering them to the analytical instrumentation, which presumably works in conditions somewhat less harsh those that characterize the ambient environment. To support these instruments, the landers require long-lived power systems that can operate at the Venus surface, as well as thermal control systems, high temperature electronics, and other high-pressure, high-temperature mechanical systems (pressure vessels, feedthroughs, bearings, etc.). An even broader range of technologies will be needed to enable future Venus sample return missions.

4.2. Research and analysis

There are currently no focused Venus Data Analysis Programs because NASA has not sent a dedicated mission to Venus in more than a decade. Basic research on this planet has evolved into a fringe activity, supported intermittently by the Planetary Astronomy, Planetary Atmospheres, and Planetary Geology and Geophysics Programs. Much of the available NASA funding supports the development of advanced theoretical modeling tools, and the application of these tools to the further analysis of data collected by earlier dedicated missions (Magellan, Pioneer Venus, Venera) and recent flyby opportunities (e.g., Galileo and Cassini). This work continues to produce advances in our understanding of this enigmatic planet, but the rate of progress has slowed in recent years, following the precipitous decline in the level of support. Significant growth in these NASA R&A programs will be needed to support future flyby opportunities, such as the planned Discovery Messenger flybys in June 2004 and March 2006, and to optimize experiments for future dedicated missions.

In contrast to the relative dearth of spacecraft observations over the past decade, new instrumentation and observational techniques becoming available at ground-based astronomical observatories have yielded a treasure trove of new data to support some areas of Venus exploration. In particular, the development of advanced near-infrared and microwave instruments has provided new opportunities to study the thermal structure, composition, and dynamics of the Venus atmosphere. Although Venus is our nearest planetary neighbor and usually the third brightest object in the sky (after the sun and the moon) these new instruments produced surprises and breakthroughs at each Venus inferior conjunction throughout the 1990's. This work has played a major role in the development of Discovery proposals targeted at Venus.

Although some of these observations have yet to be analyzed or published because of the limited support available for basic research, a number of future advances in this area are anticipated as new observing capabilities become available and are exploited by individual researchers. In addition to these individual efforts, a great deal could be learned about Venus from coordinated ground based efforts that employ global networks of observatories to obtain long-term, round-the-clock measurements. Small scale efforts of this kind were organized to support the Galileo and Cassini fly-bys of Venus. More ambitious efforts that include more global coverage and a wider range of wavelengths (UV, near IR, thermal IR, and microwave) are being considered to support the Discovery Messenger encounters on 24 June 2004 and 16 March 2006, and to support the European Venus Express and Japanese Venus Climate Orbiter (Planet C) missions. These efforts will require enhanced support from the NASA Planetary Astronomy Program, or a dedicated Venus Research and Analysis Program.

5. Opportunities for Education and Public Outreach (E&PO)

Because Venus is Earth's sister planet, comparisons between Venus, Earth, and Mars provide valuable tools for teaching physical, chemical, and geological processes that affect the environment. In particular, comparisons of Venus, Earth, and Mars can also provide insight into processes such as:

- Clouds: Why does the Earth have water clouds, while Venus has sulfuric acid clouds, and Mars has both water and carbon dioxide (dry ice) clouds?
- Seasons: Why do Earth and Mars have seasons, while Venus does not?
- Climate: Are greenhouse effects real? How much can greenhouse effects alter a planet's climate (Venus is the 300,000 times CO_2 case)?
- Planetary rotation: How does the Earth's rotation rate affect its weather and climate? What happens on a very slowly rotating planet like Venus? Why does the sun rise in the east and set in the west on Earth. Does it rise in the west on Venus?
- Plate tectonics. Terrestrial geology is dominated by plate tectonics on large scales. What geologic processes dominate on planets that have no plate tectonics, like Venus?

In short, by providing a dramatic counterexample to the Earth, Venus provides an effective tool for illustrating how various processes affect our environment.

6. Specific Recommendations:

While it is not possible to address all of the questions with a single mission, it should be possible to dramatically improve our understanding of our sister planet within the constraints of a coordinated program of small (Discovery Class), medium (New Frontiers Class), and large (~$1 B class) missions that are conducted by NASA and its international partners over the next 10 to 20 years. The NASA mission with the highest priority and greatest technological maturity is the Noble Gas/Trace Gas Explorer. The other missions listed below have comparable priorities, but they focus on different topics and have different levels of technological maturity. These missions should fly as their critical technologies become available.

6.1. The Noble Gas and Trace Gas Explorer

This has been identified as the highest priority mission because its data are vital to our understanding of the origin of Venus and its implications for our understanding of the early evolution of terrestrial planets. This small mission requires a single entry probe that can carry the state-of-the-art instruments needed to complete the noble gas and trace gas inventories between the cloud tops and the surface. These instruments include:

- State-of-the-art mass spectrometer optimized for noble gases

- Near IR and UV spectrometers for high-precision trace gas profiling
- Pressure and temperature sensors for altitude reference
- Feed-forward technology, such as a validation experiment for surface imaging and surface chemistry

This mission could be combined with the Atmospheric Composition Orbiter described below to produce a Small to Medium class mission that provides a comprehensive description of the Venus trace gas distributions.

6.2. The Global Geological Process Mapping Orbiter

This is a Small to Medium class mission that will provide global maps of the surface at horizontal resolutions of 25 to 50 m. These data are needed to identify and characterize the geologic processes that have shaped the Venus surface. This mission will fly in a global orbit and carry an advanced C and/or X-band radar designed to provide global stereo or interferometric coverage. The latter approach would require the use of a long (tens of meters) boom, like that demonstrated recently by the Shuttle Radar Topography Mission (SRTM). It might also carry instruments for identifying the atmospheric signature of volcanic activity (plumes, surface fixed sources of volcanic gases or aerosols) and discriminating these from other atmospheric processes (e.g. cloud formation, aeolean processes). This mission would require a lifetime of at least one Venus year to yield global coverage, but two Venus years would be highly desirable to search for time dependent variations in the surface associated with ongoing geologic activity. This mission could fly alone, or be combined with the Surface and Interior Explorer described below to produce a large mission that provides a much greater science return per dollar.

6.3. The Atmospheric Composition Orbiter

This is a Small mission that will carry remote sensing instruments for characterizing spatial and temporal variations in the clouds and trace gases throughout the atmosphere. It will carry the following instrumentation:

- A high resolution near-IR imaging spectrometer for characterizing clouds, lower atmosphere trace gases, and dynamics
- A UV imaging photopolarimeter for tracking cloud-top winds and characterizing the upper haze to constrain radiative processes and heterogeneous chemistry above the cloud tops
- A submillimeter heterodyne spectrometer for global measurements of middle atmosphere trace constituents and direct Doppler wind measurements
- An S-band and/or X-band radio science package allowing retrievals of: (a) Sulfuric acid vapor profiles below clouds, (b) The pressure, temperature, and density between 34 km and 100 km (c) Thermal wind profiles in this altitude range

- An instrument package for characterizing the thermospheric dynamics and exospheric mass loss, including (a) An accelerometer for thermospheric density scale heights (b) a neutral/ion mass spectrometer for in-situ exospheric mass loss measurements and (c) a Fabry-Perot interferometer or baffled mass spectrometer for thermospheric dynamics

These data will be used to characterize the radiative, chemical, and dynamical processes that are maintaining the thermal structure and composition of the present atmosphere. As noted above, this mission could be combined with the Noble Gas and Trace Gas Explorer to produce a Small to Medium mission with a much larger science return.

6.4. The Atmospheric Dynamics Explorer

This Medium mission includes both an orbiter and a network of entry probes. The four to eight entry probes will deploy twelve to twenty-four long-lived balloons at 3 to 4 altitudes between the surface and cloud tops at equatorial, mid, high latitudes. Each balloon will acquire time-resolved in-situ measurements of pressure, temperature, and the solar and thermal radiation fields over its ~1 week lifetime. VLBI tracking and radio imaging by the orbiter will be used to determine the position and velocity of each balloon over this period. This information will be analyzed to yield both the mean and transient components of the wind field between the cloud levels and the surface. The orbiter is required for communications and balloon tracking. It will use UV and near IR imaging spectrometers for tracking winds in the upper, middle, and lower cloud decks and to record the near surface static stability. It will also carry an S- and/or X-band radio science package to retrieve pressure, temperature and density at altitudes between 34 km and 100 km (Jenkins et al. 1994, Hinson and Jenkins 1995). These data will be combined with the in-situ measurements to identify the mechanisms responsible for maintaining the atmospheric superrotation.

6.5. The Surface and Interior Explorer

This Large mission will deploy three or more long-lived landers on the Venus surface. Each lander will carry a seismometer for studies of the interior structure, as well as in-situ instruments for characterizing the surface mineralogy and elemental composition. They may also carry meteorological instruments to identify mechanical disturbances associated with fluctuating winds, temperatures, and pressures that could compromise the interpretation of the seismic data. A lifetime of several months to a year is needed to address the science objectives of this mission. Because these landers must survive the harsh environment of the Venus surface, they require significant technology development for power systems, thermal control systems, electronics, and instruments.

6.6. Sample return mission

A sample return mission will eventually be needed to conduct investigations of the Venus surface and atmosphere that cannot be conducted by instruments on remote sensing platforms or on entry probes. The specific requirements for this mission are impossible to formulate at this time, but we anticipate that it will probably require a large mission and significant technology development.

7. Concluding Remarks:

Venus is our closest neighbor in this solar system, and the most Earth-like planet in size, mass, and solar distance. In spite of these similarities, and the intense scrutiny that it received early in space age, our understanding of its origin, evolution, and present environment are still shrouded in mystery. Information derived from a reinvigorated Venus Exploration Program could be combined with that provided by the ongoing Mars Exploration Program to dramatically improve our understanding of these properties of terrestrial planets. This information would be of great value in future efforts to interpret observations of extrasolar terrestrial planets, and to understand their range of habitability.

The only NASA spacecraft currently schedule to visit Venus is the Discovery Messenger mission, which will fly past this planet for a gravitational boost on 24 June 2004 at a distance of ~2500 km, and then again for a second time on 16 March 2006, at a distance of ~4300 km. While these brief encounters cannot provide answers to fundamental questions like those posed above, Messenger's instruments will provide new constraints on the enigmatic environment of our sister planet, and help to focus attention on our long-overlooked neighbor.

References

Anderson, F.S. and S.E. Smrekar 1999. Tectonic effects of climate change on Venus. *J. Geophys. Res.* **104**, 743-756.

Baker, N.L., and C.B. Leovy 1987. Zonal winds near Venus' cloud top level: A model study of the interactions between the zonal mean circulation and the semidiurnal tide. *Icarus* **69**, 202–220.

Barsukov V.L., S.P. Borunov, V.P. Volkov, V.A. Dorofeeva, M.I. Zolotov, S.V. Parotkin, I.V. Semenov, I.I. Sidorov, I.L. Khodakovskii, and A.I Shapkin 1986. Estimate of the mineral-composition of Venus soil at Venera-13, Venera-14 and Vega-2 landing sites (thermodynamic calculations). *Doklady Akademii Nauk SSSR* **287**, 415-418.

Bézard, B., C. de Bergh, D. Crisp, and J.-P. Maillard 1990. The deep atmosphere of Venus revealed by high-resolution nightside spectra. *Nature* **345**, 508–511.

Bézard B., C. de Bergh, B. Fegley, J.-P. Maillard, D. Crisp, T. Owen, J. B. Pollack, and D. Grinspoon 1993. The abundance of sulfur-dioxide below the clouds of Venus. *Geophys. Res. Lett.* **20**, 1587-1590.

Bullock, M.A., and D.H. Grinspoon 1996. The stability of climate on Venus. *J. Geophys. Res.* **101**, 752–2729.

Bullock, M.A., and D.H. Grinspoon 2001. The recent evolution of climate on Venus. *Icarus* **150**, 19-37.

Campbell, B.A. 1999. Surface Formation Rates and impact densities on Venus. *J. Geophys. Res.* **104**, 21951-21955.

Crisp, D. 1986. Radiative forcing of the Venus mesosphere I: Solar fluxes and heating rates. *Icarus* **67**, 484–514.

Crisp, D., and D. Titov 1997. The thermal balance of the Venus atmosphere. In *Venus II* (S.W. Bougher, D.M. Hunten, and R.J. Phillips, Eds.), pp. 353-384. Univ. of Arizona Press, Tucson.

Crisp, D., A.P. Ingersoll, C.E. Hildebrand, and R.A. Preston 1990. VEGA Balloon meteorological measurements. *Adv. Space Res.* **10**, 109-124.

Crisp, D., S. McMuldroch, S.K. Stephens, W.M. Sinton, B. Ragent, K.W. Hodapp, R.G. Probst, L.R. Doyle, D.A. Allen, and J. Elias 1991. Ground-based near–infrared imaging observations of Venus during the Galileo Encounter. *Science* **253**, 1538-1541.

de Bergh, C., B. Bézard, T. Owen, D. Crisp, J.-P. Maillard, and B.L. Lutz 1991. Deuterium on Venus: Observations from Earth. *Science* **251**, 547–549.

de Bergh, C., B. Bézard, D. Crisp, J.-P. Maillard , T. Owen, J. Pollack, and D. Grinspoon 1995. Water in the deep atmosphere of Venus from high-resolution spectra of the night side. *Adv. Space Res.* **15**, 79-88.

Donahue, T.M., and J.B. Pollack 1983. Origin and evolution of the Atmosphere of Venus. In *Venus* (D.M. Hunten, L. Colin, T.M. Donahue, and V.I. Moroz, Eds.), pp. 1003-1036. Univ. of Arizona Press, Tucson.

Donahue, T.M., and R.R. Hodges 1993. Venus methane and water. *Geophys. Rev. Lett.* **20**, 591-594.

Esposito, L.W., J.-L. Bertraux, V. Krasnopolsky, and V.I. Moroz 1997. Chemistry of lower atmosphere and Clouds. In *Venus II* (S.W. Bougher, D.M. Hunten, and R.J. Phillips, Eds.), pp. 415-458. Univ. of Arizona Press, Tucson.

Fegley, B., G. Klingelhofer, K. Lodders, and T. Widemann 1997. Geochemistry of surface-atmosphere interactions on Venus. In *Venus II* (S.W. Bougher, D.M. Hunten, and R.J. Phillips, Eds.), pp. 591-636. Univ. of Arizona Press, Tucson.

Florenskiy, K.P., A.T. Bazilevskiy, G.A. Burba, O.V. Nikolayeva, A.A. Pronin, A.S. Selivanov, M.K. Narayeva, A.S. Panfilov, and V.P. Chemodanov 1983. Panorama of Venera 9 and 10 landing sites. In *Venus* (D.M. Hunten, L. Colin, T.M. Donahue, and V.I. Moroz, Eds.), pp. 137-153. Univ. of Arizona Press, Tucson.

Fox, J.L., and A.J. Kliore 1997. Ionosphere: Solar cycle variations. In *Venus II* (S.W. Bougher, D.M. Hunten, and R.J. Phillips, Eds.), pp. 161-188. Univ. of Arizona Press, Tucson.

Gierasch, P.J., R.M. Goody, R.E. Young, D. Crisp, C. Edwards, R. Kahn, D. Rider, A. Del Genio, R. Greeley, A. Hou, C.B. Leovy, D. McCleese, and M. Newman 1997. The general circulation of the Venus atmosphere: An assessment. In *Venus II* (S.W. Bougher, D.M. Hunten, and R.J. Phillips, Eds.), pp. 459-502. Univ. of Arizona Press, Tucson.

Hansen, V.L. 2001. Venus diapirs, thermal or compositional? *Lunar Planet. Sci.* **XXXII**, 1036. (abstract)

Hansen, V.L. 2002. Artemis: signature of a deep Venusian mantle plume. *Geol. Soc. Amer. Bull.*, in press.

Hansen, V.L., and J.J. Willis 1998. Ribbon terrain formation, southwestern Fortuna Tessera, Venus: Implications for lithosphere evolution. *Icarus* **132**, 321-343.

Hansen, V.L., J.J. Willis, and W.B. Banerdt 1997. Tectonic overview and synthesis. In *Venus II* (S.W. Bougher, D.M. Hunten, and R.J. Phillips, Eds.), pp. 797-844. Univ. of Arizona Press, Tucson.

Hansen, V.L., B.K. Banks, and R.R. Ghent 1999. Tessera terrain and crustal plateaus, Venus. *Geology* **27**, 1071-1074.

Hansen, V.L., R.J. Phillips, J.J. Willis, and R.R. Ghent 2000. Structures in Tessera Terrain, Venus: Issues and Answers. *J. Geophys. Res.* **105**, 4135-4152.

Hauck, S.A., R.J. Phillips, and M.H. Prince 1998. Venus: Crater distribution and plans resurfacing models. *J. Geophys. Res.* **103**, 13635-13642.

Head, J.W., E.M. Parmentier, and P.C. Hess 1994. Venus–Vertical accretion of crust and depleted mantle and implications for geological history and processes. *Plan. Space Sci.* **42**, 803-811.

Herrick, R.R., and R.J. Phillips 1994. Implications of a global survey of Venusian impact craters. *Icarus* **111**, 387-416.

Hinson, D.P., and J.M. Jenkins 1995. Magellan radio occultation measurements of atmospheric waves on Venus. *Icarus* **114**, 310-327.

Hoffman, J.H., R.R. Hodges, T.M. Donahue, and M.B. McElroy 1980. Composition of the Venus lower atmosphere from the Pioneer Venus mass spectrometer. *J. Geophys. Res.* **85**, 7882–7891.

Hunten, D.M. 1995. Venus Lightning - Pros and cons. *Adv. Space Res.* **15**, 109-112.

Janes, D.M., S.W. Squyres, D.L. Bindschadler, G. Baer, G. Schubert, V.L. Sharpton, and E.R. Stofan 1992. Geophysical models for the formation and evolution of coronae on Venus. *J. Geophys. Res.* **97**, 16055-16068.

Jenkins, J.M., P.G. Steffes, D.P. Hinson, J.D. Twicken, and G.L. Tyler 1994. Radio occultation studies of the Venus atmosphere with the Magellan spacecraft: 2. Results from the October 1991 experiments. *Icarus* **110**, 79-94.

Kasprzak W.T., G.M. Keating, N.C. Hsu, I.A.F. Stewart, W.B. Colwell, and S.W. Bougher,1997. Solar activity behavior of the thermosphere. In *Venus II* (S.W. Bougher, D.M. Hunten, and R.J. Phillips, Eds.), pp. 295-324. Univ. of Arizona Press, Tucson.

Keating, G.M., and N.C. Hsu 1993. The Venus atmospheric response to solar cycle variations. *Geophys. Rev. Lett.* **20**, 2751-2754.

Lellouch, E., T. Clancy, D. Crisp, A.J. Kliore, D. Titov, and S.W.Bougher 1997. Monitoring of mesospheric structure and dynamics. In *Venus II* (S.W. Bougher, D.M. Hunten, and R.J. Phillips, Eds.), pp. 295-324. Univ. of Arizona Press, Tucson.

Limaye, S.S. 1988. Venus - Cloud level circulation during 1982 as determined from Pioneer Cloud Photopolarimeter images. 2. Solar longitude dependent circulation. *Icarus* **73**, 212-226.

Mackwell, S.J., M. E. Zimmerman, and D.L. Kohlstedt 1998. High-temperature deformation of dry diabase with application to tectonics on Venus. *J. Geophys. Res.* **103**, 975-984.

McGill, G.E., S.J. Steenstrup, C. Barton, and P.G. Ford 1981. Continental rifting and the origin of Beta-Regio, Venus. *Geophys. Res. Lett.* **8**, 737-740.

McKinnon, W.B. , K.J. Zahnle, B.A. Ivanov, and H.J. Melosh 1997. Cratering on Venus: Models and observations. In *Venus II*, (S.W. Bougher, D.M. Hunten, and R.J. Phillips, Eds.), pp. 969-1014. Univ. of Arizona Press, Tucson.

Meadows, V.S., and D. Crisp 1996. Ground-based near infrared observations of the Venus nightside: The thermal structure and water abundance near the surface. *J. Geophys. Res.* **101**, 4595-4622.

Moresi, L., and V. Solomatov 1998. Mantle convection with a brittle lithosphere: Thoughts on the global tectonic styles of the Earth and Venus. *Geophys. J. Int.* **133**, 669-682.

Moroz, V. I. 1983. Summary of preliminary results of the Venera 13 and Venera 14 missions. In *Venus* (D.M. Hunten, L. Colin, T.M. Donahue, and V.I. Moroz, Eds.), pp. 45-68. Univ. of Arizona Press, Tucson.

Nagy, A.F., and T.E. Cravens 1997. Ionosphere: Energetics. In *Venus II* (S.W. Bougher, D.M. Hunten, and R.J. Phillips, Eds.), pp. 189-224. Univ. of Arizona Press, Tucson.

Owen, T.C. 2000. The prevalence of Earth-like planets. *Acta Astronautica* **46**, 617-620.

Owen, T.C., and A. Bar-Nun 1995. Comets, impacts, and atmospheres. *Icarus* **116**, 215-226.

Owen, T.C., A. Bar-Nun, and I. Kleinfeld 1992. Possible cometary origin of heavy noble gases in the atmospheres of Venus, Earth, and Mars. *Nature* **358**, 43-46.

Parmentier, E.M., and P.C. Hess 1992. Chemical differentiation of a convecting planetary interior: consequences for a one plate planet such as Venus. *Geophys. Res. Lett.* **19**, 2015-2018.

Pepin, R.O. 1991. On the origin and early evolution of terrestrial planet atmospheres and meteoritic volatiles. *Icarus* **92**, 2-79.

Pepin, R.O. 1997. Evolution of the Earth's noble gases: Consequences of assuming hydrodynamic loss driven by giant impact. *Icarus* **126**, 148-156.

Phillips, R.J. 1993. The age spectrum of the Venusian surface: *Eos* (Supplement) **74**(16), 187.

Phillips, R.J., and M.C. Malin 1984. Tectonics of Venus. *Annu. Rev. Earth Planet. Sci.* **12**, 411-443.

Phillips, R.J., and V.L. Hansen 1994. Tectonic and Magmatic Evolution of Venus. *Annu. Rev. Earth Planet. Sci.* **22**, 597-654.

Phillips, R.J., and V.L. Hansen 1998. Geological evolution of Venus: Rises, plains, plumes and plateaus. *Science* **279**, 1492-1497.

Phillips, R.J., R.F. Raubertas, R.E. Arvidson, I.C. Sarkar, R.R. Herrick, N. Izenberg, and R.E. Grimm 1992. Impact craters and Venus resurfacing history. *J. Geophys. Res.* **97**, 15923-15948.

Phillips, R.J., M.A. Bullock, and S.A. Hauck 2001. Climate and interior coupled evolution of Venus. *Geophys. Res. Lett.* **28**, 1779-1782.

Pollack J.B., J. B. Dalton, D. Grinspoon, R.B. Wattson, R. Freedman, D. Crisp, D.A. Allen, B. Bézard, C. de Bergh, L.P. Giver, Q. Ma, and R. Tipping 1993. Near-infrared light from Venus nightside - A spectroscopic analysis. *Icarus* **103**, 1-42.

Regenauer-Lieb, K., D.A. Yuen, and J. Branlund 2001. The initiation of subduction: criticality by addition of water? *Science* **294**, 578-580.

Schaber, G.G., R.G. Strom, H.J. Moore, L.A. Soderblom, R.L. Kirk, D.J. Chadwick, D.D. Dawson, L.R. Gaddis, J.M. Boyce, and J. Russell 1992. Geology and distribution of impact craters on Venus - What are they telling us? *J. Geophys. Res.* **97**, 13257-13301.

Schubert, G., C. Covey, A. Del Genio, L.S. Elson, G. Keating, A. Seiff, R.E. Young, J. Apt, C.C. Counselman III, A.J. Kliore, S.S. Limaye, H.E. Revercombe, L.A. Sromovsky, V.E. Suomi, F. Taylor, R. Woo, and U. von Zahn 1980. Structure and circulation of the Venus atmosphere. *J. Geophys. Res.* **85**, 8007-8025.

Smrekar, S.E., W.S. Kiefer, and E.R. Stofan 1997. Large volcanic rises on Venus. In *Venus II* (S.W. Bougher, D.M. Hunten, and R.J. Phillips, Eds.), pp. 845-879. Univ. of Arizona Press, Tucson.

Solomon, S.C., S.E. Smrekar, D.L. Bindschadler, R.E. Grimm, W.M. Kaula, G.E. McGill, R.J. Phillips, R.S. Saunders, G. Schubert, S.W. Squyres, and E.R. Stofan 1992. Venus tectonics: an overview of Magellan observations. *J. Geophys. Res.* **97**, 13199-13255.

Solomon, S.C., M.A. Bullock, and D.H. Grinspoon 1999. Climate change as a regulator of tectonics on Venus. *Science* **286**, 87-90.

Squyres, S.W., D.M. Janes, G. Baer, D.L. Bindschadler, G. Schubert, V.L. Sharpton, and E.R. Stofan 1992. The morphology and evolution of coronae on Venus. *J. Geophys. Res.* **97**, 13611-13634.

Stofan, E.R., D.L. Bindschadler, J.W. Head, and E.M. Parmentier 1991. Corona structures on Venus: Models of origin. *J. Geophys. Res.* **96**, 20933-20946.

Stofan, E.R., V.L. Sharpton, G. Schubert, G. Baer, D.L. Bindschadler, D.M. Janes, and S.W. Squyres 1992. Global distribution and characteristics of coronae and related features on Venus: Implications for origin and relation to mantle processes. *J. Geophys. Res.* **97**, 13347-13378.

Stofan, E.R., V.E. Hamilton, D.M. Janes, and S.E. Smrekar 1997. Coronae on Venus: Morphology and origin. In *Venus II* (S.W. Bougher, D.M. Hunten, and R.J. Phillips, Eds.), pp. 931-968. Univ. of Arizona Press, Tucson.

Strom, R.G., G.G. Schaber, and D.D. Dawson, 1994. The global resurfacing of Venus. *J. Geophys. Res.* **99**, 10899-10926.

Taylor, F.W., R. Beer, M. T. Chahine, D.J. Diner, L.S. Elson, R.D. Haskins, D.J. McCleese, J.V. Martonchik, P.E. Reichley, S.P. Bradley, J. Delderfield, J.T. Schofield, C.B. Farmer, L.Froidevaux, J. Leung, M.T. Coffey, and J.C. Gille 1980. Structure and meteorology of the middle atmosphere of Venus: Infrared remote sensing from the Pioneer Orbiter. *J. Geophys. Res.* **85**, 7963-8006.

Weinstein, S.A., and P.L. Olsen 1992. Thermal convection with non-Newtonian plates. *Geophys. J. Int.* **111**, 515-530.

Wetherill, G.W. 1981. Solar-wind origin of Ar-36 on Venus. *Icarus* **46**, 70-80.

Zhang, S., S.W. Bougher, and M.J. Alexander 1996. The impact of gravity waves on the Venus thermosphere and O_2 IR nightglow. *J. Geophys. Res.* **101**, 23195-23205.

The Future of Solar System Exploration, 2003-2013
ASP Conference Series, Vol. 272, 2002
Mark V. Sykes, ed.

Terrestrial Analogs to Mars

T.G. Farr[1], S. Arcone[2], R.E. Arvidson[3], V. Baker[4], N.G. Barlow[6], D. Beaty[1], M.S. Bell[7], D.D. Blankenship[8], N. Bridges[1], G. Briggs[9], M. Bulmer[10], F. Carsey[1], S.M. Clifford[11], R.A. Craddock[12], P.W. Dickerson[7], N. Duxbury[1], G.L. Galford[3], J. Garvin[13], J. Grant[12], J.R. Green[1], T.K.P. Gregg[14], E. Guinness[3], V.L. Hansen[15], M.H. Hecht[1], J. Holt[8], A. Howard[16], L.P. Keszthelyi[5], P. Lee[9,17], P.D. Lanagan[5], R.C.F. Lentz[18], D.W. Leverington[12], L. Marinangeli[19], J.E. Moersch[18], P.A. Morris-Smith[20], P. Mouginis-Mark[21], G.R. Olhoeft[22], G.G. Ori[19], P. Paillou[23], J.F. Reilly II[7], J.W. Rice, Jr.[24], C.A. Robinson[25], M. Sheridan[14], K. Snook[7,9], B.J. Thomson[26], K. Watson[7], K. Williams[24], & K. Yoshikawa[27]

[1] *Jet Propulsion Laboratory, California Institute of Technology, Pasadena, CA*
[2] *Cold Regions Research and Engineering Laboratory, Hanover, NH*
[3] *Department of Earth and Planetary Sciences, Washington University, St. Louis, MO*
[4] *Department of Hydrology, University of Arizona, Tucson, AZ*
[5] *Lunar and Planetary Laboratory, University of Arizona, Tucson, AZ*
[6] *Department of Physics, University of Central Florida, Orlando, FL*
[7] *NASA Johnson Space Center, Houston, TX*
[8] *Institute for Geophysics, University of Texas, Austin, TX*
[9] *NASA Ames Research Center, Moffett Field, CA*
[10] *Joint Center for Earth Systems Technology, University of Maryland Baltimore County, Baltimore, MD*
[11] *Lunar and Planetary Institute, Houston, TX*
[12] *Center for Earth and Planetary Studies, National Air and Space Museum, Washington, DC*
[13] *NASA Headquarters, Washington, DC*
[14] *Department of Geology, University at Buffalo, Buffalo, NY*
[15] *Department of Geological Sciences, Southern Methodist University, Dallas, TX*
[16] *Department of Environmental Sciences, University of Virginia, Charlottesville, VA*
[17] *SETI Institute, Mountain View, CA*
[18] *Department of Geological Sciences, University of Tennessee, Knoxville, TN*
[19] *International Research School of Planetary Sciences, Universita d'Annunzio, Pescara, Italy*
[20] *Department of Natural Science, University of Houston Downtown, Houston, TX*

[21] *Hawaii Institute Geophysics and Planetology, University of Hawaii, Honolulu, HI*
[22] *Department of Geophysics, Colorado School Mines, Golden, CO*
[23] *Observatoire Astronomique de Bordeaux, University of Bordeaux, Floirac, France*
[24] *Department of Geological Sciences, Arizona State University, Tempe, AZ*
[25] *Center for Remote Sensing, Boston University, Boston, MA*
[26] *Department of Geology, Brown University, Providence, RI*
[27] *Institute of Northern Engineering, University of Alaska Fairbanks, Fairbanks, AK*

Abstract. It is well recognized that interpretations of Mars must begin with the Earth as a reference. The most successful comparisons have focused on understanding geologic processes on the Earth well enough to extrapolate to Mars' environment. Several facets of terrestrial analog studies have been pursued and are continuing. These studies include field workshops, characterization of terrestrial analog sites, instrument tests, laboratory measurements (including analysis of Martian meteorites), and computer and laboratory modeling. The combination of all these activities allows scientists to constrain the processes operating in specific terrestrial environments and extrapolate how similar processes could affect Mars. The Terrestrial Analogs for Mars Community Panel has considered the following two key questions: (1) How do terrestrial analog studies tie in to the Mars Exploration Payload Assessment Group science questions about life, past climate, and geologic evolution of Mars, and (2) How can future instrumentation be used to address these questions. The panel has considered the issues of data collection, value of field workshops, data archiving, laboratory measurements and modeling, human exploration issues, association with other areas of solar system exploration, and education and public outreach activities.

EXECUTIVE SUMMARY

The overarching scientific goals of the Mars program have been documented by the Mars Exploration Payload Assessment Group. They include: a) detect life or its remnants, b) determine present and past climate and its drivers, c) determine the geologic processes that have created and modified Mars, and d) prepare for human exploration. Underlying this is the fact that we are entering a new phase of Mars study: the era of global mapping is ending and studies will begin to focus on detailed analysis at the regional to outcrop level. Analog studies will play a crucial role in this transition.

Prediction of capabilities and interpretation of data from future instruments is also a critical need. By optimizing the choice of instruments and their operating parameters through tests in analog environments on Earth, we maximize our chances for success. Instruments in need of such study include low-frequency airborne or orbital radar sounders; low-frequency airborne or orbital imaging radar;

surface ground penetrating radar and EM sounders; drills; rovers; high resolution stereo imagers from orbit, lander, or microscope; high resolution topography; hyperspectral visible, near-IR, thermal IR; Raman; Mössbauer; SEM/EDAX, Laser-Induced Breakdown Spectroscopy. Accurate georeferencing or geocoding of these data needs to be developed operationally. In addition, Mars surface operations, both robotic and human, need to be developed and tested in analog settings.

Much Mars analog research has been accomplished and a great deal of activity is currently ongoing. However, more coordination among the many current individual projects could yield additional benefits. Focusing the community on a few field sites will also help build the diverse remote sensing data sets and supporting data needed to interpret their equivalents from Mars. This should particularly benefit the technology and operations development community as new instruments and techniques need testing. Laboratory measurements are another area in which ongoing efforts could be enhanced through appropriate funding.

Coordinated deployment of airborne, spaceborne, field instrumentation, and personnel to several sites to test instruments, technology, and exploration strategies intended for Mars and to provide data for ongoing studies of terrestrial geologic processes relevant to Mars is the most important area for support. The support will be used to leverage current investments by NASA codes S (Space Science), Y (Earth Science), and M (Human Exploration); NSF; US Army CRREL; and international agencies by acquiring additional data sets at important analog sites which are being studied for a variety of reasons. These activities should begin as soon as possible in order to benefit current and upcoming Mars missions.

Places where activities should commence as soon as possible include desert sites in the western US, central Australia, the Sahara Desert in Egypt, and the Dead Sea. High-priority cold regions to be studied include Antarctica as well as sites in Alaska, Canada, and Iceland. Volcanic areas of high priority include Hawaii, the Snake River Plain in Idaho, Kamchatka, Iceland, Mt. Pinatubo in the Philippines, and the East Africa Rift in the eastern horn of Africa. Young impact craters such as Meteor Crater, Arizona; Zhamanshin, Kazakhstan; Roter Kamm, Namibia; and several in Australia should also be given priority. These are all areas where significant activities are already underway, so that this program would be enhancing the data collection to improve applications to the study of Mars.

Costs for this part of the program will be modest for small enhancements to ongoing work in easily accessed areas, but high for coordinated multi-aircraft and field campaigns to places like Antarctica. These costs should be shared across agencies and it will require significant cross-agency and international coordination to make the larger projects happen.

Data archives are an important facet of this program. They would collect, document, and make available via the Web existing terrestrial data relevant to Mars studies. These could include various remote sensing and radar sounder data of important sites, laboratory data, and a comprehensive bibliography including actual articles and data files. The existing NASA Planetary Data System could host these archives. A working group is needed to identify data sets and appropriate formats for inclusion.

Field workshops should be continued and expanded to include a wider range of researchers, with guidebooks published on paper as well as Web-based for wider community access. These field workshops could also include operation of field instruments similar to those intended for Mars, modeled after the Marsokhod field tests at Silver Lake (Stoker et al. 2001). The Lunar and Planetary Institute is prepared to assist in workshop organization as it did for the Martian Highlands/Mojave Desert Workshop (Howard et al. 2001).

Laboratory measurements are ongoing, however this panel identified several areas that should be given a high priority. There is a critical need to characterize more fully various Mars analog materials, such as clathrates, palagonites, and rock coatings. Electromagnetic characterization of Martian analogs, such as ices, evaporites, clays and magnetic minerals needs to continue in support of the definition of future subsurface radar sounders.

Mars analog studies and geological field training exercises will be crucial in the human exploration of Mars because of its rich and complex geological history (perhaps even biological history). At analog sites, astronauts will be able to carry out the proper field investigations for correct geological context of the landing site as well as the surrounding environs, conduct precision landings, maximize surface mobility, and also perform intelligent sample acquisition and interpretation.

In the area of education and public outreach, the more accessible Mars analog sites especially could become "standard" sets for filmed news stories and documentaries. Demonstrations of rovers are always popular- more so in a Mars-like setting. The public can be encouraged to visit some of the sites, yielding an opportunity to describe how the site is like Mars. National Monuments and Parks such as Death Valley, Mojave Desert, and Hawaii Volcanoes may be interested in helping to support this.

REPORT

1. Introduction

It is well recognized that interpretations of Mars must begin with the Earth as a reference. The most successful comparisons have focused on understanding geologic processes on the Earth well enough to extrapolate to Mars' environment.

Several facets of terrestrial analog studies have been pursued and are continuing. These studies include field workshops, process studies at terrestrial analog sites, instrument and operations tests, laboratory measurements, and computer and laboratory modeling. The combination of all these activities allows scientists to constrain the processes operating in specific terrestrial environments and extrapolate how similar processes could affect Mars.

We attempt to summarize some of the past and current activities in these areas to set the stage for a set of Recommendations.

Figure 1. Planetary scientist (P. Lee) conducting geological field work in a "Mars analog" terrain setting on Devon Island, Canadian high Arctic, while testing out an EVA system concept comprising the Hamilton-Sundstrand concept spacesuit for advanced planetary exploration and an individual short-range mobility vehicle (a Kawasaki 220 all-terrain vehicle - ATV). Courtesy, NASA Haughton-Mars Project 2000 / Mark Webb.

2. Field Workshops

Field workshops have been recognized as a very important aspect of comparative planetology training since before the Apollo missions to the Moon. The Apollo astronauts underwent field training under the supervision of planetary and terrestrial geologists in order to better recognize and return lunar rocks of particular importance to our understanding of the geologic processes that have operated on the Moon. At present, astronaut candidate classes receive a week of field geological training in the Rio Grande rift, including geophysical data acquisition. As we continue our exploration of Mars with robotic landers and rovers and prepare for a future that likely will include human exploration, we again find that the terrestrial experience is invaluable in interpreting the remote sensing data being returned by the various Mars missions (e.g., Figure 1). Planetary geologists have organized a number of field workshops in order to train each other in the interpretation of Mars mission data. Terrestrial analog sites are identified based on their similarity to specific environments on Mars (Table 1).

Table 1. Terrestrial analogs to Mars

Mars Feature/ Environment	Terrestrial Analog Site	References
Low viscosity volcanism	Hawaii Snake River, Idaho	Greeley 1974 Greeley and King 1977
Wrinkle Ridges	Columbia River Plateau	Watters and Golombek 1989
Outflow Channels	Channeled Scablands	Baker and Nummedal 1978
Valley Networks	Colorado Plateau Devon Island	Howard et al. 1988 Lee and Rice 1999 Rice 1993
Gullies	Devon Island Greenland	Lee et al. 2001 Costard et al. 2001
Paleo-lakes	Bonneville, Utah Devon Island	Rice 1992 Lee et al. 1998
Aeolian Features	Southern California Southwest Egypt	Greeley et al. 1978 El-Baz and Maxwell 1982
Glacial/Permafrost Features	Alaska Devon Island Canada	Kargel et al. 1993 Lee et al. 1998, Lee 2000
Mars Polar Science	Iceland	
Volcano-Ice Interactions	Iceland	Gulick and Gudmundsson 2000
Spectral Analogs	Hawaii	
Fluvial, Aeolian Features	Death Valley, Mojave	Howard et al. 2001
Banded Iron Formations	Canada	Allen et al. 2001

Field workshops have been a common addition to the Planetary Geology Mappers meetings supported by NASA. Examples of some of the field trips that have been held in association with these meetings are shoreline analogs at the Great Salt Lake and Lake Bonneville, Utah (1992), glacial features in northwestern New York (1999), and volcanic features near Flagstaff, Arizona, (1998) and Albuquerque, New Mexico (2001). Several workshops have also been run by the University of Hawaii for graduate students, post-docs, and junior faculty. The workshops have focused on comparisons between Kilauea and Mauna Loa volcanoes and Mars. Remote sensing data from visible, near infrared, thermal infrared, and radar sensors over lava flows, cones, and calderas are used to help interpret Viking, Mars Global Surveyor, and Earth-based data for Mars. A comprehensive field guide and set of remote sensing examples are provided to the workshop participants. The next workshop is planned for summer 2003. Discussions are underway for a possible summer workshop in the Canadian High

Arctic in the future which will include opportunities for field visits to several sites of interest (P. Lee and W. Pollard, personal communication).

3. Process Studies

Study of potential geologic processes operating on Mars through study of similar processes at analog field sites on Earth makes the results easier to extrapolate and less site-specific. Processes of interest include those that form and modify surfaces in arid and polar regions (e.g., aeolian, flood, periglacial, mass-wasting, glacial, rock weathering, lacustrine/marine, evaporites), ice and glaciers, volcanic processes (e.g., explosive, effusive, lahars, hydrothermal), tectonic (faulting, folding), magmatic/rock-forming processes, impacts, and radiation. A number of sites have been studied or are currently being studied for their application to understanding geologic processes on Mars. The following list attempts to summarize sites at which significant work has been done or is ongoing. Future work is proposed in the Recommendations Section.

3.1. Surficial processes (aeolian, floods, mass-wasting, rock weathering, lacustrine/marine, evaporite)

Mojave/Death Valley: The Mojave Desert/Death Valley region of southeastern California is currently a dry environment but climatic conditions in the past favored wetter conditions. Many examples of dry lake beds (playas) are found in this region and serve as excellent terrestrial analogs to Martian basins proposed to contain paleolake deposits and evaporites (Baldridge et al. 2001). As the evidence continues to mount that Mars has experienced warmer and wetter periods throughout its history, areas such as the Mojave Desert and Death Valley become increasingly important as terrestrial analog sites. In addition to the playa deposits, many other geologic processes have affected this region, including aeolian (sand dunes, sand sheets, and dust devil activity), tectonism and volcanism which also have implications for the development of the Martian highlands. The importance of the southeastern California desert region to understanding Mars is demonstrated by the early field guide developed for this region (Greeley et al. 1978) and the October 2001 workshop on its applicability as a terrestrial analog to the Martian highlands (Howard et al. 2001).

Arizona: This past summer, the MATADOR team put together a field expedition to Eloy, Arizona, to pursue dust devils with about a dozen instruments. MATADOR is an atmospheric dust payload developed in response to an AO released two years ago by the HEDS organization for the planned 2003 Mars Lander. The lander has since been cancelled, and the small amount of funding was used to explore measurement strategies in this Earth analog location, chosen for the frequent appearance of vortices. MATADOR is a combination of meteorological instruments (wind, temperature, pressure), dust impact devices (triboelectric and momentum), electrostatic sensors (total E-field and radio noise, or spherics), a camera, an oxidant sensor, and lidar.

Colorado River Plateau: The small valley network systems seen throughout the Martian highlands and on the flanks of some volcanoes have been argued to result from rainfall and surficial runoff and by groundwater sapping. Many of the morphologic features of the valley networks are characteristic of sapping valleys seen on the Earth. One example of sapping valleys on Earth that is commonly used as a terrestrial analog for the valley networks of Mars are the channels found along the Colorado River Plateau in the southwestern United States. The sapping valleys are particularly well developed in southeastern Utah (Laity and Pieri 1980), where headwater retreat and hanging valleys characterize the morphology. The debate between a rainfall versus sapping origin for the Martian valley networks continues as higher resolution imagery becomes available, so such terrestrial analogs are expected to continue to play a role in the interpretation of the Martian data.

Channeled Scablands of the northwestern United States: The huge outflow channels found on Mars have a terrestrial analog in the Channeled Scablands of the northwestern United States. This area, although smaller than the Martian channels, displays the scour marks, "tadpole"-shaped islands, and other features characteristic of erosion by catastrophic floods. The Channeled Scablands were created by several episodes of catastrophic flooding associated with failures of ice dams during the Quaternary period. The last major episode of flooding extended from about 18,000 to 13,000 years ago when floodwaters from Lake Missoula spilled out over eastern Washington and down the Columbia River basin to the Pacific Ocean. Baker and Nummedal (1978) describe the similarities between the Channeled Scablands and the Martian outflow channels. The Channeled Scablands have been used as a terrestrial analog for outflow channels on many field workshops, including a pre-landing workshop to characterize the Mars Pathfinder landing site.

Central Australia: Next to Antarctica, Australia is the driest continent on Earth. Aridity began ~ 5 million years ago, but a number of climatic oscillations have occurred through the Quaternary. This climatic evolution has created many landforms that are unique to Australia, including low relief deserts with extensive duricrust plains, stony deserts, a continental-scale whorl of longitudinal dunes, drainage-aligned playas and clay pans, and a network of disorganized internal drainage systems. Although the usefulness of any of these features as potential Martian analogs is obvious, little work has been done in this regards. Bourke and Zimbelman (2001) present a compelling case that rivers in central Australia share morphologic characteristics with many Martian valley networks. This implies that many fluvial deposits on Mars may have resulted from unconfined flow, and thus should be easier to identify in orbital images.

Because of the great age of its basement rocks, central Australia contains numerous impact structures that have become type examples for understanding cratering as a geologic process. Gosses Bluff, for example established the importance of shatter cones as one of the features diagnostic of impact cratering (Milton et al. 1972).

It is also worth noting that Earth's earliest biologic record is recorded in a variety of outcrops located in central Australia. Notable examples include microfossils found in the Bitter Springs formation and stromatolites common in the Warrawoona group (Westall and Walsh 2002).

Sahara Desert: The Eastern Sahara, including Egypt's Western Desert and neighboring parts of Sudan and Libya, is the site of late Quaternary river networks buried by several meters of sand and recently revealed by Shuttle Imaging Radar (Schaber et al. 1986, 1997; McCauley et al. 1986; Abdelsalam and Stern 1996; Stern and Abdelsalam 1996). In February 2000, at a test site located in the Southern Egyptian Desert, at the Bir Safsaf oasis, these paleo-channels were mapped by French scientists using 500-900 MHz ground penetrating radar. Additional tests are planned for the coming years (see Recommendations Section).

Tunisia: The Playa deposits of Chott el Jerid and el Gharsa in Tunisia have been studied as a Mars lacustrine analog. The area is located on the southern front of the Atlas thrust and fold belt. Chott el Jerid and Chott el Gharsa are environments undergoing dramatic changes at different scales and show a combination of aeolian, fluvial and lacustrine processes (Ori et al. 2001). This combination of geological environments is very similar to what has been observed on the surface of Mars. The chotts are underlain by a thick accumulation of Cretaceous and Tertiary (Mio-Pliocene) sediments that also contain gypsum-rich ancient marine evaporites. The lacustrine evaporite formation is an ongoing process and a large quantity of new evaporite minerals are formed after major flooding events. This pulsating hydrological cycle with episodic discharge and basin filling could be a good comparison with the Martian lacustrine system. Detailed analysis of the chott area is underway using multispectral and radar data, supported by fieldwork.

Dead Sea: Evaporites have been collected recently for study of microbial populations and fossilization in this environment.

3.2. Polar processes (aeolian, permafrost, periglacial, mass-wasting, glacial)

Antarctic dry valleys: The ice free desert regions of Antarctica are no doubt the best Mars analogs in terms of mean annual temperature, aridity, winds, and remoteness. These regions also contain relevant Martian analogs such as ancient lake deposits (e.g., Rice 1992, 1994), deltas, sand dunes, cinder cones, and periglacial processes and landforms. Glacial landforms have been postulated for Mars and these regions obviously contain glacial landforms. The field investigation of Antarctic microbial life habitats (i.e., endoliths, crypto-endoliths, chasmaliths) would also provide useful training for Mars missions.

Antarctic sub-ice lakes: With the emergence of a new field of exobiology to study plausible extraterrestrial life, there has been an increasing interest in Antarctic sub-ice lakes. About 77 lakes have been identified in Antarctica under the kilometers-thick ice cover from airborne radio-echo-sounding data of the early 1970's (Oswald and Robin 1973, Siegert et al. 1996). Radio altimeter data of the ice above Lake Vostok by the European satellite ERS-1 in 1993 coincided with the revised interpretation of seismic data obtained by Russian scientists in 1964 (Kapitsa et al. 1996). These data are complemented by the more recent radar imaging of the ice surface by the Canadian Radarsat satellite (September-October 1997) and the aerial geophysical survey of Lake Vostok in

2000-2001 by the University of Texas Institute for Geophysics (UTIG) including ice-penetrating radar, laser altimetry, gravity, and magnetics (Holt 2001).

A modeling study by Duxbury et al. (2001) concluded that Lake Vostok could have survived the Antarctic glaciation without freezing completely, supporting an alternative to a basal geothermal melting origin. The numerical experiments of Duxbury et al. (2001), in combination with the recent Mars Global Surveyor Mars Orbiter Laser Altimeter (MOLA) data, also support the existence of subpolar water on Mars, even though the Martian ice caps are not thick enough to generate geothermal melting. Their model for the lake's origin relaxes the condition on the thickness of Mars's polar caps, i.e., potential subglacial lakes can exist on Mars under much thinner ice than it was thought before. As with Lake Vostok on Earth, the Duxbury et al. (2001) model has interesting implications for the search for life (either extant or extinct) on Mars: If there are lakes under the Martian polar caps, they might be the oldest preserved aquifers on Mars, and perhaps the most accessible sources of liquid water and extant or extinct life.

Antarctic ice sheets: Reliable detection and characterization of any sub-ice lakes or smaller scale water systems beneath the Mars polar caps will depend on a foundation of analogous interpretations for the Antarctic ice sheet. Most importantly, identification of any sub-ice water systems on Mars using radar sounding techniques will require detailed knowledge of the physical properties of the intervening ice. This is because attenuation of radar waves is a function of the temperature as well as the impurity content of the overlying ice column. New orbital observations of the surface elevation of Mars' northern polar cap (Zuber et al. 1998) have allowed detailed thermo-mechanical modeling of its internal properties (e.g., Larsen and Dahl-Jensen 2000). This has set the stage for calculations required to interpret future orbital radar sounding data collected over the ice caps. The process of interpreting sub-ice water from airborne radar sounding data is well established for the Antarctic ice sheet (Oswald and Robin 1973, Bentley et al. 1998). Similarly, new approaches are beginning to be used in attempts to more fully characterize sub-ice water systems in both West and East Antarctica using phase-coherent airborne radar sounding (e.g., Peters et al. 2002). Phase coherent radar surveys have recently been completed by UTIG over Lake Vostok in East Antarctica and Ice Stream D in West Antarctica.

A number of expeditions have been carried out to melt or otherwise penetrate ice sheets, a precursor to a Mars polar subsurface expedition. A group just returned from the Svalbard glacier off the coast of Norway after penetrating over 20 meters with a prototype of a Mars cryobot (Anderson et al. 2001).

Alaska/Canada: A number of regions, environments, and landforms in Alaska and Canada have potential for use in Mars analog research. A large proportion of the region is at present within the discontinuous and continuous permafrost zones, offering many opportunities for understanding the influence of permafrost processes on geological materials and features. The modern periglacial environment in the Canadian High Arctic is extensive, and surface exposure of bedrock and associated weathered materials within this polar desert is very good. Much of the region is far more accessible than similar terrain found in Antarctica. The entire landmass of Alaska and Canada has been repeatedly subjected to glacial and cold-climate conditions during the Quaternary, providing researchers today with a natural laboratory for understanding surface processes and features related to ice sheet properties and dynamics, glacio-isostacy, proglacial lakes and associated drainage systems. In addition, numerous impact features have been recognized, both on the Canadian Shield, and on geological units of the Phanerozoic (World Crater Inventory 2002, Grieve et al. 1988). A field workshop was held in Alaska in 1993 to show the similarities of terrestrial eskers, pingos, polygonal terrain, and maars to features seen on Mars (Kargel et al. 1993). Future workshops in this area are planned to address the issues of impacts into ice-rich environments. The exceptionally well-exposed and preserved Haughton impact structure on Devon Island (Robertson 1988) has been the site of rover mission simulations and simulation activities regarding human crews. The presence of both impact processes and periglacial processes makes it an especially good analog for continuing work.

Devon Island: Devon Island is located in the Canadian High Arctic Archipelago in Nunavut Territory. It bears one of the highest-latitude terrestrial impact structures known on Earth, Haughton Crater. Located at 75° N, 89° W, it is the only impact structure known to be set in a polar desert, albeit one that is less extreme than Mars. Ongoing research conducted since 1997 under the auspices of the NASA Haughton-Mars Project (HMP) has revealed a wide variety of possible parallels between features on Devon Island and possible counterparts on Mars (e.g., Lee 1997, Lee et al. 1998, Lee 2000). At Haughton Crater, analog studies may focus on impact processes, their broad range of resulting features, and the post-impact modification of an impact structure under a climate that has been predominantly cold and relatively dry since the structure's formation. Studies of how an impact record (including intracrater paleolacustrine features) may be preserved and exposed under the present polar desert setting is of unique value, in particular for helping plan the exploration of similar sites on Mars (e.g., Lee 1993, Glass and Lee 2001, Osinski et al. 2001). More generally across Devon Island (including at Haughton), a wide variety of periglacial, niveal, glacial, fluvial, aeolian, lacustrine, mass-wasting, and polar weathering features and processes can be investigated. Analog studies on Devon Island conducted by the HMP team are beginning to point to the possible key role played by cold climate niveal and glacial processes, including meltwater processes, in the formation of a variety of classically-called "fluvial" features on Mars (Lee and Rice 1999, Lee et al. 1999, Lee 2000, Lee et al. 2001). Microbiology studies on Devon Island also offer the opportunity to investigate the possibilities, adaptations and limits of life in the relatively extreme environment of the High Arctic, in particular in the context of an impact structure (Cockell and Lee 2002, Cockell et al. 2001, www.marsonearth.org).

Iceland floods: The most voluminous and catastrophic floods known to occur on Earth in historic times are the Icelandic jökulhlaups, the largest of which can have peak discharges on the order of 106 m^3/s and depths of 60-70 m. Jökulhlaups are caused by the abrupt drainage of ice dammed lakes or subglacial volcanic eruptions. The probable causes of Martian floods are massive releases of subsurface water/ice due to possible subsurface volcanic activity, mechanisms very similar to the release of jökulhlaups in Iceland. Another important aspect of jökulhlaups is that channel systems begin as point sources, at glacier termini or lacustrine breakout points, and flow is then directed along one or two primary incised channels until it emerges as unconfined outwash. This is analogous to Martian outflow channels, which also begin at point sources (chaotic terrain and box canyons), and then flow unconfined into a basin region. Erosional landforms associated with jökulhlaups include streamlined hills, extensive washed/stripped lava flows, and cataracts. Depositional landforms consist of pendant bars, boulder bars, imbricated boulder trains, transverse ribs, and giant current ripples. Many of these landforms are also found on Mars (Rice 2000b, 2001; Rice and Edgett 1997; Rice et al. 2000).

3.3. Volcanic

Hawaii: Hawaii has been used extensively for (a) developing ideas on the eruption and emplacement of lava flows (e.g., Macdonald 1953, Peterson and Tilling 1980, Hon et al. 1994); (b) analysis of the structure of basaltic volcanoes on Earth and Mars, including rift zones, calderas, and regional slopes (e.g., Rhodes and Lockwood 1995); (c) weathering processes (Farr and Adams 1984); (d) trafficability studies for Mars rovers; (e) education and public outreach. Hawaii is especially useful for the understanding of volcanic processes because of the near-continuous activity at Kilauea Volcano. No other site on the planet provides this mix of active volcanic processes and easy access. In addition to Kilauea, Mauna Loa (last erupted in 1984) is the largest volcano on Earth, providing a structural and volcanological analog to the even larger shield volcanoes on Mars. Hualalai and Mauna Kea provide excellent examples of more evolved eruptive products, which may have some relation to the "andesites" reported from Mars. Coordination with the USGS Hawaii Volcano Observatory would greatly increase the science value of future workshops in this area. Coordination with the US National Park Service and the native Hawaiian groups is also required for expanded studies.

Snake River Plain: The Snake River Plain in Idaho consists of a variety of low-viscosity basaltic lava flows emplaced primarily by fissure eruptions. These extensive volcanic plains serve as a good terrestrial analog to the volcanic plains of Mars, including the ridged plains and lava flows extending outward from the Tharsis and Elysium volcanoes. The use of the Snake River Plain as a terrestrial analog for extensive sheet lavas not only on Mars but also on Venus and the Moon has resulted in their being the destination for several workshops (Greeley and King 1977), including a Planetary Geologic Mappers meeting in 1994. As high resolution imagery continues to be collected on the volcanic plains of Mars, this region will continue to serve as an excellent terrestrial analog for the study of lava channels, aa versus pahoehoe flows, and fissure vents which are associated with such flows.

East African Rift: Tectonic and volcanic processes have been studied extensively in the East Africa Rift. Recently French scientists analyzed the composition of the basalts and weathering products and found strong similarity to Mars' composition. In particular, many iron-bearing phases were found, including ferromagnetic materials (Paillou et al. 2001).

Iceland: Iceland is a land of fire and ice, with the inevitable catastrophic floods when mixed. The largest basaltic eruption recorded by humans was the 1783-1784 Laki eruption (e.g., Thordarson and Self 1993). This immense lava flow produced a lava morphology unlike any commonly seen in Hawaii, but apparently very similar to the flood lavas on Mars (Keszthelyi et al. 2000). Iceland is the only place on Earth that eyewitness descriptions of lava flows approaching the scale of flood basalt eruptions can be examined. There are a number of other pre-historic Icelandic lava flows that are larger in size, and similar in morphology to the Laki flows (Self et al. 1998). Rootless cones, formed by the explosive interaction of wet sediments and liquid lava, were first recognized in Iceland because they are very common and easy to observe there (Thorarinsson 1953). Martian examples of rootless cones are of great interest because they are proof of water/ice within the shallow subsurface (e.g., Lanagan et al. 2001). Iceland is also the home of recent sub-glacial eruptions that produce catastrophic floods (the last major example was in 1996) as well as tuyas, km-scale mesa-like landforms (Gulick and Gudmundsson 2000). Iceland provides by far the most accessible location to study the interaction between volcanic, glacial, and fluvial processes (McEwen et al. 2001).

Peru: The objective of these studies are to improve understanding of the emplacement of thick, evolved lava flows by studying those at Sabancaya volcano in Southern Peru (Bulmer at al. 1999). Inferences have been made on the rheology and flow regime of the lava flows based on the dimensions, topography, morphology, petrology and geochemistry of the flows (Bulmer and Gregg 2000, Gregg et al. 2001). In addition, an examination has been made of the extent to which surface structure and roughness of different lava flow surfaces can be used as a diagnostic tool for accurate geologic interpretation of the emplacement styles and lava compositions from remotely sensed data on Earth and the terrestrial planets (Bulmer et al. 2001, Shepard et al. 2001).

Mt. Pinatubo: Extensive data on the erosion of the lahars from the recent eruption have been collected by the University of Hawaii and the Philippine Volcano Observatory. Lahar movement on Pinatubo over the ten years since its most recent eruption has been mapped and quantified by Evans (2001) using detailed astronaut-acquired photographs.

Queensland, Australia: A workshop held in 1996 in Queensland, Australia, sponsored by AGU was convened to examine long lava flows. Young (< 2 ma) basalt flows are well preserved there due to the climate, and furthermore represent some of the longest subaerial flows on Earth. A single lava tube, emanating from Undara volcano, can be traced for ~ 160 km. The excellent preservation and great lengths of these flows make them ideal analogs for long Martian lava flows.

3.4. Tectonic

Columbia Plateau: The Columbia Plateau in the northwestern United States has been used as a terrestrial analog site for the study of wrinkle ridges (Watters and Golombek 1989, Mege and Reidel 2001). Wrinkle ridges are seen not only on the lunar mare but also on the Venusian plains and the Martian ridged plains.

They are believed to form from compressive stresses. An excellent terrestrial example of a wrinkle ridge is the Yakima Fold Belt, found in the Columbia River Plateau of central Washington. This feature is a series of anticlinal ridges and synclinal valleys composed of basalt flows from the Columbia River Basalt Group with interbedded sediments. The area was folded and faulted under north-south directed compression, perhaps associated with the formation of the Cascade volcanic range to the west and the Rocky Mountains to the east, or with the Yellowstone hotspot in a manner possibly analogous to wrinkle ridge formation associated with Tharsis on Mars (Mege and Ernst 2001). The Yakima Fold Belt holds great promise for providing better information about the formation and characteristics of the wrinkle ridges seen in the ridged plains and intercrater plains regions of Mars.

3.5. Impact

Impact craters at a variety of sites around the Earth have been studied to learn more about the impact process itself as well as the processes that degrade the original morphology of craters. Sites that have been studied include young craters in arid regions (e.g., Meteor Crater, Arizona; Gosses Bluff and Henbury craters, Australia - Milton 1968); and in cold areas (e.g., Haughton, Devon Island; Zhamanshin, Kazakhstan; Popigai, Siberia; Elgygytgyn, Siberia).

Impacts into volatile-rich regions: There is increasing evidence that the near-surface substrate of Mars contains large quantities of ice into which the vast numbers of impact craters have excavated. The characteristics of impacts into ice-rich or water-rich targets are believed to be quite different from impacts into drier materials. Thus, there is increasing interest to identify and characterize terrestrial impact structures which formed in ice-rich or water-rich environments as well as impacts into targets containing evaporite platform sediments, such as carbonates (Agrinier et al. 2001). Impact structures such as Haughton on Devon Island (Glass and Lee 2001), Popigai and other impacts in northern Russia (Masaitis 1976), and a number of recently identified impact craters in Scandinavia are undergoing renewed study as terrestrial analogs for impacts on Mars.

Impact-induced hydrothermal activity: This may have important biological implications as hydrothermal systems in general may have provided suitable habitats where life may have originated/evolved on Earth and other planets such as Mars. For example, recent work on Devon Island has shown that impact-induced hydrothermal activity occurred following formation of the Haughton impact structure, generating a transient, warm, wet oasis (Osinski et al. 2001).

4. Instrument and Technology Tests

4.1. Remote sensing tests

Geologic Remote Sensing Field Experiment: The Geologic Remote Sensing Field Experiment (GRSFE) consisted of coordinated airborne remote sensing and field observations over geological targets in the Mojave Desert, California, and the Lunar Lake Volcanic Field, Nevada in 1989. GRSFE was funded by NASA's Planetary Geology and Geophysics Program, the Geology Program, the NASA Planetary Data System (PDS), the Pilot Land Data System, the Mars Observer Project, and the Magellan Project. The GRSFE archive is maintained and updated by the PDS Geosciences Node at Washington University. Release of the GRSFE archive on a set of CD-ROMs provided a well documented data set for research and teaching. The primary GRSFE objectives were: (a) to acquire airborne remote sensing and field data in a coordinated campaign for key geological targets, e.g., alluvial fans, dunes, lava flows, volcanoes, etc.; (b) to reduce and document the data and deliver the archive to the PDS and the Pilot Land Data System (PLDS); (c) to take advantage of the multisensor approach, allowing cross-comparison of results acquired with different wavelengths; (d) to use results to test quantitative models for the extraction of surface property information from remote sensing data for Earth, Moon, Mars, and Venus; and (e) to use data in prototype terrestrial studies focused on the nature and ages of geological features and implications for regional climatic and tectonic histories. (Arvidson et al. 1993).

Volcanology data sets: Similar to GRSFE, several sets of CDs were compiled with remote sensing, field, and laboratory data for the volcanoes of Kilauea, Mauna Loa, and Kamchatka. Also, since the early 1990's, measurements have been acquired of the sub-meter topography of Surtsey, the small volcanic islet off the coast of Iceland that formed between 1963-67. Scanning airborne laser altimetry, aerial photos, and IKONOS images have been used. In 1991 the NASA DC-8 collected AIRSAR and AVIRIS data for the island.

ASTER: The capabilities of the Advanced Spaceborne Thermal Emission Radiometer (ASTER) instrument on the Earth Observing System platform Terra, for lithologic mapping based on spectral unmixing, thermal and slope models is being tested in the Mojave Desert (Galford et al. 2001). Familiarity with the types of data collected by ASTER will help the Mars community as we begin to analyze upcoming data returns from TES and THEMIS. Methods for analyzing ASTER data will be similar to THEMIS analysis methods and ASTER analysis provides an understanding of the strengths, weaknesses and various uses that may be possible for the THEMIS data. These tests are also examining the ability of investigators to "scale up" from lander images to a regional understanding by integrating orbital remote sensing with the landed data. An airborne instrument, MASTER, is also available for multispectral visible, near infrared, and thermal infrared data collection.

Hyperspectral: Guinness et al. (2001) are using Airborne Visible and InfraRed Imaging Spectrometer (AVIRIS) data, to identify and map alteration units at the summit of Mauna Kea. The intent is to understand the processes generating the alteration. In addition to the remote data sets, they are using field work and reflectance measurements of field samples to calibrate the remote spectra and to test the derived maps.

Mass movement monitoring: The University of Maryland UMBC, USGS, University of Durham, University of Leeds, CNR-IRPI, and University of Ferrara are studying the application of RADARSAT and ERS data under NASA Code Y to investigate potential precursory phenomena of landslides. The goal is to better integrate radar interferograms, field measurements, and ancillary remote sensing at four selected landslides to obtain "calibrated" interferograms, which will provide useful geologic and geophysical information to the landslide monitoring community (Bulmer et al. 2001b, Petley et al. 2002). The results from this project have relevance to the design of techniques and methodologies to study landslides on Mars.

Exploration strategies and operations: A team of scientists and engineers from the NASA ARC, NASA JSC, University of Tennessee, Arizona State University, and the University of California at Berkeley have been testing operations strategies for scientific exploration of Mars on Devon Island. Target sites are selected prior to field deployment based on analysis of orbital images. Sites are then prioritized and further investigated using simulated autonomous aerial platforms and rovers, followed by more detailed simulated telerobotic investigation by field team members. The highest priority sites are then explored in simulated human EVA traverses, with significant interaction between the field team and the simulated Science Backroom at ARC.

4.2. Microtopography of desert surfaces

A helicopter-borne, fixed-baseline, dual-camera system for close-range stereophotography has been used in the southwest U.S. to image volcanic, alluvial fan, and playa surfaces. The resulting close-range stereo photographs and microtopographic data can provide input to geomorphic and volcanological studies and radar scattering models for the Earth, Venus, and Mars. (Farr 1992, Wall et al. 1991)

4.3. Antarctica and the Arctic

The University of Texas Institute for Geophysics (UTIG) is currently funded by NSF to develop airborne radar sounding techniques in Antarctica using a new testbed radar, funded in part by NASA. This is a chirped, coherent system that operates at a center frequency of 60 MHz, has 8 kW of peak power with variable PRF, pulse length and chirp bandwidth. Multiple acquisition systems provide different, simultaneous gain, detector and stacking options. Portions of this radar were developed by JPL in conjunction with the University of Kansas (KU) as a prototype for the Europa Radar Sounder. The remainder was developed by UTIG and the Technical University of Denmark (TUD) and has been used in Antarctica since the early 1970's (TUD) and 1994 (UTIG). The JPL/KU portion was operationally tested in Antarctica in 2000, and has since been integrated with the other components. The entire system is being operated in Antarctica during the 2001/02 austral summer by UTIG over both East and West Antarctic ice sheet targets. This testbed is being used to evaluate techniques for improved penetration, resolution, and interface characterization, and could form the basis for a new radar design for performing large-scale surveys in polar regions or adaptive surveys of specific targets. It is capable of penetrating over 4000 m of the more absorptive West Antarctic ice.

Dry valleys: The US Army Cold Regions Research and Engineering Laboratory (USACRREL) and the University of New Hampshire (UNH), in a joint NSF proposal, visited the Dry valleys in 1998 and 1999. The intent was a pilot study of the ability of ground Penetrating Radar (GPR) to penetrate the glacial deposits there and reveal stratigraphy. The possible Martian analog is that the permafrost is very cold (-18 to -23 C), possibly volcanic, and variable ice content but not ice-rich. GPR frequencies of 50, 100 and 400 MHz showed 35-40 m penetration leading to the conclusion that electromagnetic loss processes are not significant. Most importantly, the presence of volcanic minerals was not an impediment, although the local rock and soil may not necessarily contain magnetic minerals.

UTIG has operated its new testbed radar in the Dry Valleys in order to evaluate techniques for the detection of shallow subsurface water and ice. Targets included permafrost in Taylor Valley (including the USACRREL/UNH GPR sites described above), Beacon Valley (ice body beneath rock glacier), subglacial water below Taylor Glacier, and the grounding line of Ferrar Glacier. Wilson Piedmont Glacier, at the foot of the Dry Valleys, is also part of the regional targets.

West Antarctic ice cap: USACRREL and UMaine are collaborating on a proposal, funded by NSF, to determine historical accumulation rates and chemical climate records in West Antarctica. CRREL is acquiring high resolution radar stratigraphy using a very small antenna unit (400-MHz) and achieving penetration of more than 100 m in the dry snow and firn, which is about 10X greater than previously achieved by any other researcher at this frequency. The penetration is achieved by high rates of signal trace stacking; the output power is only 1.2 Watts peak, and milliwatts average.

High-resolution airborne radar studies of the ice sheet along the ice divide between the Ross Sea and the Amundsen Sea embayments have been accomplished by UTIG using an 8 kW, short pulse (60 ns) radar sounder operating at 60 MHz (Morse et al. 2002). This reveals internal layers whose thickness variations can be attributed to climatic changes (e.g., Morse et al. 1998) and complements previous large-scale West Antarctic regional surveys by UTIG of surface elevation, ice thickness, magnetics and gravity (Blankenship et al. 1993, 2001; Behrendt et al. 1994, 2002; Holt, 2001). During the 2001/02 season, UTIG is operating its new testbed radar in West Antarctica inorder to evaluate techniques for improved layer resolution, characterization of ice stream basal properties, and penetration of the deepest parts of the West Antarctic ice cap (i.e., the Bentley Subglacial Trench).

Devon Island: A number of studies have been conducted or are underway on Devon Island under the auspices of the NASA Haughton-Mars Project (HMP) to help develop some of the technologies, strategies and human factors experience needed to help plan the future exploration of Mars by robots and humans. In support of geological studies specifically, drilling was conducted in 1998 in collaboration with the U.S. Army CRREL to sample permafrosted intracrater paleolakebeds at Haughton Crater without using any liquid cooling or lubrication of the coring augers and drill bit. The collected core samples are curated at NASA Ames Research Center and are currently under study for paleoclimatic content. Ground-penetrating radar surveys were also conducted to study the distribution of ground-ice in impact-produced materials (breccia formation) and other potential Mars analog features (ice-cored mounds).

5. Laboratory Measurements and Modeling

5.1. Spectral analogs

Wyatt et al. (2001) and Hamilton et al. (2001) compared TES spectra (and mineral compositions via a deconvolution technique) of the Martian surface with terrestrial basalts and andesites, whose spectra were measured in the lab. This gave an interesting look at how well these rock types could actually be distinguished by seeing if rocks of known composition could be adequately distinguished.

A 3-year project funded by NASA's Exobiology program has begun to determine limits of detection (i.e., spatial and spectral resolutions needed) for aqueously-derived minerals in the thermal and near infrared using terrestrial analog sites. So far evaporite basins in Death Valley have been investigated using sensors MASTER (Modis/Aster airborne simulator), ASTER (Advanced Spaceborne Thermal Emission Radiometer on Terra), and AVIRIS (Airborne Visible/Infrared Imaging Spectrometer) (Moersch et al. 2001). In order to make this work more relevant as a terrestrial analog study, plans are to expand to at least one or two more evaporite basins including Salar de Atacama in Chile, the salt pans of Australia, and the Chott el Jerid in Tunisia (in cooperation with G. Ori and G. Komatsu).

5.2. Physical properties of analogs (thermal, electrical, magnetic)

Ices: Water ice, CO_2 ice, and CO_2 clathrate hydrate are the principal ices expected to be present in the polar ice caps of Mars and in Martian soil. Laboratory experiments are now giving us well resolved information on the strength of these ices under ductile flow conditions, such as during the slow spreading of an ice sheet under the influence of gravitational forces. Recent results (Durham et al. 1999, 2000) have shown that the three pure phases have very distinct rheologies. The clathrate hydrate is extremely strong and can be considered almost undeformable in comparison with other ices. CO_2 in contrast is exceedingly weak and will dominate the behavior of icy mixtures of which it is a component. The strength of water ice is intermediate to that of the other two ices.

Electromagnetic properties: A number of radar experiments are in the planning stages to map the Martian subsurface down to several kilometers, searching for subsurface liquid water reservoirs, all based on the penetration property of radio frequency waves in arid soils. The penetration depth of low frequency radar is mainly related to the electromagnetic properties of the medium, emphasizing the importance of investigations of the electrical and magnetic properties of possible Martian surface and subsurface materials. One approach is to use measurements of the electric permittivity and magnetic permeability of laboratory analogs to then construct a dielectric profile representative of the Martian subsurface (Heggy et al. 2001). Laboratory experiments have also been conducted by Williams and Greeley (2001a,b) to quantify the ability of X-band through

P-band radars to penetrate sand and Martian analog dust material. These laboratory experiments have been conducted and compared to field measurements and radar images to discuss near-surface penetration abilities of potential imaging radar missions to Mars.

5.3. Studies of Martian meteorites

Current investigations of Martian meteorites utilize a wide range of available analytical techniques including electron microprobe, SEM, TEM, SIMS, ICPMS, XRF, XRD, Raman and Mössbauer spectroscopy, and controlled crystallization experiments. The general goal of most of these studies is to measure compositions of the whole rock, major, minor, and trace elements and isotopes, both stable and radiogenic, all with the aim of understanding the conditions on Mars under which these rocks formed and what processes, if any, were involved in their later alteration. With the new focus in NASA on "Follow the water", several studies have focused on the involvement of water during formation (e.g., McSween et al. 2001, Dann et al. 2001, Lentz et al. 2001) as well as alteration (e.g., Gooding et al. 1991, Swindle et al. 1997) of the Martian meteorites, and what this may imply for the recent water budget, and current water location, for Mars (e.g., Leshin 2000, Wadhwa 2001). There have also been a few studies using terrestrial analogs for the Martian meteorites. The nakhlites were compared with Theo's Flow, an archean differentiated lava flow in Ontario, Canada (Treiman 1987, Lentz et al. 1999), in an attempt to elucidate their unusual formation. Terrestrial comparisons were also prominent in the discussions of ALH84001. Several of the observations purportedly serving as evidence of life in the rock were compared with terrestrial counterparts: organically and inorganically grown magnetites (e.g., Chang and Kirshvink 1989), carbonate rosettes (?), nanofossils (?), etc. While the jury is still out on the life in ALH84001 issue, these sorts of comparisons are critical for learning as much as possible from these small samples from Mars.

5.4. Microscopic observations of petrologic textures

Hawaiian basaltic lava flows are being frozen and sectioned at different points in their cooling history, providing information on the formation and modification of primary petrologic textures within real lava flows (Sharma et al. 2000).

5.5. Mass movement modeling

Evidence exists for various types of landsliding, including debris flows on Mars, however little is known about the slope-forming materials on Mars and the processes that govern their failure. By focusing on a terrestrial debris flow, ulmer et al. (2002) tested the application of Chezy-type modeling to provide more direct estimates of the dynamics of proposed debris flows on Mars. The model was applied to the General's slide debris flow in Madison County, Virginia. Topography along and across the debris flow was obtained using GPS equipment. Determinations of velocity were made from observations of superelevation, mudlines, clasts imbedded in trees, vegetation stripping, and H/L. Such an approach is also being used to study the Chaos Jumbles Rock Avalanche, California.

6. Key Science Questions

The overarching scientific goals of the Mars program have been documented by the Mars Exploration Payload Assessment Group (MEPAG; Greeley 2001). They include: 1) Detect life or its remnants, 2) Determine present and past climate and its drivers, 3) Determine the geologic processes that have created and modified Mars, and 4) Prepare for human exploration. Underlying this is the fact that we are entering a new phase of Mars study: The era of global mapping is ending and studies will begin to focus on detailed analysis at the regional to outcrop level. This makes it all the more important for Mars geologists of the future to be fully cognizant of geologic processes that operate on Earth which may be analogous to those on Mars.

Prediction of capabilities and interpretation of data from future instruments is a critical need. By optimizing the choice of instruments and their operating parameters through tests in analog terrains, we maximize our chances for success. Possible instruments in need of such study include low-frequency airborne or orbital radar sounders; low-frequency airborne or orbital imaging radar; surface ground penetrating radar and EM sounders; seismic instruments; drills; rovers; high resolution stereo imagers from orbit, lander, or microscope; high resolution topography; hyperspectral visible, near-IR, thermal IR; Raman; Mössbauer; SEM/EDAX, Laser-Induced Breakdown Spectroscopy. Accurate georeferencing or geocoding of these data needs to be developed operationally. These instruments should be tested in a variety of sites, listed below. In addition, Mars surface operations, both robotic and human, need to be developed and tested in analog terrains.

7. Recommendations

As the previous sections indicate, much Mars analog research has been accomplished and a great deal of activity is currently ongoing. However, more coordination among the many current individual projects could yield additional benefits. In particular, the collection and wide dissemination of terrestrial analog data could be expanded. This could take the form of enhancing already-approved field projects so that additional remote sensing and field measurements are obtained. In addition, data from these campaigns could be archived and made available via CD or the Web to a wider community, allowing broad use of the data. Both additional data collection and its subsequent reduction, calibration, and archiving will require new funding to make it happen.

Focusing the community on a few field sites could also help build the diverse remote sensing data sets and supporting data needed to interpret their equivalents from Mars. This should particularly benefit the technology and operations development community as new instruments and techniques need testing. Laboratory measurements are another area in which ongoing efforts could be enhanced through appropriate funding. Detailed recommendations are given below.

Support for data collection and dissemination for specific Mars analog field sites, however should not preclude science proposals seeking to investigate other field sites, as it is impossible to capture all features and processes of Mars in a small set of sites.

7.1. Data collection and field observations (DCFO) at terrestrial analog sites - General

Coordinated deployment of airborne, spaceborne, field instrumentation, and personnel to several sites should be undertaken to test instruments and technology intended for Mars and to provide data for ongoing studies of terrestrial geologic processes relevant to Mars. Selected scientist-astronauts could participate in field data acquisition to complement their experience in spaceborne observations; they would contribute to the research and develop capabilities for Mars exploration (http://geoinfo.nmt.edu/penguins/home.html).

This is the most important area for support. The support will be used to leverage current investments by NASA codes S (Space Science), Y (Earth Science), and M (Human Exploration); NSF; US Army CRREL; and international agencies by acquiring additional data sets at important analog sites which are being studied for a variety of reasons. These activities should begin as soon as possible in order to benefit current and upcoming Mars missions. An important aspect of these campaigns is the rapid and wide dissemination of data, which should be an explicit part of any data collection under this program.

Costs for this part of the program will probably range from less than $100K for sites requiring only a few field measurements, up to as much as $1M for a coordinated multi-aircraft, spacecraft, and field campaign to Antarctica. These costs should be shared across agencies and it will require significant cross-agency and international coordination to make the larger projects happen.

Desirable field site characteristics include easy access, good infrastructure (power, communication, roads, accommodations, etc.), easy aircraft overflight access, little to no vegetation, cold and/or dry, dominance of a few processes thought to be important on Mars, good documentation (e.g., surficial maps, geologic maps, drill logs, etc.). This work must be accompanied by field mapping and determination of the types, rates, and magnitudes of processes to couple with the interpretation of the instrumental data. Terrestrial geologists familiar with the area should be included in these studies and deployments.

Field sites proposed for initial activities under the Mars analog program are detailed in the following sections.

7.2. DCFO - Arid regions

Mojave/ Death Valley: The Mojave Desert/Death Valley region of southeastern California is currently a dry environment but climatic conditions in the past favored wetter conditions. Many examples of dry lake beds (playas) are found in this region which serve as excellent terrestrial analogs to Martian basins proposed to contain paleolake deposits and evaporites (Baldridge et al. 2001). As the evidence continues to mount that Mars has experienced warmer and wetter periods throughout its history, areas such as the Mojave Desert and Death Valley become increasingly important as terrestrial analog sites. In addition to the playa deposits, many other geologic processes have affected this region, including tectonism and volcanism which also have implications for the development of the Martian highlands. The importance of the southeastern California desert region to understanding Mars is demonstrated by the early field guide developed for this region (Greeley et al. 1978) and the October 2001 workshop on its applicability as a terrestrial analog to the Martian highlands (Howard et al. 2001). The close proximity of NASA's base of remote sensing aircraft such as AVIRIS, MASTER, and AIRSAR, make data collection even easier.

Central Australia: Wasson et al. (1988) estimate that the dune fields in Australia represent more than 38% of the world's aeolian landscape. Although the dune fields are contained in seven interconnected deserts, most dunes are longitudinal, up to 300 km long, 10-35 m high, and spaced 160-200 m apart (Croke 1997). Twidale (1972) suggests that these dunes originated downwind from debris mounds associated with playas and alluvial plains. Although the dune sand ranges in color from white to dark red, there is an ongoing debate as to whether this is a function of age or sediment source. Understanding sediment sources, transport distances, and ages of dune fields in Australia has implications for understanding similar features on Mars.

A majority of rivers in Australia are ephemeral. Because they are storm-fed, appreciable runoff typically occurs in localized areas within the much larger drainage basin. This results in only portions of the channel or tributaries containing flow at any one time. Streamflow is infrequent, has a rapid onset, and is short-lived (Mabbutt 1977, Graf 1988, Bourke and Pickup 1999). To date, most terrestrial analog studies have focused on the morphologic characteristics associated with perennial rivers or catastrophic floods. However, if Martian fluvial systems had abundant sediment supply, were subjected to oscillating climates or base levels, or were formed in an arid environment with high transmission losses, then ephemeral rivers may make better analogs. At the very least, ephemeral rivers represent an end member to the range of potential fluvial processes that may have occurred on Mars.

Australian deserts also contain vast plains of duricrust materials, particularly silcrete. The formation of these materials may have occurred as silica-rich groundwater met saline waters contained in paleolakes. They are often interpreted to represent wetter climates. Given the evidence for paleolakes on Mars and earlier, wetter conditions, formation of silcrete may be possible. Ancient (3.45 Ga) microbial mats - possible analogues for Martian biota - are available for sampling as well (Westall and Walsh 2002).

Outside of North America, Australia probably offers the best infrastructure for supporting fieldwork for Americans. The local facilities are excellent, the language is English, the climate is usually bearable, yet the area is remote enough so that land use does not create problems with interpreting deposits the way it can in the southwestern United States (e.g., Cooke and Reeves 1976). NASA aircraft instruments have already visited and there are sophisticated airborne remote sensing instruments available locally as well.

Sahara Desert: The Eastern Sahara, including Egypt's Western Desert and neighboring parts of Sudan and Libya, is floored mainly by quartzitic sandstones, interbedded shales, and conglomerates of the "Nubia Sandstone" with scattered outcrops of granite and granitic gneiss. Much of the region is surfaced by thin sheets of windblown quartz sand and a sediment blanket, centimeter to meters thick, that completely obscure the underlying geologic units, in particular late Quaternary river networks. These hidden structures, mainly composed of alluvial deposits containing calcium carbonate and gypsum, were revealed by Shuttle Imaging Radar images. Such buried geological structures could be very relevant analogs to the past history of Mars, since MOC images indicate that much of the surface of Mars has been intensely reworked by aeolian processes, and key evidence about the history of the Martian environment seems to be hidden beneath a widespread layer of sand and dust.

The French NetLander mission, to be sent to Mars in 2007, offers a unique opportunity to explore simultaneously the deep interior of the planet and the first layers of the subsurface in 4 landing sites displaying different geophysical morphologies. To take advantage of this opportunity, a ground penetrating radar (GPR) has been proposed on the 4 landers. This experiment aims at characterizing geological structures in the vicinity of the landers and at detecting possible water reservoirs in the form of ground ice or even of liquid water down to one kilometer depth. The GPR will operate at a frequency of 2 MHz (Berthelier et al. 2000).

The Siwa, Egypt site was selected to test the NetLander GPR since it presents deep groundwater (between 500 and 600 m depth) covered by homogeneous limestone layers. Good geological maps and drilling logs are available. The site can be easily accessed by road, and the Siwa oasis offers all accommodation facilities. Another site is also being considered for testing the NetLander GPR is a kind of "hematite desert" in the region of Aswan. This work is done in cooperation with the University of Cairo.

A third test site is located in the Southern Egyptian Desert, at the Bir Safsaf oasis. Previous data collection will be supplemented by dense GPR mapping (2D grid of GPR profiles over large areas) and P-band (70 cm wavelength) airborne SAR (RAMSES French facility) planned for February 2003. It may be possible to add experiments relevant to Mars: ground-based geologists for paleo-hydrology, drilling facilities, other EM sounding techniques, GPR on-board rovers, etc. The Italian Space Agency is also planning airborne experiments in Africa with a low-frequency radar.

These measurements will also support other planned radar sounder efforts such as the Mars Advanced Radar for Subsurface and Ionosphere Sounding (MARSIS), planned for Mars Express in 2003 and other possible orbital and surface ground penetrating radars being considered for 2005 and beyond.

Italy: A workshop in Sept. 2002 in Italy is currently being planned to replace the cancelled Tunisia Workshop originally scheduled for September 2001. The associated field trip will be focused on exobiology and volcanism.

Tunisia: A field workshop is also being discussed for Tunisia in 2004 and would include the first data provided by Mars Express.

Dead Sea: An evaporite basin located within the Jordan Rift Valley, it is 400 m below sea level and dry. The dominant cations (in descending order) are Mg, Na, Ca, K. In the southern end of the Dead Sea there are excellent salt deposits that are readily accessible from the road. There are excellent hotel and restaurant facilities on the Israeli side of the Dead Sea. The area is barren and relatively easy to reach. A short distance from the Dead Sea, there are solar ponds, and closer to Eilat, sabkhas. All would be excellent for evaporite analogs.

Chaos Jumbles avalanche, California: The Chaos Jumbles rock avalanche deposit in Lassen Volcanic National Park was emplaced as three separate events. These are distinguishable on the basis of morphology, size variation of large dacitic clasts, and by the matrix of the entrained blocks. The deposits have well defined margins and surface structures such as folds. The avalanche appears to have been dry when emplaced and most likely triggered by a seismic event. The avalanches traveled NW from the Chaos Crags until deflected by topography, most notably Table Mountain, where the deposit is 100 m above the valley floor. The deposit covers an area of 6.8 km^2 and traveled $\sim$ 4.5 km from the base of the Chaos Crags. The characteristics of the Jumbles make them morphologically similar to those of landslides on Mars identified in Viking and MOC images and offer insight into their triggering, movement, and emplacement.

7.3. DCFO - Antarctica

Ongoing research in Antarctica presents an opportunity to bring additional resources to the area such as airborne radar sounders, imaging radars, GPR, drills, etc. These opportunities will probably continue for many years, especially for the ice cap and ice-covered lakes. An important aspect of this program will be to collect data from past studies and archive them for easier access to raw or processed data for future workers.

Dry valleys: USACRREL and the University of Kansas have collaborated on a NASA Planetary Instrument Definition and Development Project with JPL to develop a prototype of a miniature ground-based GPR. This was recently tried out at a permafrost site in Alaska. The same radar technology will be applied to an NSF funded information technology proposal to develop robotic Synthetic Aperture ground based radar for polar ice studies. There will be trials in Antarctica and possibly Greenland.

Ice sheets and subglacial lakes: Ice sheets and glaciers of Earth can contribute to the preparation for future Mars investigations. The development of robotic systems, for both surface and subsurface operations in and on ice sheets of Earth is a direct benefit to Mars exploration. Instrumentation development and the evolution of scientific strategy for in situ investigations in Earth ice will greatly aid studies on Mars. One way to develop insight into subglacial scientific approaches is through Earth analog studies of the deep subsurface glacial environment. Such studies could address basal ice dynamics, ice sheet-bed interactions, subglacial biota, subglacial lakes and similar investigations. A systematic program to detect Antarctic sub-ice water systems including subglacial lakes, using both airborne and ground level radar sounders, is needed. Accurate detection and characterization of these water bodies will require that this program utilize modern phase-coherent radar sounders. In addition, these radar sounding data will need to be interpreted through tightly coupled thermomechanical modeling of the overlying ice column. This systematic radar study should be followed by a similarly rigorous program to model and sample sub-ice water systems. This program can be leveraged by the existing scientific ice drilling expertise developed for studies of terrestrial paleo-climate and ice stream dynamics in Antarctica.

A valuable technology development for Mars polar cap subsurface studies is thus a probe such as the thermal probe, or Cryobot, which melts its way into an ice cap. One such probe is in development and has recently been deployed for the first time in glacial ice. In-situ instruments for subsurface probes have technology issues as well. While miniaturization is underway for almost all instruments, the problems of sample handling are great for a high-pressure subsurface environment and small sample volumes.

There is a scientific benefit to undertaking parallel studies for the East and West Antarctic ice sheets. East Antarctica has a thick and comparatively stable ice column giving rise to numerous subglacial lakes, and is therefore a natural laboratory for studying lake formation and interconnections as well as the implication of analogous processes for Martian exobiology. In addition the smaller scale sub-ice water systems beneath the warmer and more ephemeral West Antarctic ice sheet are more likely a result of enhanced geothermal flux and active subglacial volcanism (e.g., Blankenship et al. 1993, 2001; Behrendt et al. 1994, 1998). In the absence of organized subglacial lakes, sub-ice water on Mars due to enhanced geothermal flux would give rise to exobiological communities that have encountered analogous selective processes.

The Italian Space Agency is also planning airborne experiments in Antarctica with a low-frequency radar to support MARSIS data interpretation.

7.4. DCFO - Other cold areas

Besides ongoing work in Alaska, Canada, and Iceland, much of which has already been discussed, there are other projects in the planning stages that should be supported.

Drill tests: A proposal has been made to the NASA Astrobiology program to field-test a Mars drill in Axel Heiberg Island, north of Devon Island in the Canadian arctic (Briggs, pers. comm.). Test objectives include: Core sampling in ice and permafrost with minimal alteration, Asceptic drilling, Autonomous operations, and Low mass and power. The work is proposed for July 2002, 2003, and 2004. Axel Heiberg provides many options for relevant Mars analog sites and sites that are of astrobiology significance including Arctic tundra polygonal ground, permafrost, ground ice in marine sediments, ground ice in diamicton, weathered bedrock, pingo, gypsum dome, and floodplain. Further, we know from previous studies that there are ice-covered lakes and subsurface hydrothermal systems in the area that are active (Pollard et al. 1999, Anderson et al. 2001). Perennial springs located on Axel Heiberg Island provide useful analogs to liquid water habitats that may have existed on Mars. The springs occur in a region with a mean annual air temperature of -17 C and thick, continuous permafrost reaching depths of 400-600 meters.

Devon Island: The ongoing Haughton-Mars project is in place to investigate the unique Polar Desert of Devon Island and to capitalize on the well preserved geologic features of the Haughton Impact feature and surrounding area.

Proposed work by University Alaska Fairbanks (Yoshikawa, pers. comm.) will involve collecting data from air- and helicopter- borne radar and coring of ground materials. Terrestrial data and drill-core data will be collected in the following permafrost and/or groundwater related sites: a saline brine layer (Barrow, Copper River basin, Alaska), a rock glacier (Wrangell mountain, Alaska), a thermokarst (Copper River basin, Alaska), an open system pingo (Tanana River basin, Alaska), closed system pingo (North Slope, Alaska), polygons (North Slope, Alaska), debris deposits (aprons) (Brooks range, Alaska), ice cap (Greenland), surging glacier (Alaska range, Alaska and Disko Island, Greenland), hot springs (Seward Peninsula, Alaska), lava field (Seward Peninsula, Alaska), and mountain permafrost (Mauna Kea, Hawaii). Studies have shown that perennial groundwater in these areas has three different sources (Yoshikawa, pers. comm.): 1) near-surface permafrost free flow, 2) deep bedrock sub-permafrost flow, and 3) intra-permafrost (mainly between sediment and weathered bedrock) flow. The intra-permafrost aquifer affects refreezing depending on the temperature and thickness of the permafrost. The deep sub-permafrost water infiltrates from permafrost-free south facing slopes.

Finally, the Canadian Space Agency is considering an initiative for Mars analog sites around the Canadian Arctic (http://www.csedi.ca/CJSEv1n1.pdf).

If these proposals are accepted, they may be a good opportunity to add other resources to some of the sites, including ground-based geologists. It may also be possible to add some sites, if resources are made available. In any case, it should also be explored whether data from the proposed studies and any other ancillary data could be archived for easier access to raw or processed data for future workers.

7.5. DCFO - Volcanic areas

Hawaii: Kilauea and Mauna Loa are active volcanoes, allowing effusive eruption processes to be studied first-hand. The rain shadow of the Kau Desert and the higher slopes of Mauna Loa and Mauna Kea provide dry areas to study surficial processes operating on lava flows of varying age.

Snake River Plain: Greeley and King (1977) proposed the term "plains volcanism" to describe the style observed at the Snake River Plains: overlapping small basaltic shields separated by lava flows commonly fed by channels or tubes. MOLA data have revealed numerous small shields within the northern plains of Mars, as well as within the Tharsis and Elysium regions. It may well be that plains volcanism is among the most common of volcanic expressions on Mars. A proposal has been made for a thorough topographic analysis of regions within the Snake River Plains to provide a quantitative basis for comparison with Mars (Sakimoto, pers. comm.).

Kamchatka: Kamchatka contains some of the world's most active explosive volcanoes. Shiveluch is in nearly constant activity with the production of an active dome and small pyroclastic flows. The 1964 debris avalanche deposit is excellently preserved and the overlying pyroclastic flows preserve many of their pristine textures. This site is accessible by helicopter or rugged military vehicles.

The Kliuchevskoi group of volcanoes, a few tens of kilometers to the south of Shiveluch contains several large explosive centers including Kliuchevskoi, Bezymianny, and Tolbachik. Kliuchevskoi is the highest volcano in Kamchatka.

It is a perfect conical stratovolcano that has erupted 68 times since its first observed activity in 1697. Bezymianny is best known for its classic horizontal blast of 1956. Since then it has produced several huge explosions with their attendant pyroclastic deposits. Tolbachik is a basaltic shield volcano that has erupted 31 times since 1730. It is best known for the giant fissure eruption of 1975, the largest eruption of this type in an arc setting this century.

In the southern part of Kamchatka is the Ksudach complex of calderas and domes. The eruption of the Stubel volcano within this caldera in 1907 was one of the largest explosive plinian events of the 20th century. The blast deposits of this volcano are well preserved as the uppermost horizon in the complex. The nested calderas and surrounding welded tuffs are also fine examples.

Mt. Pinatubo: Field analysis of Mt. Pinatubo lahar deposits provides many striking comparisons with Utopia Planitia deposits. Erosion within deep canyon systems, braided valley systems, and broad shallow fans of material at the distal ends of the channels are all seen at both Mt. Pinatubo and Utopia. Observations by PHIVOLCS and the University of Hawaii has shown the speed at which the Mt. Pinatubo lahars can evolve, and the physical characteristics of the resultant deposits bear a good similarity to the Viking Lander 2 site. Detailed grain-size analyses and field mapping, as well as the collection of a comprehensive remote sensing data set over the last decade, make it relatively easy to place the Mt. Pinatubo lahars into the context of data collected by Mars Global Surveyor.

Iceland: The recently formed island of Surtsey is important for several reasons: first, it is a closed sub-arctic landscape system composed entirely of effusive lavas, hydrothermally altered (palagonitized) basaltic tephra, and was devoid of a biological record at its time of birth. Erosional gullies from subsurface release of interlayer water has produced features geometrically similar to ones observed on Mars. Future efforts should include testing models for flood dynamics against observed floods and the identification of different fluvial features (especially at the "rover-scale") as a function of bedrock and flow regime. Spectroscopically, the palagonite resembles some of the Martian spectra and multispectral surface imaging data. While Surtsey is certainly wet and in that sense not very Mars like, it offers some unique opportunities. Other analogues in Iceland are well known and relevant to aspects of the preserved geomorphology on Mars, including the Skeidararsandur (site of the 1996 jökulhlaups), recent activity at Hekla, the mixed magmatism in the Krafla area, rootless cones, maar craters. Low lava shields in Reykjanesskagi and larger ones such as Skjaldbreidur, as well as tuyas, are also germane. The degradation of rootless cones via aeolian and fluvial processes needs to be better characterized. The characteristics of mafic pyroclastic deposits, and their facies changes away from vents, is best studied in Iceland. Iceland also provides excellent locations to test rover mobility and the trafficability of various scientifically important terrains.

East Africa Rift: The Republic of Djibouti in East Africa appears to be a good analog to the volcanic surface of Mars. It is located in the Afar depression, a triple junction between the Somalia, Arabia and Africa plates, and presents a unique emerged rift with active basaltic volcanism. Results of the analysis of basaltic rocks indicate a mineral content similar to Martian basalts analyzed by

Viking and Mars Pathfinder instruments. Basalts collected in Djibouti originate from layered lavas flows (the Afar Stratoid Series) that look remarkably similar to the stratigraphy of the upper crust of Mars as revealed by high resolution MOC imagery. Sediments produced by alteration and weathering of these basalts have strong magnetic properties and contain a high concentration of iron bearing materials. This composition is also quite similar to the Martian soil, and results of X-ray analysis of the samples show the presence of ferromagnetic materials, in particular maghemite, ilmenite and pyrite. Calcareous lacustrine deposits were also observed in several places. Moreover, the Djibouti region shows strong hydrothermal activity which produces large quantities of evaporites (halite) and gypsum. Finally, the whole region is covered with fine basaltic dust transported by wind and dust devils, as also observed on Mars. (Paillou et al. 2001).

Pyroclastic deposits: Good examples of terrestrial ash-flow sheets exist as analogs for suspected regions of Mars. The Bishop Tuff of California contains many features that are characteristic of such deposits, as does the Bandellier Tuff in New Mexico. Erosion of variably welded tuffs could provide insight into similar processes on other planets. Surface characteristics of these examples easily could be measured and compared with data from similar features on Mars. It is generally assumed that Martian volcanic products are dominantly mafic. However, the Martian paterae are most likely composed of pyroclastic deposits. To understand this important class of Martian volcanoes, we need to investigate large-scale basaltic pyroclastic deposits on Earth. Masaya Caldera, Nicaragua, experienced a large-scale Plinian eruption that generated a wide-spread, basaltic pyroclastic deposit (Gregg and Williams 1996). This deposit and the way it has eroded through time needs to be examined as a likely terrestrial analog for the deposits comprising Martian paterae.

Volcanic debris avalanche deposits: There are many good examples of well-preserved volcanic debris avalanche deposits. Many of these preserve original features and are easily accessible for study. Good sites for study include the 1964 Shiveluch deposit (Kamchatka), debris deposits from Colima Volcano (Mexico), and Socompa (Chile).

Volcanic blast deposits: Volcanic blasts are associated with many terrestrial volcanoes. Although these deposits are thin, they have textures and surface features that are distinctive. Good analog examples include the 1980 Mount St. Helens deposit, the 1956 Bezymianny beds, and the 1907 Stubel deposits, Kamchatka. Each deposit has some features in common, but other distinctive characteristics.

Peperites: MOC has revealed that the surface of Mars is replete with cycles of deposition and exhumation. Work proposed by Dr. Tracy Gregg (University at Buffalo) will examine locations within the Snake River Plains and elsewhere in the American west where lavas now on the surface were originally emplaced as shallow sills beneath wet sediments that have subsequently been eroded. These shallow sills, called peperites, have a unique morphology that can be quantified once they are exhumed. Identifying peperites on Mars would allow us to constrain deposition and erosion cycles through time.

7.6. DFCO - Impact sites

Impact craters on the Earth and other solar system bodies are an important database for evaluating the mechanics of crater formation and degradation. Study of the varying morphology of pristine impact craters and their ejecta occurring in a wide range of planetary settings can yield information for use in relating the formation of certain characteristics to the influence of key environmental parameters. Interpreting possible mechanisms responsible for the evolution of pristine crater morphology can be complicated, however, by post-formation degradation. Examination of simple, unglaciated impact craters on the Earth helps to define characteristic degradation signatures for use in placing first-order limits on the number and intensity of processes that have been active. For example, Martian impact craters display a tremendous range in pristine and degraded morphology. Some degraded craters reveal evidence of extensive crater wall incision, rim breaching, and reduction of wall slopes below the angle of repose. Refinement of terrestrial crater degradation sequences may eventually permit more quantitative evaluation of Martian craters and help to confirm the possible role and source of running water in Martian crater degradation (Grant 1999, Grant et al. 1997). Finally, attention should also be focused on the search and characterization of post-impact hydrothermal sites within impact craters, especially bearing in mind the possible astrobiological implications.

Sites for continued study include young craters in arid regions (e.g., Meteor Crater, Arizona; Gosses Bluff, Henbury, Wolf Creek, and Connolly Basin, Australia - Milton 1968; Roter Kamm, Namibia; Lonar Crater, India); in cold areas (e.g., Haughton, Devon Island; Zhamanshin, Kazakhstan; Popigai, Siberia; Elgygytgyn, Siberia); and into carbonate-rich targets (e.g., Haughton, Devon Island - Martinez et al. 1994; Ries, Germany - Hørz 1982). There are some excellent field geophysics datasets for these features and orbital as well as airborne remote sensing data (e.g., McHone et al. 2002). However, additional airborne and spaceborne radar images as well as GPR and potentially seismic data need to be acquired for many of these sites. These data, as well as field photographs, need to be compiled into an accessible database.

7.7. DFCO - Aqueous sites

Sites at Mid-ocean ridges, hot springs, and other aqueous environments on Earth have special significance for the search for life on Mars. Chemosynthetic communities at mid-ocean rifts and hydrocarbon seeps in the Gulf of Mexico and the supporting bacteria (both macro and micro) may be a potentially good modern day analog to early forms of life on Mars. At Sharks Bay, Australia, cyanobacteria exist as a modern analog to the nearby fossil sites of the Cambrian-Devonian, which may also mimic early, simple forms of life on Mars. Hot springs and geysers of Yellowstone National Park contain hyperthermophyllic organisms which may be even earlier in the evolutionary tree than the cyanobacteria.

7.8. Data archives

Collect, document, and make available via the Web existing archives of terrestrial data relevant to Mars studies. These could include GRSFE data, various remote sensing data of important sites, Devon Island results, Surtsey observations, Hawaii remote sensing and spectral analogs data, microtopographic data, raw or processed GPR and sounder data of various sites, laboratory data, a comprehensive bibliography including actual articles scanned in as PDF files. The existing NASA Planetary Data System could be the host of these archives. It already has tools for users to submit data. The availability of an archive like this would increase the efficiency of investigations, eliminating redundant work, and allowing new proposals to be made, building on the archived work.

Only modest funding would be required to acquire, format, and send the highest priority data sets to the PDS. A working group may be needed to identify data sets for inclusion.

7.9. Field workshops

Continue support of a series of workshops for established scientists, graduate students, and scientist-astronauts. The workshops should be process-oriented. Suggested processes should include aeolian, volcanic, biologic, glacial/ice, fluvial/flood, rock weathering, periglacial, impact. The current Planetary Geology Mappers workshops could be expanded to include a wider range of researchers with guidebooks published on paper as well as web-based for wider community access. These field workshops could also include operation of field instruments similar to those intended for Mars, modeled after the Marsokhod field tests. The Lunar and Planetary Institute is prepared to assist in workshop organization as it did for the Martian Highlands/Mojave Desert Workshop (Howard et al. 2001).

Funding would be modest for publishing past guidebooks on CD or the web. Adding participants and activities to ongoing field workshop series may prove more costly. Support of new workshops would probably cost on the order of $100K per workshop, depending on location.

7.10. Laboratory measurements and modeling

A number of studies are ongoing, however this panel identified several areas of laboratory measurements that should be given a high priority. There is a critical need to characterize more fully various Mars analog materials, such as clathrates, palagonites, rock coatings. Additional effort needs to go into collecting and making available new Mars analog materials such as JSC-1. Rock and mineral analogs would also be useful for instrument tests. In particular, while pure clathrates are relatively easy to produce in the laboratory, soil-clathrates are more difficult, but potentially more relevant to Mars studies.

One strategy in understanding Mars ice better is to create analog ice in the laboratory; while this is a high-priority activity, it is complicated by uncertainty on the basic method of deposition of Mars water ice and frost. A useful Mars analog laboratory study would then be the modeling of the generation of polar frost and ice in simulations of the Mars atmosphere, a challenging experimental task. Other laboratory examinations could follow, e.g., the selective ablation and concentration of dust.

Remote sensing and electromagnetic characterization of Martian analogs, such as evaporites, clays and magnetic minerals, especially at low temperatures, needs to continue in support of the definition of future subsurface radar sounders.

As many of these activities are already funded and ongoing, only modest funding is foreseen to ensure high-priority areas such as clathrate characterization are covered.

7.11. Association with other areas of solar system exploration

Results of studies of terrestrial analogs to Mars will be applicable to many studies of the other terrestrial planets such as ice deposits on the Moon and Mercury, drilling on other planets, and remote sensing data analysis. Results from ice sheet and sub ice lakes will be applicable to Europa exploration and planning. Results from ice sheet and other cold regions may also be applicable to Titan exploration.

8. Maintaining Capabilities for the Future

- Continue expanded field workshops and support of projects such as the Haughton Mars Project. Change sites/processes as Mars results come in. Continue building human observational experience and insight by involving astronaut researchers.

- Continue coordinated deployments of bread-board instruments or instruments similar to those intended for Mars on aircraft, Earth-orbiting spacecraft, and the ground. Develop relationships between NASA offices and other national and international agencies for support of these multi- disciplinary activities. For example, NASA Earth Science may be interested in ice studies or sounding for deep soil moisture; NSF is interested in polar studies as is CRREL. International interest in Martian analogs (e.g., CNES, ESA, ASI, etc.) could help support non-US sites in, for example, Africa, Tibet, Atacama Desert, Australia, Iceland, Russia, Canada.

- Continue support of data archive of terrestrial analog data. Review periodically based on latest Mars results. As Mars materials become better understood, define new Mars analog materials for laboratory study.

- Establish a few prime sites for continuing study, human training, Education and Public Outreach, etc. (see below)

9. Human Exploration

Mars analog studies and geological field training exercises will be crucial in the manned exploration of Mars because of its rich and complex geological history (perhaps even biological history). This will present new and exciting challenges to the human exploration of both the surface and subsurface of Mars. At analog sites, astronauts will be able to carry out the proper field investigations for correct geological context of the landing site as well as the surrounding environs, conduct precision landings, maximize surface mobility, and also perform intelligent sample acquisition and interpretation (Rice 2000a).

At JSC, Shuttle crews are briefed in observing/documenting/interpreting features on Earth that may be analogues for those on Mars. As a result, a growing collection of astronaut-acquired photography is being built, focused on diagnostic attributes of terrestrial sites (e.g., Channeled Scablands) for rigorous comparison with MOC and future images of Mars. An outgrowth of early efforts was an online education/outreach publication produced in collaboration with planetary scientists at JSC and LPI (Willis et al. 2000).

On Devon Island, a number of studies are currently underway under the auspices of the NASA Haughton-Mars Project to develop some of the technologies, strategies and human factors experience that will help plan the future exploration of Mars and other planets by robots and humans (www.marsonearth.org). The HMP is established as an international interdisciplinary field research project managed for NASA by the private SETI Institute, thus offering an opportunity at the same time to explore government-private partnerships in extracting the scientific and technological potential of Mars analog research.

10. Education and Public Outreach

For this application, the visual aspect of the analog site and the logistics and accessibility are the most important factors. The more accessible sites especially could become "standard" sets for filmed news stories and documentaries. Demonstrations of rovers are always popular- more so in a Mars-like setting. The public can be encouraged to visit some of the sites, yielding an opportunity to describe how the site is like Mars and how it isn't. National Monuments and Parks such as Death Valley and Mojave Desert, may be interested in helping to support this.Hawaii Volcanoes National Park already hosts annual workshops for teachers and school students who use the volcanic terrain as a planetary analog. Planning for manned and un-manned surface activities on the planets (including Mars) are included in the program.

References

Abdelsalam, M.G. and R.J. Stern 1996. Mapping Precambrian structures in the Sahara Desert with SIR-C/X-SAR radar: The Neoproterozoic Keraf Suture, NE Sudan. *J. Geophys. Res.* **101**, 23063-23076.

Agrinier, P., A. Deutsch, U. Scharer, and I. Martinez 2001. Fast back-reactions of shock-released CO_2 from carbonates: An experimental approach. *Geochim. Cosmochim. Acta* **65**, 2615-2632.

Allen, C.C., F. Westall, and R. Schelble 2001. Importance of a Martian hematite site for Astrobiology. *Astrobio.* **1**, 111-123.

Anderson, F.S., M.H. Hecht, F.D. Carsey, P.G. Conrad, W.F. Zimmerman, L.C. French, and H. Engelhardt 2001. In-Situ ice core analysis of Longyearbreen Glacier using a cryobot: Preparation for the northern polar cap of Mars. *Eos (Fall Supplement)* **82**, Abstract P22B-0548.

Arvidson, R.E., M.K. Shepard, E.A. Guinness, S.B. Petroy, J.J. Plaut, D.L. Evans, T.G. Farr, R. Greeley, N. Lancaster, and L.R. Gaddis 1993. Characterization of lava-flow degradation in the Pisgah and Cima volcanic fields, California, using Landsat Thematic Mapper and AIRSAR data. *Geol. Soc. Amer. Bull.* **105**, 175-188.

Baker, V.R., and D. Nummedal 1978. *The Channeled Scabland*, NASA Planetary Geology Program, Washington, D.C.

Baldridge, A.M., J.D. Farmer, and J.E. Moersch 2001. Remote sensing of terrestrial analogs for evaporite basins on Mars: Analysis of ground truth. *Eos (Fall Supplement)* **82**, Abstract P22B-0550.

Behrendt, J.C., D.D. Blankenship, C.A.Finn, R.E. Bell, R.E. Sweeney, S.M. Hodge, and J.M. Brozena 1994. CASERTZ aeromagnetic data reveal late Cenozoic flood basalts in the West Antarctic rift system. *Geology* **22**, 527-530.

Behrendt, J.C., C.A. Finn, D.D. Blankenship, and R.E. Bell 1998. Aeromagnetic evidence for a volcanic caldera complex beneath the divide of the West Antarctic Ice Sheet. *Geophys. Res. Lett.* **25**, 4385-4388.

Behrendt, J.C., D.D. Blankenship, D.L. Morse, and R.E. Bell 2002. Removal of subglacially erupted volcanic edifices beneath the divide of the West Antarctic Ice Sheet interpreted from aeromagnetic and radar ice sounding surveys. In *Proceedings Volume 8th International Symposium on Antarctic Earth Sciences* (J.A. Gamble, D.N.B. Skinner, S. Henrys, and R. Lynch, Eds.), Wellington, Carmel, in press.

Bentley, C.R., N. Lord, and C. Liu 1998. Radar reflections reveal a wet bed beneath stagnant Ice Stream C and a frozen bed beneath Ridge BC, West Antarctica. *J. Glaciol.* **44**, 149-156.

Berthelier J.-J, R. Ney, F. Costard, M. Hamelin, A. Meyer, B. Martinat, A. Reineix, Th. Hansen, M. Bano, W. Kofman, F. Lefeuvre, and Ph. Paillou 2000. The GPR experiment on Netlander. *Planet. Space Sci.* **48**, 1161-1180.

Blankenship, D.D., R.E. Bell, S.M. Hodge, J.M. Brozena, J.C. Behrendt, and C.A. Finn 1993. Active volcanism beneath the West Antarctic ice sheet and implications for ice-sheet stability. *Nature* **361**, 526-529.

Blankenship, D.D., D.L. Morse, C.A. Finn, R.E. Bell, M.E. Peters, S.E. Kempf, S.M. Hodge, M. Studinger, J.C. Behrendt, and J.M. Brozena 2001. Geologic controls on the initiation of rapid basal motion for West Antarctic ice streams: A geophysical perspective including new airborne radar sounding and laser altimetry results. In *The West Antarctic Ice Sheet: Behavior and Environment* (R.B. Alley and R.A. Bindschadler, Eds.), Vol. 77, Antarctic Research Series, pp. 105-121. American Geophysical Union, Washington, D.C.

Bourke, M.C., and G. Pickup 1999. Fluvial form variability in arid central Australia. In *Varieties of Fluvial Forms* (A. Miller and A. Gupta, Eds.), pp. 248-271. Wiley and Sons, New York.

Bourke, M.C., and J.R. Zimbelman 2001. The Australian paleoflood model for unconfined fluvial deposition on Mars. *Lunar and Plan. Sci.* **XXXII**, Abstract 1679.

Bulmer, M.H., and T.K.P. Gregg 2000. Fold generation on a silicic flow - A planetary analog. Lunar Planet. Sci. **XXXI**, Abstract 1469.

Bulmer, M.H., F.C. Engle, and A. Johnston 1999. *Analysis of Sabancaya volcano, southern Peru using RADARSAT and Landsat TM data.* RADARSAT ADRO Program Summary, Can. Space Agency, Montreal, http://radarsat.space.gc.ca. [CD-ROM]

Bulmer, M.H., Campbell, B.A., and J. Byrnes 2001. Field studies and radar remote sensing of silicic lava flows. *Lunar Planet. Sci.* **XXXII**, 1850. (abstract) [CDROM].

Bulmer, M.H., O.S. Barnouin-Jha, M.N. Peitersen, A.K. Johnston, and M. Bourke 2002. An empirical approach to studying debris flows - Implications for planetary modeling studies. *J. Geophys. Res.*, in press.

Chang S.B.R., and J.L. Kirshvink 1989. Magnetofossils, the magnetization of sediments, and the evolution of magnetite biomineralization. *Annu. Rev. Earth Planet. Sci.* **117**, 169-195.

Cockell, C.S., and P. Lee 2002. The biology of terrestrial impact craters: A review. *Biolog. Rev.* In press.

Cockell, C.S., P. Lee, A.C. Schuerger, L. Hidalgo, J.A. Jones, and M.D. Stokes 2001. Microbiology and vegetation of micro-oases and polar desert, Haughton impact crater, Devon Island, Canada. *Arctic Antarct. Alpine Res.* **22**, 306-318.

Cooke, R.U., and R.W. Reeves 1976. *Arroyos and Environmental Change in the American Southwest.* Clarendon Press, Oxford.

Croke, J. 1997. Australia. In *Arid Zone Geomorphology, Process, Form and Change in Drylands* (D.S.G. Thomas, Ed.), pp. 563-573. John Wiley & Sons, New York.

Dann J.C., A.H. Holzheid, T.L. Grove, and H.Y. McSween, Jr. 2001. Phase equilibrium of the Shergotty meteorite, Constraints on pre-eruptive H_2O contents of Martian magmas and fractional crystallization under hydrous conditions. *Meteor. Planet. Sci.* **36**, 793-806.

Durham, W.B., S.H. Kirby, and L.A. Stern 1999. Steady-state flow of solid CO_2: Preliminary results. *Geophys. Res. Lett.* **26**, 3493-3496.

Durham, W.B., S.H. Kirby, and L.A. Stern 2000. Comparative rheologies of solid H_2O, CO_2, and CO_2 clathrate hydrate. *Second International Conference on Mars Polar Science and Exploration.* LPI Contrib. No. 1057, Abstract 4105, Lunar and Planetary Institute, Houston, TX.

Duxbury, N.S., I.A. Zotikov, K.H. Nealson, V.E. Romanovsky, and F. Carsey 2001. A numerical model for an alternative origin of Lake Vostok and its exobiological implications for Mars. *J. Geophys. Res.* **106**, 1453-1462.

El-Baz, F. and T.A. Maxwell 1982. Desert Landforms of Southwest Egypt: A Basis for Comparison with Mars. NASA CR-3611, Washington, D.C.

Evans, C.E. 2001. Astronauts photograph Mt. Pinatubo. NASA Earth Observatory, June 15, 2001, http://earthobservatory.nasa.gov/Study/AstronautPinatubo/

Farr, T.G. 1992. Microtopographic evolution of lava flows at Cima volcanic field, Mojave Desert, California. *J. Geophys. Res.* **97**, 15171-15179.

Farr, T.G., and J.B. Adams 1984. Rock coatings in Hawaii. *Geol. Soc. Amer. Bull.* **95**, 1077-1083.

Galford, G.L., R.E. Arvidson, F.P. Seelos, B.L. Jolliff, and L.E. Griggs 2001. Lithologic mapping in the Southern Soda Mountains, Mojave Desert, using ASTER Data. *Eos (Fall Supplement)* **82**, Abstract P52B-0586.

Glass, B.J., and P. Lee 2001. Airborne geomagnetic investigations at the Haughton Impact Structure, Devon Island, Nunavut, Canada. *Lunar Planet. Sci.* **XXXII**, Abstract 2155.

Gooding J.L., S.J. Wentworth, and M.E. Zolensky 1991. Aqueous alteration of the Nakhla meteorite. *Meteoritics* **26**, 135-143.

Graf, W.G. 1988. *Fluvial Processes in Dryland Rivers.* Springer-Verlag, New York.

Grant, J.A. 1999. Evaluating the evolution of process specific degradation signatures around impact craters. *Int. J. Impact Eng.* **23**, 331-340.

Grant, J.A., C. Koeberl, W.U. Reimold, P.H. Schultz, and D. Brandt 1997. Degradation history of the Roter Kamm impact crater, Namibia. *J. Geophys. Res.* **102**, 16327-16388.

Greeley, R. 1974. *Hawaiian Planetology Conference.* NASA TMX 62362, Washington, D.C.

Greeley, R. 2001. Science Planning for Exploring Mars: Part 2, Scientific Goals, Objectives, Investigations, and Priorities. JPL Pub. 01-7, Jet Propulsion Laboratory, Pasadena.

Greeley, R. and J.S. King 1977. Volcanism of the Eastern Snake River Plain, Idaho, A Comparative Planetary Geology Guidebook. NASA Office of Planetary Geology, Washington, DC.

Greeley, R., M.B. Womer, R.P. Papson, and P.D. Spudis 1978. Aeolian Features of Southern California, A Comparative Planetary Geology Field Guide. NASA Office of Planetary Geology, Washington, D.C.

Gregg, T.K., M.H. Bulmer, and H.H. Warner 2001. Lava flow fields on Earth and Mars: Scales of comparison. *Eos (Fall Supplement)* **82**, Abstract P22D-05.

Gregg, T.K.P., and S.N. Williams 1996. Explosive mafic volcanoes on Mars and Earth: Deep magma sources and rapid rise rate. *Icarus* **122**, 397-405.

Grieve, R.A.F., C.A. Wood, J.B. Garvin, G. McLaughlin, and J.F. McHone, Jr. 1988. *Astronaut's guide to terrestrial impact craters.* LPI Tech. Rep. 88-03, Lunar and Planetary Institute, Houston.

Guinness, E.A., R.E. Arvidson, B.L. Jolliff, R.V. Morris, and D.W. Ming 2001. Mapping alteration minerals exposed on the summit of Mauna Kea Volcano using AVIRIS data: Implications for mapping Mars mineralogy. *Lunar Planet. Sci.* **XXXII**, Abstract 1659.

Gulick, V.C., and M.T. Gudmundsson, Eds. 2000. *Volcano/ice interactions on Earth and Mars*, Abstract Volume. http://wwwflag.wr.usgs.gov/USGSFlag/Land/IcelandMeeting.

Hamilton V.E., M.B. Wyatt, H.Y. McSween, Jr., and P.R. Christensen 2001. Analysis of terrestrial and Martian volcanic compositions using thermal emission spectroscopy: Application to Martian surface spectra from the Mars Global Surveyor Thermal Emission Spectrometer. *J. Geophys. Res.* **106**, 14733-14746.

Heggy, E., Ph. Paillou, G. Ruffi, J.-M. Malzieux, F. Costard, G. Grandjean 2001. On water detection in the Martian subsurface using sounding radar. *Icarus* **154**, 244-257.

Holt, J.W. 2001. Airborne surveys conducted by SOAR for geologic studies in Antarctica, 1998-2001. *Eos (Spring Supplement)* **82**, Abstract GP42A-07.

Hon, K., J. Kauahikaua, R. Denlinger, and K. Mackay 1994. Emplacement and inflation of pahoehoe sheet flows: Observations and measurements of active lava flows on Kilauea Volcano, Hawaii. *Geol. Soc. Amer. Bull.* **106**, 351-370.

Hørz, F. 1982. *Ejecta of the Ries Crater, Germany.* Geol. Soc. Amer. Special Paper 190.

Howard, A.D., R.C. Kochel, and H.E. Holt 1988. Sapping Features of the Colorado Plateau, A Comparative Planetary Geology Field Guide. NASA SP-491, Washington, D.C.

Howard, A.D., J. Moore, and J. Rice 2001, Eds. *Field trip and Workshop on the Martian Highlands and Mojave Desert Analogs.* http://www.lpi.usra.edu/meetings/martianhighlands2001.

Kapitsa, A.P., J.K. Ridley, G. de Q. Robin, M.J. Siegert, and I.A. Zotikov 1996. A large deep freshwater lake beneath the ice of central East Antarctica. *Nature* **381**, 684-686.

Kargel, J.S., J. Moore, and T. Parker 1993. Workshop on the Martian Northern Plains, Sedimentological, Periglacial, and Paleoclimatic Evolution. LPI Tech. Rep. 93-04, Lunar and Planetary Institute, Houston.

Keszthelyi, L., A.S. McEwen, and Th. Thordarson 2000. Terrestrial analogs and thermal models for Martian flood lavas. *J. Geophys. Res.* **105**, 15027-15049.

Laity, J.E. and D.C. Pieri 1980. Sapping processes in tributary valley systems. NASA Tech. Memo 81776, pp. 271-273.

Lanagan, P.D., A.S. McEwen, L.P. Keszthelyi, and Th. Thordarson 2001. Rootless cones on Mars indicating the presence of shallow equatorial ground ice in recent times. *Geophys. Res. Lett.* **28**, 2365-2367.

Larsen, J. and D. Dahl-Jensen 2000. Interior temperatures of the Northern Polar Cap on Mars. *Icarus* **144**, 256-462.

Lee, P. 1993. Briny lakes on early Mars? Terrestrial intracrater playas and Martian candidates. LPI Tech. Rep. 93-03, Part 1, p. 17, Lunar and Planetary Institute, Houston.

Lee, P. 1997. A unique Mars/Early Mars analog on Earth: The Haughton impact structure, Devon Island, Canadian Arctic. In *Conf. on Early Mars: Geologic and hydrologic evolution, physical and chemical environments, and the implications for life.* LPI Contrib. No. 916, p. 50, Lunar and Planetary Institute, Houston.

Lee, P. 2000. Selective fluvial erosion on Mars: Glacial selective linear erosion on Devon Island, Nunavut, Arctic Canada, as a possible analog. *Lunar Planet. Sci.* **XXXI**, Abstract 2080. [CD-ROM]

Lee, P., and J.W. Rice, Jr. 1999. Small valley networks on Mars: The glacial meltwater channel networks of Devon Island, Nunavut Territory, Arctic Canada, as possible analogs. In *5th International Conference on Mars.* LPI Contrib. No. 972, Abstract 6273. Lunar and Planetary Institute, Houston.

Lee, P., J.W. Rice, Jr., T.E. Bunch, R.A.F. Grieve, C.P. McKay, J.W. Schutt, and A.P. Zent 1999. Possible analogs for small valleys on Mars at the Haughton impact crater site, Devon Island, Canadian High Arctic. *Lunar Planet. Sci.* **XXX** abstract 2033. [CD-ROM]

Lee, P., T.E. Bunch, N. Cabrol, C.S. Cockell, R.A.F. Grieve, C.P. McKay, J.W. Rice, Jr., J.W. Schutt, and A.P. Zent 1998. Haughton-Mars 97 - I: Overview of observations at the Haughton impact crater, a unique Mars analog site in the Canadian High Arctic. *Lunar Planet. Sci.* **XXIX**, 1973-1974. (abstract)

Lee, P., C.S. Cockell, M.M. Marinova, C.P. McKay, and J.W. Rice, Jr. 2001. Snow and ice melt slope flow features on Devon Island, Nunavut, Arctic Canada, as possible analogs for recent slope flow features on Mars. *Lunar Planet. Sci.* **XXXII**, Abstract 1809. [CD-ROM]

Lentz, R.C.F., G.J. Taylor, and A.H. Treiman 1999. Formation of a Martian pyroxenite: A comparative study of the nakhlite meteorites and Theo's Flow. *Meteor. Planet. Sci.* **34**, 919-932.

Lentz, R.C.F., H.Y. McSween Jr., J.G. Ryan, and L.R. Riciputi 2001. Water in Martian magmas: Clues from light lithophile elements in shergottite and nakhlite pyroxenes. *Geochim. Cosmochim. Acta* **65**, 4551-4565.

Leshin, L.A. 2000. Insights into Martian water reservoirs from analysis of Martian meteorite QUE94201. *Geophys. Res. Lett.* **27**, 2017-2020.

Mabbutt, J.A. 1977. *Desert Landforms*, The MIT Press, Cambridge.

Macdonald, G.A. 1953. Pahoehoe, aa, and block lava. *Am. J. Sci.* **251**, 169-191.

Martinez, I., P. Agrinier, U. Scharer, and M. Javoy 1994. A SEM-TEM and stable isotope study of carbonates from the Haughton impact structure, Canada. *Earth Planet. Sci. Lett.* **121**, 559-574.

Masaitis, V.L. 1976. Astroblemes in the USSR. *Int. Geol. Rev.* **18**, 1249-1258.

McCauley, J.F., C.S. Breed, G.G. Schaber, W.P. McHugh, B. Issawi, C.V. Haynes, M.J. Grolier, and A.E. Kilani 1986. Paleodrainage of the Eastern Sahara - the radar rivers revisited. *IEEE Trans. Geosci. Remote Sens.* **24**, 624-648.

McEwen, A., D. Burr, J. Hardardottir, A. Hoskuldsson, L. Keszthelyi, P. Lanagan, A. Snorrason, and T. Thordarson 2001. Icelandic analogs for volcanic and fluvial processes on Mars. abs. *Eos (Fall Supplement)* **82**, Abstract P32D-0573.

McHone, J.F, R. Greeley, K.K. Williams, D.G. Blumberg, and R.O. Kuzmin 2002. Space Shuttle observations of terrestrial impact structures using SIR-C and X-SAR radars. *Meteor. Planet. Sci.*, in press.

McSween , H.Y., Jr., T.L. Grove, R.C.F. Lentz, J.C. Dann, A.H. Holzheid, L.R. Riciputi, J.G. Ryan 2001. Geochemical evidence for magmatic water within Mars from pyroxenes in the Shergotty meteorite. *Nature* **204**, 487-490.

Mege, D. and R.E. Ernst 2001. Contractional effects of mantle plumes on Earth, Mars, and Venus. In *Mantle Plumes: Their Identification Through Time* (R.E. Ernst and K.L. Buchan, Eds.), pp. 103-140. Geological Society of America Special Paper, Vol. 352.

Mege, D. and S.P. Reidel 2001. A method for estimating 2D wrinkle ridge strain from application of fault displacements scaling to the Yakima folds, Washington. *Geophys. Res. Lett.* **28**, 3545-3548.

Milton, D.J. 1968. *Structural Geology of the Henbury Meteorite Craters Northern Territory, Australia.* U.S. Geological Survey Prof. Paper 599-c, Washington, DC.

Milton, D.J., B.C. Barlow, R. Brett, A.R. Brown, A.Y. Glikson, E.A. Manwarning, J.J. Moss, E.C.E. Sedmick, J. Van Son, and G.A. Young 1972. Gosses Bluff impact structure, Australia. *Science* **175**, 1199-1207.

Moersch, J.E, J. Farmer, and A. Baldridge 2001. Remote sensing of evaporite minerals in Badwater Basin, Death Valley, at varying spatial scales and in different spectral regions. In *Field trip and Workshop on the Martian Highlands and Mojave Desert Analogs.* (A.D. Howard, J. Moore, and J. Rice, Eds.), Abstract 4023. http://www.lpi.usra.edu/meetings/martianhighlands2001.

Morse, D.L., E.D. Waddington, and E.J. Steig 1998. Ice Age Storm Trajectories Inferred from Radar Stratigraphy at Taylor Dome, Antarctica. *Geophys. Res. Lett.* **25**, 3383-3386.

Morse, D.L., D.D. Blankenship, E.D. Waddington, and T.A. Neumann 2002. A Site for Deep Ice Coring in West Antarctica: Results from Aerogeophysical Surveys and Thermo-kinematic Modeling. *Ann. Glaciol.* **35**, in press.

Ori, G.G., G. Komatsu, and L. Marinangeli, Eds. 2001. *Field trip guidebook for Workshop on Exploring Mars' Surface and its terrestrial analogues,* http://irsps.sci.unich.it/education/tunisia.

Osinski, G.R., J.G. Spray, and P. Lee 2001. Impact-induced hydrothermal activity within the Haughton impact structure, Arctic Canada: Generation of a transient, warm, wet oasis. *Meteor. Planet. Sci.* **36**, 731-745.

Oswald, G.K.A., and G. de Q. Robin 1973. Lakes beneath the Antarctic Ice Sheet. *Nature* **245**, 251-254.

Paillou, Ph., G. Grandjean, J.-M. Malzieux, G. Ruffi, E. Heggy, D. Piponnier, P. Dubois, and J. Achache 2001. Performances of ground penetrating radars in arid volcanic regions: Consequences for Mars subsurface exploration. *Geophys. Res. Lett.* **28**, 911-914.

Peters, M.E., D.D. Blankenship, and D.L. Morse 2002. A new radar sounding method for improving the characterization of subglacial interfaces applied to Ice Streams B and C, West Antarctica. *J. Glaciol.*, submitted.

Peterson, D.W. and R.I. Tilling 1980. Transition of basaltic lava from pahoehoe to aa, Kilauea Volcano, Hawaii, Field observations and key factors. *J. Volcanol. Geotherm. Res.* **7**, 271-293.

Petley, D., M.H. Bulmer, and W. Murphy 2002. Patterns of movement of landslides in rotational and translational landslides. *Geology,* in press.

Pollard, W.H., C. Omelon, D.T. Andersen, and C.P. McKay 1999. Perennial spring occurrence in the Expedition Fiord area of Western Axel Heiberg Island, Canadian High Arctic. *Can. J. Earth Sci.* **36**, 1-16.

Rhodes, J.M. and J.P. Lockwood, Eds. 1995. *Mauna Loa Revealed: Structure, Composition, History, and Hazards.* AGU Geophysical Monograph 92, American Geophysical Union, Washington, DC.

Rice, J.W., Jr. 1992. Antarctic Lakes (above and beneath the ice sheet): Analogs for Mars. In *MSATT Workshop on the Polar Regions of Mars: Geology, Glaciology, and Climate History.* LPI Tech. Rep. 92-08, Part I, pp. 23-24. Lunar and Planetary Institute, Houston.

Rice, J.W., Jr. 1993. The Martian valley networks: Origin by Niveo-fluvial processes. In *Workshop on Early Mars, MSATT.* LPI Tech. Rep. 93-03, pp. 22-23. Lunar and Planetary Institute, Houston.

Rice, J.W., Jr. 1994. Terrestrial Polar Beach Processes: Martian Paleolake Analogs. *Lunar Planet. Sci.* **XXV**, 1125-1126. (abstract)

Rice, J W., Jr. 2000a. Mars analog field training of astronauts. *Concepts and Approaches for Mars Exploration.* LPI Contrib. No. 1062, pp. 261-262. Lunar and Planetary Institute, Houston.

Rice, J.W., Jr. 2000b. Flooding and Ponding on Mars: Field Observations and Insights from the Polar Realms of the Earth, *Lunar Planet. Sci.* **XXXI**, 2067. (abstract) [CDROM]

Rice, J.W., Jr. 2001. High Latitude Terrestrial Lacustrine and Fluvial Field Analogs for the Martian Highlands. In *Field trip and Workshop on the Martian Highlands and Mojave Desert Analogs.* (A.D. Howard, J. Moore, and J. Rice, Eds.), Abstract 4010. http://www.lpi.usra.edu/meetings/martianhighlands2001.

Rice, J.W., Jr., and K.S. Edgett 1997. Catastrophic Flood Sediments in Chryse Basin, Mars, and Quincy Basin, Washington: Application of Sandar Facies Model. *J. Geophys. Res.* **102**, 4185-4200.

Rice, J.W., Jr., A.J. Russell, F.S. Tweed, Knudsen, M.J. Roberts, P.M. Marren, R.I. Waller, E.L. Rushmer, H. Fay, and T.D. Harris 2000. Field investigations of Icelandic jökulhlaups as an analog to floods on Mars. In *Second International Conference on Mars Polar Science and Exploration.* LPI Contrib. No. 1057, Abstract 4098. Lunar and Planetary Institute, Houston.

Robertson, P.B. 1988. The Haughton impact structure, Devon Island, Canada, Setting and history of investigations. *Meteoritics* **23**, 181-184.

Schaber, G., J.F. McCauley, C.S. Breed, and G.R. Olhoeft 1986. Shuttle Imaging Radar, Physical controls on signal penetration and subsurface scattering in the Eastern Sahara. *IEEE Transactions on Geoscience and Remote Sensing* **GE-24**, 603-623.

Schaber, G.G., J.F. McCauley, and C.S. Breed 1997. The use of multifrequency and polarimetric SIR-C/X-SAR data in geologic studies of Bir Safsaf, Egypt. *Rem. Sens. Env.* **59**, 337-363.

Self, S., L. Keszthelyi, and Th. Thordarson 1998. The importance of pahoehoe. *Annu. Rev. Earth. Planet. Sci.* **26**, 81-110.

Sharma K, L. Keszthelyi, C. Thornber, and S. Self 2000. Variations in the cooling, crystallization textures, and glass chemistry during emplacement of Kilauea pahoehoe lobes, as constrained by in situ field experiments. *GSA Annu. Meeting, Online Abstracts* #51433.

Shepard, M.K., B.A. Campbell, M.H. Bulmer, T.G. Farr, and L. Gaddis 2001. The roughness of natural terrain: A planetary and remote sensing perspective. *J. Geophys. Res.* **106**, 32777-32795.

Siegert, M.J., J.A. Dowdeswell, M.R. Gorman, and N.F. McIntyre 1996. An inventory of Antarctic sub-glacial lakes. *Antarct. Sci.* **8**, 281-286.

Stern, R.J. and M.G. Abdelsalam 1996. The origin of the Great Bend of the Nile from SIR-C/X-SAR Imagery. *Science* **274**, 1696-1698.

Stoker, C.R., N. Cabrol, T. Roush, J. Moersch, J. Aubele, N. Barlow, E.A. Bettis III, J. Bishop, M. Chapman, S. Clifford, C. Cockell, L. Crumpler, R. Craddock, R. De Hon, T. Foster, V. Gulick, E. Grin, K. Horton, G. Hovde, J.R. Johnson, P.C. Lee, M.T. Lemmon, J. Marshall, H.E. Newsom, G.G. Ori, M. Reagan, J.W. Rice, S.W. Ruff, J. Schreiner, M. Sims, P.H. Smith, K. Tanaka, H.J. Thomas, G. Thomas, and R.A. Yingst 2001. The 1999 Marsokhod rover mission simulation at Silver Lake California: Mission overview, data sets, and summary of results. *J. Geophys. Res.* **106**, 7639-7663.

Swindle T.D., J.A. Grier, B. Li, E.D. Olsen, D.J. Lindstrom and A.H. Treiman 1997. K-Ar ages for the near-surface liquid water on Mars in the last few hundred million years. *Lunar Planet. Sci.* **XXVIII**, 1403-1404. (abstract)

Thordarson, Th., and S. Self 1993. The Laki (Skaftr Fires) and Gramsvntn eruptions in 1783-1785. *Bull. Volcanol.* **55**, 233-263.

Thorarinsson, S. 1953. The Crater Groups in Iceland. *Bull. Volcanol.* **14**, 3-44.

Treiman A.H. 1987. Geology of the nakhlite meteorites, Cumulate rocks from flows and shallow intrusions. *Lunar Planet. Sci.* **XVIII**, 1022-1023. (abstract)

Twidale, C.R. 1972. Evolution of sand dunes in the Simpson Desert, Central Australia. *Trans. Inst. Brit. Geograph.* **56**, 77-109.

Wadhwa, M. 2001. Redox state of Mars' upper mantle and crust from Eu anomalies in Shergottite pyroxenes. *Science* **291**, 1527-1530.

Wall, S.D., T.G. Farr, J.-P. Muller, P. Lewis, and F.W. Leberl 1991. Measurement of surface microtopography. *Photogramm. Eng. Rem. Sens.* **57**, 1075-1078.

Wasson, R.J., K. Fitchett, B. Mackey, and R. Hyde 1988. Large-scale patterns of dune type, spacing and orientation in the Australian Continental Dunefield. *Austral. Geograph.* **19**, 89-104.

Watters, T.R. and M.P. Golombek 1989. *MEVTV Workshop on Tectonic Features on Mars.* LPI Tech. Rep. 89-06, Lunar and Planetary Institute, Houston.

Westall, F., and M.M. Walsh 2002. Fossil biofilms and the search for life on Mars. In *Fossil and Recent Biofilms* (W.E. Krumbein, T. Dornieden, and M. Volkmann, Eds.), Kluwer, Amsterdam, in press.

Williams, K.K., and R. Greeley 2001a. Radar attenuation by sand: Laboratory measurements of radar transmission. *IEEE Trans. Geoscience Remote Sensing* **39**, 2521-2526.

Williams, K.K. and R. Greeley 2001b. Laboratory measurements of radar transmission through dust with implications for radar imaging on Mars. *Lunar Planet. Sci.* **XXXII**, 1264. (abstract) [CDROM]

Willis, K., P.W. Dickerson, and B.H. McRay 2000. *Canyons, Craters and Drifting Dunes - Terrestrial Analogues on Earth's Moon and Mars.* http://eol.jsc.nasa.gov/newsletter/planetary/sld001.htm.

World Crater Inventory 2002. http://www.unb.ca/passc/inventory/

Wyatt M.B., V.E. Hamilton, H.Y. McSween Jr., P.R. Christensen, and L.A. Taylor 2001. Analysis of terrestrial and Martian volcanic compositions using thermal emission spectroscopy: 1. Determination of mineralogy, chemistry, and classification strategies. *J. Geophys. Res.* **10**, 14711-14732.

Zuber, M.T., and 20 colleagues 1998. Observations of the north polar region of Mars from the Mars Orbiter Laser Altimeter. *Science* **282**, 2053-2060.

The Future of Solar System Exploration, 2003-2013
ASP Conference Series, Vol. 272, 2002
Mark V. Sykes, ed.

Lunar Exploration Manned and Unmanned

P.D. Spudis[1], S.W. Asmar[2], D.B.J. Bussey[3], N. Duxbury[2], L.J. Friesen[4], J.J. Gillis[5], B.R. Hawke[3], G. Heiken[6], D. Lawrence[6], J. Manifold[7], M.A. Slade[2], A. Smith[8], G.J. Taylor[3], & R.A. Yingst[9]

[1] *Lunar and Planetary Institute, Houston, TX*
[2] *Jet Propulsion Laboratory, California Institute of Technology, Pasadena, CA*
[3] *Hawaii Institute of Geophysics and Planetology, University of Hawaii at Manoa, Honolulu, HI*
[4] *Division of Natural and Applied Sciences, University of Houston at Clear Lake, Houston, TX*
[5] *Department of Earth and Planetary Sciences, Washington University, St. Louis, MO*
[6] *Los Alamos National Laboratory, Los Alamos, NM*
[7] *Applied Space Resources, Inc., Hicksville, NY*
[8] *American Physical Society, Washington, DC*
[9] *Natural and Applied Sciences, University of Wisconsin, Green Bay, WI*

Abstract. The past decade has seen two global reconnaissance missions to the Moon, Clementine and Lunar Prospector, which have mapped the surface in multiple wavelengths, determined the Moon's topography and gravity fields, and discovered the presence of water ice in the permanently dark regions near the poles. Although we have learned much about the Moon, many key aspects of its history and evolution remain obscure. The three highest priority questions in lunar science are: 1) the Moon's global composition, particularly the abundance of aluminum and magnesium; 2) the extent, composition, and physical state of polar deposits, including the extent, purity, and thickness of ice, the elemental, isotopic, and molecular composition of polar volatiles, the environment of the polar regions; and 3) the cratering chronology of the Moon and the implications of a possibly unique history, such as a cataclysm, for our understanding of other Solar System objects. Answering and addressing these questions require a series of new missions, including an orbiter (carrying XRF, imaging radar, and other instruments), the deployment of surface network stations equipped with seismometers and heat flow probes, selected robotic sample return missions from geologically simple areas (e.g., youngest lava flow or crater melt sheet), and complex geological field work, conducted by human explorers. Because the Moon is a touchstone for the history and evolution of other rocky bodies in the solar system, we believe that these questions are of very high scientific priority and that lunar missions should receive much more serious attention and detailed study than they have in the past by the NASA Office of Space Science.

REPORT

1. Current State of Knowledge

The decade of the 1990's saw two missions to the Moon, the first since the 1970's. Both the Clementine and Lunar Prospector missions mapped the entire Moon. Clementine obtained a global image cube of the Moon in 11 wavelengths in the visible and near-IR, mapped the topography of the Moon using laser altimetry, and detected water ice in the permanently dark areas near the south pole of the Moon. Lunar Prospector mapped the global concentration of selected elements (Th, Fe, and Ti) in the Moon's surface, the locations of surface magnetic anomalies, determined the size of the lunar core, and confirmed deposits of ice in the dark regions at both poles. We now have a global surface reconnaissance that permits us to understand the global context of lunar rocks, crustal structure and evolution, and volatile history.

Despite these advances, our understanding of the Moon and its evolution is incomplete. The Moon is the nearest planetary body and is a likely destination for humans traveling beyond low Earth orbit. It's processes and history provide us with a template for the deciphering and understanding of the processes and histories of all of the terrestrial planets. As an example, our current understanding of the stratigraphic chronology of the planets is strongly dependent on our understanding of the cratering record of the Moon, even though recent research suggests that the Moon has several unique aspects that may make its use as a planetary touchstone questionable. Because of the Moon's importance to comparative planetology, we must fully understand its origin and evolution to better comprehend the history of the other terrestrial planets.

We now think that the Moon was created as a consequence of a giant, planetary-scale impact 4.6 Ga ago. A hypothetical, Mars-sized object hit the Earth a glancing blow, hurling vaporized debris into Earth orbit. This material later coalesced into the Moon. The rapid timescale of lunar accretion resulted in nearly total melting (the lunar "magma ocean") and this global differentiation resulted in a plagioclase-rich crust (composed of anorthosite and related rocks), about 70 km thick. Considerable crust-forming magmatism continued for the next 500 Ma, with crustal and mantle remelting, the production and transport of magma, and the emplacement of igneous bodies into the crust, creating a lunar crust of surprising structural complexity. For some reason, large amounts of radioactive, incompatible trace elements were concentrated on what is now the western near side of the Moon; although it had been thought to be the site of a huge impact basin, it now appears that this Procellarum depression is somehow related to internal, magmatic processes.

The Moon underwent an intense, heavy bombardment early in its history, but the timing, intensity, and consequences of this bombardment have recently become more, rather than less, obscure. Clementine revealed the presence of huge, multi-ring basins that cover the Moon's surface, including the immense

South Pole-Aitken basin, at 2600 km across and over 12 km deep, the largest, deepest impact crater known in the solar system. This impact served to create major compositional provinces on the far side of the Moon and it is somewhat surprising that the impact which created it did not shatter the Moon into pieces. A key question in lunar chronology is the exact nature of the early bombardment. Impact melt samples from the Apollo collections all date from a very narrow time interval, between 3.8 and 3.9 Ga ago. Do these ages represent the entire cratering history of the lunar highlands? Or, are these ages representative of only a few, or in the extreme interpretation, one impact event? The nature of this lunar cratering "cataclysm" calls into question the use of the lunar example as a guide to the interpretation of impact cratering records and ages on the other planets.

The Moon underwent extensive volcanic resurfacing between 3.9 and 3 Ga ago and volcanic eruptions may have continued well past the age of the youngest Apollo mare basalts at 3.15 Ga. The lavas are all rich in iron, but have a "bimodal" distribution of titanium content. The new remote sensing data show that the extremely high Ti mare basalts returned by Apollo 11 and 17 are, in fact, not common among mare basalts on a global basis. Most mare lavas are moderate to low in Ti content, on both near and far sides, and tend to be relatively thin, on the order of a few tens to a few hundreds of meters - thicker than earlier, post-Apollo estimates that suggested meter-scale thickness, but much less than the kilometer-thick stacks estimated before the Apollo missions.

After the cessation of mare volcanism, the Moon has been relatively inert, witnessing only the occasional large impact, creating spectacular rayed craters such as Tycho, 108 Ma ago, and the constant bombardment of micrometeorites, which grind the surface rocks into a powder and churn the upper surface of the Moon constantly. It was somewhat of a surprise to find the existence of deposits of water ice in the dark regions of the Moon's poles. This water is not indigenous to the Moon but is probably derived from comets and water-bearing meteoritic debris that has landed on the Moon over the past 3 billion years. Such water is not stable on the Moon's surface, but the presence of permanent zones of darkness near the poles, which might have temperatures as low as 50 degrees above absolute zero, provide a locality where water molecules, and other volatile species, are stable and get "trapped" for geological time scales. Current estimates suggest that as much as 10 billion tons of water ice might exist near the poles of the Moon. Such water would be a valuable resource to future inhabitants of the Moon, both to sustain human life and to provide rocket propellant to power a space transportation infrastructure. Thus, the Moon possesses the resources needed to allow humanity to bootstrap its first off-world community.

2. Key Science Questions

Although we have learned much about the Moon over the last decade, many key aspects of its history and processes remain elusive. Key questions about the Moon that remain after the global mapping missions of the 1990's are summarized in Table 1. In the following paragraphs, we attempt to rank these questions in general priority order and suggest specific measurements and their required fidelity that will allow us to answer these questions. How such measurements and the instruments required to obtain them are grouped together as missions will be covered in the next section, "recommendations" (see below).

2.1. Lunar composition: Completing the assessment

Although the two polar-orbiting missions of the last decade, Clementine and Prospector, provided us with a global maps of selected mineral and chemical properties, the need remains for additional compositional data for key elements of geochemical interest. Such knowledge is necessary for studies dealing with the origin and evolution of the Moon, its thermal and volcanic history, and an accurate assessment of its resource potential.

For global compositional study from orbit, the two key missing elements are aluminum and magnesium. Both of these rock-forming elements are extremely important for a variety of chemical and geological studies and are currently only estimated to within an order of magnitude. Aluminum can be estimated because its concentration varies inversely with iron (although it is important to test this relation globally), but magnesium requires direct measurement. Magnesium is extremely important as an indicator of igneous fractionation processes and measurement of Mg and its variation would yield great insight into the igneous evolution of the Moon's crust and mantle.

The best method for estimating Al and Mg is by X-ray fluorescence spectrometry from a polar-orbiting spacecraft. As the sun illuminates the lunar surface, incident solar X-rays excite atoms of low atomic number in surface materials and the fluorescent X-rays carry information on the concentration of these elements. This technique was proven during the late Apollo orbital missions and it is now time to complete the global, orbital compositional survey.

In addition to these two key elements, it is also important to understand the composition of the interior of the Moon, both for comprehending the Moon's bulk composition and for deciphering the interior structure of the Moon. The best methods of determining such properties is through geophysical investigation, particularly seismometry and heat-flow measurements. The Apollo missions laid out a geophysical network on the near side that operated for nearly seven years. From these data, we found that the Moon has a crust and mantle, determined its approximate thicknesses, and estimated the refractory element budget of the Moon from two heat flow measurements (lunar heat flow is caused by the decay of radioactive elements). Although important, more data of much better quality are needed to resolve critical issues in lunar composition and evolution. A surface network of seismic stations and heat flow probes deployed globally could give us greatly improved knowledge of crustal thickness variations, deep structure, approximate compositions, and refractory element content of the bulk Moon. Coverage of the Moon's far side gravity field will allow us to better assess variations in lunar crustal structure and thickness.

Finally, our knowledge of lunar compositions will always remain rudimentary until we can obtain more samples from the lunar surface. The wide variety of scientific topics that can be addressed via sample collection ensure that it will always be a high priority. Sample collection is the preferred method for determining certain critical properties, such as the absolute age of the emplacement of lunar geological units, like impacts and volcanic eruptions. Although sample collection provides the greatest fidelity data, the role of in-situ measurements,

Table 1. Moon Decadal Strategy

Knowledge Requirements	Data or Measurements Needed	Instruments and Missions
Composition and structure of the lunar crust - lateral and vertical heterogeneity - origin of hot spot, SPA	a. global Mg, Al maps b. min./chem. composition of unusual terranes c. samples of far side, limb, anomalous areas d. trace-element composition for selected regions e. composition as function of depth in crust f. crustal layering and subsurface structure g. estimates of bulk crust, mantle from seismic and heat flow data	X-ray spect. / orbiter (a,b,e) MS imaging + XRF in situ / landers (b,d,e) Regolith and rocklets / simple sample return (b,c,d,e) Full element comp. of selected spots / active GRS (Double Eagle) (d,e) Remote, in situ, and samples of central peaks, basin rings (b,c,d,e,f) Global seismic and heat flow network (e,f,g)
Impact and cratering history of the Earth-Moon system - cataclysm or not? - Post-3 Ga Flux history - Spikes and periodicity in impact flux	a. radiometric ages of rocks from key units: oldest basin on the Moon (SPA), Nectaris, Imbrium basins, and stratigraphic units from a variety of times b. impact melts of key basins, craters to characterize nature and sources of projectiles	in situ radiometric dating / landers, rovers (a,b) rocklets and regolith / simple sample return (a,b) selected rocks and field work / complex sample return (a, b) in situ MS imaging + XRF / landers (b) radiochemistry of impact melts / sample return (in situ??) (b)
Nature of the Impact Process - crater evolution; scaling with size - origin of central peaks, basin rings - regolith formation and evolution; origin and evolution of foreign debris in regolith - role of lateral mixing of crust, ancient and modern	a. composition of ejecta, walls, peaks, rings b. reconstruction of impact targets; stratigraphy of crater subsurface c. composition (and variations) of impact melt sheet and ejected blobs d. characterization of quantities and compositions of exotic debris in regolith e. nature of compositional gradients at mare-highland boundaries	exploration of crater geology / orbiters, landers, rovers (a-e) in situ analysis, MS imaging + XRF / rovers (a,e) sample characterization / simple and complex sample return (a,c,d) trenching and coring of regolith / rovers, human missions (d,e) ejecta radial traverse and characterization / rovers, human missions (d,e)
Origin, evolution, and fate of lunar volatiles	a. nature, extent, and composition of polar deposits b. composition of lunar outgassing c. volatile coats on pyroclastic gases d. trapped volatiles in vesicles? e. Time-varying volatile emission	radar mapping of polar regions / orbiter (a) temperature of cold traps / orbiter, landers (a) char. of ices, Mass spect. / landers (a,b,e) char. volatiles in pyroclastics / landers, simple sample return (b,c,d) outgassed volatiles compos. / orbiter (b,c,e)
Thermal and tectonic history of the Moon - Timing, duration and extent of mare and non-mare volcanism - near side-far side dichotomy - origin and evolution of Procellarum province	a. absolute ages of lunar volcanic units b. ages of crustal deformation (via stratigraphic evidence) c. isostatic state of craters and basins d. composition, morphology of volcanic deposits	in situ radiometric dating / landers, rovers (a,b) rocklets and regolith / simple sample returns (a,b,d) hires topography / radar, laser, stereogram. Orbiters (b,c,d) digital stereoimage model (5 m/pixel) / orbiter (b,c.d)
Origin of the Moon	a. bulk composition of the Moon	complete global survey for all elements / orbiter global network, seismic and heat flow / surface network, (microwave radiometery? / orbiter) mantle samples; SPA floor, xenoliths in pyroclastics / orbiter, rover, complex sample return

particularly in regard to new technical developments, needs to be evaluated as well. In-situ measurements can provide detailed compositional information without the high mission costs of providing a return stage for samples.

In sum, we need additional information on lunar compositions from a variety of techniques. In priority order, we should complete the global reconnaissance from orbit by measuring Al and Mg concentrations. We should next emplace a global surface network to improve our understanding of crust and mantle structure, composition, and evolution. We should plan carefully targeted landed missions both to make in-situ measurements of surface chemical and mineralogical composition and to collect and return surface samples for analysis back on Earth. These approaches ensure a robust and diverse set of compositional data which will shed new light on the origin and evolution of the Moon.

2.2. Polar volatiles: Origin, evolution, and fate

The startling discovery of ice in the permanently dark areas near the poles of the Moon by Clementine and Lunar Prospector has raised many more questions about the nature, composition, and origin of these deposits. As the Moon has virtually no indigenous volatiles, this material must have extra-lunar origins, probably as remnants of cometary and meteoritic debris that has impacted the Moon over geological time. The preservation of such deposits has important scientific and utilization implications and we must learn more about the origin, nature, and fate of these deposits.

The first step in characterizing these features is to map their distribution, assess their quantity and physical state, and better characterize their environment of deposition. To a first order, such a characterization can be accomplished from lunar orbit. We can use imaging radar to map the distribution of the ice deposits, their approximate thickness and purity, their areal extent, and the nature of the surrounding environment. Temperature measurements from orbit can determine the actual thermal conditions in the lunar cold traps, which have never been directly measured. Detailed, high resolution altimetry can map the topography of the poles and these data can be compared with continuous images over the course of a lunar year to reveal the extent of permanent darkness and any zones of permanent or near-permanent solar illumination.

After this orbital reconnaissance is completed, we must have surface measurements of the polar deposits. We need to understand the elemental, isotopic, and molecular composition of the ices, the nature of any clathrates, other volatiles species present and their abundance. Although there are several ways to obtain these data, the simplest is to make in-situ measurements of the ice using a laser flash heater and a mass spectrometer. This instrumentation can be placed on a roving vehicle for multiple measurements or, multiple instruments can be placed on multiple, small probes that may be, in part, semi-hard landers. The return of a sample of lunar polar ice to Earth would allow the most detailed measurements to be performed, but would require great care to preserve the very cold environment that exists in the polar cold traps during return of the sample to Earth.

As the Moon has apparently been accumulating cometary ice for at least the last 2 billion years, detailed characterization of the polar deposits of the Moon will allow us to unravel the volatile history of the inner solar system. But even more important than their scientific value is their potential for strongly leveraging an Earth-Moon transportation infrastructure. Lunar polar ice can be mined for many purposes, but the production of rocket propellant (hydrogen and oxygen) will greatly reduce the costs of routine transport between low Earth orbit and the Moon - and by extension, geosynchronous orbit. The creation of a transportation infrastructure in cislunar space serves scientific and national strategic goals in that such a system will be of great value for future, large missions to the planets, including human missions.

2.3. Impact chronology: Implications for the history of the Moon and other planets

The Moon remains the principal guide to our interpretation of the early geological histories of all the terrestrial planets. We use in particular the impact chronology derived from the study of lunar highland rocks to understand the phase of planetary history call the "early heavy bombardment." In brief, this paradigm envisions that after accretion on all of the terrestrial planets, a period of intense impact flux occurred, churning up planetary crusts with huge basin-forming impacts, creating enormous bodies of shock melted material, some of which might have subsequently differentiated into igneous complexes, and pulverizing the outer few kilometers of planetary crusts. Moreover, the chaotic environmental disruptions caused by basin impacts might have precluded any initial biosynthesis, suggesting that life may have originated (and then immediately became extinct) many times over.

In recent years, a new paradigm for lunar origin and evolution has called into question the suitability of using the Moon as a planetary touchstone. Because the origin of the Moon was the result of a singular event, namely, the collision of two proto-planetary objects early in solar system history, the Moon's early thermal and impact history may have been likewise singular. Global melting and subsequent bombardment, inferred for all the planets by analogy to the Moon, may have been quite different for different objects. On the other hand, early heavy bombardment as the tail-end of accretion is predicted by models of planetary formation and growth, so perhaps the Moon does offer substantial guidance in the interpretation of planetary histories.

Because of its importance, we must understand early lunar cratering history much better than we do now. Specifically, recent work demonstrates that the Moon may have undergone a significant increase in the rate of impact around 3.8 Ga ago, a phase of lunar history referred to as the "cataclysm." This spike in the rate of large body impacts is not predicted by normal models of planetary accretion and special circumstances must be invoked to account for it.

But there is considerable controversy about whether there really was a cataclysm or not. Because of the importance of the Moon and its cratering history to the interpretation of the histories of the other planets, it is important to resolve this issue. Understanding the cataclysm is also important to fully understand lunar history, the interplay between cratering and crustal magmatism, lunar volcanic history, and the nature of the outer few kilometers of the Moon (the "megaregolith").

At the other end of the time scale, the post-mare cratering record of the Moon is important because it complements the "missing" pages of Earth's history. Recently, evidence has been gathered that suggests that the Earth is periodically showered with impact debris, possibly from comets perturbed into the inner solar system by either galactic plane crossings or some unseen, massive object outside our solar system. These putative periodic showers have been tentatively linked to mass extinctions found in the fossil record. Such a connection has profound implications - perhaps the history of life on Earth is fundamentally linked to the impact flux. But such an idea cannot be tested on Earth because the dynamic processes of erosion and tectonism have destroyed much of earth's cratering record. However, the Moon's cratering record for the last 3 Ga is clearly and plainly preserved and Earth and Moon are inexorably linked together in their joint journey in space.

So the Moon's cratering history serves both to illuminate the early evolution of the planets, by presenting a paradigm for early bombardment and the later evolution of life on Earth during the post-mare era. Because of this importance, a better understanding of the Moon's impact history is a high priority goal in planetary exploration. Two principal scientific problems in the cratering history of the Moon need to be addressed. First, did the cataclysm occur and if so, what was its nature and magnitude? Second, is there periodicity in the impact flux in the Earth-Moon system during the last few billion years?

Answering the first question is easy in principle and difficult in practice. Basically, we need to know the absolute age of the oldest basin on the Moon, the South Pole-Aitken basin. As all other basins post-date this one, determining its absolute age could resolve the nature of the cataclysm; if it is as old as 4.3 Ga, there was no "cataclysm" as such, but merely a spike in the cratering rate. If SPA is 3.9 Ga old, we had a cataclysm of the most extreme variety possible. Seldom in science is an issue so clearly and definitively resolvable.

Unambiguous samples of SPA basin impact melt will be difficult to recognize prior to selection. This makes a successful robotic sample return from this basin unlikely. Such a mission would obtain basin floor material, some of which might contain basin impact melt. But we would not be certain that such melt comes from SPA basin - the younger age of all other basins means that, theoretically, impact melt from any basin on the Moon might be returned. To give ourselves the best chance of obtaining SPA impact melt, we must focus on the basin massifs of SPA and concentrate on impact melt units, boulders, and ledges that comprise the fabric of these massifs. Although such a strategy does not guarantee direct samples of SPA melt, that is where our best hope lies. Such a sample return mission is likely to be successful only with people. Hence, we believe that this task, important as it is, is likely to be successfully resolved only with the advent of a human return to the Moon.

Deciphering the late cratering history of the Moon is a problem that might be tractable with simple robotic sample return missions. The basic approach is to obtain unambiguous impact melt samples of a large number ($\sim$100) small (D$\sim$500 m or smaller), post-mare craters. As melt tends to be concentrated in the very bottoms of crater floors, small rovers could traverse into these craters, recognize glassy in-situ melt deposits, and obtain small fragments for radiometric dating. Ideally, such samples would be returned to Earth for analysis, but the technology of in-situ K-Ar dating should be investigated. If such a technique is feasible, it would greatly reduce mission cost. A mission like this could provide direct evidence of the impact flux and its variations in the Earth-Moon system during the last 3 billion years.

Ultimately we need many different absolute ages for a variety of units across the Moon. Lava flows, basins, craters, and pyroclastic deposits need to be sampled and dated. We are particularly thin for calibration tie-points in middle lunar history, i.e., the late Eratosthenian- early Copernican periods (3-0.5 Ga ago). Because of the importance of this cratering history to our interpretation of the ages on all of the terrestrial planets, it makes sense to use the nearby Moon as a natural planetary laboratory and "witness plate" to the impact flux in the inner solar system.

3. Recommendations

To address our scientific questions requires a variety of different measurements, instruments, and missions (summarized in Table 1). We here outline a few possible missions, their possible payloads, sequence, and operations designed to address the principal scientific goals described above.

The most pressing need is to complete the orbital surveys of the Moon. A polar-orbiting spacecraft could complete the chemical survey of the Moon by carrying an X-ray spectrometer to measure lunar Al and Mg. The same spacecraft could also carry a payload designed to resolve the first order questions about the polar volatiles - their extent, purity, total abundance, and physical state. These measurements could be made with an imaging radar (to map the ice and obtain a detailed topographic map of the poles), a thermal infrared bolometer (for temperature determination), an imaging camera (to observe the polar lighting conditions through a seasonal cycle). This payload, combined with an X-ray instrument and possibly one or two others (e.g., a mass spectrometer to characterize the lunar atmosphere) would provide much of the critical data we need to complete the orbital reconnaissance of the Moon. A small, subsatellite would allow us to map the gravity field in detail on the lunar far side, an important data set to assess variations in crustal thickness.

Next, we need to plan a series of surface missions. The first of these could land near the dark areas of the polar regions to obtain direct, in-situ measurements of the chemical and isotopic composition of the polar ice. We also need a further series of small, network stations, delivered by separate landers or a long-distance rover, to begin the next generation survey of the lunar interior - each package should contain a 3-axis seismometer and heat flow probe at a minimum; a third instrument (such as a magnetometer) could be added at little additional cost. The surface network should be designed to operate for at least 2-4 years and ideally, an RTG or more likely, a rechargeable battery/fuel cell power source, should be provided for it.

The next set of surface missions should plan for extended exploratory traverses, designed to collect samples of impact melt from small craters to assess the variations in the post-mare impact flux of the Moon. Although a return of these samples to Earth is highly desirable, the development of in-situ radiometric dating is a technological challenge that should be carefully assessed; if such an instrument is feasible, the commercial spin-off possibilities might reflect very favorably on the Code S exploration program.

Sample return from the Moon is highly desirable from many scientific perspectives and will most likely be required in order to assess the nature of the early lunar cataclysm. We support the return of humans to the Moon for many reasons, but their contributions to the lunar science questions outlined here will be immense. Human field work will be needed to obtain the carefully selected samples necessary to date the oldest basins on the Moon. Moreover, after the initial characterization of the polar ice deposits, people will be needed to emplace and align surface experiments, including advanced geophysical instrumentation to better determine lunar interior composition and structure. Human field work will be needed to carefully decipher the complex geological history and processes of the Moon.

The orbiter mission can be flown for a cost on the order of \$100M. The small landers and rovers to characterize the poles are in the \$200-300M cost category. More capable landed missions to emplace geophysical stations or deploy a sampling rover are on the order of \$500 M each. An initial human return to the Moon is likely to cost a few to tens of billions of dollars, but of course, such missions will not be undertaken solely for science and hence, will not be solely budgeted to Code S. The lunar science community should stand ready to support a human return to the Moon with carefully thought out strategies of scientific investigation, including goals and possible tactical approaches to human exploration. It is entirely possible that humans will be on the Moon by the end of the coming decade and it is not too early to begin to think about innovative exploration plans for that eventuality.

4. Maintaining Capabilities for the Future

The lunar community is still assimilating the return of global remote sensing data from the missions of the last decade. Although we still do not fully understand all of these data, significant progress is being made. Programmatic considerations suggest that we should make efforts to maintain a base of expertise in lunar science and lunar mission technology in the coming decade. Right now, graduate theses are being undertaken using the data of Clementine and Lunar Prospector. NASA Research and Analysis funds should be budgeted to support this continuing work so that these students and post-docs will be ready to fly the lunar missions of the coming decade and interpret the scientific results.

We need to develop new instruments and instrument technology, particularly in small lightweight chemical analysis labs, mass spectrometers, geophysical instruments, and compact, rechargeable power supplies. The development of instruments for in-situ radiometric dating should be an important priority. Special sample environmental containers, designed to preserve the cryogenic conditions of the polar dark areas, should be developed for the eventual return of polar ice samples from the Moon for analysis on Earth. It is highly desirable to develop an inexpensive means for the retrieval of surface grab samples ($\sim$1 kg) from a variety of reconnaissance targets all over the Moon. The use of space tethers to accomplish this has been examined and found to be tentatively feasible - such work in innovative technologies applied to small missions should continue.

We believe that the Moon is an important scientific object, both in its own right and in its character as a touchstone for the interpretation of the histories of all the terrestrial planets. Our outline of lunar missions and experiments could easily be expanded to twice the current length - we have attempted here to focus our attention on the most pressing, most scientifically significant problems, the bulk of which can be addressed in the coming decade with small robotic spacecraft. Ultimately, we need to return to the Moon with people in order to unravel its complex history in detail. This exploration plan gives us the knowledge we need to make that exploration productive and profitable.

The Future of Solar System Exploration, 2003-2013
ASP Conference Series, Vol. 272, 2002
Mark V. Sykes, ed.

Determinative Mineralogy: An Essential Component of Planetary Exploration

D.L. Bish[1], D.T. Vaniman[1], D.F. Blake[2], J.R. Green[3], C.T. Johnston[4], B.A. Kelly-Serrato[5], D.W. Ming[5], J.J. Papike[6], A.S. Yen[3], & M.E. Zolensky[5]

[1] *Los Alamos National Laboratory, Los Alamos, NM*
[2] *NASA Ames Research Center, Moffett Field, CA*
[3] *Jet Propulsion Laboratory, California Institute of Technology, Pasadena, CA*
[4] *Department of Agronomy, Purdue University, West Lafayette, IN*
[5] *NASA Johnson Space Center, Houston, TX*
[6] *Institute of Meteoritics, University of New Mexico, Albuquerque, NM*

Abstract. Mineralogy is a fundamental characteristic of extraterrestrial bodies because different mineral assemblages can be used to characterize present and past conditions of the atmosphere, the surface, the crust, and the deep interior of a planet. Many tools are used in space exploration, and the exploration of extraterrestrial mineralogy is no exception. The tools that are needed include remote observations, in-situ analysis with landed instruments, and analysis of returned samples in laboratories on Earth. In a balanced exploration program, all three sources of data are necessary. Likewise, no one method will provide the information needed to understand the makeup and history of extraterrestrial bodies. This paper is intended as a starting point for the development of truly integrated mineralogic instrumentation efforts that will combine the diverse capabilities of many analytical methods.

EXECUTIVE SUMMARY

One of the most informative characteristics of a planet is its mineralogy. Suites of surface minerals can be used to characterize past and present climatic conditions, sedimentary weathering processes or hydrothermal activity. More than elemental data, mineralogical data are linked to surficial conditions and processes and can be used to elucidate present and past conditions of the atmosphere, the surface, the crust, and occasionally the deep interior of a planet. Minerals have known ranges of stability and paragenetic relationships. Even when minerals persist out of their stability fields (as is common in sedimentary rocks), disequilibrium associations, or the presence of relict unstable phases, can be used to unravel detailed sedimentary or diagenetic histories that can be linked to internal evolution, surface processes, or climate history.

The importance of mineralogical studies extends beyond the terrestrial planets and rocky moons. For example, the colored fracture zones of Jupiter's moon Europa are thought to be due to hydrated salts. An evaluation of the mineralogy of the ices and salts present in these zones would shed light on the nature of the proposed ocean deep beneath Europa's icy crust. The mineralogic composition of comets is equally important. The need for in-situ mineralogical studies is particularly pressing for these icy bodies, for it is very unlikely that samples can be collected from them and returned to Earth unaltered.

Our existing knowledge of extraterrestrial mineralogy has been gained largely from analyses of returned or collected samples - returned samples and meteorites from the Moon, meteorites from Mars, and the more extensive set of meteorites that represent early solar system bodies. However, there is a large gap between what we have learned from these samples and what is inferred from remote planetary exploration. This gap can be closed by multidisciplinary efforts to extend the capabilities of remote analysis, to provide a better correlation between remotely obtained data and mineralogic verification, and to use the remotely obtained data to carefully select landing sites and to collect samples for return to Earth.

There are many tools to be used in space exploration, and the exploration of extraterrestrial mineralogy is no exception. The tools that are needed include remote observations, in-situ analysis with landed instruments, and analysis of samples in laboratories on Earth. In a balanced exploration program, all three sources of data are necessary.

What instruments do we need? This is a question that has been the focus of much debate and the resolution of this debate is beyond the scope of this paper. Rather, this paper is intended as a starting point for the development of truly integrated mineralogic instrumentation efforts that will combine the diverse capabilities of many analytical methods. A list of instruments that should be considered is provided below, but this list is based mainly on current research directions and is not a final and definitive list.

- X-ray Diffraction
- Reflectance and Emittance Spectroscopy
- Thermal Emission Spectroscopy
- Mössbauer Spectroscopy
- Laser Raman Spectroscopy

The most common method of phase identification in the past eight decades has been X-ray diffraction, which relies on the fact that all crystalline substances diffract X-rays in a known, predictable fashion, producing a pattern that can be used much like a fingerprint. Because of this, X-ray diffraction has been the standard for producing unambiguous mineral identifications for more than 75 years. Indeed, other identification methodologies, such as spectroscopic and chemical, routinely use X-ray diffraction to provide baseline information. Other methods such as reflection and emission spectroscopy, Raman spectroscopy, and Mössbauer spectroscopy are sensitive to specific attributes of minerals and organic materials and in many cases also provide fingerprint identifications. Each of these methods has specific strengths and weaknesses, and when used in combination can provide a complementary and more complete assessment of mineralogy. Indeed, a well-designed payload for study of extraterrestrial mineralogy should consist of several instruments, operating together to provide the most complete mineralogical picture possible.

What are the relative roles of surface analysis and remote sensing in extraterrestrial mineralogical studies? Remote spectral analysis techniques have been applied to many objects within our solar system, with the result that in general terms, we know the compositions of the solid surfaces that we can image. In some instances, high-resolution spectral data are available as is the case for the Moon, Mars, and Europa, among others. However, even these high-resolution images and analyses have lateral spatial resolutions of tens to thousands of meters, many orders of magnitude larger than the scale length of the phases in surficial deposits that comprise the images. Remote sensing is principally useful in the formulation of hypotheses that can later be investigated by in-situ analysis using techniques that can provide more unambiguous mineralogical information. Mineralogical analyses by landers provide not only a scale of measurement (mm to m) more amenable to mineral identification, but they allow analysis by methods that are particularly powerful for mineral identification but require physical contact with the sample (e.g., X-ray diffraction). The most successful exploration program will not deny one source of data at the expense of another, but it will build supporting systems of remote analysis, analysis by landers, and selected sample return to unlock the fundamental information contained in the minerals that comprise most of the solids in our solar system.

REPORT

1. Introduction

Our existing knowledge of extraterrestrial mineralogy has been gained largely from analyses of returned or collected samples - returned samples and collected meteorites from the Moon, meteorites from Mars, and the more extensive set of meteorites that represent early solar system bodies. However, there is a large gap between what we have learned from these samples and what is inferred from remote planetary exploration. This gap can be closed by multidisciplinary efforts to extend the capabilities of remote analysis, to provide a better correlation between remotely obtained data and definitive mineral determinations, and to use the remotely obtained data to carefully select landing sites and to collect samples for return to Earth.

There are many tools to be used in space exploration, and the exploration of extraterrestrial mineralogy is no exception. The tools that are needed include remote observations, in-situ analysis with landed instruments, and analysis of samples in laboratories on Earth. In a balanced exploration program, all three sources of data are necessary. Remote sensing covers large portions of planetary bodies and places mineralogy in an integrated whole-body context. Landers with instruments can provide many on-site analyses for ground-truth data on a scale that approaches the "footprint" of the orbital instruments. If appropriate instruments are used on landers, accurate information can be obtained in-situ on chemistry and mineralogy, and it is likely that a combination of landed diffraction, chemical, and spectroscopic instruments can provide unambiguous identifications and detection limits comparable to those obtained during routine analyses on Earth.

Naturally, mineralogy studies, whether lander based or done with returned samples, will be no more representative than the samples chosen for analysis. However, returned samples (or fortuitous meteorites) provide the ultimate assurance of mineral identity and composition, for the ability to dissect and probe and re-analyze samples by human researchers in their laboratories will not be matched by remote methods. Clearly, sample return will be required to gain information on mineral zonation, the existence of nanofossils or indirect mineralogical evidence of biological processes, to see features such as the tiny reduced Fe metal grains in lunar samples, or to detect new phases such as yoshiokaite, which is a product of extreme volatilization associated with impact processes on the surface of the Moon. The sequencing and emphasis placed on these three tool sets (remote sensing, landers, and samples in hand) will depend on the bodies being studied, but all three tool sets have important contributions to make in understanding extraterrestrial mineralogy and in using knowledge gained from minerals to understand planetary processes.

In this report, a definitive identification of an individual solid phase having a specific crystal structure constitutes confirmation of a specific mineral. On Earth, this definitive identification is typically gained using methods that are sensitive to the crystal structure of a solid phase, including optical methods, vibrational spectroscopy, and diffraction methods. The most common and unambigious method of phase identification in the past eight decades has been X-ray diffraction (XRD), which relies on the fact that all crystalline substances diffract X-rays in a known, predictable fashion, producing a pattern that can be used much like a fingerprint. Other methods such as reflection and emission spectroscopy, Raman spectroscopy, and Mössbauer spectroscopy are sensitive to specific attributes of minerals and organic materials and in many cases also provide fingerprint identifications. Each of these methods has specific strengths and weaknesses, and when used in combination can provide a complementary and more complete assessment of mineralogy. Indeed, a well-designed payload for study of extraterrestrial mineralogy should consist of several instruments, operating together to provide the most complete mineralogical picture possible.

2. Goals of Extraterrestrial Mineralogic Studies

There is a wide variety of objects in our solar system for which mineralogic studies are needed to provide true understanding:

- The rocky planets Mercury, Venus, and Mars,
- Rocky planetesimals, including asteroids and planetary moons,
- Icy planetesimals, including Kuiper belt objects, Oort cloud comets and icy moons

What would we like to learn by studying these objects?

- Origin of the Solar System. Asteroids and comets have crystallization ages which predate the formation of the planets. An understanding of their mode of formation will shed light on early solar nebula processes and planet formation.

- Comparative planetology. By analyzing planetary materials from Mars, Mercury, Venus, Earth, and the Moon, information on early differentiation, plate tectonics, upper mantle and crustal processes, and the presence/absence of magnetic fields over time can be determined and compared.

- Natural history of volatiles and biogenic elements. Water and other volatiles, including the entire inventory of the biogenic elements H, C, O, N, S and P may have been deposited onto the surfaces of the rocky planets as a thin veneer during the late bombardment of the early solar system. What is the source of the volatiles? What is their distribution over space and time?

- Did life evolve there? Earth harbors the only example of life as we know it, but we have reason to believe that life could originate wherever appropriate conditions existed (liquid water, a source of simple carbon-containing molecules and energy). Such conditions apparently existed on Mars in the distant past, and perhaps in the water ocean thought to exist below the ice-covered surface of Europa.

- Can we live there? Human missions to Mars are being planned, perhaps as early as 2018. Permanent colonization will require utilization of local resources for food and shelter, in addition to knowledge of the environmental hazards associated with surface dusts.

Much of this information is contained in the mineralogy of planetary surfaces. We have learned a great deal from remote sensing techniques, but definitive answers, particularly where mineralogy is involved, will require in-situ instruments or samples returned to Earth.

XRD has been the workhorse and the standard for determinative mineralogy for many decades. However, improvements in the instrumentation and understanding of other spectroscopic methods that are sensitive to mineralogy and the development of extensive data bases and cross reference to the extensive XRD databases has increased their use in mineralogical studies. The development of CCD detectors and miniaturization of components has made some of the spectroscopic methods especially suitable for high-resolution mapping from orbital platforms. The rapid pace of improvements in laser technology has led to increased applications of some methods.

3. Instrumentation in Planetary Exploration

3.1. X-ray diffraction

XRD is based on the fact that crystalline substances, which have periodic crystal structures, diffract X-rays in a well-defined manner. Measured diffraction data are directly related to mineral structure in a way that is well understood and can be calculated from first principles. This direct link to crystal structure makes XRD the benchmark identification tool, providing typically unambiguous results, even on multicomponent mixtures. In addition, XRD has the potential to achieve detection limits well below 1% and it can provide quantitative information on fine-grained and disordered materials such as clay minerals.

Between 1960 and 1968, considerable effort was expended in designing and building a practical XRD for use on the Moon (e.g., Dunne and Nickle 1968). The first proposal for a planetary XRD was from William Parrish of Philips Electronic Instruments in 1960. A series of breadboard and prototype instruments was constructed at JPL in the following years. The final flight prototypes, which were Bragg-Brentano geometry instruments utilizing a Cu X-ray tube, Soller slits and Xe gas proportional detectors, achieved a performance equal to laboratory diffractometers. The instruments suffered from relatively unreliable power supplies, and, at 20 kg, were heavy for spacecraft instruments. The lunar XRD program was ended in 1968 without ever deploying an instrument on a lunar mission.

A number of prototype XRD instruments were built and evaluated during the years 1964-1968. Combined XRD/XRF (X-ray fluorescence) instrument concepts were investigated as well, and it was concluded that such capabilities were feasible in a flight instrument using then-current technology. In addition to the basic instrumentation, simple drills and sample delivery devices were devised to provide powder samples to the instrument. The U.S. Geological Survey utilized one of the prototype XRD instruments as a field-operable system and performed hundreds of diffraction analyses during two field tests.

There is a renewed interest in XRD instrumentation for space exploration. The power of XRD in mineralogical analysis is compelling because it remains the most versatile and reliable method for characterization of materials with crystalline structure (the presentation of definitive XRD data is still the key requirement in acceptance of a new mineral description by the International Mineralogical Association). Vaniman et al. (1998) present a case for XRD (combined with XRF) in the search for deposits that may host evidence of life on Mars. Sarrazin et al. (2000) provide a broader analysis of the combined XRD/XRF method and its potential applications in determination of extraterrestrial mineralogy. Significant advances in the past 20 years have led to precise and accurate determination by XRD of mineral abundances in complex rocks and soils. As with several other methods described below, XRD (combined with XRF) has benefited from and has yet-undeveloped potential based on new CCD technology.

3.2. Reflectance and emittance spectroscopy

Reflectance and emittance spectroscopy rely upon the interaction of electromagnetic radiation at visible to near infrared wavelengths with particulate materials. These interactions involve a combination of reflection, absorption, scattering, and emission processes, dependent on geometry of the interaction, physical character of the surface, and the wavelength of radiation (Hapke 1993). Reflectance and emittance spectroscopy are sensitive to both crystalline and amorphous materials. A key advantage is that they can be used in the laboratory as well as remotely (e.g., from orbit). Some of the disadvantages are broad, overlapping spectral features and sensitivity of these features to small changes in the chemistry and structure of a material. Variations in material composition commonly cause shifts in the position and shape of absorption bands. Thus, with the great variety of chemistry typically encountered in natural materials, spectral signatures can be quite complex. However, recent improvements have increased knowledge of the natural variation in spectral features and the causes of the shifts. As a result, the previous disadvantage can and has been turned into an advantage, and extensive libraries of a wide variety of materials of known composition and mineralogy are now available.

With the advances in computer and detector technology, the new field of imaging spectroscopy is developing (reviewed by Clark 1999). Imaging spectroscopy is a technique for obtaining a spectrum in each position of a large array of spatial positions so that any one spectral wavelength can be used to make a recognizable image. The image might be of a rock in the laboratory, a landing site from a descent image or a lander or rover on the surface, or a whole planet from an orbiter. By analyzing the spectral features, and thus specific chemical bonds in materials, one can map where those bonds occur, and thus map materials, e.g., using spectral feature analysis. Even if end-member minerals (components) are not known, variations and material diversity can be mapped in heterogeneous materials such as rocks.

3.3. Thermal emission spectroscopy

Thermal emission spectroscopy, which is currently in use in Mars orbit (Mars Global Surveyor - TES instrument), senses infrared radiation emitted by objects in the mid to far IR wavelength region. This range covers vibrational absorption features that are diagnostic of mineral structure and composition. Each spectrum is produced by fundamental vibrations of major covalent ionic groups, lattice vibration bands, overtones, and combination modes. In principle, mineral identification can be accomplished by direct comparison of a measured spectral pattern with spectra of pure minerals, given a suitable spectral database (e.g., Christensen et al. 2000). The strongest IR bands of most oxyanionic minerals occur in slightly different wavelength regions, and these can be used for first-order determination of the mineral type (carbonate, sulfate, phosphate). For silicates having different degrees of polymerization (e.g., orthosilicates, single-chain, double-chain, layer, and framework silicates), the major fundamental vibrations of the SiO_4^{4-} group (asymmetric stretching and bending modes) show characteristic regularities in shifts of their peak positions. Such shifts can be used to classify these silicates and to estimate the rock types. A decomposition calculation is essential in order to make mineral phase identifications from the spectra of mixtures (rocks and soils), mainly due to the broad widths and complicated shapes of thermal emission spectral bands. Such a procedure has been developed by the TES science team and has been used for MGS-TES data and terrestrial rocks (Wyatt et al. 2001). Making measurements in the thermal IR provides the capability to penetrate through dust coatings that are common, for example, on the Martian surface and can present problems for remote sensing observations.

Linear decomposition methods have been developed for thermal emission spectroscopy that can effectively discriminate some mixtures of minerals, for example, the members in a series of igneous rocks (e.g., basalt-andesite-dacite).

Such methods rely on the assumption that a thermal emission spectrum of a mixture is a linear combination of the spectra of its constituent minerals and that the spectrum of a solid solution is a linear combination of the spectra of end-members, which is approximately true only for fundamental vibrations. Decomposition of thermal emission spectra is more complicated than that for Mössbauer spectra, for which the shape of spectral peaks is well understood.

3.4. Mössbauer spectroscopy

^{57}Fe Mössbauer spectroscopy is based upon the interactions between the Fe nucleus and its surrounding electromagnetic field, which is determined by the symmetry of the crystal framework and the types of atoms and bonds in the minerals in which it occurs. Iron atoms of different valence states (Fe^0, Fe^{2+}, Fe^{3+}) and different spin states, and located in different crystallographic sites in different mineral phases, yield Mössbauer spectra with distinctive patterns. ^{57}Fe Mössbauer spectra of pure minerals are linear combinations of these patterns. The number of components of each pattern in a spectrum of a mineral depends on the number of different types (valence and spin) of Fe in that mineral and the number of crystallographically equivalent sites where Fe occurs. From the Mössbauer spectrum of a mineral or rock, the valence, electronic configuration, coordination symmetry, site occupancy, and magnetic state of Fe cations in their host crystal structures may be determined. The relative contributions of different forms and crystallographic environments of Fe are proportional to the ratios of their Mössbauer peak areas. Mössbauer spectroscopy is thus well suited to the study of Fe-minerals including oxides, hydroxides, sulfides, silicates, and nanophase Fe metal, and should be very useful on the surface of Mars where Fe-rich mineralogy is expected on the basis of the SNC meteorites and previous remote and in-situ studies (Morris et al. 2000).

A significant challenge in the use of reflectance, emission, and Mössbauer spectroscopies is the investigation of mixtures (e.g., rocks and soils) because of the substantial overlap of spectral features of different components. It is thus usually not possible to identify different phases from mere inspection of raw spectra of rocks and soils (although in some cases, a rough estimate of assemblage can be made from the shape of a mixed spectrum). Profile decomposition procedures are well developed for Mössbauer spectra. Spectral peak overlap among some minerals, however, is severe enough that they cannot be distinguished in a mixture even with the most sophisticated decomposition procedures.

3.5. Laser Raman spectroscopy

Laser Raman spectroscopy is used to study the inelastic scattering phenomena in analyzed samples stimulated by a monochromatic light source. The wavelength differences between the Raman scattered radiation and laser wavelength are totally determined by the molecular properties of the target, i.e., its structural and compositional features. Measurement of the wavelength difference yields the so-called Raman shift, and for a given mineral, many different molecular vibrational modes contribute to a "fingerprint" pattern. Laser-produced, monochromatic light of ultraviolet, visible, or infrared frequency can be used as an excitation light source to produce a Raman signal. A visible laser is normally used for instrumentation convenience and higher Raman efficiency. Raman spectroscopy provides similar molecular vibrational information as mid-IR (i.e., TES) does, but with different selection rules. Raman spectra of minerals contain mainly fundamental vibrational modes and lattice modes; the overtones and combination modes are 1-2 orders of magnitude weaker. Thus, Raman spectral peaks are typically very narrow, sharp, and non-overlapping for well-crystallized minerals and many organic compounds. The strongest Raman peaks of oxyanionic minerals shift systematically according to their molar mass and bonding strength; the Raman spectral patterns of silicates change with degrees of polymerization. Thus phase identification can often be made from the inspection of raw spectra. For poorly crystalline or amorphous material, peaks may be subdued or broad, but peak positions still reflect the dominant molecular components and their structures. The Raman effect is best for molecular structures that have high degrees of covalent bonding (oxyanionic compounds), thus Raman spectroscopy is well suited for the identification of silicates, carbonates, sulfates, phosphates, nitrates, borates, oxides, sulfides, and hydroxides, as well as organic compounds. Because of the difference in selection rules, graphitic carbon (and molecules like O_2, N_2) cannot be detected by IR techniques, but can be detected down to 50 ppm using the Raman technique. Variations in composition within mineral groups typically result in predictable shifts in peak positions, but not in changes to the overall spectral pattern unless the compositional variation is accompanied by a significant change in mineral structure or symmetry (e.g., Wang et al. 2001). Because the incident laser beam can be designed to provide a very small laser spot size and can be used over a significant focal range, the spectra of individual mineral grains of many multiphase materials such as rocks can be determined and interpreted without recourse to complex spectral decomposition (e.g., Haskin et al. 1997). Analysis can generally be done without any preparation of the target material. For rocks and soils, that means characterizing their properties as they are found in nature.

4. In-Situ Mineralogical Analysis of the Surfaces of Extraterrestrial Bodies

Suites of surface minerals can be used to characterize past and present climatic conditions, sedimentary weathering processes or hydrothermal activity. More than elemental data, mineralogical data are linked to surficial conditions and processes and can be used to elucidate present and past conditions of the atmosphere, the surface, the crust, and occasionally the deep interior of a planet. Minerals have known ranges of stability and paragenetic relationships. Even when minerals persist out of their stability fields (as is common in sedimentary rocks), disequilibrium associations or the presence of relict unstable phases can be used to unravel detailed sedimentary or diagenetic histories that can be linked to climate history. In addition to surface mineralogy, the determination of mineralogy in extruded or detrital fragments of igneous or metamorphic rocks can be used to determine much about planetary interiors.

The importance of mineralogic studies extends beyond the terrestrial planets and rocky moons. XRD has been proposed for landed missions to icy bodies. The possibility of a Discovery-class mission to one of the poles of the Moon to look for water ice or hydrated minerals has also been discussed (Sarrazin et al.

2000). XRD, combined with XRF in a single instrument as proposed by Sarrazin et al. (2000), would be equally suited to the task of evaluating the mineralogy of the colored fracture zones of Jupiter's moon Europa, thought to be due to hydrated salts (McCord et al. 1999). An evaluation of the mineralogy of the ices and salts present in these zones would shed light on the nature of the proposed ocean deep beneath Europa's icy crust. The mineralogic composition of comets is equally important. The need for in-situ mineralogic studies is particularly pressing for these icy bodies, for it is very unlikely that samples can be collected from them and returned to Earth unaltered.

5. Remote Sensing of Solar System Objects

Remote spectral analysis techniques have been applied to many objects within our solar system, with the result that in general terms, we know the compositions of the solid surfaces that we can image. In some instances, high-resolution spectral data are available as is the case for the Moon, Mars, and Europa, among others. However, even these high-resolution images and analyses have lateral spatial resolutions of tens to thousands of meters, many orders of magnitude larger than the scale length of the phases in surficial deposits (e.g., minerals) that comprise the images. The elemental / chemical information obtained by remote sensing, although informative, has spawned a cottage industry of studies of analog materials and a myriad of interpretations. Remote sensing is principally useful in the formulation of hypotheses that can later be investigated by in-situ or returned-sample analysis using techniques that can provide more unambiguous mineralogical information.

5.1. The Moon

The Earth's moon has been visited by us a number of times, and during the Apollo program, several hundred kilograms of rocks and soil were returned to the Earth for analysis. However, it is surprising to some to learn that we are still very much interested in returning to that body to analyze additional materials. The Neutron Spectrometer on the Lunar Prospector mission identified excess hydrogen within craters at the poles of the Moon, thought to be water ice (?) about 40 cm beneath the surface in perennially shaded regions (Feldman et al. 1998). This water is in all likelihood a vestige of the bombardment of the inner solar system by icy planetesimals. The water is presumably from the same outer solar system reservoir of icy objects that must have bombarded the Earth, and we would like to know whether the Earth's oceans are indeed a legacy of cometary influx. This original hypothesis - that the world's oceans and other volatiles were delivered to a barren and rocky, volatile-free early Earth, has been challenged by recent remote analyses of a number of cometary bodies which suggest that cometary water is too high in deuterium to have contributed significantly to the Earth's ocean.

The nature of the water reservoir at the lunar poles is uncertain. The Lunar Prospector fast and epithermal neutron data indicate as much as 40 cm of desiccated regolith above water-ice reservoirs in permanently shadowed polar craters. It is unlikely that this reservoir is a simple accumulation of ice within regolith pore space. Limited data on the lunar geotherm (four measurements from the

Apollo missions) indicate temperatures of ~245-255 K at regolith depths of 40 cm, well below the freezing temperature for pure water ice but within a range where brines may exist as liquids. These factors indicate that any surface exploration of the lunar poles for water should be provided with (1) an ability to drill to depths of >40 cm and (2) the capability to determine the mineral and chemical constituents of the samples obtained. The samples obtained by such a system could be readily analyzed by XRD/XRF. X-ray diffraction analysis, coupled with X-ray fluorescence can explicitly determine not only the presence or absence of hydrous alteration phases such as clays or zeolites, but can also identify the structural variants or types of clay or zeolite present (e.g., well-ordered versus poorly ordered smectite; chabazite versus phillipsite). Finally, if coring and analysis is performed during the lunar night, XRD can provide information on any crystalline ices that might occur in the regolith samples. Beyond an abiding scientific interest in the presence of water and its source, the identification of a significant source of water on the Moon vastly improves the chances for success of any lunar colony.

Evidence for the presence of water at the lunar poles follows many years of speculation regarding lunar polar ice, assumed to be accumulated as cold-trapped volatiles released from cometary impacts (Watson et al. 1961, Arnold 1979). Before the Lunar Prospector data were acquired, the presence of any water in lunar regolith was highly suspect because of the complete absence of evidence for water-rock interactions in any lunar samples. This evidence ranges from the absence of aqueous drivers in lunar pyroclastic eruptions (where carbon species, sulfur, and halogens dominate) to the well-documented absence of hydrous minerals in lunar samples. Of course, no samples have been obtained from the lunar poles. The Lunar Prospector data have rekindled the interest in sampling these regions.

These factors indicate that any surface exploration of the lunar poles should be provided with (1) an ability to drill to depths of > 40 cm and (2) the capability to determine the mineral and chemical constituents of the samples obtained. It is important to note that successful robotic drilling operations have already been fielded on the Moon; three Soviet Luna missions were able to drill successfully to depths of 35, 27, and 160 cm into the lunar regolith. The last and deepest robotic sample (Luna 24) was based on a sophisticated but easily replicated drilling system that utilized a core-lining membrane. The samples obtained by such a system could be analyzed by a number of methods to determine whether hydrous minerals have formed in icy regolith at the lunar poles from the Moon's own heat or from the heat of regolith-gardening processes.

5.2. Mars

Mars occupies a special place in solar system exploration because of widespread interest in extraterrestrial life. The importance of Mars to our exploration activities is illustrated by the fact that NASA is planning two missions to Mars, an orbiter and a lander, at each launch opportunity (about every two years). The principal emphasis of this robotic exploration initiative is first, to ascertain whether life ever existed on Mars, and second, to pave the way towards human exploration in the early-to-mid 21st century.

Mineralogical analysis could lend insight into the early history of Martian volatiles, could establish the presence and lateral extent of hydrothermal systems, and could reveal the locations of rock types (silica sinter, travertine, etc.) that may harbor evidence of liquid water, prebiotic organic material or even extinct life. Once mineralogical surveys are completed, follow-on studies could perform in-situ mineralogical analyses of rock outcrops or select the most likely candidate rock samples for return to Earth. Categories of paleoclimatological/exobiological research interest are listed below along with a description of their distinctive mineralogical characteristics.

The presence and lateral extent of hydrothermal systems: Abundant morphological evidence exists for early and extensive volcanic activity on the surface of Mars, and for the presence of liquid water. The juxtaposition of these features is compelling evidence that hydrothermal systems once (or have always) existed on Mars. Isotopic evidence from the SNC meteorites also suggests that there was an exchange of hydrogen and oxygen between the crust and the atmosphere. The depletion of volatiles such as sulfur and carbon is puzzling and these elements may now be contained within minerals precipitated in hydrothermal systems. A knowledge of the presence and distribution of sulfur- and carbon-containing mineral phases is important because first, they are storage sites for volatiles, and second, hydrothermal systems might have contributed significantly to the Mars atmosphere over time.

Ancient hydrothermal systems could have been eroded or exhumed, exposing minerals and mineral assemblages at the surface that were formed at depths inaccessible in presently active systems. The mineralogical characterization of such a system would provide an evaluation of the role hydrothermal processing has played in modifying the early Martian atmosphere and in altering deep-seated igneous rocks. Hydrothermally altered rocks, dissected and exposed at the surface by erosion or impact cratering, may contain evidence of the nature of the early Mars atmosphere and of the fate of its volatiles.

Evidence of prebiotic organic chemistry: On the Earth, all evidence of prebiotic organic chemistry has been erased. In the earliest terrestrial rocks for which conditions of formation and of subsequent metamorphism would permit it, evidence of life is present. Therefore, even if life never originated on Mars, it would be exceedingly valuable to find some evidence in the geologic record of Martian prebiotic organic chemistry. This would be possible on Mars more than on the Earth, because extensive regions of ancient terrain exist on Mars that have not been subject to metamorphism due to the apparent absence of plate tectonics. Current concepts of prebiotic organic chemistry suggest that many important reactions have occurred in hydrothermal systems where some energy could have been provided by mineral hydration reactions. Hydrothermal systems also provide a means for gas exchange with the atmosphere and transport of reactants to the sites of reactions.

It is often the case that hydrated minerals and their anhydrous counterparts are compositionally very similar. Thus, elemental analyses would not easily distinguish one from the other. Mineralogic analysis, however, comprising both compositional and structural information, can provide a definitive result. Once hydrothermal terrains are identified by their distinctive mineralogy, areas or samples can be chosen for more comprehensive organic analyses or sample return.

Evidence of extinct life: Evidence of life on the Earth occurs in the earliest rocks which could have preserved signs of its presence ~3.5 billion years ago (recent reports indicate that life may have been present 3.85 billion years before the present, Mojzsis 1996). However, much of the geologic record from the earliest sedimentary sequences has either been heated to the extent that metamorphism would have removed evidence of life, or has simply been destroyed by burial and subduction. Because of the apparent lack of plate-tectonic activity on Mars, a great deal of the sediment deposited on the ancient Martian surface probably still exists and was probably not heated to as high a temperature as equivalent-age sediments on the early Earth. Impact heating could still have obliterated ancient surface material, but orbital mapping should be able to identify old, undisturbed sediments. Therefore, it is likely that if life originated early in Martian history, some record of its existence would be manifest in the earliest geologic record accessible to surface landers.

Life as we know it requires liquid water, a source of energy, and nutrients. Good preservation requires early burial or entombment in fine-grained clastic sediments or chemically precipitated sedimentary materials such as silica or carbonate. In the search for evidence of extinct life on Mars, we can be guided by our knowledge of the existence, characteristics and style of preservation of organisms on the early Earth. We know that for 4/5 of Earth history, life consisted of nothing more than single-celled microbes living in lacustrine or marine environments. The presence of these microbes is typically revealed by the observation of sedimentary structures (i.e., stromatolites) or trace fossils, not by the organisms themselves.

Absolute determination of the presence of prebiotic organic material or of evidence of extinct or extant life will most likely require the return of a sample to Earth. The purpose of in-situ sample analysis on the Martian surface would be to survey rocks or sediments, to ensure that the very best candidates for fossiliferous strata are collected, and to provide a global or regional context for the samples brought back.

History of volatiles on Mars: Mars is presently a dry eolian planet with a tenuous atmosphere of ~7 millibars, dominated by CO_2. However, ancient surface morphological features such as stream channels and other fluvial and lacustrine features provide compelling evidence that liquid water existed on the surface of Mars in large quantity. This apparent abundance of liquid water implies that early Mars was once wetter and warmer, and had a denser atmosphere.

The ultimate fate of atmospheric CO_2 is of interest to paleoclimatologists and exobiologists because this compound figures prominently in prebiotic organic chemistry and because a full inventory of the CO_2 sinks is required to provide a balanced volatile budget for the planet. A major reservoir for the carbon dioxide could be in the form of carbonates deposited as chemical sediments or as hydrothermal precipitates. Alternatively (or additionally), CO_2 could be stored in the form of clathrate hydrates beneath the surface or as solid carbon dioxide at the poles, in addition to that CO_2 that is seasonally deposited there. Each of these phases can be unequivocally distinguished and characterized by diffraction techniques. In the case of carbonates, the crystal symmetry (rhombohedral or orthorhombic), the extent of cation solid solution (Ca, Mg, Fe, Mn, etc.), cation order/disorder and cation stoichiometry can all be determined through mineralogical analysis. Each of these distinguishing characteristics provides information about the specific origin of the mineral phase and its environment of formation.

The total quantity of water that apparently existed on the surface of Mars early in its history cannot be accounted for by the polar caps alone. In addition to that water lost to space, it is likely that hydrated phases exist either as a consequence of their direct deposition from aqueous solution or as products of the reaction of anhydrous igneous minerals with water. The quantity, type and degree of crystallinity of clays, micas and other hydrated phases can be determined by mineralogic analysis, and their known stability relationships can constrain the conditions under which they formed.

6. The Future

Our current best knowledge of extraterrestrial mineralogy is based on samples in hand from the Moon and meteorites. Many of the questions we wish to answer will still require samples returned for very detailed analysis (witness the recent efforts devoted to Martian meteorites). Although the need to analyze returned samples will remain as long as advanced analytical methods evolve on Earth, the state of the art in remote analytical methods is progressing at a very rapid pace. In reference again to the experience from Martian meteorite studies, it is evident that no single method of mineralogical analysis is sufficient, whether in a laboratory on Earth or from a spacecraft. Combined methods gain far more than isolated analyses and the technical capability of combining mineralogical, chemical, thermal, electromagnetic, optical, and other analyses will add a power to space exploration that holds vast potential.

References

Arnold, J.R. 1979. Ice in the lunar polar regions. *J. Geophys. Res.* **84**, 5659-5668.

Clark, R.N. 1999. Spectroscopy of rocks and minerals, and principles of spectroscopy. In *Manual of Remote Sensing* (A. Rencz, Ed.), pp. 3-58. John Wiley and Sons, New York.

Christensen, P.R., J.L. Bandfield, V.E. Hamilton, D.A. Howard, M.D. Lane, J.L. Piatek, S.W. Ruff, and W.L. Stefanov 2000. A thermal emission spectral library of rock-forming minerals. *J. Geophy. Res.* **105**, 9735-9739.

Dunne, J.A., and N.L. Nickle 1968. *The lunar and planetary X-ray diffraction program: A summary.* NASA Technical Report 32-1248, Jet Propulsion Laboratory, Pasadena.

Feldman, W.C., S. Maurice, A.B. Binder, B.L. Barraclough, R.C. Elphic, and D.J. Lawrence 1998. Fluxes of fast and epithermal neutrons from Lunar

Prospector: Evidence for water ice at the lunar poles. *Science* **281**, 1496-1500.

Hapke, B. 1993. Combined theory of reflectance and emittance spectroscopy. In *Remote Geochemical Analysis: Elemental and Mineralogical Composition* (C.M. Pieters and P.A.J. Englert, Eds.), pp. 31-42. Cambridge University Press, Cambridge.

Haskin, L.A., A. Wang, K.M. Rockow, B.L. Jolliff, R.L. Korotev, and K.M. Viskupic 1997. Raman spectroscopy for mineral identification and quantification for in-situ planetary surface analysis: A point-count method. *J. Geophys. Res.* **102**, 19293-19306.

McCord, T.B., G.B. Hansen, D.L. Matson, T.V. Johnson, J.K. Crowley, F.P. Fanale, R.W. Carlson, W.D. Smythe, P.D. Martin, C.A. Hibbitts, J.C. Granahan, and A. Ocampo 1999. Hydrated salt minerals on Europa's surface from the Galileo near-infrared mapping spectrometer (NIMS) investigation. *J. Geophys. Res.* **104**, 11827-11851.

Mojzsis, S.J., G. Arrhenius, K.D. McKeegan, T.M. Harrison, A.P. Nutman, and C.R.L. Friend 1996. Evidence for life on Earth before 3,800 million years ago. *Nature* **384**, 55-59.

Morris, R.V., D.C. Golden, J.F. Bell III, T.D. Shelfer, A.C. Scheinost, N.W. Hinman, G. Furniss, S.A. Mertzman, J.L. Bishop, D.W. Ming, C.C. Allen, and D.T. Britt 2000. Mineralogy, composition, and alteration of Mars Pathfinder rocks and soils: Evidence from multispectral, elemental, and magnetic data on terrestrial analogues, SNC meteorites, and Pathfinder samples. *J. Geophys. Res.* **105**, 1757-1817.

Sarrazin, P., D. Blake, D. Bish, D. Vaniman, S. Chipera, S.A. Collins, and S.T. Elliott 2000. In situ investigation of ices and hydrous minerals at the lunar poles using a combined X-ray fluorescence and diffraction instrument. *J. Phys. IV France* **10**, 343-352.

Vaniman, D., D. Bish, D. Blake, S.T. Elliott, P. Sarrazin, S.A. Collins, and S. Chipera 1998. Landed XRD/XRF analysis of prime targets in the search for past or present Martian life. *J. Geophys. Res.* **103**, 31477-31489.

Wang, A., B.L. Jolliff, L.A. Haskin, K.E. Kuebler, and K.M. Viskupic 2001. Characterization and comparison of structural and compositional features of planetary quadrilateral pyroxenes by Raman spectroscopy. *Am. Mineral.* **86**, 790-806.

Watson, K., B.C. Murray, and H. Brown 1961. The behavior of volatiles on the lunar surface. *J. Geophys. Res.* **66**, 3033-3045.

Wyatt, M.B., V.E. Hamilton, H.Y. McSween Jr., P.R. Christensen, and L.A. Taylor 2001. Analysis of terrestrial and Martian volcanic compositions using thermal emission spectroscopy: 1. Determination of mineralogy, chemistry, and classification strategies. *J. Geophys. Res.* **106**, 14711-14732.

The Future of Solar System Exploration, 2003-2013
ASP Conference Series, Vol. 272, 2002
Mark V. Sykes, ed.

Near-Earth Objects: Discovery, Tracking, and Characterization

D.K. Yeomans[1], E. Asphaug[2], W.F. Bottke[3], P.G. Brown[4], A. Cellino[5], R.A. Fevig[6], U. Fink[6], C.W. Hergenrother[6], A.R. Hildebrand[7], S.M. Larson[6], J.-L. Margot[8], & D.J. Tholen[9]

[1] *Jet Propulsion Laboratory, Pasadena, CA*
[2] *Earth Sciences Department, University of California, Santa Cruz, CA*
[3] *Southwest Research Institute, Boulder CO*
[4] *Department of Physics and Astronomy, University of Western Ontario, Canada*
[5] *Osservatorio Astronomico di Torino, Pino Torinese, Italy*
[6] *Lunar and Planetary Laboratory, University of Arizona, Tucson, AZ*
[7] *Department of Geology and Geophysics, University of Calgary, Calgary, Canada*
[8] *Division of Geological and Planetary Sciences, California Institute of Technology, Pasadena, CA*
[9] *Institute for Astronomy, University of Hawaii, Honolulu, HI*

Abstract. The rationale for the efforts to discover, track, and characterize Near-Earth Objects (NEOs) are outlined. While great strides have taken place over the last few years with regard to NEO research, there still is a lack of large telescopes in the search process, a less than comprehensive effort to coordinate the search efforts, an under representation of search facilities outside the U.S., and efforts need to be increased to physically characterize the NEO population. Recommendations are provided to address these issues.

REPORT

1. Current State of Knowledge

The rationale for the study of near-Earth objects (NEOs) includes their being relatively primitive bodies left over from the early solar system formation process, their future utilization as raw materials for building structures and habitats within the inner solar system, and their hazards because of their infrequent, but potentially catastrophic, collisions with Earth. It is primarily this last topic that is addressed here. We note that NASA's interest in this topic is not driven by scientific justification alone. This interest includes the very real threat to life on Earth. As a result of a robust search for NEOs, the scientific community not only ensures that there will be no surprises from Earth threatening objects but there will be new discoveries and future close Earth approaches of these objects that can be studied via ground-based programs including radar "imaging" characterizations. In addition, these NEO discoveries will provide future mission targets that are among the most accessible in terms of spacecraft energy requirements. Indeed, there are many near-Earth asteroids (NEAs) for which it would be easier to effect a spacecraft rendezvous and landing than it would for the moon itself.

About four billion years ago, perhaps as a result of the outer planets formation process, there began a period of intense impacts by comets and asteroids upon the Earth. This period has been termed the "late heavy bombardment" and while it may have been part of Earth's formation process, conditions on the surface of the Earth were far too hot to allow the necessary building blocks of life to survive (i.e., liquid water and organic molecules). It was only after the vast number of asteroids and comets in the inner solar system thinned out as a result of planetary agglomeration and gravitational scattering that conditions improved to a point where they were suitable for the pooling of water and organic molecules. As a result, about 3.8 billion years ago, the late heavy bombardment drew to a close and life developed rather quickly. Subsequent asteroid and cometary impacts, while not frequent enough to completely frustrate the development of life on Earth, did punctuate its evolution, allowing only the most adaptable species to evolve further. Thus in some sense, we mammals owe our preeminent position atop the world's food chain to ancient Earth impacts that wiped out our principal competition, including the dinosaurs. If we wish to maintain this position, we must deal with potential Earth impactors before they have a chance to relegate us to the evolutionary trash heap.

The rough dividing line between an impactor that would have global catastrophic consequences and one that would have only local consequences (e.g., tsunamis), is about 1 kilometer in size. An Earth impactor with a diameter larger than about one kilometer would be expected to wipe out a significant fraction of life on Earth since the effects would be global (Chapman and Morrison 1994, Toon et al. 1997). Impactors less than a kilometer in size, while extremely serious threats in some cases, will likely cause only local damage and large scale loss of life can be avoided via evacuations if the threat is identified with sufficient advance warning.

Thus the Spaceguard Goal of NASA, as articulated in 1998 by Dr. Carl Pilcher before the House Subcommittee on Space and Aeronautics, is to discover 90% of the near-Earth asteroids larger than one kilometer within 10 years. It is this Spaceguard Goal that is driving much of today's research into, and funding for, the near-Earth objects (NEOs). Assuming an average albedo of about 11% for near-Earth asteroids (NEAs), the dividing line between one kilometer and larger-sized objects and those that are smaller has been taken to be absolute magnitude 18. Toward the end of 2001, about 550 NEAs brighter than, or equal to, absolute magnitude 18 have been discovered out a total estimated population of about 1000 large NEAs (Rabinowitz 2000, Bottke et al. 2000, Stuart 2001, D'Abramo et al. 2001). While at first look this progress seems very impressive, when the facility with which new objects are detected in the first few years of an intensive search effort is taken into account, the current discovery rate does not seem sufficient to meet the Spaceguard Goal.

Currently, NASA supports five NEO search teams (Table 1). The LINEAR program, near Socorro NM, employs two Air Force GEODSS telescopes of 1-

Table 1. Current NEO Search Programs

LINEAR	Lincoln Laboratory Near-Earth Asteroid Research, MIT (Grant Stokes, P.I.)
NEAT	Near Earth Asteroid Tracking, JPL (Eleanor Helin, P.I.)
LONEOS	Lowell Observatory Near-Earth Object Search (E. Bowell, P.I.)
Catalina	Catalina Sky Survey, LPL, Univ. of Arizona (Steve Larson, P.I.)
Spacewatch	LPL, University of Arizona (Bob McMillan, P.I.)

meter aperture and the NEAT program uses an Air Force telescope of 1.2 meters aperture on Mt. Haleakala, Maui. NEAT also has a 1.2-meter aperture telescope operational at Palomar mountain. The LONEOS and Catalina teams currently use 0.6 m aperture telescopes while the Spacewatch group uses 0.9 meter and 1.8 meter aperture instruments.

The discovery of NEOs, the necessary follow-up astrometric observations required to secure their orbits, and the integration of their motions into the future to check for Earth encounters are critically important steps, but these activities are not the entire effort required. We must also understand the physical characteristics of NEOs including the nature of their rotations, structures, sizes, masses and chemical compositions. The general characterization of NEO members by optical, IR, and radar techniques are especially valuable. Physical characterization studies, including the determination of an objects' spectral class, are necessary steps toward determining a NEO's composition and structure - key bits of information should it be necessary to deflect any of these objects from an Earth threatening trajectory.

2. Key Science Questions and Recommendations

Q: How can the current search efforts to reach the Spaceguard Goal be optimized? How can this optimization be achieved within each team and how might cooperative efforts between teams improve the discovery rate?

R: Re-form NEO Science Working Group to assess current progress toward meeting Spaceguard Goal (i.e., by 2008, discover 90% of NEAs whose diameters are larger than 1 km) and suggest how the entire search efforts could be optimized. This group should consider and recommend strategies for:

- Optimizing efforts of current search teams.
- Further extending sky coverage to the southern hemisphere and closer to the sun.
- Coordinating search efforts between teams (to the extent possible).

- Identifying what new equipment/techniques/resources are required to meet the Spaceguard Goal?
- Providing for an appropriate support level for the Minor Planet Center (Cambridge, MA) so this vital clearing house for NEO data is operationally robust.
- Providing for the efficient searching of data archives for NEO prediscovery observations
- Using the data from Earth orbital wide field surveys (e.g., SIRTF, GAIA) to significantly improve the NEO discovery rate and our understanding of their physical characteristics.
- Identifying under-utilized wide-field, large-aperture telescopes that could be employed to discover NEOs in the northern and southern hemispheres.
- Improving our knowledge of the size distribution for the near-Earth asteroids
- Dealing with the long period comets whose Earth impacts, while far more rare than impacts by asteroids of comparable size, can occur at much faster velocities and with much shorter warning times.

Q: What are the critically important studies for characterizing the physical nature of NEOs and how might these studies be made a more integral part of the NASA NEO Observations Program?

R: The annual NASA Research Opportunities in Space Science (ROSS) announcements have recently called out the search and follow-up of NEO to be of higher priority than physical characterization studies. In FY2000, the very limited resources available to NEO observations went to support only those proposals offering search and astrometric follow up programs. There was a complete lack of FY2000 funding for NEOs within the Near-Earth Object Observation Program. While the majority of the funding should go toward the search and follow-up activities, a certain fraction of the available funding should be guidelined for NEO characterization studies to maintain a balanced NEO research program. Just as the NEO discovery and follow-up facilities themselves require an appropriate amount of infrastructure support, the physical characterization facilities (including the planetary radars at Arecibo and Goldstone) also require this infrastructure support.

Efforts to provide a balanced NEO research program will require additional funding to be added to the NEO program since the realization of the Spaceguard Goal will not likely be achieved within the current level of funding and the current level includes insufficient allocations for the necessary physical characterization studies.

Q: What is the role of in-situ spacecraft studies for the physical characterization of near-Earth asteroids and comets?

R: While the proximity of NEOs to the Earth from time to time make them prime targets for ground-based physical characterization studies, there are a host of important measurements that require in-situ measurements. Knowledge of an object's mass, chemical composition, bulk density, porosity, internal structure and spin characteristics will require rendezvous and landing missions carrying the necessary imagers, spectrometers, landers, active seismology experiments and spacecraft tracking equipment. A detailed analysis of an object's chemical composition will require the study of a returned surface sample in Earth-based laboratories. At a minimum, rendezvous missions should be sent to C-type and M-type NEAs to allow comparisons of these objects with the S-type asteroid Eros which was closely scrutinized by the NEAR spacecraft.

We recommend that a conference or panel address the specific science requirements for the mitigation of Earth threatening near-Earth objects (Ahrens and Harris 1994). This group should address what are the specific physical parameters of a potential impactor that must be known in advance of an effective mitigation campaign.

References

Ahrens, T., and A. Harris 1994. Deflection and fragmentation of near-Earth asteroids. In *Hazards Due to Comets and Asteroids* (T. Gehrels, Ed.), pp. 897-927. Univ. of Arizona Press, Tucson.

Bottke, W.F., R. Jedicke, A. Morbidelli, J.-M. Petit, and B. Gladman 2000. Understanding the distribution of near-Earth asteroids. *Science* **288**, 2190-2194.

Chapman, C.R., and D. Morrison 1994. Impacts on the Earth by asteroids and comets: Assessing the hazard. *Nature* **367**, 33-39.

D'Abramo, G., A.W. Harris, A. Boattini, S.C. Werner, and A.W. Harris 2001. A simple probabilistic model to testimate the population of near-Earth objects. *Icarus* **153**, 214-217.

Rabinowitz, D., E. Helin, K. Lawrence, and S. Pravdo 2000. The number of kilometer-sized near-Earth asteroids. *Nature* **403**, 165-166.

Stuart, J.S. 2001. A near-Earth asteroid population estimate from the LINEAR Survey. *Science* **294**, 1691-1693.

Toon, O.B., R.P. Turco, and C. Covey 1997. Environmental perturbations caused by the impacts of asteroids and comets. *Rev. Geophys.* **35**, 41-78.

The Future of Solar System Exploration, 2003-2013
ASP Conference Series, Vol. 272, 2002
Mark V. Sykes, ed.

Near-Earth Asteroid Sample Return

D.W.G. Sears[1], C.C. Allen[2], D.T. Britt[3], D.E. Brownlee[4], A.F. Cheng[5], C.R. Chapman[6], B.C. Clark[7], B.G. Drake[8], R.A. Fevig[9], I.A. Franchi[10], A. Fujiwara[11], S.P. Gorevan[12], H. Kochan[13], J.S. Lewis[9], M.M. Lindstrom[8], K. Nishiizumi[14], M.S. Race[15], D.J. Scheeres[16], E.R.D. Scott[17], G.J. Taylor[17], & H. Yano[11]

[1] *Arkansas-Oklahoma Center for Space and Planetary Sciences, University of Arkansas, Fayetteville, AR*
[2] *NASA Johnson Space Center, Houston, TX*
[3] *Department of Geological Sciences, University of Tennessee, Knoxville, TN*
[4] *Department of Astronomy, University of Washington, Seattle, WA*
[5] *Johns Hopkins Applied Physics Laboratory, Baltimore, MD*
[6] *Southwest Research Institute, Boulder, CO*
[7] *Lockheed Martin Astronautics, Denver, CO*
[8] *NASA Johnson Space Center, Houston, TX*
[9] *Lunar and Planetary Laboratory, University of Arizona, Tucson, AZ*
[10] *Planetary and Space Sciences Research Institute, The Open University, Milton Keynes, United Kingdom*
[11] *Planetary Science Division, Institute of Space and Astronautical Science, Kanagawa, Japan*
[12] *Honeybee Robotics, New York, NY*
[13] *German Aerospace Center, Köln, Germany*
[14] *Space Sciences Laboratory, University of California, Berkeley, CA*
[15] *SETI Institute, Mountain View, CA*
[16] *Department of Aerospace Engineering, University of Michigan, Ann Arbor, MI*
[17] *Hawaii Institute of Geophysics and Planetology, University of Hawaii, Honolulu, HI*
[18] *Planetary Science Division, Institute of Space and Aeronautical Science, Kanagawa, Japan*

Abstract. The NEA Sample Return Community Panel for the NRC Decadal Study was assembled from selected participants of the Near-Earth Asteroid Sample Return Workshop held at the Lunar and Planetary Institute in December 2000 and summarized in presentations at the 33rd Lunar and Planetary Science Conference and the American Institute of Aeronautics and Astronautics' Space 2001 meeting in Albuquerque. Detailed views on the scientific value of NEA sample return of the many of the individual panel members prior to panel discussions can be obtained in the abstract volume for the workshop. This paper represents efforts of the panel between June and November 2001 and was presented to the NRC Primitive Bodies Discipline Panel on 25th October 2001.

EXECUTIVE SUMMARY

A unique source of information about the early solar system, the formation of the planets, and the connection between stars and our Sun, are the meteorites and asteroids, yet studies of both are hindered by a lack of unequivocal and detailed information linking the meteorites and asteroids. Meteorites are rock samples of unknown provenance. We have no information about the geological context of the source of meteorites. They are also highly non-representative sampling of primitive solar system material because the terrestrial meteorite population is dominated by the ejecta of stochastic impacts and because the atmosphere filters out all but the toughest rocks. Without sample return, asteroids are not amenable to the depth and breadth of techniques available in the laboratory, yet the NEAR images indicate that there are many processes occurring on asteroids - or that could have occurred in the past - that we must understand if the meteorite data are ever to yield a clear image of early solar system processes. Technical developments of the last few years - the success of the NEAR-Shoemaker and Deep Space 1 missions, the progress of the technology development mission MUSES C, and the availability of sample collection and containment apparatus and the discovery of large numbers of NEA - mean that sample return from multiple asteroids is now within small mission capability. NEA sample return missions lend themselves well to Education and Public Outreach efforts, present few and well understood planetary protection issues, have a well established research and management infrastructure already in place, are excellent opportunities for international cooperation and pooling resources, integrate well into existing mission plans, have a part to play in the human exploration and development of space effort, and, most importantly, have the highest scientific priority.

REPORT

1. Current State of Knowledge

1.1. Primitive meteorites and the objects from which they come (asteroids and comets) offer unique insights into a wide variety of early solar system processes.

Primitive solar system bodies (comets and asteroids) have the potential to provide unique information about the early solar system and the material it contained. Meteorites and cosmic dust are samples of these bodies that are falling naturally to Earth. They have been intensely studied (Kerridge and Matthews 1988, Hewins et al. 1996). While much can be learned from cosmic dust, considerably more has been learned from meteorites, which can be studied on the macroscopic scale by a wide variety of very sophisticated techniques. Primitive meteorites have ages comparable to the age of the solar system, they have bulk compositions very similar to that of the Sun, and they have unique textures, sometimes being referred to as "cosmic sediments". The major components are chondrules and refractory inclusions, metal and sulfide grains, and a fine-grained matrix. It has been argued that trace components in primitive chondrites, such as graphite, diamond, silicon carbide, and alumina, probably have an interstellar origin.

Detailed chemical and physical studies of primitive chondrites enables subdivision into a number of discrete classes, the largest of which are the ordinary chondrites, the H, L and LL chondrites, but especially significant scientifically are the rare carbonaceous chondrites, some of which can be up to 20% water by volume (Sears and Dodd 1988). The classes show subtle but significant deviations in composition from those of solar abundances. The existence of these classes and the physical and chemical trends they represent are important clues to processes occurring in the early solar system. One such process is the separation of silicates and metal. Another is volatile loss. Yet another is associated with the formation of the chondrules, glassy silicate droplets containing conspicuous crystal structures. Early solar system processes also resulted in variations in elemental abundance and isotopic proportions of oxygen. It is not clear what caused these variations in property or how they relate, but it is clear that they represent fundamental processes in the early solar system.

A variety of dating techniques have not only shown the antiquity of primitive meteorites, but have made it possible to resolve a great many events, some of them involving small time intervals for events occurring many years ago, such as the time interval between the end of nucleosynthesis and meteorite formation. Other dating techniques have identified the times of major and lesser breakup events. We return to this below. Near-Earth Asteroids (a few of which might be extinct comets) are the immediate parent bodies of primitive meteorites and returned samples will provide new insights into processes witnessed by the meteorites.

It seems clear that the asteroids are material that was prevented from accreting into planet by gravitational interaction with Jupiter and Saturn and that they are thus primitive solar system material (Binzel et al. 1989, SSB 1998). Size distributions and families of asteroid sharing similar orbits suggest that, except for the largest few, they are fragments resulting from multiple collisional breakups. A few have satellites and many appear bifurcated, consisting of two lobes. The spectra of sunlight reflected from the asteroids indicates that they are compositionally very diverse, ranging from carbonaceous material not unlike the carbonaceous chondrites through silicate rich material superficially resembling the ordinary chondrites to metallic asteroids resembling iron meteorites (Binzel et al. 1989). In a few cases, it is possible to link meteorite classes with asteroid types by a convincing match of their spectra, especially if allowance is made for alteration of the surface by possible space weathering effects.

It is thought that nearly 1600 asteroids are in orbits that bring them close to the Earth (i.e., perihelia$<$1.3 AU), and nearly 350 pass close enough to the Earth to be considered as potentially hazardous ($<$1.3 AU, diameter$>$200m or H$<$22). The NEA are in intrinsically unstable orbits with lifetimes on the order of 108 years and the population must be continually replenished from the main belt or from sources deeper in the solar system. In terms of their spectral classification, the population of NEA is very similar to that of the main belt.

Figure 1. A pebble beach at the foot of the White Cliffs of Dover on the southeast coast of England (Corbis.com, with permission). Meteorites have the potential to tell us much about the origin and history of the early solar system, but they are cosmic jetsam, brought to Earth after a number of selection and alteration effects. Attempts to learn about the early solar system from them are somewhat analogous to learning about the geology of southern England from the pebbles on the beach instead of visiting the chalk cliffs themselves.

Many new insights have been obtained over the last one hundred years or so of primitive body research, but many questions have arisen and unequivocal answers to some fundament questions - like the origins of the meteorite and asteroid classes and their characteristic properties - are still lacking. Our panel suggests that missions to return samples from near-Earth asteroids will provide major opportunities to for advancement in both meteorite and asteroid research and therefore in our understanding of some of the most fundamental questions in planetary science, the origin and early evolution of the solar system and the processes by which interstellar material becomes new planetary life-bearing systems.

1.2. Returned samples I - New kinds of primitive material, or "Why bring back asteroid samples when we have meteorites?"

Meteorites are, in effect, cosmic jetsam. The meteorite samples reaching the Earth are biased towards the fragments of bodies that recently broke up and towards materials strong enough to survive the rigors of reaching Earth. We are, in effect, trying to determine the formation of a mountain from stream wash deposits rather than visiting the outcrop. A terrestrial analogy might be the chalk cliffs of southern England (Figure 1). At the base of every cliff is a pebble beach of flint nodules that have weathered out of the chalk. How difficult it would be to determine the geological history of the cliffs, and the large regions of England they represent, by looking only at the pebbles and knowing nothing about the chalk.

The Yarkovsky effect - uneven radiation pressure on a spinning object in space - ensures that small ($\sim$10 cm) samples of most material in the inner part of main belt reaches the vicinity of the Earth. In addition, the largest and most famous meteorite classes come from a few bodies that were broken up by impact relatively recently. They may have been located near dynamically favorable locations in the solar system, such as near resonances with Jupiter and Saturn. The existence of this event is demonstrated, and its age apparent, from preferred values in their cosmic ray exposure ages, in the case of the H chondrites (Figure 2), and their Ar-Ar ages (Figure 3), in the case of the L chondrites. We do not see similar preferred ages, or even sensible cosmic ray ages for the more primitive carbonaceous meteorites. Such fragile material is does not survive impacts but is continually abraded in space so that a single meaningful exposure age is not apparent.

Probably the major process removing the most primitive meteorites from the Earth's meteorite collection is passage through the Earth's atmosphere. Only tough material reaches the surface of the Earth while the fragile water-rich and particularly primitive materials are destroyed during atmospheric passage (Figure 4). Quantitative modeling by aerospace engineers interested in spacecraft reentry show that primitive carbonaceous meteorites suffer 1000-fold greater mass attrition during atmospheric passage compared with the tough ordinary chondrites that are the dominant meteorite classes on Earth (Figure 5). It seems almost certain that many kinds of currently unknown primitive materials exist in the asteroid belt. Perhaps relevant to this are the measured densities of asteroids that are lower than those of ordinary chondrites and only the largest of which are comparable with carbonaceous chondrites. Whether this reflects abundant volatiles or an unusual internal texture is unclear, but it does indicate that meteorites are giving an incomplete story about the nature of primitive solar system material.

Tantalizing evidence that there is primitive material not surviving atmospheric passage are a few spectacular falls, like Revelstoke and Tagish Lake. These are meteorites that produced large amounts of dust in the atmosphere, and huge visual and audible effects, but deposited very little in the way of macroscopic meteorites on the surface of the Earth. If reports had not been widely disseminated and search teams quickly on the spot, it is doubtful that any material would have been recovered. The small samples that were recovered were found to be very friable, carbon and water-rich meteorites. Every year there are thousands of bright fireballs that deposit no material on the Earth. Are these samples of new kinds of primitive material, full of undreamed of information about the solar system's formation and early history?

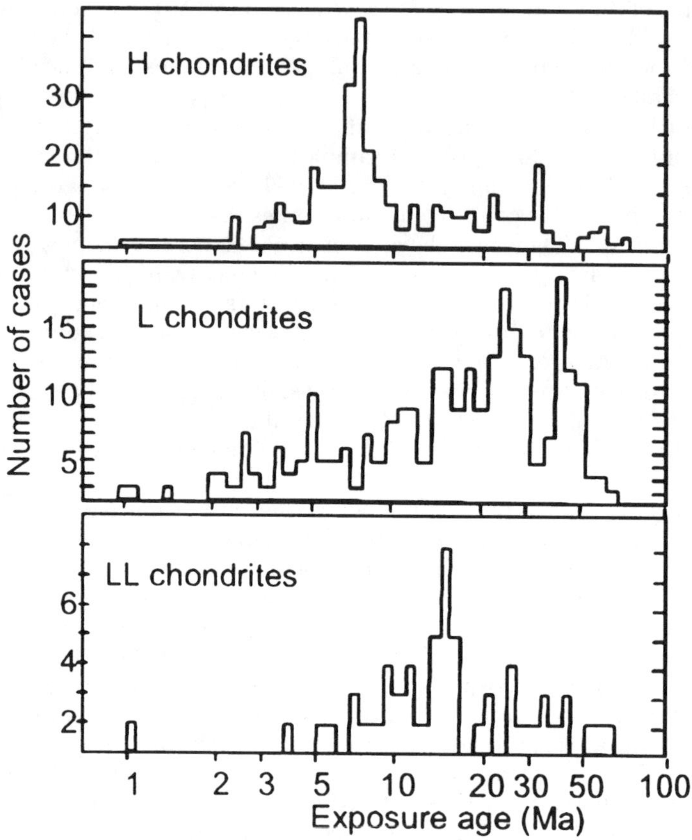

Figure 2. Cosmic ray exposure ages for the ordinary (H, L and LL) chondrites (Marti and Graf 1992). The peak in the distribution of cosmic ray ages for H chondrites at 8 Ma suggests that large numbers of these meteorites became exposed to cosmic radiation at this time by fragmentation into meter-sized chunks. The LL chondrites show a similar peak at 17 Ma. Some authors have argued that the L chondrites show evidence for a number of such major fragmentation events. Apparently, a large fraction of the meteorites in these classes shared a common parent body until very recently in solar system history. The carbonaceous chondrites do not show preferred values in their cosmic ray exposure ages which is commonly interpreted to mean that these friable rock continually break up during their passage through space. They also contain large amounts of inert gases trapped in their primary textures, so it is difficult to detect the cosmic ray produced gases over this large background.

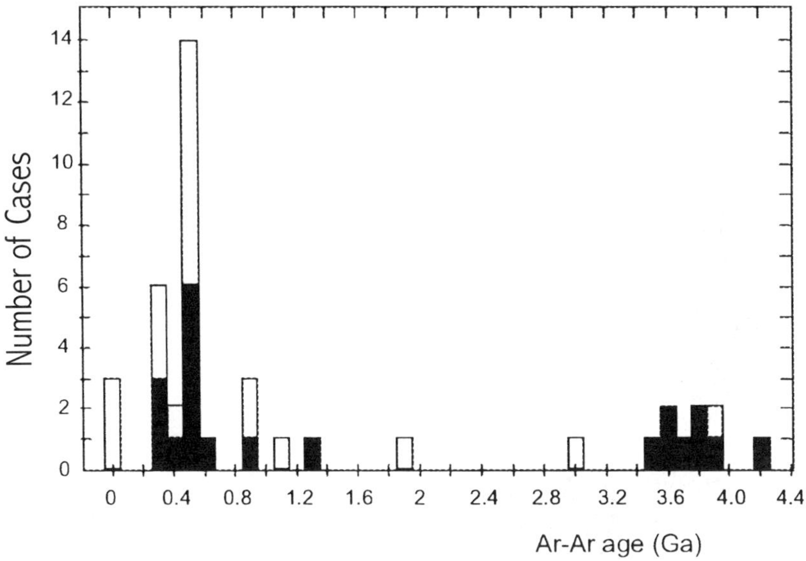

Figure 3. Argon-argon ages for L chondrites (Bogard 1995). The peak in the distribution of these ages at 500 Ma indicates that a major disruption of a common parent body (asteroid) at this time that heated the meteorites and caused the loss of accumulated Ar and thus a resetting of the chronometer. Meteorites plotting in the 500 Ma peak show a number of petrographic symptoms of shock heating.

1.3. Returned samples II - The value of context, or "Why bring back asteroid samples when we have meteorites?"

To return to our chalk cliffs of England, aside from knowing that there was chalk in the cliff, in addition to the pebbles of the beach, a terrestrial geologist would need to know whether the cliff made mainly of pebbles, or whether the pebbles are rare nodules that preferentially survive weathering, albeit somewhat altered. If we suspected the chalk existed, without ever seeing it, would we have any way of knowing whether the flint is spread uniformly through the chalk, or in narrow time horizons? Studying cosmic jetsam also means that not only do we not know what type of asteroid the meteorites are from, but we do not know the geological context from which the samples came. We do not know whether the samples are from inside a crater, from the crater rim, from the ejecta blanket of a crater, from bedrock, from the surface, from depth, from rare veins of particularly tough material, or from some other undreamed-of geological feature. No terrestrial geologist would think of discussing the origin and history of a rock without knowing its geological context. The NEAR-Shoemaker spacecraft presented a whole new world of structures and features and a number of scenarios in which meteorites could have formed (Figure 6). There were craters and ejecta blankets, linear structures and grooves, boulder strewn fields, regions of uninterrupted regolith (Veverka et al. 2001) and curious flat regions of apparently fine-grained

Figure 4. A painting of the fall of the Sikhote Alin meteorite in eastern Siberia in 1947. Sikhote Alin is an iron meteorite, yet it produced enormous amounts of dust in its trail. Carbonacous meteorites that are weakly agglomerated and contain ~20% water are almost entirely destroyed by atmospheric passage and produce large amounts of dust in the atmosphere and very little survives to reach the surface of the Earth. It is to be expected that anything more friable and more water-rich like primitive material is expected to be will not survive at all.

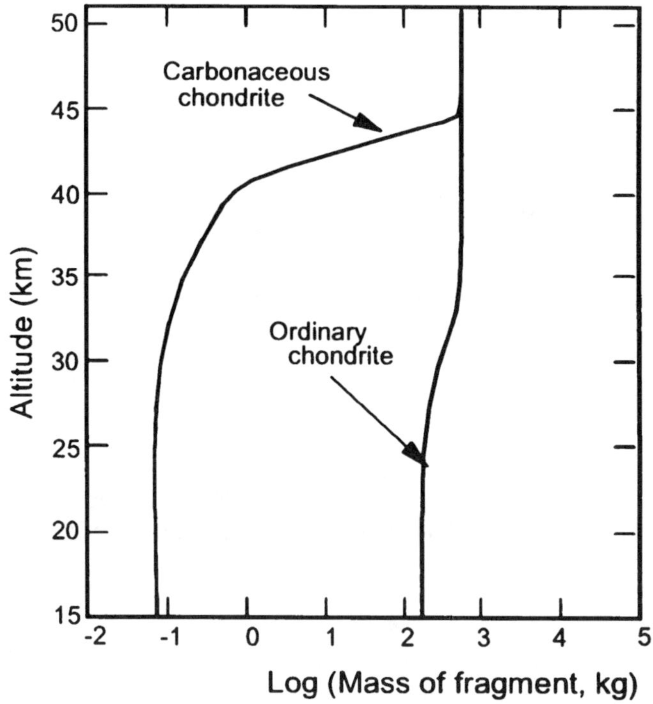

Figure 5. Altitude in the atmosphere against the mass of the surviving fragment of a body initially 1000 kg (Baldwin 1971). The fragmentation of carbonaceous meteorites (CI and CM chondrites) is approximately 1000 times greater than that of ordinary chondrites (H, L and LL chondrites) leading to a 1000-fold bias in favour of ordinary chondrites on Earth.

material that was raised relative to other local features often, but not always, in crater bottoms that have come to be known as "ponds" (Robinson et al. 2001). The NEAR-Shoemaker spacecraft revealed a wide range of environments in which meteorites might have formed.

There are a great many examples of how returned samples from known context might resolve long-standing questions in meteorite and asteroid studies and thus our understanding of conditions and processes in the early solar system. Two of the most fundamental questions in meteorite studies are (1) how did chondrules form, and (2) how were the various metal to silicate ratios produced. These questions relate to the formation of the chondrite classes in as much as every class has unique combination of metal-silicate ratio and chondrule and metal abundances and sizes (Glavin et al. 2000). Many researchers have argued that chondrules are impact melt droplets produced when a meteorite impacted the parent body, while others have argued that a process in the nebula produced the chondrules, perhaps lightning, perhaps one of many other proposed mechanisms. If it were found that samples rich in crater ejecta were rich in chondrules, while

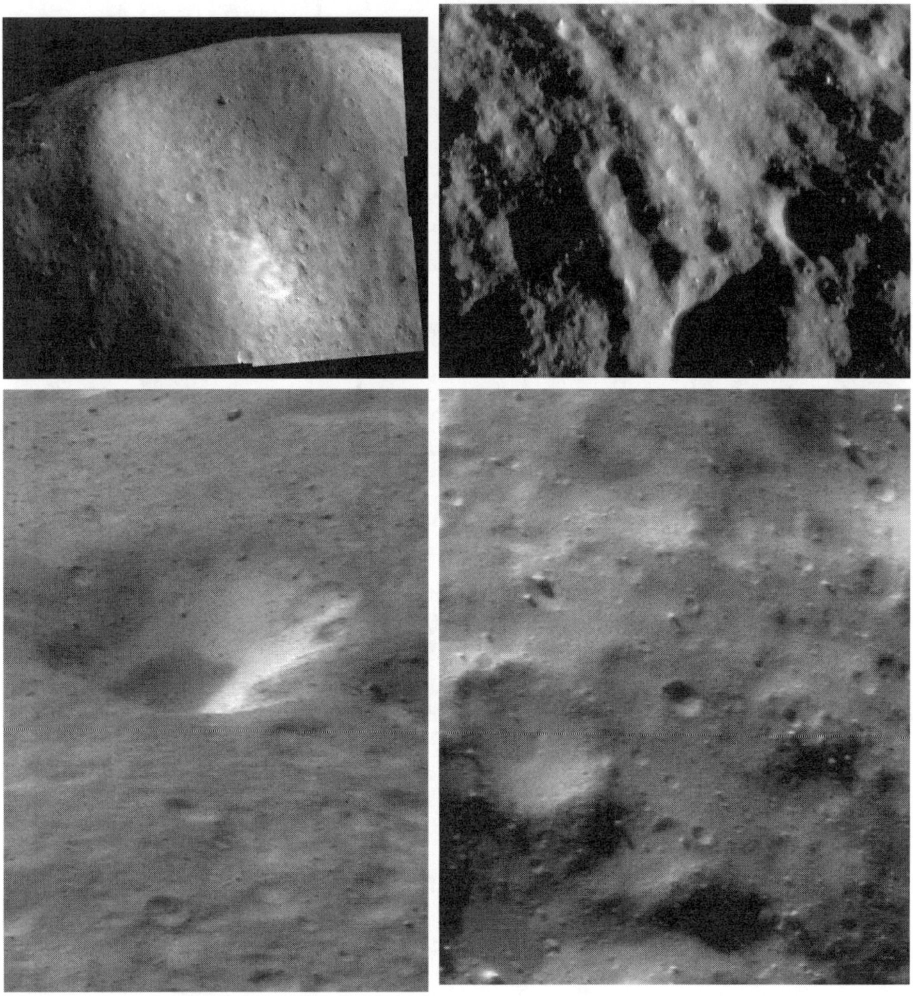

Figure 6. Surface textures on Eros, boulders, grooves, craters, and regolith (NASA). In addition, NEAR discovered flat areas of fine-grained material referred to as "ponds".

samples from the interior or inter-crater plains were free of chondrules, then it would be clear that chondrules were impact melts. Until we do fieldwork on asteroids and bring back samples we will not know for sure if chondritic asteroids have interiors made of weak unconsolidated dirt or whether they contain chondrules throughout. Similarly, many authors have argued that metal silicate ratios reflect some unknown process that occurred in the solar nebula. However, if metal-silicate ratios on asteroid surface samples varied in some way, say with depth in the regolith or distance from major impact sites, then we might conclude that processes on the surface of the asteroids caused the metal silicate ratios. These are just two examples of specific questions that would be addressed with fieldwork and returned samples from chondritic asteroids. These are just two examples of specific questions that we would be addressed with fieldwork and returned samples from chondritic asteroids. There are many more. Consider what was learnt from fieldwork and laboratory study of returned lunar samples. Without these tools we should still be arguing about whether volcanic action had created any of the lunar features that we now recognize to be impact craters and ejecta blankets. Until we do comparable field work on asteroids and study returned samples from selected sites we will not know which come from bodies that were heated to form cores, mantles and crusts. We will not know which S asteroids if any were partly melted or wholly melted and remixed by impacts.

Samples from the surface of asteroids will also enable us to characterize space weathering on asteroids, just as lunar samples enabled us to understand space weathering on the Moon. While one might expect certain similarities, differences due to the higher impact rate and velocity on asteroids and differences in target chemistry, especially the presence of volatiles, are to be expected. In fact, space weathering effects in samples returned from asteroids of different classes could be compared to such effects for lunar samples in an excellent example of comparative planetology. A fundamental understanding of space weathering would facilitate the interpretation of spectra for all asteroids. We return to this below.

1.4. The value of returned samples, or "Why return asteroid samples when we have in-situ techniques?"

The depth and breadth to which asteroid samples can be studied in terrestrial laboratories will always be many times greater than will be available from in-situ techniques as even the most cursory glance at the literature will demonstrate. The first report of the Tagish Lake meterorite in the technical literature included data for 78 elements using 10 techniques - most of them requiring sophisticated procedures that could only be performed on Earth (Table 1) (Brown et al. 2000). By comparison, in-situ analysis by the Pathfinder mission to Mars yielded data from 7 elements using one technique (Table 2) (Rieder et al. 1997). Aside from the volume of the data, the quality of data obtained in the laboratory was superior.

Not only can better data be obtained in the laboratory, but there are certain kinds of measurement that cannot yet be performed on robotic spacecraft but which are crucial to an understanding of the samples. The classic example is dating. About ten kinds of age can be determined by laboratory methods, while as of today none are available for in-situ measurements. Depending on the type of

Table 1. Analytical data for the Tagish Lake meteorite (Brown et al. 2000).

	Method	Concentration (ppm)		Method	Concentration (ppm)
H	Prt-Gm	1.5 ± 0.3 wt%	Cd	AQ-ICPMS	0.82 ± 0.16
Li	TD-ICPMS	2.5 ± 0.2	In	AQ-ICPMS	0.075 ± 0.005
Be	TD-ICPMS	0.052 ± 0.003	In	F-ICPMS	0.060 ± 0.007
B	Prt-Gm	0.8 ± 0.1	Sn	F-ICPMS	0.92 ± 0.09
C	C-IRA	3.6 ± 0.2 wt%	Sb	LL-INAA	0.17 ± 0.03
Na	LL-INAA	4450 ± 60	Te	AQ-ICPMS	1.5 ± 0.3
Mg	WRA-ICP	10.8 ± 0.5 wt%	I	SL-INAA	<0.2
Al	WRA-ICP	0.99 ± 0.03 wt%	Cs	F-ICPMS	0.146 ± 0.006
Si	WRA-ICP	11.4 ± 0.4 wt%	Ba	F-ICPMS	3.6 ± 1.2
P	TD-ICP	927 ± 50	Ba	TD-ICPMS	7.4 ± 0.8
S	C-IRA	3.8 ± 0.2 wt%	La	F-ICPMS	0.31 ± 0.02
Cl	SL-INAA	560 ± 90	La	LL-INAA	0.33 ± 0.03
K	TD-ICP	650 ± 50	Ce	F-ICPMS	0.81 ± 0.06
Sc	LL-INAA	7.2 ± 0.3	Pr	F-ICPMS	0.111 ± 0.007
Ti	WRA-ICP	520 ± 56	Nd	F-ICPMS	0.58 ± 0.03
V	F-ICPMS	57 ± 3	Sm	F-ICPMS	0.19 ± 0.02
V	SL-INAA	51 ± 1	Sm	LL-INAA	0.20 ± 0.02
Cr	LL-INAA	2840 ± 150	Eu	F-ICPMS	0.072 ± 0.004
Mn	WRA-ICP	1450 ± 150	Gd	F-ICPMS	0.24 ± 0.02
Mn	SL-INAA	1530 ± 77	Tb	F-ICPMS	0.049 ± 0.005
Fe	LL-INAA	19.3 ± 0.9 wt%	Dy	F-ICPMS	0.30 ± 0.04
Co	LL-INAA	517 ± 9	Ho	F-ICPMS	0.064 ± 0.006
Ni	LL-INAA	1.16 ± 0.08 wt%	Er	F-ICPMS	0.20 ± 0.02
Cu	TD-ICP	116 ± 5	Tm	F-ICPMS	0.032 ± 0.002
Zn	TD-ICP	253 ± 9	Yb	F-ICPMS	0.203 ± 0.009
Ga	F-ICPMS	8.4 ± 0.3	Yb	LL-INAA	0.21 ± 0.02
Ge	F-ICPMS	30 ± 2	Lu	LL-INAA	0.034 ± 0.006
As	LL-INAA	1.74 ± 0.06	Hf	F-ICPMS	0.18 ± 0.02
Se	LL-INAA	14.3 ± 0.4	Ta	F-ICPMS	0.022 ± 0.006
Br	LL-INAA	2.8 ± 0.2	Re	AQICPMS	0.056 ± 0.004
Rb	F-ICPMS	2.0 ± 0.2	Os	FA-INAA	460 ± 18 ppb
Sr	F-ICPMS	9.4 ± 0.5	Ir	FA-INAA	547 ± 10 ppb
Y	F-ICPMS	1.7 ± 0.1	Pt	FA-INAA	1.22 ± 0.05
Zr	F-ICPMS	6.0 ± 1.3	Au	LL-INAA	0.19 ± 0.03
Nb	F-ICPMS	0.31 ± 0.15	Tl	TDICPMS	0.090 ± 0.004
Mo	AQ-ICPMS	1.13 ± 0.09	Pb	AQICPMS	2.9 ± 0.8
Ru	FA-INAA	1.08 ± 0.09	Bi	AQICPMS	0.09 ± 0.02
Rh	FA-INAA	0.25 ± 0.02	Th	F-ICPMS	0.040 ± 0.008
Pd	FA-INAA	0.98 ± 0.09	U	F-ICPMS	0.008 ± 0.004

Table 2. Analytical results for the Mars surface rocks as determined by instruments on the Pathfinder spacecraft (Rieder et al. 1997)

	Method	Concentration
Na2O	APXS	3.2 ± 1.3 wt%
MgO	APXS	3.0 ± 0.5 wt%
Al2O3	APXS	10.8 ± 1.1 wt%
SiO2	APXS	58.6 ± 2.9 wt%
SO3	APXS	2.2 ± 0.4 wt%
Cl	APXS	0.5 ± 0.1 wt%
K2O	APXS	0.7 ± 0.1 wt%
CaO	APXS	5.3 ± 0.8 wt%
TiO2	APXS	0.8 ± 0.2 wt%
FeO	APXS	12.9 ± 1.3 wt%
Sum		*92.7 wt%*

material, the dating method, and certain interpretations, laboratory techniques can determine the time interval the end of nucleosynthesis and agglomeration, the duration of agglomeration, time of accumulation, crystallization ages, time since the onset of inert gas retention, the age of major heating and degassing events, the time of metamorphism, time of aqueous alteration, the duration of exposure to cosmic radiation, the time on Earth, and for Antarctic meteorites the duration of burial and the time on the surface of the ice.

Returned samples also have the potential to be archived for future reference, pending new techniques and new ideas. The Apollo lunar samples archived at the Johnson Space Center have been the subject of a great many studies using equipment not dreamed of when the samples were collected. Most notable are probably micrometer scale variations in isotopic properties.

It is sometimes argued that one needs only to collect data of a quantity and quality needed to address specific questions and that if in-situ data are adequate for the problem then the additional expense of obtaining better data is not justified. This is true, and there are many science questions for which in-situ data has been sufficient to make major advances. But this is not true of questions relating to the composition, mineralogy, petrology, and isotopic properties of primitive solar system materials where the full arsenal of data is required. The problems are complex and diverse, and not amenable to a few very simple measurements made by simple automated instruments. This has been demonstrated repeatedly in meteorite studies - in-situ techniques could never have discovered extinct nuclides or complex diversity in elemental and isotopic properties of chondrules, for instance. The NEAR Shoemaker spacecraft carried instruments to characterize the mineralogy and chemistry of Eros from close orbit but was not able to answer the fundamental questions, such as the Mg/Si or Fe/Si ratio of the surface, the presence and type of chondrules, the ratio of metal to silicates, the oxygen content or the oxygen isotopic ratios of the surface materials (McCoy et al. 2001). It is doubtful that instruments specifically designed for in-situ measurements on the surface would have faired much better.

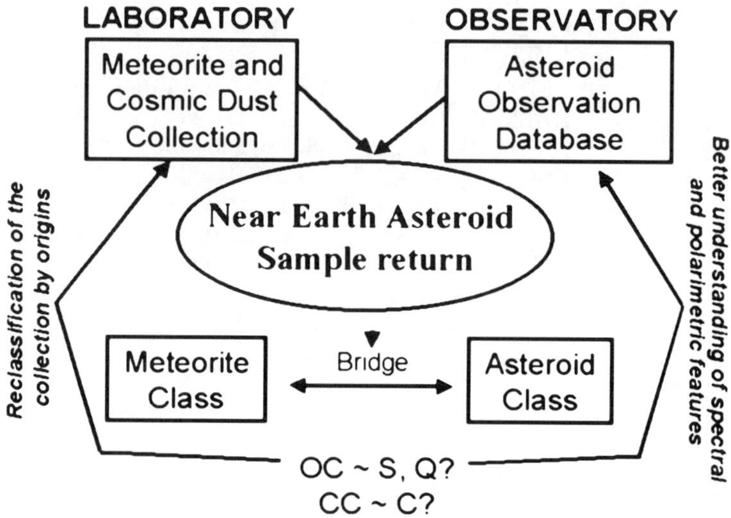

Figure 7. Near-Earth asteroid samples will provide information about the asteroids from which they came, but additionally will provide new understanding that will feed back into the whole of meteorite and asteroid studies.

1.5. A cultural problem and in-situ analysis - do what you do best regardless of the demands of the Science.

There is a cultural problem that must be overcome in placing an appropriate value on asteroid sample return, which is that because sample return has only recently become viable the planetary science community has a long history of finding ways to get round this. There are a great many techniques for in-situ analysis that have been painstakingly developed over the last thirty years that we would like to see fly even though they are not the way to do good science. This tends to cloud the issue and for the difficulties of sample return to be exaggerated.

1.6. The multiplying effect

While the samples returned from near-Earth asteroids will be of great scientific value in themselves, there will actually be a "multiplying effect" in the science return since they will form a bridge between rock samples investigated in great detail in the laboratory and the astronomical objects that have until now only been observed from great distances through a telescope. Thus the data from returned samples will provide new insights into meteorite formation and history that will improve the interpretation of all meteorite data, and data from the returned samples will also help in our interpretation of astronomical spectra for asteroids (Figure 7). Only sample return from asteroids previously studied by remote observations and in-situ measurements can bridge the gap between ground observations of minor bodies and laboratory analyses of meteoritic samples. There will be an influx of new data and a refreshing reevaluation of ideas on a scale that has not been seen since the return of lunar samples by the Apollo program.

1.7. A natural "next step" in solar system exploration

Perhaps the strongest argument for an NEA sample return mission is that it is the natural next step in solar system exploration in the logical and systemic investigation of the building blocks of the solar system. After centuries of telescopic observation of these primitive solar system objects, numerous fly-bys, a mission to orbit an asteroid, and now a landing, sample return is the inevitable next step. Consider the well tried and tested traditional method of scientific exploration of a geological field site. The area is mapped, making use of aerial photographs and perhaps even remote sensing data, sometimes geophysical data obtained, then the geologist goes into the field to obtain detailed mapping, and then finally the geologist takes samples that can be brought back to the laboratory for detailed study. The only equipment they take into the field is normally a hand lens and a note book, and perhaps a magnet and a bottle of acid. The quipment of modern science is best kept in the laboratory. The analogous line of investigation for the examination of asteroids is to obtain astronomical spectra (map), fly rendezvous and orbital missions (fieldwork), and then take samples. We are entering the decade of sample return. The Stardust and Genesis missions are scheduled to return samples. The Johnson Space Center has a fully operational Astromatrials Branch equipped to process the samples and pass them into the science community, and the NASA Cosmochemistry Program has generated a large community of sample analysts equipped with some of the most sophisticated equipment on Earth.

While we can be confident that there is a good chance the samples will shed new light on fundamental questions, exploring new terrains always stirs new insights and new questions that could not have been predicted. The return of samples from the Moon resulted in a complete overturn of ideas about the origin of the Moon and its history. The existence of a magma ocean was not predicted and few would have predicted the widespread acceptance of an impact origin for the Moon prior to the return of Apollo samples. Thus while it is difficult to write it into budget requests, the best reason to return samples might be that we can expect the unexpected.

1.8. So can it be done?

It is often said that sample return from asteroids is "high science, high risk". In other words, the scientific value of returning new samples of primitive solar system material from known context on known asteroids is beyond question, but that the mission is a challenge to current technology. The NEA Sample Return Panel suggests that while this was true a few years ago, events of the last year or two have changed this. The events are:

Figure 8. The Deep Space 1 technology development mission demonstrated the reliability of solar electric propulsion and the required automatic navigation techniques in deep space (NASA). Using a new technologies, the spacecraft performed a flyby mission of asteroid Braille and then comet Borelly. Useful images of Braille were not obtained, but the first image of a comet nucleus since that of Halley's Comet fifteen years ago was obtained. Solar Electric Propulsion has been in use in low-Earth orbit for many years, but this was the first use in deep space.

- The success of the Deep Space 1 mission and the new confidence it places in Solar Electric Propulsion. As a result of the end of the primary mission phase of Deep Space 1, JPL recommended and NASA Headquarters approved new specifications for the NSTAR SEP thrusters that bring a number of asteroid missions into the capability of current technology using small missions (Figure 8).

- The spectacular rate of discovery of near-Earth asteroids. Almost 80% of the 1600 known asteroids in near-Earth orbits were discovered in the last two years (Figure 9). Among them are about 30 in orbits that have lower Δv than the Moon. In a pilot study for the Hera mission, Leon Gefert has identified over 60 trajectories (Sears et al. 2001) that would take a NEAR-type spacecraft powered by SEP thrusters to three asteroids and return to Earth.

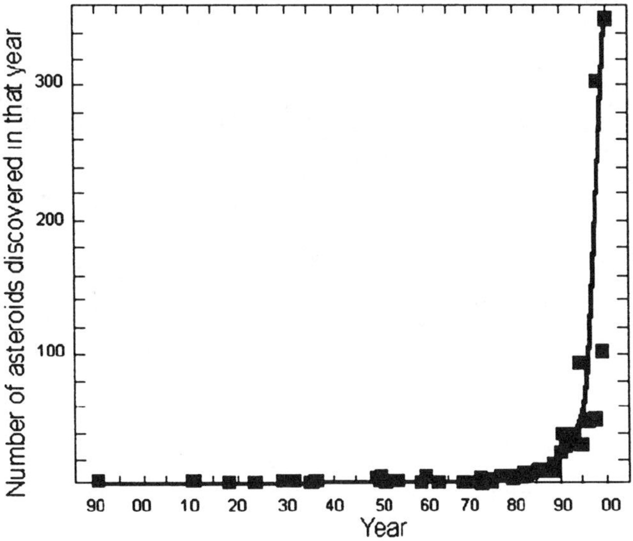

Figure 9. Rate of discovery of near-Earth asteroids has increased considerably in recent years. A large number of low-energy targets now exist for a near-Earth asteroid sample return mission.

- The spectacular success of the NEAR-Shoemaker mission. In many respects, NEAR was a dry run for an NEA sample return mission, accomplishing many crucial operations flawlessly. These were, going into orbit around an asteroid, maintaining a stable orbit for a year, maneuvering repeatedly with high precision while in orbit, and finally landing (Figure 10) (Williams et al. 2000, Fujiwara et al. 2000).

- The pending launch of the technology development mission MUSES-C. The Japanese MUSES-C mission is not a science driven mission but it will return a few grams of sample from a NEA. It will rendezvous with NEA 1998 SF36, station keep, descend to the surface momentarily, fire a projectile into the surface, collect the ejecta in a cone that will channel it into a container inside a sample return capsule, return to Earth for a recovery in the USAF Utah test range (Fujiwara et al. 2000). The published budget for MUSES C is \$150M, but this does not include personnel costs, however it is clear that the mission would fit within the Discovery Program cap of \$250M.

The crucial steps in asteroid sample return are (1) getting to the asteroid, (2) maneuvering in the vicinity of the asteroid, (3) taking the sample, (4) returning the sample to Earth. The first is not a problem given the large numbers of NEA and the availability of SEP. We now have considerable experience and confidence in maneuvering in the vicinity of asteroids through the NEAR experience, and while this was not exactly the same as hovering to take a sample, work by Scheeres and others associated with the NEAR mission and Japanese

Figure 10. The NEAR-Shoemaker spacecraft about to land on Eros (NASA). Although the spacecraft was intended as an oribiter, after a year in which a large number of maneuvers were performed the NEAR team were able to make a soft landing on the asteroid.

colleagues working on MUSES-C mission has resulted in algorithms for controlling a spacecraft during hovering operations (Williams et al. 2000, Scheeres et al. 2000, Fujiwara et al. 2000). Sample return from space has been commonplace since the film canisters from spy satellites were captured in the air over the Utah test range. More recently, sample return procedures have been developed for Stardust and Genesis missions and workers at NASA Langley have developed even simpler methods of sample return from deep space by direct reentry.

The most challenging aspect of near-Earth asteroid sample return is taking the sample. There are a number of sample collection techniques with flight heritage, such as the automated drill cores of the Luna missions and the trowels of Viking and Surveyor. For human missions there are a number of techniques developed by the Apollo mission. Of course, the MUSES-C team have already developed a technique for sample collection which is flight-ready. In addition to this, the Lockheed Martin Astronautics company have developed a large collector which relies on a fly wheel and screw and Honeybee Robotics have developed a collector that uses two counter-rotating auger bits on the end of a flexible rod that can be withdrawn to haul the sample into the spacecraft (Figure 11) (Allen and Lindstrom 2000, Yano et al. 2000, Nygren 2000, Rafeek and Gorevan 2000, Rafeek et al. 2000). The 750 g collector can reliably pick up 5-10 g and a 5 kg version could probably pick up 500 g.

Any sample return mission must address planetary protection issues to avoid cross contamination from possible living entities contained on or in returned samples. Samples falling naturally to Earth will have received sufficient radiation dose to sterilize them, and this is probably true of samples obtained

Figure 11. Three methods for collecting samples from an asteroid that do not involve landing. (left) Honeybee Robotics has designed a sampling head which involves counter-rotating drill heads attached to a rod which bounces across the surface. (middle) Lockheed Martin Astronautics have designed apparatus that involves a large screw and fly-wheel that are jettisoned after sample collection. (right) The Muses C collector fires a cannon into the surface and collects the ejecta.

from the surface of most asteroids, but it will not be true of samples taken from depth. Fortunately, NRC panels have considered this matter at great length and suggested that sample return from most asteroids will not require any special containment or handling based on planetary protection concerns (NRC 1994). However, special handling and containment is warranted for P- and D-type, and perhaps for F-type asteroids (the asteroids probably containing free organic compounds) until such time that more information is available about them. In general, samples returned from asteroids are not likely to pose a serious threat to Earth. Nevertheless it will be necessary to be sensitive to planetary protection issues during the initial design of any mission returning samples to Earth (Clark 2000).

1.9. Why sample NEAs, rather than main belt asteroids?

Clearly, the energetics of reaching NEA with spacecraft are less demanding than main belt asteroids and thus will be accessible to scientific research sooner than main belt asteroids. As already mentioned, some are easier to get to than the Moon. Available evidence suggests that NEA are representative of the main belt, at least the distribution of asteroids over the spectral classes is the same for NEA as it is for the main belt so the potential science returns are greater. NEA have suffered an event not experienced by the main belt, namely transfer from the main belt to the near-Earth vicinity, but there is no reason to expect that this has changed their mineralogical or chemical properties and they are still pristine material from the earliest days of solar system history. Since they are fragments of primary bodies, they expose the interior of that object.

However, there are actually reasons why we would prefer to explore NEA before going to the main belt. They are part of the near-Earth space environment, and NASA's current plans for exploring the solar system with robotic and human missions involve a steady progression outwards, from low-Earth orbit, to the Moon, to Mars, to main belt asteroids, to the outer solar system. NEA exploration fits neatly between the exploration of the Moon and the exploration of Mars. Missions to NEA would have shorter durations than missions to Mars, and would be technically less demanding. Second, NEA could ultimately provide local and relatively cheap resources (water, for instance) for the International Space Station or lunar colonies (Lewis 2000). Third, NEAs include potential Earth-impactors and there is widespread interest in identifying and characterizing asteroids that could potentially impact Earth.

2. Key Science Questions

- Primitive solar system processes. What are the processes occurring in the primitive solar system and accompanying planet formation? How can we distinguish meteorite parent from nebular processes? How does this information relate to existing data and prevailing ideas based on meteorites? What are the lessons to be learned about interpreting data for meteorite classes whose parent body look-alikes have not been visited. If there is new primitive material in the asteroids that is not surviving atmospheric passage, then exciting surprises are to be expected.

- Presolar grains and the relation between the Sun and adjacent stars. New kinds of primitive materials means a whole new range of possibilities for discovering new kinds of presolar material: new types of pre-solar grains, new types of interstellar molecules. To date, interstellar materials have been limited to refractory minerals and compounds have been restricted only to the most primitive of the meteorites. Presolar materials in primitive meteorites are the only direct link we have between stars in the vicinity of the solar system and the Sun and the solar system.

- What are the organic materials in as yet unsampled primitive materials and are there lessons for our understanding of the origin and distribution of life in the solar system? There will almost certainly be volatile organic compounds present in the new types of primitive solar system materials discovered by NEA sample return missions. Characterizing these will yield information about the formation and evolution of prebiotic molecules necessary for life on planets. Does this information require reevaluation of current conclusions based on meteorite organics?

- How did asteroids and meteorite classes form and have their present properties? How do the major elemental, mineralogical and isotopic properties asteroid samples (1) vary with asteroid class and (2) compare with the meteorite classes. What does this say about the origin of meteorites and their relevance to the formation and history of the solar system and processes occurring there?

- How do the elemental, mineralogical and isotopic properties of the asteroid samples vary with geological context on the surface? How uniform is the asteroid, and if it is inhomogeneous, how do the variations relate to variations in meteorite properties? What can be learned from studying the asteroid samples in context about possible parent body effects in meteorites? What does this imply for nebula conditions inferred from meteorites?

- The space environment. What processes, physical, mineralogical, elemental and isotopic, can be identified as happening on the surface of these small airless bodies as a result of exposure to the space environment. What are the nuclear effects and what is the exposure age of the surface, what is the gardening rate, what are the cosmic ray dose rates, how do these compare with data for Apollo lunar samples? What do these data imply for human radiation doses in the 1 AU region of the solar system? What do these data imply for space weathering on different types of surface, and how does the process compare with those on the Moon? Can we see both radiation effects and mineralogical effects on all bodies? What do these observations imply for the interpretation of astronomical spectra for asteroids?

- What is the internal nature of asteroids? Can anything be learned from the nature of asteroid surface materials about the nature of the interior and therefore the low bulk densities of the asteroids, one of the major problems in asteroid science?

- What can the science community contribute to the human exploration and development of space and mitigation of the impact hazard?

3. Findings and Recommendations

3.1. Supporting facilities

NASA already has in place a facility for astromaterials curation at the Johnson Space Center (Allen and Lindstrom 2000). The facility has over thirty years experience of handling extraterrestrial materials in a way acceptable to the scientific community and society-at-large. They are currently working with teams for the Genesis and Stardust missions, and they are responsible for handling Antarctic meteorites and Apollo samples. In addition, information developed for Mars sample return missions (Race et al. 2001) can provide extensive information in preparation for handling and analyzing future returned samples from asteroids.

3.2. Costs

Judging from the cost of the Japanese MUSES-C mission, asteroid sample return from multiple asteroids in a single mission could be accomplished within the Discovery price cap, especially if there is there is a modest increase in the cap or an augmentation to the funds by overseas partners.

3.3. Supporting programs

The Panel believes that the near-Earth asteroid sample return should be considered primarily as a science-driven project suitable for the Discovery program. However, it could also be considered as a technology development mission and submitted to the New Millennium Program or as an impact mitigation mission for submittal to the United States Air Force who have been given the mandate

NASA Space Science Enterprise Goals

Science Goals	1	2	3	4	5	6	7	8	9	10	NEA Sample Return
1. Understand how the structure of our Universe (galaxies, stars and planets) emerged from the Big Bang.	●				●			●		●	
2. Test physical theories and reveal new phenomena throughout the Universe, especially through the investigation of extreme environments.	●	●			●				●	●	
3. Understand how both dark and luminous matter determine the geometry and fate of the Universe					●	●		●		●	
4. Understand the dynamical and chemical evolution of galaxies and stars in the exchange of matter and energy among stars and the interstellar medium					●	●		●	●	●	
5. Understand how stars and planetary systems form together					●	●	●	●			●
6. Understand the nature and history of our Solar System, and what makes Earth similar to and different from its planetary neighbors.	●		●	●	●						●
7. Understand the mechanisms of long- and short-term solar variability, and the specific processes by which Earth and other planets respond.	●	●							●		●
8. Understand the origin and evolution of life on Earth				●							●
9. Understand the external forces, including comet and asteroid impacts, the affect life and the habitability of Earth	●		●								●
10. Identify locales and resources for future habitation within the Solar System	●			●							●
11. Understand how life may originate and persist beyond Earth			●	●		●	●				●

Figure 12. Figure from NASA's 1997 "Space Science Enterprise Strategic Plan" illustrating the goals of the space science program and how each mission addresses them (NRC 1997). To this diagram we have added the contributions of a mission to take and return samples from near-Earth asteroids (NEA Sample Return). Other missions include 1) Solar Terrestrial Probes, 2) Bohr Probe, 3) Europa Orbiter and Pluto/Kuiper Express, 4) Mars Surveyor, 5) Next Generation Space Telescope, 6) Space Interferometer Mission, 7) Terrestrial Planet Finder, 8) Far-Infrared Submillimeter Telescope, 9) Gamma-ray Large Array Space Telescope, and 10) Constellation X-ray.

by congress to consider impact mitigation. NASA has the mandate to characterize potential earth impactors, but does not have a specific mission program for that purpose.

The community responsible for extraterrestrial material (meteorites, cosmic dust, Apollo samples) research is funded primarily by NASA's Cosmochemistry program. Another argument for near-Earth asteroid sample return is that a large, well-coordinated and well-funded community already exists for the analysis of the returned samples and existing resources would be adequate for their characterization. Lagging, however, is major equipment, and a sample return mission should assign resources for updating the aging instrumentation in the field.

3.4. Priority relative to other activites

The NEA Sample Return Panel argues that the return of samples from near-Earth asteroids is a mission of highest priority relative to other solar system activities currently under study as part of the NRC decadal review. The scientific value of studies of these small solar system objects has been recognized by NASA in the "Space Science Enterprise Strategic Plan" (Figure 12) (NRC 1997). Seven of the 11 goals laid out in the strategic plan can be uniquely addressed by sample return. These are:

a. Understand processes occurring during planet formation from evidence contained within primitive asteroids.

b. Understand stellar evolution and the relationship between stars and planet formation from pre-solar grains they contain.

c. Shed light on the origin of molecules necessary for life from organic compounds.

d. Detect chemical processes that preceded life on Earth from chemical trends in the samples. This can help us also understand the possibilities of life on other planets.

e. Determine the record of solar activity for bodies in known orbits, from solar wind and solar energetic particles trapped in these surface materials.

f. Enable the design of devices to deflect potentially hazardous objects and predict the effects should they reach Earth's atmosphere from both small body sample return and the data from the encounters.

g. Identify asteroid resources to facilitate human exploration and the development of space.

In addition, another NASA document, "Mission to the Solar System: A Mission and Technology Roadmap", which lays out technology development and missions that are required to accomplish the strategic plan, advocates sample return from small solar system objects (Figure 13) (NRC 1996) in fact, of the 18 "portrait" missions listed in the Roadmap, all but one have either been flown or their goals are being addressed in some way. The exception is a mission to recover samples from small solar system bodies.

Missions to recover samples from near-Earth asteroids, the only feasible targets for macroscopic sample return with currently technology in the small mission budgets, addresses some of the most fundamental questions in planetary science - questions relating to the origin of the solar system and all materials in it - which underpin many of the activities currently under consideration.

3.5. Specific recommendations

- Samples should be obtained from NEAs with two specific objectives in mind. (a) To collect primitive material not represented in the terrestrial meteorite flux. To this end, emphasis should be on C-type asteroids. (b) To collect samples from a variety of geological contexts that are likely to

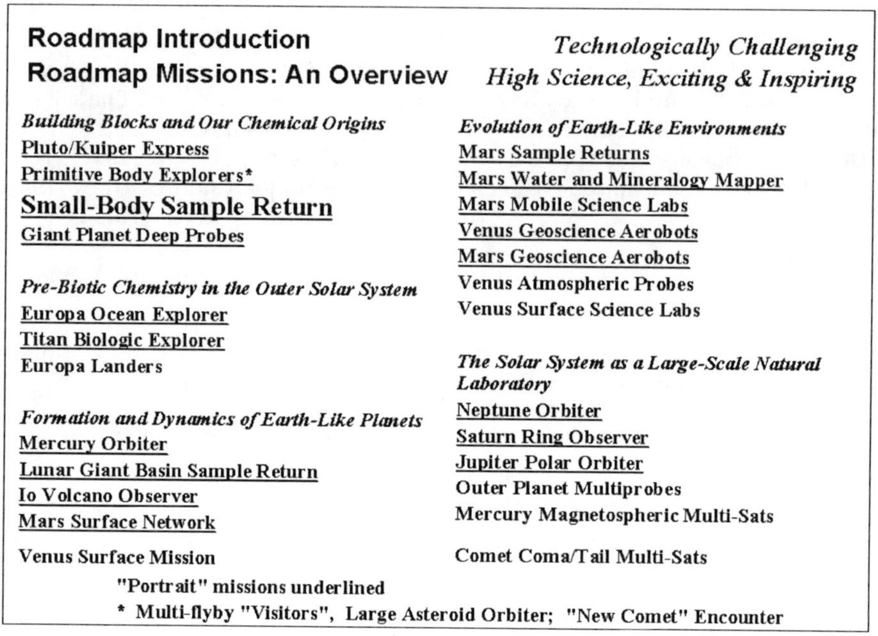

Figure 13. An overview of missions intended to address the major questions in NASA's space science program as described in the document "Mission to the Solar System: A Mission and Technology Roadmap" (NRC 1996).

shed light on the origin of the chondrite and asteroid classes. To this end, emphasis should be on S-type asteroids, probably those that most closely resembling the ordinary-chondrites, namely the S(IV) asteroids.

- Sufficient samples should be obtained from a number of asteroids so as to reasonably bracket the material in the main belt, i.e. include the major classes. We suggest that three asteroids is appropriate for an initial mission, but recognize that in the long run a program of systematic exploration of the asteroids with multiple missions and multiple programs will be necessary.

- Sufficient total mass of sample should be obtained so that the full armory of terrestrial techniques can be applied and still have sufficient material for long-term archiving. This means that about 500 g - 1 kg of each asteroid should be returned.

- Samples should be obtained from all the scientifically significant sites on the asteroids so that the effects of processes occurring on the asteroid surface can be identified.

- Material should be distributed to the scientific community through a process that recognizes the depth and breadth of techniques available and sensitive to newly emerging techniques.
- Material should be archived on the assumption that the asteroid will never be visited again and in such a way its scientific value is not compromised, yet it should remain accessible to researchers with scientific justification.
- A systematic program of NEA sampling should be international in scope and involve multiple missions. An excellent start has been made by the technology development mission MUSES-C. It would be appropriate to follow this up by a science driven mission to multiple asteroids.

4. Maintaining Capabilities for the Future

Samples need to be subjected to sufficient preliminary examination to ensure that information reaches appropriate researchers and that appropriate research is done.

Samples need to be need to be distributed to the scientific community in a way that ensures maximum scientific return from the samples. Thus while care should be taken to ensure that the samples are not used irresponsibly, an effort should be made to see that all scientists with the appropriate credentials receive samples.

A significant fraction of the material (one-quarter?) should be placed in long-term storage pending further developments and instrumentation.

NEA sample return is an excellent example of how a single mission can make major contributions to the space science strategic plan, but how integrations with other NASA missions and the international programs can yield a total science return that is greater than the sum of the return from the individual missions. In this sense, the achievements of one mission will be perpetuated through future missions. The MUSES C mission will visit one asteroid and return less than a gram of material, yet it will have demonstrated the technology and provide first-look samples of an S asteroid. NASA could contribute the next step by funding a science driven mission to return amounts of material that will enable the full range of measurements to be made, possibly visiting several asteroids. The European Aurora program is an example of a future program that could eventually return samples. These three missions could represent an excellent opportunity of international collaboration for optimum science by coordinating targets and minimum risk and cost, by sharing technology and expertise.

5. Human Space Activities

The current vision for the orderly exploration of the solar system favored by NASA involves an orderly growth outwards from low Earth orbit, to the Moon, to Mars, the Main Asteroid Belt and then the outer planets. There is a big technological gap between missions to the Moon and Mars which can usefully be filled by missions to NEA. Gene Shoemaker advocated manned missions to NEA two decades ago (Shoemaker and Helin 1978), and at the NEA workshop Bret Drake described recent studies by mission planners at JSC of NEA missions as a useful way to test Mars-bound technologies (Drake 2000). The missions would be of shorter duration (about one year), simpler and cheaper no gravity well or atmosphere at the target, greater launch opportunities to numerous potential NEA targets, and lower energy would be required than for a mission to Mars. The International Space Station might make such missions even more economic. Robotic missions are the precursors of all manned missions and NEA sample return missions would be robotic precursor missions for the human explorations and development of space.

6. Education and Public Outreach

The Public Outreach possibilities of NEA missions are considerable because of the relevance to the impact hazard and popularization of the topic by its relevance to the termination of the dinosaurs and several recent movies. On the strategic and political front there is considerable interest in impact prediction and mitigation and society has every right to expect that the scientific community will make a significant contribution to these efforts. We can do so, while still addressing the demands of fundamental planetary science, by near-Earth asteroid sample return (Drake 2000).

The Education possibilities of integrating a Near-Earth asteroid sample return mission into K-12 education curricular are also numerous. The orbital mechanics involves knowledge of gravitational theory and simple solar system structure, which can be used to help teach basic mathematics and astronomy at many grade levels. The nature of the samples can be integrated into seventh and eighth grade earth science courses, while the impact hazard aspects of the mission can be integrated into history of life on Earth and concepts of time. Solar electric propulsion can be used as a means of teaching 11th and 12th grade physics of moving charges in electric fields and solar cells are an excellent introduction into solid state physics. There are few missions that relate to such diverse aspects of the K-12 school science and mathematics curriculum.

References

Allen, C.C. and M.M. Lindstrom 2000. Near-Earth asteroid sample curation. Near Earth Asteroid Sample Return Workshop, December 11-12, LPI Contrib. 1073, pp. 1-2. Lunar and Planetary Institute, Houston.

Baldwin, B.S. 1971. Ablation and breakup of large meteorids during atmospheric entry. *J. Geophys. Res.* **76**, 4653-4668.

Binzel, R.P., T. Gehrels, and M.S. Mathews, Eds. 1989. *Asteroids II.* Univ. of Arizona Press, Tucson.

Bogard, D.D. 1995. Impact ages of meteorites - A synthesis. *Meteoritics* **30**, 244-268.

Brown, P.G., A.R. Hildebrand, M.E. Zolensky, M. Grady, R.N. Clayton, T.K. Mayeda, E. Tagliaferri, R. Spalding, N.D. MacRae, E.L. Hoffman, D.W. Mittlefehldt, J.F. Wacker, J.A. Bird, M.D. Campbell, R. Carpenter, H. Gingerich, M. Glatiotis, E. Greiner, M.J. Mazur, P.J.A. McCausland, H. Plotkin, and T.R. Mazur 2000. The fall, recovery, orbit, and composition of the Tagish Lake meteorite: A new type of carbonaceous chondrite. *Science* **290**, 320-325.

Clark, B.C. 2000. Sample containment options for planetary protection. Near Earth Asteroid Sample Return Workshop, December 11-12, LPI Contrib. 1073, pp. 8. Lunar and Planetary Institute, Houston.

Drake, B. 2000. Implications of NEA sample return for human exploration and development of Space. Near Earth Asteroid Sample Return Workshop, December 11-12, LPI Contrib. 1073, (Oral presentation). Lunar and Planetary Institute, Houston.

Fujiwara, A., T. Mukai, M. Abe, H. Yano, J. Kawaguchi, and K. Uesugi 2000. MUSES-C: Mission scenario and science. Near Earth Asteroid Sample Return Workshop, December 11-12, LPI Contrib. 1073, pp. 14. Lunar and Planetary Institute, Houston.

Glavin, D.P., P. Ehrenfreund, O. Botta, G. Cooper, and J.L. Bada 2000. Amino acid chemistry as a link between small solar system bodies and carbonaceous chondrites. Near Earth Asteroid Sample Return Workshop, December 11-12, LPI Contrib. 1073, pp. 17-18. Lunar and Planetary Institute, Houston.

Hewins, R.H., R.H. Jones, and E.R.D. Scott 1996. *Chondrules and the Protoplanetary Disk.* Cambridge Univ. Press, Cambridge.

Kerridge, J.F. and M.S. Matthews, Eds. 1988. *Meteorites and the Early Solar System.* Univ. of Arizona Press, Tucson.

Lewis, J.S. 2000. Utilization of resources from NEA. Near Earth Asteroid Sample Return Workshop, December 11-12, LPI Contrib. 1073, pp 25. Lunar and Planetary Institute, Houston.

Marti, K., and Graf T. 1992. Cosmic ray exposure history of ordinary chondrites. *Annu. Rev. Earth Planet. Sci.* **20**, 221-241.

McCoy, T.J., T.H. Burbine, L.A. McFadden, R.D. Starr, M.J. Gaffey, L.R. Nittler, L.G. Evans, N. Izenberg, P. Lucey, J.I. Trombka, J.F. Bell III, B.E. Clark, P.E. Clark, S.W. Squyres, C.R. Chapman, W.V. Boynton, and J. Veverka 2001. The composition of 433 Eros: A mineralogical-chemical synthesis. *Meteor. Planet. Sci.* **36**, 1673-1695.

National Research Council (NRC) 1994. *Evaluating the Biological Potential in Samples Returned from Planetary Satellites and Small Solar System Bodies.* National Research Council, Washington, DC.

National Research Council (NRC) 1996. *Mission to the Solar System: A Mission and Technology Roadmap.* National Research Council, http://sse.jpl.nasa.gov/site/nasa/ .

National Research Council (NRC) 1997. *Space Science Enterprise Strategic Plan.* National Research Council, Washington, DC.

Nygren, W.D. 2000. Rapid remote sample collection mechanism for near-Earth asteroid sample return. Near Earth Asteroid Sample Return Workshop, December 11-12, LPI Contrib. 1073, pp. 32-33. Lunar and Planetary Institute, Houston.

Race, M.S. 2000. Planning for NEA sample return: Planetary protection controls and non-scientific factors. Near Earth Asteroid Sample Return Workshop, December 11-12, LPI Contrib. 1073, pp. 35. Lunar and Planetary Institute, Houston.

Race, M.S., D.L. de Vincenzi, J.D. Rummel, and S.E. Acevedo, Eds. 2001. *Mars Sample Handling Protocol Workshop Series.* NASA Conference Procedings, in press.

Rafeek, S. and S. Gorevan 2000. NEA Touch-and-Go Surface Sampler (TGSS). Near Earth Asteroid Sample Return Workshop, December 11-12, LPI Contrib. 1073, pp. 36-37. Lunar and Planetary Institute, Houston.

Rafeek, S., K.Y. Kong, S. Shadick, and C.C. Porter 2000. NEA multi-chamber sample return container with hermetic sealing. Near Earth Asteroid Sample Return Workshop, December 11-12, LPI Contrib. 1073, pp. 38-39. Lunar and Planetary Institute, Houston.

Rieder, R., T. Economou, H. Wänke, A. Turkevich, J. Crisp, J. Brückner, G. Dreibus, and H.Y. McSween Jr. 1997. The chemical composition of Martian soil and rocks returned by the Mobile Alpha Proton X-ray Spectrometer: Preliminary results from the X-ray mode. *Science* **278**, 1771-1774.

Robinson, M.S., P.C. Thomas, J. Veverka, S. Murchie, and B. Carcich 2001. The nature of ponded deposits on Eros. *Nature* **413**, 396-400.

Scheeres, D.J. 2000. Trajectories in close proximity to asteroids. Near Earth Asteroid Sample Return Workshop, December 11-12, LPI Contrib. 1073, pp. 42-43. Lunar and Planetary Institute, Houston.

Sears, D.W.G. and R.T. Dodd 1988. Overview and classification of chondrites. In *Meteorites and the Early Solar System* (J.F. Kerridge and M.S. Matthews, Eds.), pp. 3-31. Univ. of Arizona Press, Tucson.

Sears, D.W.G., L. Gefert, and D. J. Scheeres 2001. Asteroid constraints on multiple near-Earth asteroid sample return. *Meteor. Planet. Sci.* **39**, A186-187. (abstract)

Shoemaker, E.M. and E.F. Helin 1978. Earth-approaching asteroids as targets for exploration. In *Asteroids: an Exploration Assessment* (D.Morrison and M.C. Wells, Eds.), pp. 245-256. NASA Conf. Publ. 2053.

Space Studies Board (SSB) 1998. *The Exploration of Near-Earth Objects.* Committee on Planetary and Lunar Exploration, National Academy of Sciences, Washington, DC.

Veverka, J., M. Robinson, P. Thomas, S. Murchie, J. F. Bell III, N. Izenberg, C. Chapman, A. Harch, M. Bell, B. Carcich, A. Cheng, B. Clark, D. Domingue, D. Dunham, R. Farquhar, M.J. Gaffey, E. Hawkins, J. Joseph, R. Kirk, H. Li, P. Lucey, M. Malin, P. Martin,1 L. McFadden, W.J. Merline, J.K. Miller, W.M. Owen Jr., C. Peterson, L. Prockter, J. Warren, D. Wellnitz, B.G. Williams, and D.K. Yeomans 2001. NEAR at Eros: Imaging and Spectral. *Science* **289**, 2088-2097.

Williams, B.G., J.K. Miller, P.G. Antreasian, J.J. Bordi, E. Carranz, S.R. Chesley, C.E. Helfrich, W.M. Owen, and T.C. Wang 2000. Orbit operations at 433 Eros: Navigation for the NEAR Shoemaker mission. Near Earth Asteroid Sample Return Workshop, December 11-12, LPI Contrib. 1073, pp. 53. Lunar and Planetary Institute, Houston.

Yano, H., A. Fujiwara, S. Hasegawa, M. Abe, K. Higuchi, J. Kawaguchi, K. Tsuchiya, K. Okano, and K. Yoshida 2000. MUSES-C's impact sampling device for small asteroid surfaces. Near Earth Asteroid Sample Return Workshop, December 11-12, LPI Contrib. 1073, pp. 54-55. Lunar and Planetary Institute, Houston.

The Future of Solar System Exploration, 2003-2013
ASP Conference Series, Vol. 272, 2002
Mark V. Sykes, ed.

The Next Giant Leap: Human Exploration and Utilization of Near-Earth Objects

T.D. Jones[1], D.R. Davis[2], D.D. Durda[3], R. Farquhar[4], L. Gefert[5], K. Hack[5], W.K. Hartmann[2], R. Jedicke[6], J.S. Lewis[6], S. Love[7], M.V. Sykes[8], & F. Vilas[9]

[1] *Consultant, Oakton, VA*
[2] *Planetary Science Institute, Tucson, AZ*
[3] *Southwest Research Institute, Boulder, CO*
[4] *John Hopkins Applied Physics Laboratory, Baltimore, MD*
[5] *NASA Glenn Research Center, Cleveland, OH*
[6] *Lunar and Planetary Laboratory, University of Arizona, Tucson, AZ*
[7] *NASA Johnson Space Center, Houston, TX*
[8] *Steward Observatory, University of Arizona, Tucson, AZ*
[9] *NASA Headquarters, Washington, DC*

Abstract. Small Near-Earth Objects offer an opportunity for short-duration, low-Δv missions as the next logical step in human space activities beyond Earth orbit. Population characterization by groundbased observations using facilities including a Large Synoptic Survey Telescope is a critical precursor activity.

EXECUTIVE SUMMARY

Planetary science plays a unique role as the pathfinder for future human space activities beyond the International Space Station. It can also provide the rationale for the first human departure from LEO since the Apollo program. We examine potential manned missions to a new class of small near-Earth objects (typically tens of meters in diameter), passing close by the Earth-Moon system with very low Δv. Preliminary estimates suggest there may be many thousands of these objects. To demonstrate the feasibility of such missions we simulate a mission to 1991 VG during 1991/1992 when it passed within 0.004 AU of the Earth. This mission takes a total of 60 to 90 days, including a 30 day stay time at the asteroid. Such accessible targets provide an opportunity to perform complex, in-situ research on NEO via an extended visit with human explorers, develop and test systems needed for longer duration interplanetary missions to Mars, and to engage in precursor space resource utilization activities. Manned missions would be preceded by robotic probes to test acquisition, rendezvous, and local operations while returning significant new science from target objects. An effort to expand the known members of this population is needed. Early discovery, orbit determination, and target characterization should be pursued, and the required space-based and ground-based systems are assessed. As we look beyond ISS over the next five years, we must reinvigorate a mutually supportive relationship between our human spaceflight and solar system exploration goals. The scientific and exploration rationale for sending humans to NEO's must be mature in time to meet NASA's decision-making window for operations beyond near-Earth space.

Our recommendations are as follows:

1. We strongly endorse the Large Synoptic Survey Telescope as an effective means of increasing both the known numbers and the degree of characterization of our target population. This discovery effort must include adequate funding for conducting an asteroid search and characterization program capable of identifying and extracting observations of our target population. Follow-up observations should be supported by a dedicated telescope in the 3- to 4-m class.

2. NASA's Code M should direct its Exploration Office to include NEO mission capability in its studies of human activities beyond low Earth orbit, and direct a study of a family of deep space mission scenarios that include round-trip missions to NEO, among other destinations in Earth-Moon space. This effort should be conducted by a joint Code M/Code S working group, tasked to evaluate target discovery, orbit determination and characterization efforts, to define precursor robotic missions, and to identify priority operations and science goals for human crews operating on and near NEOs.

3. We endorse continued support for radar orbit determination of NEOs as an essential precursor activity for robotic and manned missions.

REPORT

1. Introduction

In the next decade, the International Space Station will be completed, and planning will begin in earnest for human activities beyond Earth orbit. Planetary science should play a crucial role in both justifying and preparing for such missions, and planetary research will be the top activity of human crews as they expand our presence beyond the Earth-Moon system. Such an opportunity for the discipline has not been seen since the Apollo era, when planetary science took advantage of a Cold War-driven technology program. Now the planetary science community can play an active role in shaping the decisions that will drive our scientific activities in space for decades. We identify a class of manned missions and their supporting planetary science activities that can be precursors to human missions to Mars, and open the door to space resource utilization activities.

On Dec 5, 1991, asteroid 1991 VG came within 0.005 AU of Earth (the mean orbital radius of the Moon is 0.003 AU). Other than the Moon, near-Earth objects (NEO) are Earth's closest neighbors in space. More importantly for the future of space exploration, spacecraft can rendezvous with this class of objects from Low Earth Orbit (LEO) using energies even lower than those required for a lunar landing. It was of particular interest that 1991 VG had the lowest Δv of all known asteroids (Figure 1). Subsequent discoveries of additional small asteroids

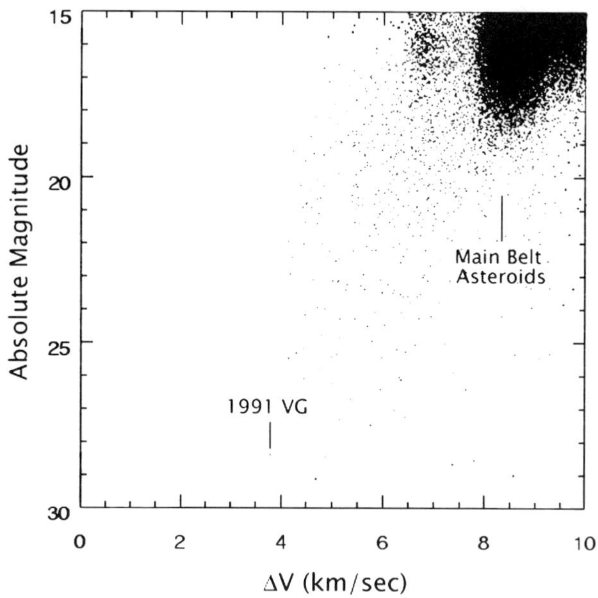

Figure 1. Δv for two-impulse Hohmann transfer orbits of asteroids. 1991 VG has an absolute magnitude of 28.4, corresponding to a spherical diameter of 9 meters, assuming a geometric albedo of 0.1.

coming near the Earth-Moon system suggest there may be a large number of these objects. Many targets per year may be dynamically available, offering extremely attractive mission opportunities of modest duration (8-12 weeks) and low energy. Given their dynamical accessibility, inherent scientific value, and the prospect for providing useful amounts of natural resources, such NEOs are excellent targets for the next step in human exploration beyond ISS.

2. Characterizing the Target Population

The distribution of orbital elements and absolute magnitudes of NEOs has been recently modeled by Bottke et al. (2000, 2001) and predicts roughly 1000 of them with $H<18$. Figure 2 shows the predicted distribution from this model of NEO semimajor axes and eccentricities in the case where $\Delta v<4.2$ km/sec. The apparent 'squares' in the data are artifacts of the sampling of Spacewatch observational data to which the model was constrained. The population of NEOs increases rapidly as one goes to smaller and smaller sizes (increasing absolute magnitude) and is illustrated in Figure 3 (left), based on Bottke et al. (2000) for $H<24$ and Rabinowitz et al. (2000) for $H>24$. According to this model, there are more than 107 NEOs larger than 1991 VG. Using the above information, the cumulative number of NEOs having a Δv above a given limit can be derived

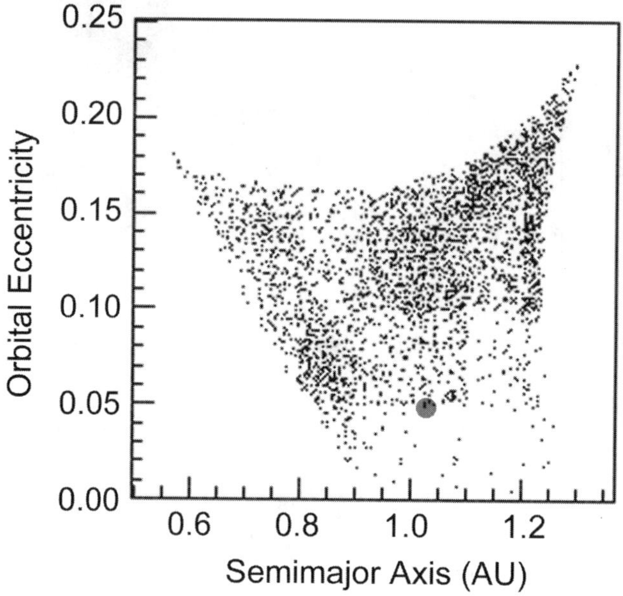

Figure 2. Orbital element distribution of NEOs having $\Delta v<4.2$ km/sec. 1991 VG is indicated by the filled circle.

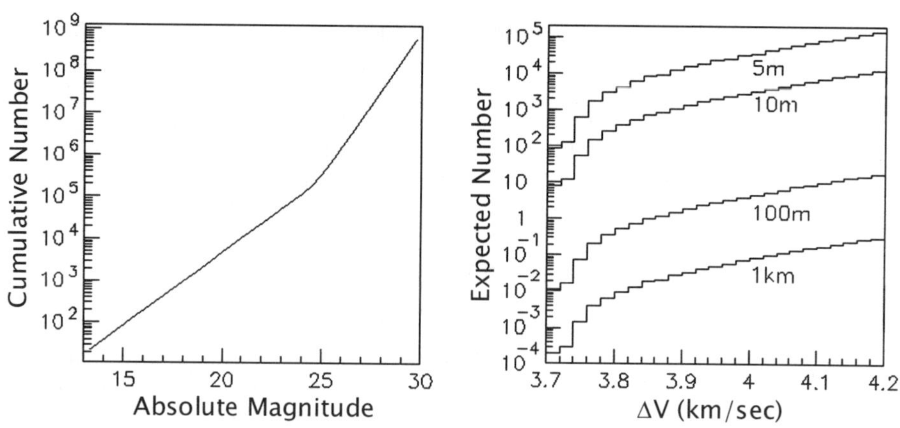

Figure 3. Model NEO absolute magnitude cumulative number (left). Expected cumulative number of NEOs having maximum Δv for a minimum diameter (right).

and is shown in Figure 3 (right). There may be more than 10,000 NEOs with diameters greater than 10 m and $\Delta v<4.2$ km/sec. This suggests a reasonable number of target opportunities over the course of a year.

3. Key Science Questions

- What compositional classes are represented among NEOs in the size range of our paradigm mission target?
- To what extent does space weathering affect various materials on asteroid surfaces, and what is the specific process(es) responsible for the alteration?
- Based on previous asteroid science results, what is our state of understanding of the internal structure of such objects, i.e., the nature of the transition (gradual, abrupt, composition-dependent?) from coherent, monolithic objects at diameters smaller than about 100-300 meters to gravity-dominated aggregates with negligible tensile strength ("rubble piles") at larger sizes?
- What is the compositional range of asteroidal material across the surface of an NEO?
- How will a significant return of NEO material, from one or more objects, enhance our understanding of solar nebula composition and processes, and the formation of the terrestrial planets and asteroids?

4. Key Operational and Engineering Questions

- What NEO orbit types offer the best set of low-energy trajectories for round-trip missions?
- What obstacles will regolith and dust environment present when we put landers and people on NEO surfaces?
- What field exploration strategies should be pursued on such objects - sampling, coring, geophysical data collection, etc.?
- What programmatic advantages will accrue from piloted missions to NEO, for both conducting human operations in deep space, and for achieving the goal of sending human beings to Mars?

5. Detecting Potential Mission Targets

Our target objects are located near the Earth's orbit (Figure 4), but small asteroids are faint, and are bright enough to be observed only when they are within proximity to the Earth. Two scenarios for identifying target NEOs are considered: (1) identify a large enough fraction of the population of small, low-Δv NEOs that targets can be predicted far in advance of their optimal Earth approach, and (2) discover NEOs a short time prior to their entering a dynamical

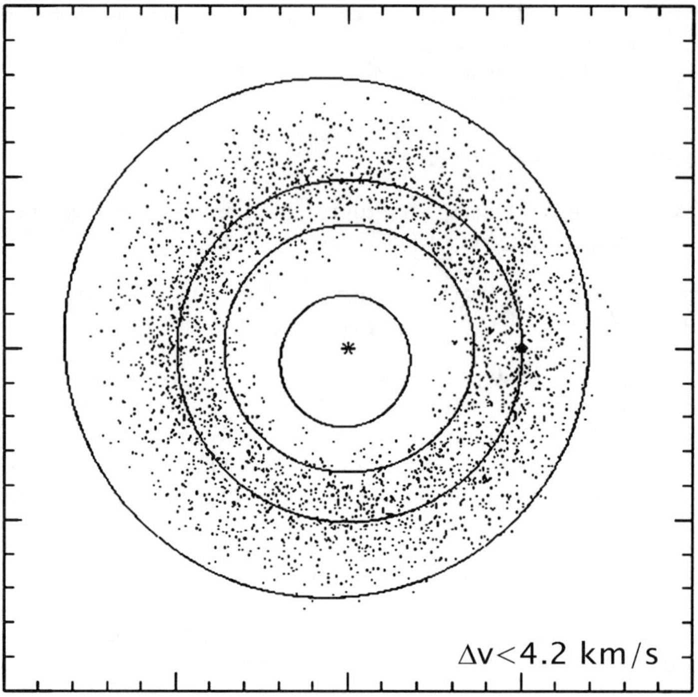

Figure 4. Positional distribution of more than 3000 model target NEOs. The orbits of Mercury, Venus, Earth, and Mars are shown for scale.

'box' defining a target to which a mission will be launched. To assess these scenarios and our ability to detect these small objects, we consider three means: a spacebased system, operating in the thermal infrared, having capabilities similar to the Space Infrared Telescope Facility (SIRTF); the Arecibo radar; and the proposed Large Synoptic Survey Telescope (LSST).

The thermal behavior of these small asteroids is assumed to be consistent with the Near-Earth Asteroid Thermal model of Harris (1998). The infrared system is presumed to detect a 1 milli-Jansky source at 10 μm. Ranges in solar elongation are arbitrarily set to be slightly larger than that of the polar orbiting Infrared Astronomical Satellite (IRAS). The results are shown in Figure 5 and assume a geometric albedo of 0.18 and a phase integral calculated from Bowell et al. (1989) assuming a slope parameter, G=0.15. Such a system would have discovered 1991 VG at a distance of only 0.02 AU (about 8 Earth-Moon distances) about a month before perigee (and at a solar elongation, outside the diagram, > 135°).

For objects around the size of 1991 VG and smaller, radar can detect them at greater distances from the Earth (assuming the capabilities of the recently upgraded Arecibo facility - Ostro, private communication). However, such radar systems are not designed to scan large volumes of space, but are rather useful

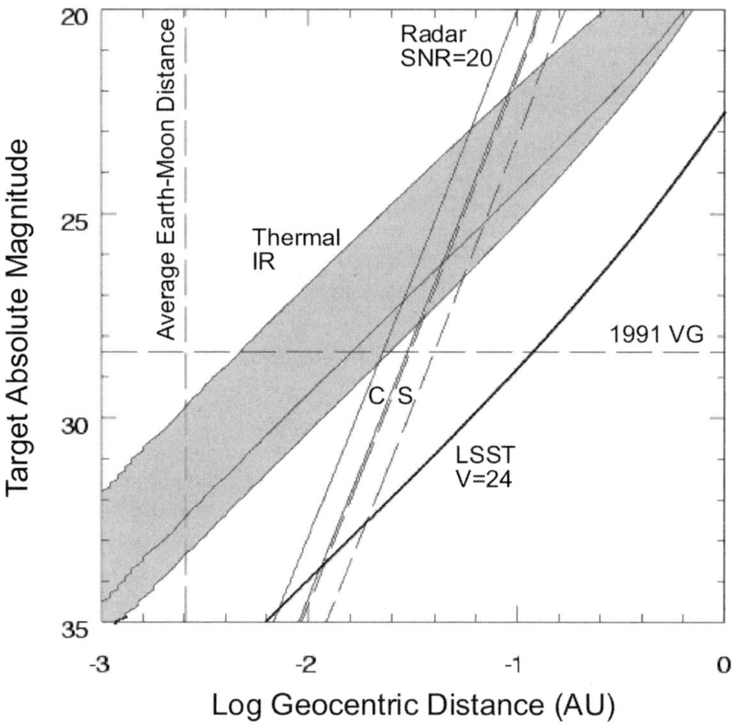

Figure 5. Detecting small NEOs using a spacebased thermal infrared system (gray), radar (unfilled), and a groundbased optical system (LSST, solid heavy curve). The ability to detect paradigm target 1991 VG is shown. For the thermal system a detection limit of 1 mJy at 10 μm is used for solar elongation angles of 45°, 90°, and 135° (curves top to bottom). For radar, the detection of C and S type asteroids are shown for slow and fast rotators (see text).

for targeted observations. The usefulness of radar in this context is its ability to greatly reduce the uncertainties in the orbital elements of the target object, resulting in very accurate predictions of its location with time (Yeomans et al. 1987) - necessary for any manned or unmanned missions (we need to know where we are going). The ability of radar to refine an orbit is a complex function of the apparent motion and location of the target relative to the Earth, but it can be coarsely stated that Doppler and range observations at SNR=20 with a time baseline of a week may ensure future recovery of the target object at future apparitions and provide precise locations in the near-term (Ostro, private communication). A conservative estimate of radar sensitivity in Figure 5 assumes a return signal having SNR=20 for both C ($P_v = 0.05$) and S-type ($P_v = 0.20$) objects that are both "slow" (2 hour) and "fast" (0.2 hour) rotators, observed at a declination of 8° (gain at Arecibo is about constant between 3° and 30°).

Using a spacebased thermal system for discovering target objects, radar could be used almost immediately for orbital refinement. In the context of our nominal mission, however, with a two-week outbound trip time, discovering targets only one month out would provide very little time under the best of circumstances.

The situation is substantially improved by considering the capabilities of the proposed Large Synoptic Survey Telescope. LSST (an 8-m f/1.25 telescope) is supposed to obtain a good detection of a "V=24" object in 10 seconds, and survey the entire available sky within one week. Such a facility would have detected 1991 VG at a geocentric distance of 0.13 AU - about 3.5 months before perigee. This would provide sufficient lead time to enable either robotic or manned rapid deployment missions. Based on the model used to generate Figures 1-3, LSST should discover 2038 potential targets within 10 years.

6. Paradigm Mission Scenario: 1991 VG

6.1. Assumptions

Initial state (Bowell orbital element file):

Semimajor axis	1.02685668
Eccentricity	0.04913590
Inclination	1.445833
Ascending node	73.965429
Argument of perihelion	24.480684
Mean Anomaly	56.900547
Epoch of osculation	2001/07/10 12:00:00.0

This state was propagated back to the 1991 close approach to Earth using the GRC n-body integration code SNAP. SNAP used the DE 403 JPL ephemeredes for the propagation. The resulting earth relative state was found to be:

Semimajor axis	-501781.317
Eccentricity	2.72063465
Inclination	137.618615
Ascending node	275.174114
Argument of perigee	315.064251
Mean anomaly	0
Epoch of osculation	1991/12/13 12:27:59.736

Using this Earth relative state a round trip mission analysis was performed. A 30 day stay at the asteroid was assumed, such that rendezvous would occur 15 days before closest approach and departure would occur 15 days after closest approach. The outbound trip time and return trip times were assumed to both be 15 (or 30) days. Given these transfer times an outbound and inbound transfer hyperbolas are calculated using Gauss's solution to Lamberts problem. The characteristic velocities (C3) of the departure and return hyperbola is 1.08 km^2/s^2 (0.216 km^2/s^2). The rendezvous and departure Δv's are 1.315 km/s (675 m/s).

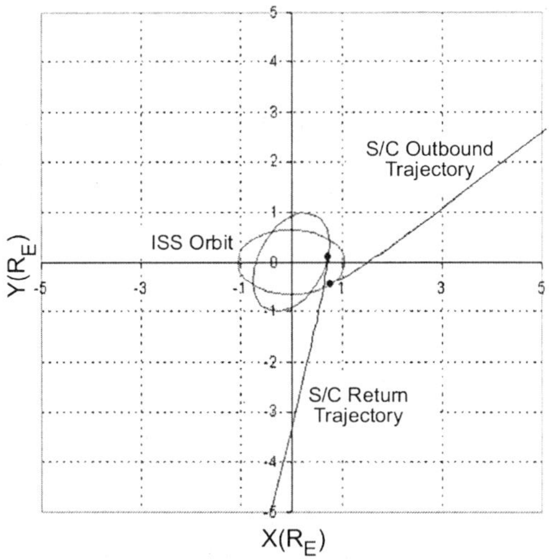

Figure 6. Earth-centered outbound and return trajectories for the paradigm mission. The projection of the International Space Station (ISS) orbit is shown on the equatorial plane of the Earth. Filled circles represent the points at which burns occur to leave and return to ISS orbit.

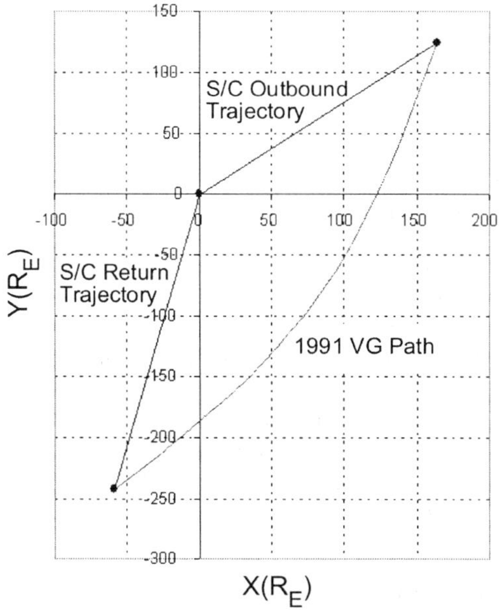

Figure 7. Earth-centered trajectory plot showing outbound and return legs and the path of the target asteroid over the course of the mission. Relative to the Earth the asteroid is traveling retrograde.

6.2. Trajectory plots (for 90 day mission)

The mission begins from ISS orbit (LEO). Two burns are assumed in each of the outbound and return legs of the trip (Figure 6), one to leave ISS orbit, the other to rendezvous with the target asteroid, then one burn to head back to ISS orbit, and the other to rendezvous. The total Δv for a 60-day roundtrip mission from LEO to 1991 VG is 6.1 km/sec, nearly the same as a one-way trip to the lunar surface. For a 90-day mission, the total Δv is just 4.9 km/sec. These numbers assume aerobraking to LEO on the return leg. Over the duration of the mission, the astronauts stay within 4.5 lunar orbital radii of the Earth (Figure 7).

7. Considerations for Actual Missions

7.1. Libration point staging for transporation beyond Earth-Moon space

Any piloted mission launched from the surface of Earth would be prohibitively expensive to keep on "standby" while waiting for a suitable target to be identified. Farquhar (2000) has explored a family of trajectories in Earth-Sun space that provide access to a variety of useful destinations. In particular, the Earth-Sun L2 libration point, on the far side of Earth from the Sun, can serve as a staging point for vehicles and cargo for spacecraft and telescope servicing, interplanetary departures, and NEO missions. An NEO mission might be readied at L2 for a departure "launch window" lasting several weeks, awaiting suitable target discovery. The L2 location provides energetically cheap access to properly chosen NEO targets, and can take advantage of Earth swing-by's on departure. These advantages should be further explored in the context of NEO mission opportunities.

7.2. Departure

For piloted missions to NEO, there appear to be no advantages to using the ISS (with its 51.6° inclination) as an assembly point for departure. In fact, lifting a spacecraft to that inclination from the Cape or from Baikonur incurs a substantial payload penalty. We have learned from ISS experience that it's cheaper and less risky to integrate and test a spacecraft on the ground before launch than to perform major assembly in orbit, with the challenges of EVA and delicate robotics work. Any spacecraft ultimately bound for an NEO should be packaged in as few pieces as possible for launch; ideally, the mission should depart from Earth in a single launch, fully tested and integrated on the ground (like Apollo). Given the spacecraft size and mass, some simple assembly in orbit or at L2 may be required (ideally limited to docking operations); the eventual option will depend on whether the space shuttle or a heavy-lift successor is the workhorse Earth launch vehicle.

7.3. Infrastructure requirements

A piloted NEO spacecraft should have the attributes of a "space tug", designed to be modular and reusable. Such reusable spacecraft will enable a variety of piloted operations in Earth-Moon space, and the NEO spacecraft should be designed for refurbishment at an orbital logistics node. For multiple missions to

NEO from HEO or L2, it will be useful to operate a transfer vehicle to move people and material from LEO to the staging point. These cargo transfer functions, not time-sensitive, can take advantage of high-ISP engine concepts. Such a vehicle is not essential to mount a single piloted NEO mission, but its existence will make possible repeated visits to NEO, and develop the infrastructure for more ambitious voyages. The NEO missions discussed here can drive us along an evolutionary path to achieving the goal of putting scientists and explorers on Mars

7.4. Spacecraft requirements

For piloted missions beyond LEO, our next-generation spacecraft should be capable of a number of mission profiles in Earth-Moon and deep space. Such mission scenarios include servicing missions to telescope facilities at the Earth-Sun L1 and L2 Langrange points, lunar orbit operations, and expeditions to NEOs (Farquhar 2000). The spacecraft should be reusable (see above), using clustered, replaceable fuel pallets or orbital refueling techniques to enable a family of such missions. ISS, Mir, and shuttle have provided us good information on the habitation space needed for a mission of this duration. For mission durations of 60 days or less, the habitable volume of the spacecraft can be quite modest. For example, the initial pressurized volume for the inaugural ISS crew was the equivalent of a couple of Spacelab modules, with interior dimensions of roughly 2 x 2 x 20 m. The Expedition One crew easily handled a four-month stay in such tight confines. Longer expeditions of six months to a year would require the equivalent of several ISS modules for living and laboratory space; finding NEO targets that offer fast round-trip missions provides great programmatic cost leverage, because such accessible targets can greatly reduce the spacecraft size and mass.

7.5. Precursor missions

The best way to pave the way for human expeditions beyond LEO, and to demonstrate the accessibility and scientific potential of NEO targets, would be to execute a series of scientific precursor missions. Search programs already have discovered several promising NEO targets, and a first step would be a quick one-way rendezvous mission to conduct physical and chemical characterization of a small NEO. We should then follow up with a simple rendezvous and return mission, including orbital reconnaissance and characterization of the NEO, followed by a touchdown and grab sample return. Such a sample return will not only yield first-rate science, but also furnish the navigation and operations experience needed for the first human expedition to an NEO (expanding on the success of NEAR's orbital and landing operations).

8. Maintaining Capabilities for the Future

8.1. Challenges

We already know a great deal about the technical challenges facing a piloted mission to an NEO. Taking stock, we have already been to the Moon (30 years ago), which requires a greater Δv, and includes much riskier descent, landing, ascent, and rendezvous phases. ISS mission durations of 4-6 months are routine; NEO missions are of comparable length. The largest challenges are liable to be programmatic, e.g. regaining the capability to throw big masses out of LEO; funding a "critical mass" of space infrastructure; and committing to a long-term program of human exploration beyond LEO. There are technical challenges, of course: protecting the crew against radiation on long-duration flights outside the Earth's magnetosphere; developing better reliability and autonomy of spacecraft systems, given 10-second round-trip communication delays; and ensuring a robust abort and crew-return capability.

8.2. Benefits

There are two categories of benefits stemming from a piloted mission to an NEO. The first is scientific. NEO flights promise to add a totally new area of scientific return to our solar system exploration program (Davis et al. 1993, Jones et al. 1994). Missions to NEO will not only expand our space flight experience base, but will also reap a scientific bonanza. Putting a scientific team on an NEO surface for a month offers an outstanding opportunity to collect the full range of material on the NEO surface, addressing major questions about the materials in the solar nebula and the processes that led to terrestrial planet and asteroid formation. The expedition would be able to conduct detailed geophysical measurements of the NEO interior, including seismic, radar, and gravity surveys. Interior samples of the NEO would be obtained from craters and from coring experiments. The scale and data return of such investigations dwarf the capabilities of any robotic probe or series of probes. The clues to early solar system history held by NEOs are highly complementary to the "evolved planet" studies that are the focus of Mars research. Given the dearth of detailed knowledge about the composition and origin of NEOs, a human expedition to even one would produce ground-breaking results well in advance of the first Mars field expedition.

Included in the scientific return from an NEO mission are detailed assessments of the resource potential of such objects. Utilization of space resources from NEOs is the key to long-term economic activity in space (Lewis et al. 1993), and practical experiments on resource exploitation should be a priority for NEO surface expeditions.

The second category of benefits is technological. Reaching an NEO with a piloted mission will produce space-qualified hardware with direct application to both lunar and Mars exploration. Only enhancements to propulsion and life support capability must be added to turn a crewed NEO spacecraft into one capable of operating in Mars orbit. Thus, a human mission to an NEO is the logical precursor to the first piloted flight to Mars, just as the Apollo 8 and 10 lunar orbital missions prefaced the first lunar landing on Apollo 11 (Durda 2001).

In addition, development of an asteroid visit capability will give us much more confidence in the physical and chemical makeup of NEOs; this detailed characterization, possible only with an extended field expedition, will enable development of credible diversion or destruction technologies from any threatening objects that may yet be discovered.

9. Conclusion

Before ISS completion (scheduled for 2006), planning will begin on the scope and destinations for human expeditions beyond LEO. Our goals for scientific exploration of the solar system should be an integral part of planning to send humans to Earth's nearest planetary neighbors. A very feasible, and attractive direction for this next human step into the solar system is toward the family of small NEOs that offer favorable (low energy) opportunities for fast round-trip visits. Such missions provide scientist/explorers direct access to materials and small bodies that date from the earliest era of solar system formation, giving a strong scientific rationale and set of goals for human explorers on these varied surfaces.

We recognize that decisions on future human space activity will be driven by more than scientific considerations alone. However, hard-won knowledge from our scientific exploration of the solar system, and the potential for future discoveries, are critical to choosing the next step in that activity. We have identified a family of targets which allow for intermediate destinations (in both energy and time) between LEO and Mars or (in energy) the surface of the Moon. By maximizing the synergy between science goals and other programmatic, operational, and technological factors in human exploration, we ensure the maximum benefit to all.

10. Recommendations

- We strongly endorse the Large Synoptic Survey Telescope as an effective means of increasing both the known numbers and the degree of characterization of our target population. This discovery effort must include adequate funding for conducting an asteroid search and characterization program capable of identifying and extracting observations of our target population. Follow-up observations should be supported by a dedicated telescope in the 3- to 4-m class.

- NASA's Code M should direct its Exploration Office to include NEO mission capability in its studies of human activities beyond low Earth orbit, and direct a study of a family of deep space mission scenarios that include round-trip missions to NEOs, among other destinations in Earth-Moon space. This effort should be conducted by a joint Code M/Code S working group, tasked to evaluate target discovery, orbit determination and characterization efforts, to define precursor robotic missions, and to identify priority operations and science goals for human crews operating on and near NEOs.

- We endorse continued support for radar orbit determination of NEOs as an essential precursor activity for robotic and manned missions.

References

Bottke, W.F., R. Jedicke, A. Morbidelli, J-M. Petit, B. Gladman 2000. Understanding the distribution of Near-Earth asteroids. *Science* **288**, 2190-2194.

Bottke, W.F., A. Morbidelli, R. Jedicke, J-M. Petit, and H. Levison 2001. Debiased orbital and size distributions of the NEOs. *Bull. Am. Astr. Soc.* **32**, 705.

Bowell, E., B. Hapke, D. Domingue, K. Lumme, J. Peltoniemi, and A.W. Harris 1989. Application of photometric models to asteroids. In *Asteroids II* (R. Binzel, T. Gehrels, and M. Matthews, Eds.), pp. 524-556. Univ. of Arizona Press, Tucson.

Davis, D.R., A.L. Friedlander, and T.D. Jones. 1993. The Role of Near-Earth asteroids in the Space Exploration Initiative. In *Resources of Near-Earth Space* (J. Lewis, M.S. Matthews, and M.L. Guerrieri, Eds.), pp. 619-655. Univ. of Arizona Press, Tucson.

Durda, D.D. 2001. Stepping stones to Mars. *Astronomy* **29**, 44-49.

Farquhar, R. 2000. The Role of the Sun-Earth Collinear Libration Points in Future Space Exploration. *Space Times* **39**, 9-12.

Harris, A.W. 1998. A thermal model for Near-Earth asteroids. *Icarus* **131**, 291-301.

Jones, T.D., D.B. Eppler, D.R. Davis, A.L. Friedlander, J. McAdams, and S. Krikalev. 1994. Human exploration of Near-Earth asteroids. In *Hazards Due to Comets and Asteroids* (T. Gehrels, Ed.), pp. 683-708. Univ. of Arizona Press, Tucson.

Lewis, J.S., M.S. Matthews, and M.L. Guerrieri, Eds. 1993. *Resources of Near-Earth Space.* Univ. of Arizona Press, Tucson.

Rabinowotz, D.L., E. Helin, K. Lawrence, and S. Pravdo 2000. A reduced estimate of the number of kilometre-sized Near-Earth asteroids. *Nature* **403**, 165-166.

Yeomans, D.K., S.J. Ostro, and P.W. Chodas 1987. Radar astrometry of near-earth asteroids. *Astron. J.* **94**, 189-200.

The Future of Solar System Exploration, 2003-2013
ASP Conference Series, Vol. 272, 2002
Mark V. Sykes, ed.

Mars: Its Place in Solar System Exploration

J.M. Moore[1], L.A. Leshin[2], D.D. Bogard[3], C.K. Shearer[4], N.G. Barlow[5], A.F.C. Haldemann[6], R.J. Wilson[7], B.P. Weiss[8], A.H. Treiman[9], R.N. Clark[10], & S.M. Clifford[9]

[1] *NASA Ames Research Center, Moffett Field, CA*
[2] *Department of Geological Sciences, Arizona State University, Tempe, AZ*
[3] *NASA Johnson Space Center, Houston, TX*
[4] *Department of Earth and Planetary Sciences, Institute of Meteoritics, University of New Mexico, Albuquerque, NM*
[5] *Department of Physics, University of Central Florida, Orlando, FL*
[6] *Jet Propulsion Laboratory, California Institute of Technology, Pasadena, CA*
[7] *Geophysical Fluid Dynamics Laboratory, National Oceanic & Atmospheric Administration, Princeton, NJ*
[8] *Division of Geological and Planetary Sciences, California Institute of Technology, Pasadena, CA*
[9] *Lunar and Planetary Institute, Houston, TX*
[10] *US Geological Survey, Denver, CO*

Abstract. We concur with the NRC Committee on Planetary Exploration that sample return should be the primary goal of the Mars Exploration Program (MEP) and that several missions should be flown. However, we have become convinced that to execute sample return, there needs to be significant investment in technology development either implicitly (e.g., Smart Lander) or explicitly as a line item in the MEP budget. We also strongly recomment that Scout missions should be given their own "sheltered" funding line. A minority of us note that it would be very desirable to place a Synthetic Aperture (Imaging) Radar in orbit about Mars.

REPORT

1. Current State of Knowledge

Mars studies have resulted in a vast wealth of knowledge that cannot be meaningfully summarized in a few paragraphs. The forthcoming NRC Committee on Planetary Exploration (COMPLEX) report "Assessment of Mars Science and Mission Priorities" has several chapters that thoroughly cover the present state of knowledge regarding Mars. Therefore we believe that it is unnecessary to repeat that material here.

2. Key Science Questions

We concur with COMPLEX (which served as the NRC Mars Discipline Panel for the decadal study) that the key science questions regarding Mars derive from the overarching science-driving theme: "What is the evolution of an Earth-like planet" and the sub-themes:

- What is the potential of Mars as a past or present abode of life?
- What is the history of climate and water on Mars?
- What is the structure and evolution of the interior of Mars ?

3. Recommendations

- We concur with COMPLEX that Sample Return is the primary goal of the Mars Exploration Program (MEP). For instance prioritization and baseline capabilities of Mars Reconnaissance Orbiter (MRO) and the "Smart Lander" are most easily justifiable if seen as essential, partially technology development-driven, prerequisites to Sample Return.

- We concur with COMPLEX that several (very bare minimum of 3) Sample Return missions must be flown to successfully address the science objectives of that activity. Though a minority of us argues for a single MEP-culminating Sample Return mission on programmatic grounds (see Appendix A).

- The Community Panel's discussion with Mars Program Systems Engineering Team (MPSET) convinced us that significant technology development is necessary to execute Sample Return and significant investment in technology development would have to come implicitly (e.g., Smart Lander) or explicitly as a line item in the MEP budget. Technology development alone pushes Sample Return missions into the next decade.

- We strongly recommend that Scout Missions should be given their own "sheltered" funding line. Otherwise it will be very difficult to effectively address significant aspects of the Sub-Themes with a program exclusively focused on Sample Return

- A minority of us note that it world be very desirable to place a Synthetic Aperture (Imaging) Radar in orbit about Mars. Such a mission would probably be difficult to place in a Scout Mission format and would probably have to be treated as a higher cost mission. Such an Imaging Radar mission could potentially reveal extent and character of shallowly buried morphologies indicative of a wetter and warmer early Mars, and the integrated relationship of these putative morphologies to each other.

4. The Place of the Mars Program in the Context of Missions to Other Planetary Objects

We note that presently approximately half of all SSE funding for planetary missions goes to Mars, and that this funding proportionality is currently planned to continue though this decade.

The Mars Community panel believes this level is appropriate and justifiable when measured against the overarching themes of planetary exploration: (1) What is the origin and what factors contribute to the initial configuration of the Solar System (Origins); (2) How far and in what forms does organic chemistry advance toward life, and what is the frequency, duration, and level of complexity of life in the Solar System (Life); and (3) What processes have operated to shape the evolution of the worlds of the Solar System throughout their histories up to their present configurations (Evolution)?

The Mars Community Panel concludes that the Scientific Goals of the MEP (including the Scout Program) bear directly and potentially most fruitfully on the "Life" theme, which is considered by both the Scientific Community and the General Public as perhaps the most interesting, thus important. But also the Martian NAS theme of "What is the evolution of an Earth-like planet" places MEP goals as a major component of the "Evolution" theme. Again the evolution of an Earth-like planet is viewed by both the Scientific community and the General Public as perhaps the most interesting, thus important.

Appendix A

Reservations About the Scientific and Programmatic Impact of Sample Return Missions on the Mars Exploration Program

Steve Clifford

No one questions the desirability of studying samples from Mars, the merits of which are discussed at some length in this report. However, there is substantial controversy within the Mars community regarding the relative priority, number, and programmatic impact of sample return missions that is underrepresented in this report's summary conclusions.

With an estimated price tag of $2 billion dollars per mission (latest JPL estimate), a sample return mission costs the same as sending 4-6 orbiters or 2-3 advanced landers/rovers to Mars. For this reason, in the context of the present Exploration Program budget, committing to even a single sample return mission places a considerable strain on the resources and opportunities available to pursue other first-order scientific investigations. While the potential return from such a mission appears sufficient to warrant the inclusion of a single, well-targeted sample return effort sometime during the next 20 years, the decision to pursue 3 such missions in quick succession (on a time scale any shorter than the next 35-40 years) would have a devastating financial and scientific impact on our ability to conduct a well-rounded Mars exploration program.

While the financial stress the program would experience from a commitment to multiple sample return missions appears obvious, the scientific impact has been less well discussed. The single, highest rated (by a considerable margin) goal of the Mars Exploration Program, as identified by MEPAG, is to identify the present 3-dimensional distribution and state of water on Mars. Approximately 5-10% of the planet's present inventory of H_2O is thought to reside in the polar caps, while the remainder is thought to reside in the subsurface as groundwater and ground ice. Although the analysis of returned samples might well provide information relevant to the past occurrence of water at a single location (or even the amount of water that may now reside in a chemically-bound state within the crust), it would contribute virtually nothing to our understanding of where the vast bulk of the planet's unbound inventory resides today.

Even after 35 years of robotic exploration, there are still many first-order investigations about the global character of the planet that have not been conducted – chief among these being meteorological and geophysical network investigations, which promise to dramatically improve our understanding of the Martian atmosphere, climate, and internal structure. Until such critical elements of our initial global reconnaissance of Mars are complete, the commitment of the exploration program's limited resources to support multiple, consecutive sample returns missions appear to be scientifically and fiscally ill advised.

The Future of Solar System Exploration, 2003-2013
ASP Conference Series, Vol. 272, 2002
Mark V. Sykes, ed.

Exploring Main Belt Asteroids

M.V. Sykes[1], E. Asphaug[2], J.F. Bell[3], R.P. Binzel[4], W. Bottke[5], S.J. Bus[6], A. Cellino[7], P. Clark[8], D.R. Davis[9], M.C. De Sanctis[10], D.D. Durda[5], J. Emery[12], R.A. Fevig[12], U. Fink[12], J. Granahan[13], A.W. Harris[14], W.K. Hartmann[9], R. Jedicke[12], M. Kelley[15], S.M. Larson[12], D.J. Lien[16], C. Magri[17], S.J. Ostro[14], K.L. Reed[18], A.S. Rivkin[4], D.W.G. Sears[19], A. Storrs[20], D.J. Tholen[21], R. Walker[22], R. Whiteley[12], & H. Yano[23]

[1] *Steward Observatory, University of Arizona, Tucson, AZ*
[2] *Earth Sciences Department, University of California, Santa Cruz, CA*
[3] *Hawaii Institute of Geophysics and Planetology, University of Hawaii, Honolulu, HI*
[4] *Department of Earth, Atmospheric, and Planetary Sciences, Massachusetts Institute of Technology, Cambridge, MA*
[5] *Southwest Research Institute, Boulder, CA*
[6] *Institute for Astronomy, University of Hawaii, Hilo, HI*
[7] *Osservatorio Astronomico di Torino, Pino Torinese, Italy*
[8] *NASA Goddard Space Flight Center, Greenbelt, MD*
[9] *Planetary Science Institute, Tucson, AZ*
[10] *Istituto di Astrofisica Spaziale e Fisica Cosmica, CNR, Rome, Italy*
[11] *Southwest Research Institute, Boulder, CO*
[12] *Lunar and Planetary Laboratory, University of Arizona, Tucson, AZ*
[13] *BAE SYSTEMS Missions Solutions, San Diego, CA*
[14] *Jet Propulsion Laboratory, California Institute of Technology, Pasadena, CA*
[15] *NASA Johnson Space Center, Houston, TX*
[16] *Department Of Physics, Oklahoma State University, Stillwater, OK*
[17] *Department of Natural Sciences, University of Maine, Farmington, ME*
[18] *Raytheon Corporation, Goleta, CA*
[19] *Arkansas-Oklahoma Center for Space and Planetary Sciences, University of Arkansas, Fayetteville, AR*
[20] *Department of Physics, Astronomy, and Geosciences, Towson University, Baltimore, MD*
[21] *Institute for Astronomy, University of Hawaii, Honolulu, HI*
[22] *Vanguard Research, Scotts Valley, CA*
[23] *Planetary Science Division, Institute of Space and Astronautical Science, Kanagawa, Japan*

Abstract. The asteroid belt is filled with fragments of disrupted protoplanets, of which only a few survive. It has been dynamically stirred and suffered substantial mass loss. Samples exist on the Earth as meteorites. Putting these pieces back together provides a window into an early chapter of solar system history, during and after its formation, capturing the earliest epoch of planet formation. Groundbased observations, laboratory analyses of meteorites, and theoretical studies, in addition to flyby and rendezvous missions are all required to progress in this area.

EXECUTIVE SUMMARY

The asteroid belt is a site of interrupted terrestrial planet formation and of complex collisional and dynamical activity. Our investigations of this region of the solar system can be reduced to three key questions:

A. What was the compositional gradient of the asteroid belt at the time of initial protoplanetary accretion?

   Addressing this question requires an understanding of the population and compositional structure of the main asteroid belt today, how surface modification processes affect our ability to determine this structure, how dynamics and collisions modify this structure over time, as well as the physical properties of asteroids.

B. What fragments originated from the same primordial parent bodies, and what was the original distribution of those parent bodies?

   This requires an understanding of what asteroid fragments are associated dynamically, and what objects are geochemically linked (suggesting a common origin).

C. What are the early steps in planet formation and evolution?

   This problem requires knowledge of the compositions and structures of the surviving protoplanets. The cratering records on their ancient surfaces combined with the cratering record of younger surfaces provide insights into the primordial size distribution of objects in the asteroid belt. In addition, available hand samples - meteorites - provide detailed information about the formation and evolution processes of their parent bodies, though we need to improve our understanding of the extent to which meteorites sample the asteroid belt.

These questions cannot be addressed by a single spacecraft mission. Asteroids are characterized by their diversity. Consequently, the most important priority for the next decade is healthy funding support for NASA Research and Analysis programs supporting groundbased observations, meteorite studies, and theory programs. In particular, there should be support for studies using modest aperture telescopes. Asteroid missions addressing the key science questions cannot be planned nor provide a maximum return without a credible and continuing investment in earthbased remote observation programs, meteorite studies, and theory programs. These programs have suffered at NASA in recent years as funding has shrunk, particularly for physical studies of main belt asteroids.

Our recommendations for more expensive facilities/missions include the Large Synoptic Survey Telescope (designed and operated in a manner allowing for the detection of all small mainbelt asteroids), a rendezvous mission to Vesta or Vesta and Ceres, and a Solar System Infrared Survey mission (to survey all mainbelt asteroids having diameters > 1 km). In parallel, a modest technology development program should be supported to investigate the potential for large number of multiple asteroid flybys using microsatellites and possible advanced propulsion systems such as solar sails that would allow for multiple flybys per spacecraft.

REPORT

1. Perspective

Planet formation in our solar system followed a variety of paths, leading in most cases to finished planets: rocky bodies in the inner solar system, gas giants in the middle region, and icy planets in the outer solar system. However, the planet-forming process was interrupted in two regions - the Kuiper Belt (with the exception of Pluto) and the asteroid belt - leaving a vast population of small bodies instead of a single planet or planets. Accretion formed numerous protoplanets in the region between Mars and Jupiter, certainly Ceres-sized bodies formed, and perhaps even Mars-sized bodies. These large bodies would have gravitationally perturbed the local population, increasing collision speeds, and terminating the planetary formation process. Fragmentation and disruption commenced. Jupiter was also forming, accreting (in addition to rocky material) great amounts of ices and gases that were unable to condense in the warmer region of the inner solar system. Jupiter's large mass and orbital location created strong resonances in the asteroid belt that quickly removed any body unlucky enough to move into those resonances. This depletion mechanism removed most of the primordial mass in the asteroid belt and finally cut off the material needed there for further planetary growth.

Today, main belt asteroids (Figure 1) consist of a few unshattered protoplanets (e.g., Ceres, Pallas, Vesta) and the remains of smaller, disrupted primordial bodies whose pieces were shattered and dispersed, then further shattered and dispersed leaving the hundreds of thousands of fragments that dominate the main asteroid belt today (e.g., Gaspra, Ida) as well as feed into the Near-Earth asteroid population (e.g., Eros). While the boundary between the two populations may have some fuzziness (containing perhaps small shattered protoplanets that were not dispersed), the smaller of these fragments (d<50km) tend to be relatively homogeneous, while the surviving protoplanets are more complex (e.g., preserving basaltic eruptions on Vesta and the geological context of ancient water mobilization on Ceres).

The pot has been stirred dynamically and collisionally over the age of the solar system, yet it retains evidence of initial conditions and planet formation

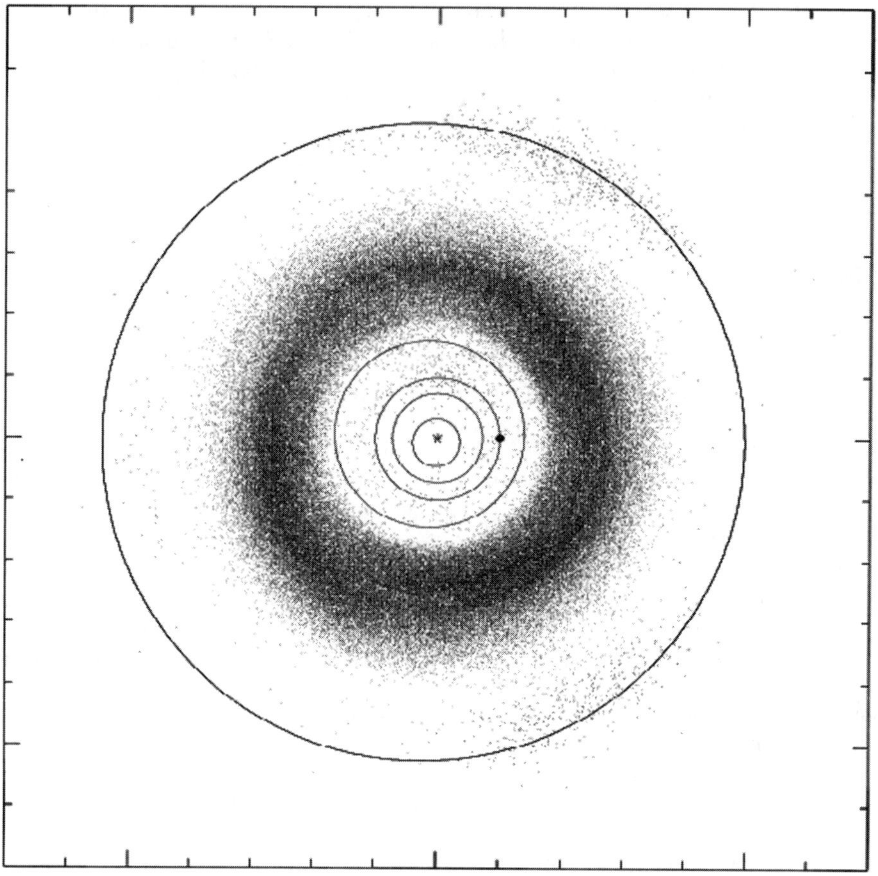

Figure 1. Instantaneous positions of more than 125,000 known asteroids are shown. The orbits of Mercury, Venus, Earth, Mars, and Jupiter are shown for scale. The main asteroid belt resides between Mars and Jupiter. Asteroids are gravitationally perturbed into orbits interior to Mars and the Trojan clouds are observed at the Jovian Lagrange points.

and evolution. Some asteroids have shifted into Earth-crossing orbits, causing concerns about the hazard of impact and interest in their potential as resources for human activity in space. The primordial outer main belt may have been the source of Earth's water (Morbidelli et al. 2000), thus critical to the existence of life on Earth. Destructive asteroid collisions have also been a source of interplanetary dust, generating much of our zodiacal cloud. The detection of non-primordial dust disks around other stars suggests that asteroids and the collisional processes we observe in our solar system occurs elsewhere in the universe.

2. Key Science Questions

There is a hierarchy of questions that motivate almost all of asteroid science which are rooted in our knowledge of the present. In earlier decades asteroid science focused primarily on the gathering of basic data from which derives our modern understanding of the processes operating in the asteroid belt at present and in the past. This is a never-ending process. Our continuously expanding capabilities to gather a wider range of more detailed data over a larger number of objects allows for more sophisticated, detailed, and overarching models, as well as more challenges to them. Asteroid science is dynamic.

On the basis of what we know today, the asteroid belt is a site of interrupted terrestrial planet formation and complex collisional and dynamical activity. There are three fundamental questions that we consider to be key. They are listed below along with the next level of questions that derive from them:

A. What was the compositional gradient of the asteroid belt at the time of initial protoplanetary accretion?

   - What is the population and compositional structure of the main asteroid belt today?
   - How do surface modification processes affect our ability to determine this structure?
   - How do dynamics and collisions modify this structure over time?
   - What are the physical properties of asteroids?

B. What fragments originated from the same primordial parent bodies, and what was the original distribution of those parent bodies?

   - What asteroid fragments are associated dynamically, suggesting a common origin?
   - What objects are geochemically linked?

C. What are the early steps in planet formation and evolution?

   - What are the compositions and structures of surviving protoplanets?
   - What does their cratering record combined with the cratering record of younger surfaces reveal about the primordial size distribution of objects in the asteroid belt?
   - What do meteorites tell us about formation and evolution processes of these bodies? How well do they sample the asteroid belt?

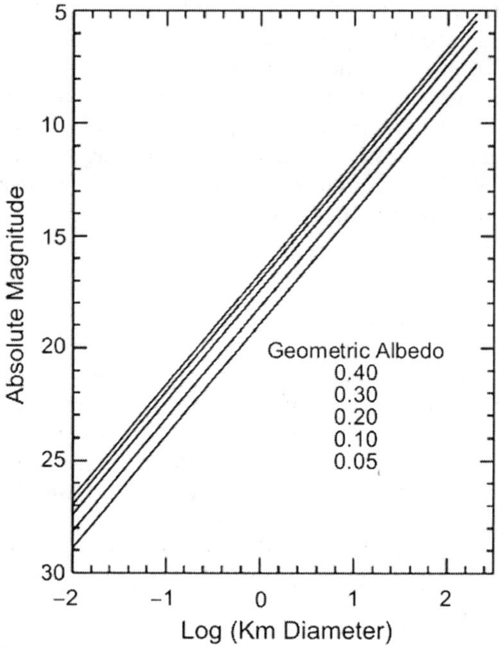

Figure 2. Relating absolute magnitude to diameter and geometric albedo.

2.1. How Can the Science Questions be Addressed?

Discovering Asteroids Basic to main belt asteroid studies is the ongoing discovery of more and more asteroids. The discrete material in the main asteroid belt extends from objects many hundreds of kilometers in radius down to the size of dust particles. Mass is continuously being redistributed from larger to smaller sizes through collisions, then ultimately lost via resonances or radiation forces. Because of their much greater numbers, smaller asteroids provide better tracers of dynamical and collisional processes. Smaller asteroids also differ in their distribution of compositions compared to large asteroids (e.g., Tedesco et al. 1989). In general, asteroids of all sizes are characterized as much by their compositional and physical diversity as their similarities, addressing our key science questions requires the study of objects from the largest (surviving protoplanets) to fragmented rubble piles to objects small enough to have retained physical coherence (tens to possibly hundreds of meters - Pravec and Harris 2000).

Current programs designed to discover Near-Earth objects (e.g., LINEAR) are discovering many main belt asteroids in the process (more than 100 times more). Such programs covering large areas of sky trade off sensitivity (V 19.5) for area. This significantly limits the population of small main belt asteroids that can be discovered. The brightness of an asteroid is a function of its size and albedo (Figure 2).

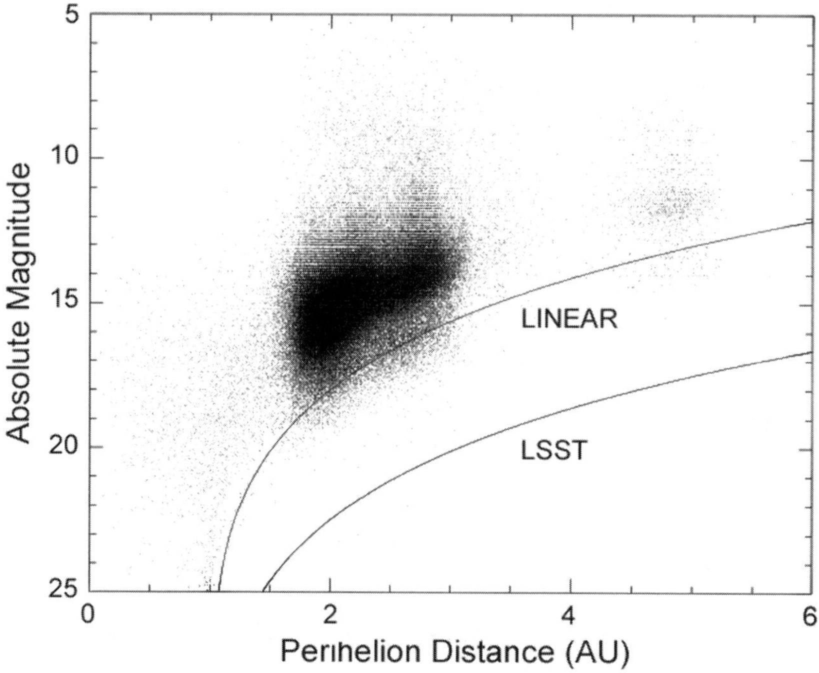

Figure 3. The sensitivity of LINEAR and LSST for discovering asteroids.

The proposed Large Synoptic Survey Telescope (LSST) covers the entire available sky to V=24 in one week. The sensitivity of an LSST survey is compared to LINEAR in Figure 3. The smallest asteroids are most likely to be discovered when they are brightest - at perihelion - as demonstrated in the plot. Most main belt asteroids for which there are orbits were discovered by LINEAR. Through the main belt, LSST would be far more sensitive than LINEAR, detecting asteroids having diameters between 125 and 250 meters. In the Trojan population, asteroids having diameters of 2 km would be detected.

Albedos and Diameters (Addressing A and B) Diameter is a fundamental physical characteristic of all asteroids. Albedo is indicative of bulk composition (e.g., dark carbonaceous, or bright and rocky). Together they determine the absolute magnitude of an asteroid (Figure 2). They are traditionally obtained through occultation studies and radiometric observations, as well as via polarimetry. Occultations require accurate predictions of when the line of sight to a background star is intercepted by an asteroid. These observations are useful because they are the most accurate measure of a specific dimension along the occultation path and they offer an opportunity to acquire the projected shape of an asteroid normal to the line of sight if a number of appropriately separated chords can be obtained. These are target of opportunity measurements. Once a diameter

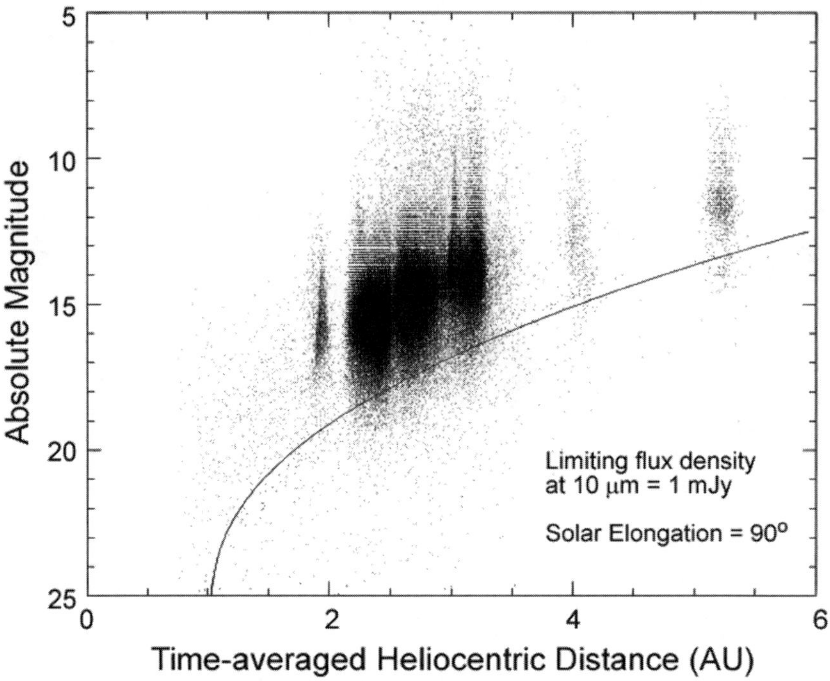

Figure 4. Detecting main belt asteroids at 10 μm using a polar orbiting space-based system at a limit of 1 mJy.

is obtained, geometric albedo can also be determined from the absolute magnitude of the asteroid. Polarimetric studies require a number of observations spread over different phase angles in order to determine albedo from the polarization phase curve and then diameter (as above) (e.g., Dollfus and Zellner, 1979). Radiometry combined with visual magnitude provide perhaps the most efficient means of determining albedo and diameter for an asteroid, requiring only a single observation of the former and knowledge of the latter at that time (e.g., Lebofsky and Spencer 1989). The difficulty of making such observations from the ground is that one must remove the significant contribution to the thermal flux of the atmosphere and telescope itself. Making such observations from space allows all such sources to be removed. The Infrared Astronomical Satellite made the largest number of asteroid radiometric observations in 1983, providing diameters and albedos for 1884 of 7309 numbered or multi-apparition asteroids (Tedesco et al. 1992). Today the number of available asteroid orbits has increased by more than an order of magnitude and the increased sensitivity of thermal detectors make most of these detectable. A thermal detector with SIRTF-like sensitivity (1 mJy at 10 μm) in an IRAS-like polar orbit and scanning the celestial sphere with a 0.5 degree FOV at a constant solar elongation of 90 degrees as the Earth orbits the Sun, would after one year would have scanned 96% of main belt asteroids and detect almost all asteroids for whom orbits have

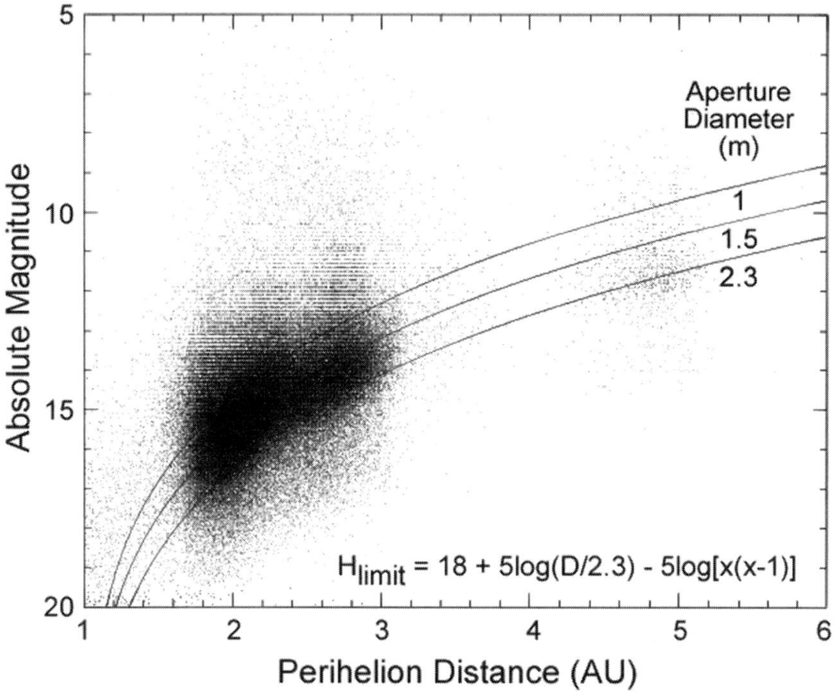

Figure 5. An assessment of modest aperture telescopes for doing lightcurve and CCD spectral studies of asteroids at opposition. For lightcurve studies, SNR=50 in ten minutes is required, for spectral studies, SNR=20 in one hour is required.

been determined (Figure 4), increasing the IRAS database by more than an order of magnitude and providing virtually complete coverage down to km diameters throughout the main asteroid belt and down to around 15 km in the Trojan population. If a visible wavelength channel were added to a spacebased thermal system, relative motions compared to the scan pattern would allow for simultaneous, if partial, thermal and visible lightcurve studies from which would be derivable other physical properties such as thermal inertia. Another byproduct might be detection of debris trails associated with asteroids, which would be directly related to modern collisional activity and dust production from crater ejecta.

Lightcurves (Addressing A and B) Asteroid lightcurves give insight into collision processes by providing information on shape and rotation state. As a consequence of apparent motions, lightcurves for asteroids must be obtained one at a time. The rate at which such information can be acquired is self-limiting. There are relatively few large-aperture and spacebased facilities that may provide the highest signal-to-noise data or data on the faintest objects -

and there is fierce competition for those resources. Acquiring lightcurves for large numbers of objects in a reasonable period of time requires the devotion of a number of facilities operating in parallel. As with asteroids, the smaller the telescope the greater their number.

Even modest aperture telescopes are capable of doing lightcurve studies of tens of thousands of asteroids (Figure 5). This suggests an opportunity for small institutions to engage in such studies, perhaps leveraging basic science education opportunities for students while contributing to very basic knowledge about asteroids.

Asteroids having a diameter greater than 0.5 km are seen to have rotation periods greater than two hours (Pravec and Harris 2000), hence a ten-minute sampling frequency should provide sufficient points to get a baseline lightcurve. However, several small asteroids tens of meters in diameter have been discovered having periods ~10 minutes. In these cases, their intrinsic faintness exclude their observation in the main asteroid belt even by existing large aperture facilities, requiring their study in the closer Near-Earth asteroid population by those facilities.

Spectroscopy and Colorimetry (Addressing A and B) At present, our ability to reconstruct the compositional gradient of the primordial asteroid belt is limited. We have only the coarsest models of the compositional structure of the main asteroid belt today. The 2000 or so asteroid spectra that exist in the literature in 2001 are very inhomogeneous in orbital element space, tending to focus on specific regions. There is significant compositional diversity in the asteroid belt, ranging from primitive carbonaceous objects dominating the outer belt and Trojans to basaltic and metallic fragments in the inner region of the main belt. The "dew" line across which water becomes stable transects the outer main belt (around 3 AU). The composition of all these fragments suggest a significant primordial gradient, an early heating event affecting some but not all asteroids, processes involving differentiation and volcanism, processes involving water mobilization, and significant dynamical processes. For the richness of the key science to be addressed in the main asteroid belt, we are extremely data poor.

As with lightcurves, asteroid spectra must be obtained one at a time. The largest apertures may be utilized to obtained spectral information on a relatively small number of fainter objects of interest, but as with lightcurve studies, modest aperture telescopes are shown in Figure 5, to be capable of acquiring data on tens of thousands of asteroids.

Broadband filter observations allow for more colorimetry obtained much more quickly than CCD spectra - spectral resolution and degraded compositional information is sacrificed for greater signal to noise. However, such information can play a useful role in that it allows for large numbers of asteroid colors to be obtained in a survey, such as the Sloan Digital Sky Survey and (in the near-infrared) the Two-Micron All Sky Survey. This allows for a coarse but relatively unbiased sampling of the compositional structure of the asteroid belt to the magnitude limits of the given survey. Unfortunately, existing such surveys have filters selected for addressing astrophysical issues and are not optimized for asteroidal research, thereby limiting their potential usefulness. Furthermore, lightcurve variations as mentioned above can potentially complicate the trans-

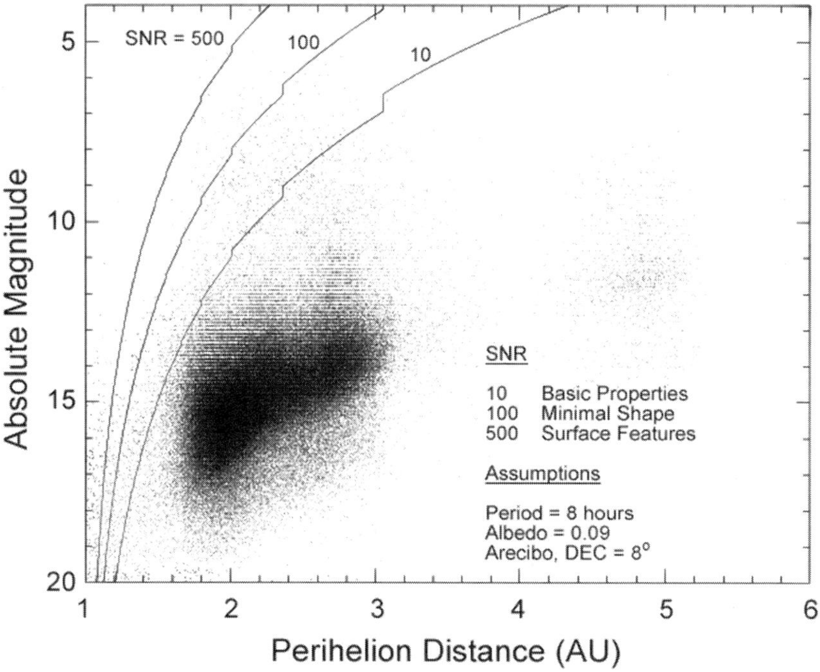

Figure 6. Radar preferentially detects nearby and large asteroids, just sampling the main asteroid belt.

formation of serendipitous but nonsimultaneous color data to taxonomic types. Nevertheless, the use of colorimetry in classifying asteroids is a long-established practice, and the major asteroid classes separate relatively well in both UBV and JHK color diagrams (Zellner et al. 1975, Bowell et al. 1978, Hahn and Lagerkvist 1986).

Relating observations to compositions requires an understanding of how spectra relate to specific mineralogies and how surface scattering properties effect that relationship. In addition, the effects of surface impacts and space weathering over time must be understood. This requires laboratory studies of meteorites and terrestrial analogs of other interplanetary materials.

Radar Studies (Addressing A and B) Radar can provide information on surface properties, reflectivity, surface roughness, spin state, and morphology. It can help to distinguish between degenerate spectral M type asteroids, some of which are metallic and others composed of some hydrated, possibly carbonaceous material.

The recent upgrades at Arecibo allow for the use of this technique to improve our knowledge of main belt asteroids. Figure 6 shows that radar can probe the larger asteroids of the main belt and Figure 7 shows that these targets are

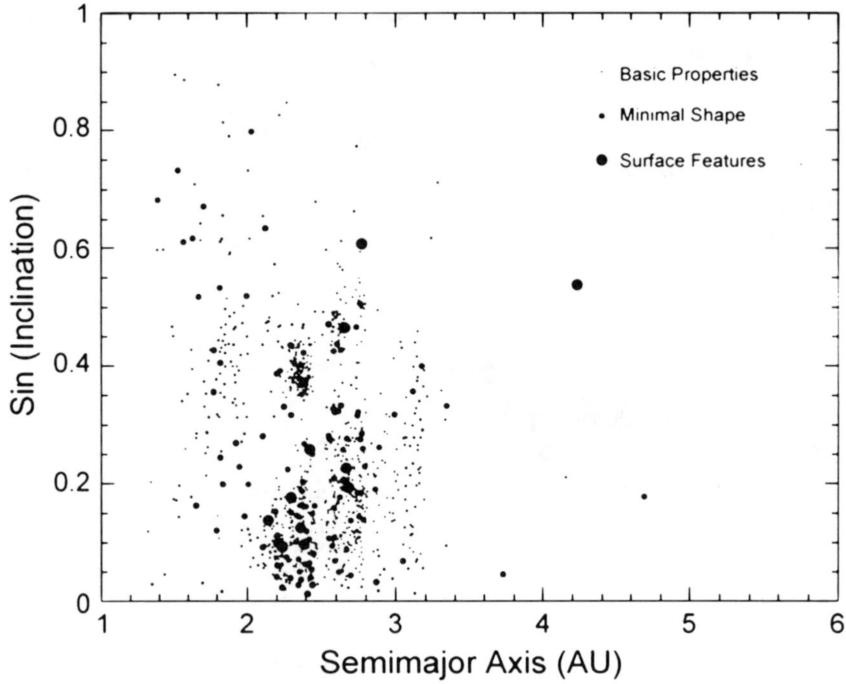

Figure 7. Radar is able to detect asteroids in all dynamical regions of the main belt.

distributed throughout the major regions of the main belt, though not as far as the Trojans.

Dynamical Studies (Addressing A and B) Even given detailed information on the modern compositional structure of the asteroid belt, inferring its primordial structure requires an understanding of the various dynamical processes acting on asteroid orbits, such as gravitational perturbations, chaotic scattering, Yarkovsky effects, collisions, etc. It also requires an understanding of collision mechanics and the preferential erosion of some compositions with their associated physical properties from the modern population of objects.

Such studies also serve to identify asteroids having discernable dynamical relationships (families), targeting them for followup physical and spectral studies to determine whether a common origin makes cosmochemical sense.

Sometimes the identification of asteroids that seem to be spectrally related, such as the Vestoids which lead from Vesta to resonance regions allowing for transport of such material into Earth-crossing orbits and explaining the HED meteorites, calls for more dynamical studies to understand how in detail this actually occurred and if that interpretation of apparent transport is correct.

High Resolution Observations (Addressing A, B, and C) Adaptive optics (AO) and speckle interferometry are now able to discover binary asteroid systems and directly measure asteroid sizes and shapes (at least for the large end of the size distribution. This allows for the determination of mass (from the binary orbit) and density (from the size and shape information combined with mass).

Collision Mechanics (Addressing A and B) Understanding how collisions affect the internal structure and physical properties of asteroids over time requires sustained support for experimental and theoretical work. Challenges for the near future include:

- A better understanding of the response of low-density, high-porosity objects to cratering and catastrophic impacts. NEAR/Shoemaker images of several extremely large craters on 253 Mathilde and the existence of asteroid satellites predominantly around primitive asteroids (i.e., dark taxonomic types like C, D, etc.) indicate the importance of understanding the response of such bodies to collisions. We believe laboratory experiments and hydrocode modeling involving impacts into low-density, high-porosity targets should be given high priority in this area.

- A merger of collisional and dynamical models allowing simultaneous treatment of both mechanisms on asteroid populations. To understand the cratering history of individual objects (e.g., Eros) as well as the collisional evolution of the main belt population, we need to create increasingly sophisticated models that include physical mechanisms like collisions, Poynting-Robertson drag, Yarkovsky thermal drag, resonances, and other related phenomena.

- A link between accretional, dynamical, and collision models in order to produce a seamless, self-consistent model of main belt evolution. The current main belt is but a tiny remnant of the primordial main belt, which may have contained several Earth masses of material and possibly multiple Vesta-like bodies. The sudden disappearance of these bodies, the existence of unusual meteorites like pallesites and mesosiderites, and the overall orbital structure of the main belt are all intimately tied to collisional and dynamical events which took place in the first few tens of Myr of solar system history. Until we understand this epoch better, we will be unable to set the context for interpreting modern-day asteroids.

Meteorite Studies (Addressing A, B, and C) Samples of asteroids are present on the Earth in the form of (the non-lunar/non-Martian) meteorites. Both the absolute number of meteorites and the number of different types of meteorites have grown in recent years as meteorite concentration areas like Antarctic glaciers and deflation surfaces in deserts have been systematically searched by groups like ANSMET. Through study of meteorites, knowledge of asteroidal properties such as compositions, density/porosity, and thermal and aqueous alteration histories can be either directly measured or inferred in a straightforward manner. However, the only current way of linking specific asteroids or asteroid classes with specific meteorites or meteorite classes is through the interpretation and comparison of meteorite and asteroid spectral properties.

Leveraging composition from spectra may be limited when the bulk composition of an asteroid are not optically active minerals or their spectra is masked (e.g., by low-albedo carbonaceous material). Furthermore, regolith processes may change asteroid spectra so that it no longer has a close resemblance to its best meteorite analog. Unfortunately, not all asteroid compositions are thought to be represented in meteorite collections due to biases introduced at every stage of meteorite delivery from impact and fragmentation to survival through the Earth's atmosphere to likelihood of being discovered on the ground. This situation might be improved with samples from dozens of targetted asteroids, which may be a goal beyond this next decade.

The association of specific asteroids or asteroid families with specific meteorite groups remains a high priority for asteroid spectroscopists. This will allow those asteroids to be analyzed as though they were the targets of a sample return, and allow a large store of meteoritical data to be applied directly to asteroidal problems. While great progress has been made in the last decade toward this goal (Gaffey and Gilbert 1998), new findings pertaining to temperature effects on mineral spectra and spectral maturity ("space weathering") effects need to be fully understood and considered.

The Role of Flyby Missions (Addressing A, B, and C) Flybys provide information on shape, density, compositional heterogeneity, and cratering records. Flybys of many members of a particular family would reveal cratering records that could be used to determine relative ages of family members, identifying multi-generational collisional fragmentation histories. Flybys can also help determine whether densities obtained are primordial or arise from collision processes. Similarly, flybys of protoplanetary survivors would reveal cratering records extending back to the earliest epoch of the solar system. Unfortunately, such observations are primarily target of opportunities for spacecraft having other principal targets (e.g., Jupiter for Galileo), and even in the rich target environment of the main belt even a spacecraft dedicated to asteroid flybys would make few observations per year. It is also a limited science return for the expense of the kinds of missions (e.g., Discovery class) that are being sent to or through the main belt, even considering the cost of a launch vehicle (>\$40M for Delta class launcher).

Technology innovation may increase the role for flyby missions by making them substantially cheaper. The "Cube-sat" program is currently building 1 kg satellites of dimension 10 cm^3 having a power of several Watts. The cost of such a satellite is approximately US\$50,000. Two are currently being built for launch: Ffize, which has 6 retroreflectors on board to be used for laser ranging and a test of the theory of relativity, and Rincon, a fully functional but miniaturized standard satellite, having solar cells, a power and battery charge regulator board, a microprocessor board, a transmitter board and a receiver board. They are to be launched from a Russian SS18 rocket in May of 2002.

These Cube-sats are designed and built by teams of faculty mentors and students under the University of Arizona student satellite program. The next generation satellites will study the effects of radiation on electronic components (supported by Alcatel), the development of micro-reaction wheels (supported by Honeywell), a low power 1Kx1K CMOS camera, a small CCD spectrometer with a mass of about 150g a spectral range from 200-1000 nm and a resolution of about 1nm, and a more powerful microprocessor capable of handling and processing spectra and images (all University of Arizona student developments). They are also looking into a joint industry development of ion engine propulsion to be coupled to the small satellites.

Another program being explored is ANTS (Autonomous Nan-Technology Spacecraft) at NASA Goddard Space Flight Center. These are 1 to 10 kg spacecraft utilizing solar sail propulsion to explore the asteroid belt. Most ANTS would contain single instruments and the idea is to launch hundreds to thousands of these to individual or numerous targets. When a large number target a single object, a smaller number would be manager/supervisors whose job would be to schedule/supervise small groups of ANTS taking observations. A few ANTS would analyse measurements and select targets.

It may be possible to launch a large number of such Cube-sats or ANTS containing simple instrument sets - perhaps a single imager - from a single vehicle and provide them with enough Δv capability to make a targeted swath through the main belt, obtaining basic information on at least as many asteroids as their are satellites launched. Many issues regarding data rates and operations have yet to be contemplated, but the possibilities are intriguing.

Rendezvous missions (Addressing B and C) Surviving protoplanets such as Ceres, Pallas, and Vesta are obvious targets for rendezvous missions. The utilization of solar electric propulsion technology allows for missions to any individually (e.g., Binzel and Clark 1999) or even a multiple rendezvous mission to Vesta and Ceres within a Discovery class cost envelope (Russell, private communication).

Only a rendezvous mission can provide detailed information on surface composition, morphology, surface structures, interior structure, magnetic fields, and core formation (or not) needed to probe the earliest stages of planetary formation.

Rendezvous missions to asteroidal fragments can provide important information about the primordial parent bodies and the processes they experienced prior to disruption, but in order to maximize the scientific return from such missions, target selection must be based on an understanding based on a large amount of groundbased data and theoretical studies.

Sample return missions (Addressing A, B, and C) Our principal source of asteroid samples are meteorites. However, meteorites are not thought to sample all regions of the main asteroid belt. They provide extremely detailed information about processes experienced by a particular portion of a single parent body over a range of time. A sample return from a single main belt object would not tell us anything about the asteroid belt as a whole. However, they might serve to provide important information regarding objects or classes of objects for which there are no meteorite samples and allow us to better make mineralogical interpretations from our spectroscopic data.

Multiple sample returns from family fragments showing spectral variations may provide important detailed geochemical information about the formation and internal evolution of the parent body prior to disruption. These might be accomplished to a more limited extent by multiple rendezvous to family fragments.

3. Maintaining Capabilities for the Future

The Key Scientific Questions for main belt asteroid studies are not going to be conclusively answered over the next decade. To ensure that the capacity to address these questions is maintained, strong and steady support for the underlying NASA Research and Analysis programs is required. These programs address these questions directly while creating the necessary bases for future asteroid missions.

4. Data Archive and Access Issues

Data associated with asteroid studies is archived by the NASA Planetary Data System. The data is peer reviewed as part of the ingestion process to ensure quality, consistency, and adequate documentation. To maximize usefulness of this data requires the creation of an interface that would allow researchers to find desired data. This might include a listing of all holdings regarding a particular object, or the identification of those objects meeting SQL criterion across all asteroid data sets.

While the PDS was designed to archive data sets from spacecraft missions, it has growing groundbased data holdings as well, particularly for small bodies. Challenges:

- All groundbased as well as spacebased asteroid data should be captured.
- Relevant laboratory data on meteorites should be captured
- All archive products should be available to remote query

5. Utilization Potential

The pursuit of basic science in the main asteroid belt may provide information that is useful to the utilization of asteroids in near-earth orbits to support future human space activity. For instance, the efficiency at which asteroids of given sizes from given (compositional) regions are removed from the main belt into near-earth orbits is useful as is the growth and physical properties of regolith of carbonaceous to metallic objects. See Jones et al. (this volume, pg. 141).

6. Advanced Technology

The continued improvement of solar panels and electric propulsion thrusters create the capability for multiple rendezvous and flyby missions that are not practical with chemical-based propulsion systems. For multiple flyby missions, solar sail technology may be ideal.

7. Education and Public Outreach

The advent of remote observational facilities coupled with the large numbers of asteroids, make asteroids a wonderful target for seasonally independent education. Programs could be designed in part to acquire data needed to address the key scientific questions - such as spectra and lightcurves. For instance, a student (high school or college) could acquire a spectrum or lightcurve on an asteroid (most might have been previously unobserved), and then fit a simple spectral model (2 or 3 component) to the spectrum, or triaxial ellipsoid to the lightcurve. By comparing the final result to the data, students could learn that science is an iterative process - and that there are no final answers, just better understanding.

8. Recommendations

Our top priority for the next decade is to ensure healthy funding support for NASA Research and Analysis programs supporting groundbased observations of asteroids, meteorite studies, and theory programs. In particular, there should be support for studies using modest aperture telescopes.

Asteroid missions addressing the key science questions cannot be planned nor provide a maximum return without a credible and continuing investment in earthbased remote observation programs, meteorite studies, and theory programs. These programs have suffered at NASA in recent years as funding has shrunk, particularly for physical studies of main belt asteroids. While programs to steadily increase our database of asteroid spectra and lightcurves may not garner headlines, they are critical to our ability to design cost effective missions that maximize the science returned for dollar spent. Success should be measured by the number and distribution (in orbital element space) of asteroids for which such observations are obtained and the rate at which it is obtained. This includes radar observations of main belt asteroids.

8.1. Discovery/Midex class and larger missions

- We endorse the Large Synoptic Survey Telescope provided it is designed and operated in a manner which allows for a determination of all main belt asteroids referred to in Figure 3 and that such a survey is supported. We note that this will place a large burden on the IAU Minor Planet Center (at Harvard) and that funding should be made available to MPC in the event that it processes LSST observations for orbits and links with prior observations.
- A rendezvous mission to Vesta or Vesta and Ceres.
- Solar System Infrared Survey - nominally, a polar orbiting, IRAS-like mission to determine albedos and diameters for all main belt asteroids having $D>1$ km. Must have non-thermal visible channel, may limit thermal channels to 10 and 20 μm (with a limiting flux density of 1 mJy), simplifying the experiment relative to IRAS. Broadband visible channels may allow for coarse taxonomic determination of the main belt.

8.2. Technology development - modest cost

Investigate the potential for large number of multiple asteroid flybys using microsatellites and possible advanced propulsion systems such as solar sails that would allow for multiple flybys per spacecraft.

References

Binzel, R., and B. Clark 1999. New World Explorer Mission to Vesta: The most ancient terrestrial world. *Lunar Planet. Sci.* **XXX**, 1218. (abstract) [CDROM]

Bowell, E., C.R. Chapman, J.C. Gradie, D. Morrison, and B. Zellner 1978. Taxonomy of asteroids. *Icarus* **35**, 313-335.

Dollfus, A., and B. Zellner 1979. Optical polarimetry of asteroids and laboratory samples. In *Asteroids* (T. Gehrels, Ed.), pp. 170-183. Univ. of Arizona Press, Tucson.

Gaffey, M.J., and S.L. Gilbert 1998. Asteroid 6 Hebe: The probable parent body of the H-Type ordinary chondrites and the IIE iron meteorites. *Meteor. Planet. Sci.* **33**, 1281-1295.

Hahn, G., and C.-I. Lagerkvist 1988. Physical studies of asteroids. XVII - JHK photometry of selected main-belt and near-earth asteroids. *Icarus* **74**, 454-471.

Lebofsky, L., and J. Spencer 1989. Radiometry and thermal modeling of asteroids. In *Asteroids II* (R.P Binzel, T. Gehrels, and M.S. Matthews, Eds.), pp. 128-147. The Univ. of Arizona Press, Tucson.

Morbidelli, A., J. Chambers, J.I. Lunine, J.M. Petit, F. Robert, G. Valsecchi, and K.E. Cyr 2000. Source regions and time scales for the delivery of water to Earth. *Meteor. Planet. Sci.* **35**, 1309-1320.

Pravec, P., and A.W. Harris 2000. Fast and Slow Rotation of Asteroids. *Icarus* **148**, 12-20.

Tedesco, E.F., D.L. Matson, and G.J. Veeder 1989. Classication of IRAS asteroids. In *Asteroids II* (R.P Binzel, T. Gehrels, and M.S. Matthews, Eds.), pp. 290-297. Univ. of Arizona Press, Tucson.

Tedesco, E.F., Ed. 1992. IRAS Minor Planet Survey. Philips Laboratory Technical Report No. PL-TR-92-2049, Hanscom AFB, MA.

Zellner, B., W.Z. Wisniewski, L. Andersson, and E. Bowell 1975. Minor planets and related objects. XVIII - UBV photometry and surface composition. *Astron. J.* **80**, 986-995.

The Future of Solar System Exploration, 2003-2013
ASP Conference Series, Vol. 272, 2002
Mark V. Sykes, ed.

Planetary Atmospheres

D.L. Huestis[1], N.G. Adams[2], S.K. Atreya[3], K.H. Baines[4], R.F. Beebe[5], S.J. Bolton[4], S.W. Bougher[6], A. Coustenis[7], S.G. Edgington[4], A.J. Friedson[4], M. Galand[8], C.A. Griffith[6], S.L. Guberman[9], H.B. Hammel[10], M.D. Hofstadter[4], A.P. Ingersoll[11], J.I. Lunine[6], M. Mendillo[8], J. Moses[12], I. Mueller-Wodarg[13], G.S. Orton[4], K.A. Rages[14], T.G. Slanger[1], D.V. Titov[15], A.R. Vasavada[16], A.-S. Wong[11], & R. Yelle[6]

[1] *Molecular Physics Laboratory, SRI International, Menlo Park, CA*
[2] *Department of Chemistry, University of Georgia, Athens, GA*
[3] *Department of Atmospheric Ocean and Space Sciences, University of Michigan, Ann Arbor, MI*
[4] *Jet Propulsion Laboratory, California Institute of Technology, Pasadena, CA*
[5] *Astronomy Department, New Mexico State University, Las Cruces, NM*
[6] *Lunar and Planetary Laboratory, University of Arizona, Tucson, AZ*
[7] *Observatoire de Paris-Meudon, Meudon, France*
[8] *Department of Astronomy, Boston University, Boston, MA*
[9] *Institute for Scientific Research, Winchester, MA*
[10] *Space Science Institute, Ridgefield, CT*
[11] *Division of Geological and Planetary Sciences, California Institute of Technology, Pasadena, CA*
[12] *Lunar and Planetary Institute, Houston, TX*
[13] *Atmospheric Physics Laboratory, University College London, London, United Kingdom*
[14] *SETI Institute, Mountain View, CA*
[15] *Max-Planck-Institut für Aeronomy, Katlenburg-Lindau, Germany*
[16] *American Association for the Advancement of Science, Washington, DC*

Abstract. Observation, characterization, and understanding the atmospheres of our neighboring planets and moons in the solar system provide essential inputs for addressing broad science goals. These goals include identifying the conditions that are favorable for producing and supporting biological activity, managing the effects of human activity on the Earth's atmosphere, and planning and evaluating observations of extra-solar planets. Here we review the current state of knowledge, pose key science questions, recommend enhancements in the research infrastructure for investigation of planetary atmospheres, and propose new space missions with specific science objectives.

Executive Summary

Observing, characterizing, and understanding planetary atmospheres are key components of solar system exploration. A planet's atmosphere is the interface between the surface and external energy and mass sources. Understanding how atmospheres are formed, evolve, and respond to perturbations is essential for addressing the long-range science objectives of identifying the conditions that are favorable for producing and supporting biological activity, managing the effects of human activity on the Earth's atmosphere, and planning and evaluating observations of extra-solar planets.

Current knowledge, based on very few observations, indicates that the planets and moons in the solar system have diverse atmospheres with a number of shared characteristics. Comparing and contrasting solar system atmospheres provides the best near-term means of addressing the broad scientific goals. Additional space missions with specific atmospheric objectives are required. At the same time, investment of additional resources is needed in the infrastructure of observation and interpretation of planetary atmospheres.

Current observational characterization of planetary atmospheres is roughly comparable to what had been learned about the Earth's atmosphere after the first rocket and satellite measurements in the 1950s and 1960s. From telescope observations and planetary missions we have determined the principal atmospheric constituents and the altitude profiles of pressure and temperature. We are able to classify the atmospheres of many of the larger solar system planets and moons into four groups:

1. Nitrogen atmospheres (Earth, Titan, Triton, Pluto)
2. Carbon dioxide atmospheres (Venus, Mars)
3. Hydrogen gas giants (Jupiter, Saturn, Uranus, Neptune)
4. Thin atmospheres, with three subgroups:
    - Rocky surfaces (Mercury, Moon)
    - Volcanic (Io)
    - Icy surfaces (Europa, Ganymede, Callisto)

Interpretative studies of radiative transport and collisional processes in the atmospheres of Venus and Mars have helped us understand the "greenhouse effect" and the impact of continued release of carbon dioxide into the Earth's atmosphere. Characterization of the composition of the atmospheres of the gas giants provides guidance about how planets and their atmospheres originate and how to interpret observations of extrasolar planets. Exploration of the current and historical abundance and state of water in the atmospheres, surfaces, and subsurfaces of Mars, Europa, Venus, and the Moon will provide important clues about photochemical stability of planetary atmospheres and the production of prebiotic chemistry.

Unfortunately, even with an increasing volume of observational data, planetary atmospheres are still grossly undersampled. For example, at the relevant altitudes in the atmospheres of Mars and Venus we have no observations of the minor chemical species (HO_x, ClO_x, SO_x) that models suggest are responsible for the stability of CO_2 atmospheres (as a result of catalytic recombination of CO and O_2) and for catalytic depletion of ozone in the Earth's atmosphere. Thus far we have sampled only a portion of the atmosphere of Jupiter. Without knowledge of the abundance of the heavier elements C, N, and O in the deep atmosphere, little can be said about whether the gas giant planets reflect the initial elemental composition of the solar system. The nitrogen/hydrocarbon atmospheres of Titan, Triton, and Pluto can provide important clues about photochemical formation of complex organic molecules in the early atmosphere of the Earth.

In addition, investigations of the Earth's atmosphere show that significant unpredictable variations occur on time scales of hours, vertical scales of a few kilometers, and horizontal scales of hundreds of kilometers. The atmospheres of many planets reveal structure and variation with respect to latitude, longitude, and season. Everything changes with solar cycle. Atmospheric models are very complicated. Many of the underlying chemical and physical processes are still poorly characterized. We think that we can produce useful explanations, but we do not have the data needed to ensure confidence that models can make quantitative predictions.

Four key questions/science themes were identified by the panel:

- *Understanding Atmospheres.* The historical attempts to understand planetary atmospheres have emphasized identification of the underlying chemical and physical processes responsible for the many fascinating observations. It is appropriate that the focus should now shift toward comparative interpretation of what the atmospheric observations and discoveries on multiple planets can teach us about broader scientific goals.

- *Learning by Exploring Planets and Moons.* Atmospheres are different each time we look at them. All future planetary mission campaigns should include explicit atmospheric components. Increased availability of observing time for planetary astronomy is essential on ground-based and near-Earth space-based telescopes.

- *Providing the Required Research Infrastructure.* Visiting planets is only one of the objectives. Lasting value comes from analyzing, interpreting, and using the data to establish broader implications, supported by independent programs for laboratory experiments, fundamental theory, modeling, and reanalysis of historical observations.

- *Assimilating Space and Planetary Science with Earth Science.* From our near neighbors in the solar system we hope to acquire additional hints about our origins and steps we should take to preserve our life-supporting environment. Better coordination between Earth science and space and planetary science can contribute to shared science goals, and justification and mobilization of additional funding resources for both disciplines.

The recommendations of the Planetary Atmospheres Community Panel fall into two broad categories. Recommendations that apply to multiple planets include establishing a mechanism for secure funding for analysis and interpretation of mission data, access to space- and ground-based telescopes dedicated to planetary observations, and creation of a new initiative for Comparative Understanding of Planetary Atmospheres. Recommendations for specific planetary missions with atmospheric goals include three Jupiter mission concepts, a Mars atmospheric explorer mission, a Post-Cassini/Huygens atmospheric/surface mission to Titan, and concepts for Neptune/Triton missions. Venus missions are also needed, but are not described here, under the assumption that they will be covered by the Venus community panel.

Other issues discussed by the panel included the public fascination with planets and the issue of international collaboration. Planetary observations make good news and well-watched television. Unfortunately, atmospheres look too much like chemistry and plasma physics. Other than the Jupiter impact of Shoemaker-Levy 9, neat colorful pictures are rare. U.S. citizens are better educated and intelligent than we suppose. Atmospheric scientists can do much more to explain why what we do is interesting, understandable, and important. The European and Japanese space agencies are launching capable planetary missions and planning future ones. A high level of coordination with these efforts is needed in order for the U.S. program of solar system exploration to reach its ambitious science goals.

REPORT

1. Current Knowledge and Key Science Questions

1.1. Introduction to planets and atmospheres

This document focuses on Planets, defined as the large objects orbiting the Sun: including Mercury, Venus, Earth, Mars, Jupiter, Saturn, Uranus, Neptune, and Pluto, along with their associated moons and companions.

The specific emphasis is on Atmospheres, defined as the interface between the planetary interior and the interplanetary medium: beginning with the top cm of the planetary surface; including atoms, molecules, ions, electrons, and cloud particles bound by the planet's gravitational field; also including planetary magnetic fields; and extending to the limit of the planet's non-gravitational influence on the interplanetary medium.

It is generally believed that the atmospheres of the small "rocky" planets and moons (e.g., Earth, Venus, Titan, etc.) are relatively young, having been created largely by outgassing as the surface cooled following planetary accretion; supplemented by later additions from impacting meteoroids, asteroids, and comets; and depleted by gradual escape of light elements to space. The "giant" or "gaseous" planets (Jupiter, Saturn, etc.) consist mostly of an atmosphere that is thought to roughly reflect the initial condensation of interplanetary atoms, molecules, and dust. Significantly, we know of only one planetary atmosphere with a large fraction of molecular oxygen, which is believed to have been formed on Earth by photosynthesis.

The surfaces and atmospheres of most planets and moons in the solar system receive more energy from external sources (usually sunlight and solar wind, supplemented on moons by tidal forces) than from upwelling from the planetary interior (e.g., original accretion energy and decay of radioactive elements). In general, the temperature of the planetary surface, and the altitude profile of temperature in the atmosphere, are controlled by the absorption of energy from the sun, reflection of visible radiation back to space, infrared emission by the surface (and clouds, if any), which is partially absorbed by the atmosphere, and eventually reemitted to space. This radiative transport problem defines the "greenhouse effect," the understanding of which is essential to predicting the impact of the increasing carbon dioxide abundance in the Earth's atmosphere resulting from combustion of carbon as an energy source for human activity.

1.2. Goals and objectives

The principal goals and objectives of the investigation of planetary atmospheres are briefly outlined below.

- Understanding the origin, history, composition, motion, and stability of planetary atmospheres: including formation during planetary accretion, by surface outgassing, or post-accretion deposition; modification of top-surface and atmospheric composition by external energy and mass sources; vertical and horizontal transport; clouds, winds, and storms; and loss of mass to space by surface ejection or exospheric escape.
- Characterizing the chemical and physical processes responsible for import of energy and mass from the sun and the interplanetary medium, response of atmospheres to external inputs, and release of energy and mass back to space.
- Identifying key observables for future planetary space missions and near-earth telescopic observations.
- Comparing the diverse planetary atmospheres in the solar system to learn what each can teach us about the others; to better understand the potential impact of human modifications of the Earth's atmosphere; to characterize the processes in the atmospheres or top surfaces that could generate molecules of prebiotic significance; and to identify what signatures might be useful for characterizing the atmospheres of extra-solar planets.

1.3. Who studies planetary atmospheres?

Three communities of scientists collaborate in the investigation, understanding, and interpretation of planetary atmospheres. "Observers" record atmospheric "data" using direct- sampling instruments on planetary probes, remote sensing instruments on Earth-orbiting satellites and planetary orbiters, and ground-based spectrometers, radar facilities, and telescopes. "Modelers" attempt to explain atmospheric observations and make predictions about which future observations would have the greatest impact. They use simulations based on microscopic processes that hopefully are well known from laboratory investigations, but if necessary, plausible numerical parameters are inferred by reproducing field observations. "Laboratory investigators" quantitatively characterize the underlying microscopic processes.

1.4. How we observe planetary atmospheres

Most observations of planetary atmospheres are derived from remote sensing. Many of these are from ground-based telescopes, although these observations are subject to limitations imposed by atmospheric transparency and seeing. Airborne, earth-orbiting and near-earth instrument platforms supplement the ground-based observations by accessing parts of the planetary spectra unavailable to ground-based observers; they can also provide spatial resolutions not bound by atmospheric seeing and not achievable through adaptive optics technology in ground-based telescopes. Near-planet remote-sensing observations that are enabled by interplanetary missions fill the important niche that not only accesses the entire spectrum but also achieves the highest possible spatial resolution. The high spatial resolution of interplanetary instruments provides horizontal discrimination of field parameters such as temperature and cloud opacity, and it enables vertical resolution through sensing of the planetary limb. Near-planet spacecraft also access geometries not available from the earth, e.g., high phase angles of illumination.

In-situ observations of planetary observations also access limited regions of the atmosphere but provide a means to measure properties that are not possible through remote sensing, such as spectrally inactive constituents or portions of the atmosphere that are too deep to be accessed by remote sensing. They can also achieve vertical resolutions usually unachievable from remote sensing,

1.5. How we understand planetary atmospheres

Below we list a sequence of questions that illustrates the roughly historical progression of observation, inference, and understanding in the study of planetary atmospheres.

- What is the Nature of the Observables?
  - Assignment and measurement of strength and line widths of spectroscopic absorbers and emitters
  - Measurement of cloud reflectivity as a function of wavelength, emission angle, and solar incidence angle; determination of Bond albedo
  - Cloud reflectivity/absorptivity measurements by atmospheric probes
  - Determination of thermal output, bolometric and in spectral regions of well-mixed constituents for thermal structure
  - Temperature vs pressure measurements by atmospheric probes
  - In-situ chemical assessment of gases and particulates (e.g., mass spectrometry, gas chromatography, spectroscopy, specific heat/acoustic wave speed) of gases and particulates
  - Local wind speed measurements by time-lapse imaging of clouds
  - Wind speed component detection by Doppler experiments on atmospheric probes

  - Radio occultation of spacecraft signals, visible/uv occultation of stellar signals
  - Radio reflection and ionospheres

- What are the Energy Sources?
  - Sunlight
  - Internal heat and tidal forces
  - Solar wind
  - The role of planetary magnetic fields

- What are the Underlying Chemical and Physical Processes?
  - Radiative emission, absorption, scattering, and transport
  - Particle impact excitation, dissociation, ionization, and scattering
  - Neutral and charged chemical reactions
  - Excited state quenching and energy transfer
  - Coagulation, condensation and sedimentation of particulates
  - Molecular and eddy diffusion
  - Large-scale vertical and horizontal wind generation

- What are the Numbers?
  - Atmospheric composition and thermal structure
  - Intensities of atmospheric ultraviolet, visible, and infrared emitted and scattered light, including X-ray and microwave/radio-frequency radiation
  - Solar spectrum
  - Composition of the solar wind
  - Rates and cross sections

- Why do Observables Vary in Space and Time?
  - Solar cycle and coronal mass ejections
  - Winds, waves, and transport
  - Seasons and global patterns
  - Storms and lightning
  - Atmospheric regions: The "spheres"

- What can be Learned from Systematic Observations?
  - Expose vulnerabilities and uncertainties in atmospheric models
  - Learn about possible signatures of "interesting" atmospheres on extrasolar planets
  - Long-term changes due to human perturbation (ozone depletion, global climate change)

- ◦ Predict sporadic short-term interference with human technology (space weather)

In the earliest stage we attempt to explain the macroscopic observables, such as colors of aurorae being due to atomic and molecular emissions and that an ionized atmosphere can reflect radio waves. In the second stage we infer what sources of energy could produce the observed perturbations of the atmosphere. Next we attempt a microscopic description of the specific processes that could be used to construct a quantitative model. But the model will not work unless we have accurate numerical values for the starting conditions, energy inputs, and the rate parameters for energy deposition and chemical transformation. As this microscopic local model begins to be trusted we back off from a local or point description and attempt to understand how variations in energy sources generate atmospheric dynamics. Finally we come to what atmospheric scientists tell the general public are the reasons why their field is important.

1.6. Specific needs by category and object

- Comparative Understanding Needs
  - Formation, evolution, stability, and structure of atmospheres:
    - All: Non-thermal exospheric escape
    - Venus/Mars: Stability of CO_2 atmospheres
    - Giants: Planetary elemental composition
  - Atmospheric motion: coupling between atmospheric regions, vertical and horizontal transport, mixing, and diffusion:
    - All: Origin of super-rotation
    - Mars: Coupling of condensation and release of polar volatiles, dust storms, atmospheric dynamics, thermal structure, and chemistry
    - Giants: Origin of long-lived powerful storms and the role of the latent heat of water
  - Planetary magnetic fields: differences between and implications of interactions of the solar wind with the atmospheres of planets with magnetic fields (e.g., Earth and Jupiter) compared to those without (e.g., Venus, Mars, Titan)
- Observational Needs
  - All: Repeated systematic observations: every 20 years is not enough
  - All: Signatures of winds and transport
  - All: Direct measurements of exospheric escape
  - All: Measurements of airglow emissions (UV, visible, IR, especially on nightside of Mars and outer planets)
  - All: Measurements of auroral emissions and correlation with the plasma environment
  - Venus/Mars: Minor species composition below 120 km

- Venus: Mid-level and surface winds
- Venus: Middle-atmospheric cloud structure and variability
- Giants: Elemental abundances of H, He, N, O, etc. in the deep atmosphere.
- Giants: Measurements of deep atmospheric winds
- Giants: Measurements of gravitational moments
- Giants: Measurements of horizontal and vertical thermal waves
- Giants: Measurements of horizontal winds over several scale heights
- Giants: Measurements of birth and decay of large- and small-scale features: particulate, tracers, temperatures
- Giants: Long-term measurements of horizontal winds
- Giants: Long-term measurements transport tracers
- Giants and Titan: Measurements of horizontal and vertical variability of temperatures, aerosols, cloud particulates, tracer gases
- Giants and Titan: Horizontal and vertical variability of CH_4-derived constituents
- Giants and Titan: High-precision abundances of isotopes
- Giants and Titan: Characterization of seasonal and other external forcing
- Jupiter: Measurements of auroral activity and correlation with solar wind
- Jupiter: Measurements of polar regions: establishing the relationships between processes in the charged and neutral atmosphere, establishing the existence and maintenance of a polar vortex.
- Titan: Temperature and chemical abundances in the troposphere and at the surface

• Modeling Needs

- Venus/Earth/Mars/Titan: Are general circulation models evolving toward a unified description that explains how planetary parameters control energy and momentum budgets?
- Venus: What can be learned from nightglow variability?
- Earth/Jupiter: What can be learned from auroral emissions?
- Titan: a unified description combining GCM, dynamics, and photochemical models.

• Laboratory and Theory Needs

- All: variability of non-Lorentzian line shapes with temperature
- All: characterization of ice index of refraction spectra, both pristine and irradiated
- Mercury/Moons: Trapping of volatiles in the top surface, radiation and impact induced chemistry and desorption

- Venus/Earth: Relaxation of O_2 excited states
- Venus/Earth/Mars: Rate of $CO_2(000)$ + $O(^3P)$ $\longrightarrow$ $CO_2(010)$ + $O(^3P)$
- Venus/Mars: Yields of $O(^1S,^1D)$ from e + $O_2^+(v>0)$
- Venus/Mars: Energetic H/H^+ collisions with O
- Giants: Equations of state, solubility, and molecular diffusion in H_2/He at low temperature and high density
- Giants: $CH_4/CH_3/CH_2/CH$ photochemistry
- Giants and Titan: Relaxation of CH_4 and hydrocarbon exited states, to characterize stratospheric non-LTE conditions
- Giants and Titan: Quantum or empirical ground-state potential energy surface, rovibrational energies, and transition probabilities for individual visible and near-infrared transitions
- Giants and Titan: measurement of low-temperature gas absorption coefficients in the visible and ultraviolet where quantum theory is intractable
- Giants and Titan: shape and scattering properties of condensed particulates
- Giants: non-Lorentzian line shapes in visible, e.g., K I, that are evident in brown dwarfs and probably influence the radiative transport process at depth
- Jupiter: Ammonia isotopic fractionation
- Jupiter: Determination of the full index of refraction for NH_4SH
- Jupiter and Saturn: determination of the temperature dependence of the submillimeter line wing absorption by NH_3
- Titan/Triton: CH_4 condensation and polyacetylene/nitrile photochemistry. More supporting laboratory measurements on aerosols, polymers, tholins and other organic material.

- Maintaining Future Capabilities
  - Justifying space missions
  - Justifying Research and Analysis (R&A) programs
  - Space telescope capabilities for planetary astronomy
  - Competition for time on large telescopes
  - Effective collaboration with Earth science programs
  - Coordinated collaboration with European and Japanese space agencies
  - Political support for NASA and NSF
  - Communicating with the public and Congress
  - Enhancing laboratory research
  - Career prospects in planetary atmospheres

2. Recommendations

The discussion above illustrates that planetary atmospheres are important, interesting, and complicated. We have learned quite a bit, but our partial understanding leads to many new questions. The following goals and themes may facilitate prioritizing the numerous science needs and mission possibilities.

- *Broad Science Goals.* Understanding how atmospheres are formed, evolve, and respond to perturbations is essential for addressing the long-range science objectives of identifying the conditions that are favorable for producing and supporting biological activity, managing the effects of human activity on the Earth's atmosphere, and planning and evaluating observations of extra-solar planets.

- *Understanding Atmospheres.* The historical attempts to understand the atmospheres of the Earth and other solar system objects have emphasized identification of the underlying chemical and physical processes responsible for the many fascinating observations. After decades of exciting discoveries, and with anticipated future discoveries of no less interest, it is appropriate that the focus should shift toward comparative interpretation of what the atmospheric observations and discoveries on multiple planets can teach us about broader scientific goals.

- *Learning by Exploring Planets and Moons.* Atmospheres evolve, move, change, and vary from place to place. A single observation is never enough. Planning of all future planetary mission campaigns should include explicit atmospheric components with specific scientific objectives emphasizing the need to fill gaps in our understanding. For example, the current strong Mars program is weak in atmospheric observations. The resource of ground-based and near-Earth space-based telescopes has been very productive historically, but observations of solar system atmospheres are currently not rated highly by time allocation committees, nor are current priorities for future space telescopes.

- *Providing the Required Research Infrastructure.* Visiting planets is only one of the objectives. Lasting value comes from analyzing, interpreting, and using the data to establish broader implications. Funding for post-flight analysis and interpretation of mission data needs to be reserved in advance and secured against escalation of hardware costs. In addition, well-funded independent Research and Analysis (R&A) programs, including laboratory experiments, fundamental theory, modeling, and reanalysis of historical observations, are essential contributors to the impact of the study of planetary atmospheres.

- *Assimilating Space and Planetary Science with Earth Science.* The broad science goals for planetary science are actually inward looking. From our near neighbors we hope to acquire additional hints about our origins and the steps we should take to preserve our life-supporting environment. In contrast, the current organization of research programs at NASA and NSF suggest a strong distinction between Earth science and space and planetary science, demarked by a boundary about 50 km above the Earth's surface. While this may reflect a perception of separate communities of researchers, it presents a barrier for effective communication, contribution to shared science goals, and justification and mobilization of funding resources for both disciplines.

The majority of the Decadal Survey Community Panels address individual solar system objects, or collections thereof. For many of these, the atmosphere is only one of several important topics of scientific interest. In contrast, the Planetary Atmospheres Community Panel is one of the few that focus on science themes that apply to many solar system objects.

Below we have grouped our recommendations into two categories: (1) programmatic and infrastructure improvements that apply to multiple solar system objects and (2) recommendations for specific planetary missions. The first group also includes issues that are broader than atmospheres, but are included here because we are passionate about them. In the second group we include three missions to Jupiter (the Giant Planets community panel joined with Planetary Atmospheres), an atmospheric mission to Mars (the Mars panel is concentrating on surface missions), an atmospheric/surface mission to Titan, and concepts for missions to Neptune and Triton (these recommendations might be adopted by the Titan and Neptune panels). We had originally planned to suggest an atmospheric mission to Venus, but defer to the Venus panel that has a strong atmospheric emphasis.

2.1. Programmatic and infrastructure recommendations

Secure funding for mission data analysis and interpretation: Successful Solar System Exploration missions generate vast amounts of observational data, only a portion of which can be analyzed and interpreted within the mission duration. In addition, planned funding for data product generation and archiving, as well as analysis and interpretation, is often consumed in advance to cover escalation of hardware and operation costs. This problem has been avoided in part by the Discovery Program, in which DA is not included in the mission funding cap. However, Discovery Program guidelines of only 1-3% of Phase C/D is not adequate for missions having diverse instruments and science objectives, allowing only one or two investigations per year per instrument, and excluding the broader community of researchers. The problem is greatly magnified when the design and generation of mission data products are also underfunded.

A healthy R&A program is necessary for the definition of well-focused planetary missions and to provide the knowledge base needed to maximize the science return to cost ratio for those missions. Having completed the initial reconnaissance of the solar system (excepting Pluto-Charon), missions are no longer in the "let's see what is there" category and these programs are even more important now than in the past to mapping out the next phase of solar system exploration. The health of Planetary R&A programs is being eroded by the fact that they are the only means by which scientists can fund post-mission data analysis (after the conclusion of a modestly funded one to two year mission-specific data analysis program) and the body of mission data is growing rapidly. This is in stark contrast with NASA Astrophysics, which funds the Astrophysics Data Program and Long Term Space Astrophysics Program for post-mission data analysis out of the OSS Data Analysis budget line.

RECOMMENDATIONS: Mission data product generation and archiving budgets should not be included in the mission cap. The budget should be justified separately and negotiated with both the mission and the NASA Planetary Data System and include margin. The requirement that Data Archive and Management Plans be developed and signed off prior to launch should be enforced. The quality of mission data products is the ultimate measure of mission success. Transparency, usefulness, and accessibility are critical.

- DA programs for every mission should not be included in the mission cap and should be funded at a level that reflects the complexity of the mission and its science objectives. A benchmark of several investigations per year per instrument should be used. DA programs should begin within a short period of time after data products are archived and available in the NASA PDS. These programs should extend for at least three years beyond the end of mission. All DA awards should be peer-reviewed.

- Planetary Data Analysis funding should be augmented to support the creation of a Planetary Data Program and a Long-Term Planetary Space Research Program. These programs would be parallel to the corresponding Astrophysics Data Program and Long-Term Space Astrophysics Program, funding planetary research utilizing data from all past planetary missions beyond their mission specific DA programs. This should also include planetary research using data from past astrophysics missions (explicitly excluded in the ADP and LTSAP). Funding levels of these programs should be equivalent to their Astrophysics counterparts. The purpose of this recommendation is in part to relieve DA pressure from R&A programs, therefore it is important that Planetary R&A program funding not be reduced or cancelled to pay for these new Data Analysis programs.

Dedicated telescopes for planetary astronomy: A major portion of our knowledge of planetary atmospheres comes from observations made from telescopes on the ground, on aircraft, or in near-Earth orbit. These telescopes provide opportunities for more scientists to participate in planetary research as well as for repeated systematic observations of how planetary atmospheres evolve.

In spite of this productivity, cost effectiveness, and growing telescope capability, observing time for solar system observations is actually becoming more difficult to acquire.

Requests for planetary observation time on large ground-based telescopes have very low approval rates. Time allocation committees (TAC) tend to be dominated by galactic and extra-galactic science interests. Requests for partial nights further reduce the approval prospects because they require extra TAC effort to schedule the remaining time. As deployment of adaptive optics and interferometry increases, the bias against planetary astronomy will get even worse. NASA controls only a small fraction of the large ground-based telescope assets: principally IRTF and one sixth of Keck.

The Hubble Space Telescope is viewed as a good platform for planetary observations and allocation of time is thought to be more balanced. In addition to providing higher spatial resolution, observations from outside the Earth's atmosphere extend the achievable wavelength regions and permit the observation of faint objects near bright ones. Unfortunately, Hubble will be decommissioned as soon as it can be replaced. NGST appears to be targeted exclusively toward extra-solar-system observations. Issues include loss of ultraviolet and visible instruments and capability to track moving objects.

RECOMMENDATIONS:

- NASA should plan, launch, and operate a new Hubble-class space telescope with UV, visible, and IR imaging spectrometers and other instrumentation appropriate to planetary astronomy.
- NASA should increase its financial support for ground-based telescopes, including existing telescopes, and join consortia for new telescopes. Even telescopes as small as 2 m can be useful for planetary observations if they are equipped with capable instrumentation. One possible approach might be to create a new grant program for telescope instrumentation, through which a telescope consortium could request NASA funds for equipment purchases, in return for allocating a fraction of the telescope time for outside planetary observers. NSF has implemented a similar approach, the Telescope System Instrumentation program (TSIP), based on the recommendations in the National Research Council report *Astronomy and Astrophysics in the New Millennium.*
- Observing time should be allocated by open competitions, as is currently done by NASA for Hubble, Keck, and IRTF (and by NSF for NOAO, etc), but the emphasis should be on planetary science.

Comparative understanding of planetary atmospheres (CUPA): Current knowledge indicates that the atmospheres of the Earth, planets, and moons are quite diverse; yet share a number of characteristics, from which each can teach us something about the others. Comparing and contrasting planetary atmospheres provides the best near-term means of addressing the broad scientific goals of identifying the conditions that are favorable for producing and supporting biological activity, managing the effects of human activity on the Earth's atmosphere, and planning and evaluating observations of extra-solar planets.

In contrast to this unifying "comparative approach" the communities of atmospheric scientists, the "sections" of professional societies, and the organization of programs at research funding agencies (NASA and NSF) tend to emphasize a separation into three separate categories: planets and moons, space physics (Earth upper atmosphere above 50 km), and atmospheric chemistry (Earth lower atmosphere below 50 km). There is also a fourth category called astrophysics, whose research topics resemble atmospheric science, but at much lower or much higher densities and temperatures. These separations have understandable historical foundations, based in part on differences in observational techniques, and amplified by independent justifications used in competition for research funding. At the same time, these separations actually weaken both scientific productivity and prospects for research funding for all atmospheric scientists.

NSF has developed approaches for attracting increased funding for multidisciplinary programs without requiring reorganization of existing program management organizations. Recent examples are the Information Technology Research (ITR) and Nanotechnology programs. NASA's "Living with a Star" initiative has been similarly successful.

RECOMMENDATIONS:

- NASA and NSF should establish a multi-agency multi-program multi-year long-term initiative for Comparative Understanding of Planetary Atmospheres (CUPA).
- Additional funding from Congress would be required, which the benefits of the comparative theme would justify. CUPA is a way to explain the importance of R&A programs.
- The initiative must be supported by strong endorsement from NAS/NRC. The planetary atmospheres research community must contribute to the campaign.
- Annual program solicitations should be issued with coordinated proposal evaluations. The resulting grants would be administered by the most relevant existing program.
- The solicitations might have an annual theme, but the main criterion should be that the research be truly "comparative."
- Much can be learned from the largely unsuccessful 1998 attempt to launch a similar joint NASA/NSF program. The justification was weak, lower atmosphere program managers were not convinced, and the solicitation required proposers to focus on Mars rather than explicit comparisons between planets.

2.2. Mission recommendations

Jupiter Microwave Sounder: As discussed elsewhere in this report, the composition and dynamics of the deep troposphere of Jupiter (down to pressures of $\sim$100 bars or more) are central to many of the questions we have about the giant planets. The abundances of oxygen and nitrogen, and the large-scale circulation pattern are of particular interest: the former placing constraints on models of planetary and atmosphere formation, and the latter telling us about how deep atmospheres organize themselves in response to solar and internal heating.

Microwave observations are particularly sensitive to the reduced forms of oxygen and nitrogen (water and ammonia), allowing centimeter to decameter wavelength data to tell us their abundances in the deep, well-mixed portion of the Jovian troposphere. Furthermore, microwave remote sensing can collect observations globally, allowing us to determine horizontal and vertical variations in abundances which in turn can be used as tracers to infer the general circulation of the atmosphere and the nature of convective and cloud-forming processes. Microwave sounding is therefore a uniquely suited remote-sensing technique to address key questions about giant planet atmospheres.

While in-situ measurements (such as provided by an entry probe) can more directly address some of the above questions, there are scientific, technological, and financial reasons to fly a passive microwave sounder as a precursor to a full-up multiprobe mission. Scientifically, probes are limited in number and cannot achieve global coverage. Entry probes can also be targeted more effectively if we already have some critical information on the deep atmosphere. From a technological point of view, conditions in the deep atmosphere must be known before a probe's design (particularly the communication system) can be optimized. Finally, a multi-probe mission is much more challenging and expensive than a relatively simple radiometer.

To achieve its objectives, the microwave sounder will most likely have to fly within Jupiter's radiation belts on a polar trajectory, allowing it to avoid confusion from synchrotron emission while sampling all latitudes. A study made for a Discovery Mission proposal has determined that a fly-by trajectory can achieve all the science objectives, while minimizing cost and the radiation hazard. Furthermore, even within the constraints of a Discovery mission, supporting instruments can be flown to provide more information on the interior of Jupiter, the cloud-top dynamics, and the fields and particles environment. This microwave mission serves as a model for future missions to all of the gas giants, particularly since a fly-by trajectory may allow multiple planets to be observed with a single spacecraft.

Deep penetration probes to determine elemental compositions of giant planet atmospheres: The Galileo Probe data on the elemental abundances of Jupiter has challenged our views of the formation of the giant planets and the subsequent evolution of their atmospheres. Contrary to expectations, the Probe measurements revealed, for the first time, that in the deep well-mixed regions of Jupiter's atmosphere, "all" of the measured heavy elements, C, N, S, Ar, Kr, and Xe are enriched relative to their solar proportions by a factor of 2-3. A plausible explanation is that the heavy elements were delivered to Jupiter largely by icy planetesimals. The volatiles containing many of these elements can be trapped in amorphous water ice only at temperatures of 30 K or lower. This implies that the planetesimals must have their origin in the pre-solar nebula nascent interstellar cloud. Another hypothesis argues for the trapping of the volatiles in clathrate hydrates. The two scenarios predict vastly different abundances of water, hence the importance of the oxygen elemental ratio on Jupiter.

Unfortunately, the elemental abundances measured by the "single" Galileo Probe may or may not be representative of the entire planet. To complicate matters, the Probe also entered a meteorologically anomalous region known as a (5 micron) hot spot, where downwelling is expected to alter the distribution of the volatiles, especially the condensable volatiles. Moreover, the abundance of the carrier of the heavy elements, water, continues to be a mystery. The water mixing ratio in the deep well-mixed part of the atmosphere–where it should be uniformly mixed–could not be measured. This is due to the fact that in the hot spot where the Probe made it measurements, the region of the uniformly mixed water must lie well below the deepest level probed, 21 bars, since the water vapor mixing ratio was found to be still increasing at this depth.

If the first of the above hypotheses of the heavy element enhancement on Jupiter is correct, the water abundance, hence the oxygen elemental ratio, must have an enrichment similar to that of the other heavy elements. The clathrate-hydrate scenario predicts at least three times greater enrichment in the oxygen elemental ratio relative to the other heavy elements. Still other scenarios may be possible.

In order to understand the formation of Jupiter and the evolution of its atmosphere, it is thus imperative that ALL heavy elements, including oxygen, be determined accurately in the deep well-mixed regions of Jupiter at several different latitude/longitude locations, and for comparison on Saturn. A comprehensive understanding of the formation of the giant planets and their atmosphere would be crucial also for modeling the formation of the extrasolar planets and the origin of their atmospheres.

In addition to the elemental abundance measurements, it would be important to study the dynamics of the deep atmosphere, including the questions of atmospheric stability and the depth to which the zonal winds extend and how they vary with latitude at great depths.

In summary, cleverly instrumented deep multiprobes into Jupiter, followed by Saturn, and eventually, Uranus and Neptune are recommended. In addition, it would be highly desirable to explore the possibility of precursor missions that could determine by remote sensing at least the N and O elemental abundances, as they would help with much more intelligent and sophisticated instrumentation and planning of the comprehensive elemental abundance measurements with the multiprobes (see the Jupiter microwave sounder description above).

Jupiter Polar Orbiter: A number of processes in Jupiter's atmosphere are centered in its polar regions, which have not been scrutinized directly since the high-latitude flyby of Pioneer 11 in 1973, although oblique views of polar areas have been obtained by Voyager, Galileo and Cassini spacecraft. In addition, a number of atmospheric properties that were to have been determined by Galileo were not accomplished because of telecommunications problems, and were addressed only briefly by the Cassini flyby.

One goal of this mission is to determine the relationship between the properties of Jupiter's electromagnetic field, together with its intrinsic time variability and that forced by solar wind variations, and the auroral discharge that is present in its upper atmosphere. In order to accomplish this, in-situ measurements of Jupiter's electric and magnetic field would be coupled with remote sensing of its auroral properties from X-ray and ultraviolet through near-infrared emission, primarily from hydrogen.

A second goal of this mission is to determine the relationship between the charged and neutral portions of Jupiter's atmosphere. A relationship exists between auroral discharge, the energetics of charged particles that are constrained by the magnetic field, and properties of the neutral atmosphere. Among the properties of the neutral atmosphere are the growth of particulates in the stratosphere at high latitudes and the existence of a region of high temperature at high altitude. Portions of the latter are characterized by anomalous abundances of hydrocarbons. Thermal measurements of temperature, a spectroscopic inventory of constituents in and out of these "auroral-related" regions, and a determination of the concentration of aerosols and their properties from ultraviolet/visible/near-infrared imaging/spectroscopy within the well-known polar haze are needed. These should include limb sensing to determine stratification of temperatures, chemicals and aerosols, to within scale-height resolution.

A third goal of the mission is to determine the meteorology and neutral energetics of the polar regions, and elsewhere on the planet. The first glimpse of the details of polar circulation were obtained by Cassini imaging and revealed surprising organization. Better details are possible with observations that are not nearly so oblique. In addition, there is a possible polar vortex that is characterized by a cold airmass that may be detectable in both the troposphere and the stratosphere but cannot be characterized well from ground-based or spaceborne facilities because of the limitations of diffraction-limited angular resolution. It is not known whether the entire polar region that appears to entrain aerosols is subject to a general upwelling, as is true for other relatively cold regions in Jupiter (its zones and anticyclonic vortices). Measurements of waves from limb sensing will also measure the amplitude and characteristic length of vertically propagating waves. These properties can also be measured at lower latitudes, as well, where the relationships between vertical winds, temperatures, aerosols, condensates, and other minor constituents (e.g., PH_3 and para-H_2), can be determined down to the size of the deformation radius and better, and tracked in some cases as a function of time.

Such a mission would be a valuable adjunct to either a multi-probe mission or to the Discovery "Inside Jupiter" mission, if either were to be approved. However, it is of sufficient merit to stand on its own. It may be somewhat greater in cost than a Discovery-class mission.

Mars Atmospheric Explorer: In many ways Mars is the most Earth-like of solar system objects and has much to teach us about the origin of life-supporting environments and possible significant signatures that might be observed on extrasolar planets. Especially intriguing are previous observations that show the presence of water in the current Martian atmosphere and polar caps. In addition, the observed surface topography provides hints that liquid water may have been present in the past and that substantial amounts may lie buried below the surface. Understanding the historical evolution of the Martian climate, and especially the history of water, requires detailed knowledge of the Martian atmosphere and the coupling between the surface and the atmosphere. Sustained repeated observations are needed to follow the geographical differences and diurnal, seasonal, and solar-cycle variations.

Unfortunately, the Martian atmosphere is the least studied of the 3 terrestrial planets. We have insufficient information to compare the chemistry of the Martian atmosphere with those of Earth and Venus. Yet, the Martian atmosphere is likely to be affected by chemistry to a greater degree than any other. Some key science questions are listed below.

- CO_2 is photochemically unstable on Mars, yet it exists. The postulated chemical explanation involves photolysis of atmospheric water, but we have no observations of the altitude profiles of the key minor chemical species (H_2O, H_2O_2, HO_2, OH, H, H_2, O, O_2, O_3, CO, and CH_4) in the middle atmosphere.

- The interaction of the Martian atmosphere with the solar wind and the escape rate of the atmosphere have never been investigated observationally. The evolution of the Martian climate and the history, stability, and loss of water are tightly coupled with the rates of exospheric escape of H and O atoms. We have some information about the H-corona but know almost nothing about the exospheric density and rate of escape of O atoms.

- The relationships among condensation and release of polar volatiles, dust storms, atmospheric dynamics, thermal structure, and chemistry are likely to be important. Almost nothing is known about the actual wind speeds in the atmosphere of Mars, other than from model simulations. Limited information about the Mars thermospheric density is being derived from aerobraking measurements by Mars Global Surveyor (1997-1999) and Mars Odyssey (2001-2002). These data show a tight coupling of the Mars lower (0-80 km) and upper (80-200 km) atmospheres. Both hydrostatic (inflation/contraction of the entire atmosphere with the seasons and dust events) and dynamical (tidal waves propagating up from the surface to the lower thermosphere) coupling processes (and their impacts) have been observed.

Recommendations for components of a Mars Atmospheric Explorer would include:

- An orbiter (similar to Pioneer Venus) with ion and neutral mass spectrometers as well as UV, visible, and IR spectrometers to measure the composition of the upper atmosphere, ionosphere, and exosphere, including dayglow and nightglow. A Fabry-Perot instrument could measure wind speeds in the upper atmosphere. It would also be useful to measure the charge-transfer depletion of O^+ ions from the solar wind, and the corresponding fast neutral O-atoms.

- Sacrificial descent probes for mass spectrometry measurements in the middle atmosphere.

- Balloons for lower atmosphere measurement of wind speeds as well as mass spectrometry.

Mars atmospheric science currently has a very low priority in the U.S. Mars Exploration Program agenda and upcoming NASA spacecraft missions. Some of the key science questions outlined above will be addressed by the upcoming Nozomi and Mars Express missions (2004-2005). A strong U.S. program of future missions could make especially valuable contributions by observing the Martian atmosphere near solar maximum (2010-2012).

Post-Cassini/Huygens atmospheric/surface mission to Titan: Ground-based and satellite observations of Titan are powerful means to study many of the processes occurring in this exciting planetary object. They are not, however, a substitute for in-situ measurements. Although the Cassini/Huygens mission to Titan will undoubtedly bring valuable answers to current problems regarding the satellite from 2004 and for 4 years, it will by the same token give rise to new questions that will require further exploring by more sophisticated missions in the future.

Thus, with the heritage of the Voyager, ISO and Cassini/Huygens data, new mission(s) to Titan, affording higher spectral and spatial resolutions, greater sensitivity and longer time spans could address items that remain unknown with regard to the atmosphere and the surface inventory of the satellite.

In the atmosphere, the nature and the spatial distribution of condensed particulates (such as aerosols, clouds, solids, etc) especially in the lower atmosphere need to be defined with emphasis on their involvement in the general organic chemistry. The degree of complexity attained by the latter and its connection with prebiotic chains remains to be established. The major composition, temperature structure and methane cycle are to be investigated by Cassini; there may however be seasonal variations that require further observations (a season on Titan is 7.5 yrs). The photochemistry, dynamics and circulation processes in the atmosphere require long-term measurements with technologically improved instruments aboard a future mission.

One of the most important remaining uncertainties with respect to Titan is, however, the nature of its surface. Even if the Cassini/Huygens probe will explore the ground, this study is limited to only one location. We have, however, indications that the landscape on Titan is variable and even with complementing observations from the orbiter, all the different facets of the surface material will not be investigated by Huygens. Hence, while the Cassini/Huygens data will provide clues as to where one should be looking, we will undoubtedly require in the future a mobile lander (balloons, aerobots, airplanes, helicopters, or other airborne platforms have been suggested) with laboratory capacities to investigate the composition of the surface and the sub-surface of Titan at different locations (preferably more than two). If such a probe could be combined with the capability to measure the tropospheric properties (as defined in the previous paragraph), then a sophisticated orbiter could be dispensed with, provided there is some means of relay to the Earth made available.

Neptune and Triton: In spite of (perhaps due to) Voyager's success at Neptune and subsequent studies with HST, many questions about Neptune remain unanswered.

- Atmospheric dynamics and structure: What powers the winds, and why are the winds and thermal structure similar to those of Uranus, though the internal heat sources differ? How deep does the zonal structure go? NEED: visible imaging and thermal mapping at various phase angles with scales down to 10 km; occultations of radio telemetry signals to probe atmosphere down to $\sim$2 bar.

- Atmospheric chemistry: What is the composition of discrete features (bright and dark), and of the atmosphere as a function of altitude? NEED: UV occultations to measure density, scale height, temperature and composition; compositional mapping at near-IR wavelengths.

- Planetary interior and magnetic field environs: Why are the magnetic fields much more asymmetric in ice giants than in gas giants? NEED: measurements of magnetic field and magnetospheric particles at a variety of latitudes and longitudes.

Short of exploring Pluto, exploring Triton may provide our best opportunity to examine the surface and atmosphere of a Kuiper Belt Object analog. Open questions regarding Triton include:

- Atmospheric structure and composition: What is Triton's atmospheric composition and structure, and how has it changed since Voyager? NEED: radio occultations for atmospheric size/structure; high phase and high-resolution (100-300 m) limb imaging for hazes/plumes; UV occultations (density, scale height, temperature, composition); atmospheric sampling (fly-through).

- Surface geology and composition: Is there evidence for "recent" solid-state convective activity in an icy mantle? How does composition vary between/within surface features? What causes geologic structures on Triton's surface? Has the geyser distribution changed since Voyager? Have atmospheric changes modified the surface? NEED: UV to near-IR global imaging (< 100 m); high-resolution imaging (10-30 m) of selected locales; thermal (50 and 100 microns) mapping; global 1-km imaging spectroscopy at 1-5 microns with R=300.

Numerous questions persist about the Neptune rings and small satellites: Are the ring arcs of Neptune a "major ring system waiting to happen"? Is a resonant model for arc stability correct? If not, how do arcs remain stable? Do Neptune's inner satellites show the effect of extreme tidal stress? NEED: low-phase 100-m scale imaging of arcs to find embedded bodies; high-phase 1-km scale imaging to detect new rings/arcs and to characterize ring/arc morphology; spectroscopic capability to determine composition.

Missions to Neptune to address these questions would include the following:

- Neptune orbiter: The orbiter is the core of the mission, providing a remote sensing platform, in-situ probes of the magnetic field and environs, and primary data links. An integrated imaging package would include: visible imager, IR imaging spectrometer, and UV imaging spectrometer. Other remote sensing devices are a thermal IR spectrometer and a microwave radiometer. Space physics detectors might include a magnetometer (and perhaps other instruments). Radio science instruments would also be necessary.

- Atmospheric multi-probe: Multi-probes are an essential part of an investigation of the deep ($\sim$100 bar) atmospheric structure and chemistry on Neptune. However, significant technology advances would be required to enable high S/N transmission from depth in a cost-effective manner. An optimal probe package would include a main probe (GCMS; sensors for temperature, pressure, and acceleration; solar and IR radiometers; nephelometer) and at least three mini- probes (GCMS; temperature, pressure, and acceleration sensors) to sample diverse atmospheric regions.

- Triton lander: A stretch goal would be a miniature surface lander to make in-situ studies of the satellite's lower atmosphere and surface geology/composition.

- Technological challenges: Recent studies indicate a Neptune mission with these capabilities is feasible given innovative technologies: high-power lightweight SEP and solar sails; qualified aeroshells; aerocapture; autonomous spacecraft communications; advances in miniaturization; lightweight power generation systems; temperature-tolerant electronics (~50K); lightweight structures. These technology drivers are required for many outer planet missions; their solutions will be broadly applicable.

- Alternative Neptune and Kuiper-Belt Flyby: If the technology challenges prove to be too difficult to overcome in the next decade, an intelligently designed Neptune flyby mission might be able to address a significant fraction of the science objectives. It might also conclude with a flyby to a Kuiper-Belt object.

Suggested Reading

Atreya, S.K. 1986. *Atmospheres and Ionospheres of the Outer Planets and their Satellites.* Springer-Verlag, New York-Berlin.

Atreya, S.K., S.G. Edgington, Th. Encrenaz, and H. Feuchtgruber 1999. ISO Observations of C_2H_2 on Uranus and CH_3 on Saturn. In *The Universe as Seen by ISO* (P. Cox and M. F. Kessler, Eds.). ESA-SP 427 1999.

Atreya, S.K., J.B. Pollack, and M.S. Mathews, Eds. 1989. *Origin and Evolution of Planetary and Satellite Atmospheres.* Univ. of Arizona Press, Tucson.

Atreya, S.K., M.H. Wong, T.C. Owen, P.R. Mahaffy, H.B. Niemann, I. de Pater, P. Drossart, and Th. Encrenaz 1999. A comparison of the atmospheres of Jupiter and Saturn: Deep atmospheric composition, cloud structure, vertical mixing, and origin. *Planet Space Sci.* **47**, 1243-1262.

Beatty, J.K., C.C. Peterson, and A. Chaikin, Eds. 1999. *The New Solar System,* 4th Ed. Sky Publishing, Cambridge, MA.

Bergstralh, J.T., E.D. Miner, and M.S. Matthews, Eds. 1991. *Uranus.* Univ. of Arizona Press, Tucson.

Bougher, S.W., D.M. Hunten, and R.J. Phillips, Eds. 1997. *Venus II–Geology, Geophysics, Atmosphere, and Solar Wind Environment.* Univ. of Arizona Press, Tucson.

Burns, J.A., Ed. 1977. *Planetary Satellites.* Univ. of Arizona Press, Tucson.

Burns, J.S., and M.S. Matthews, Eds. 1986. *Satellites.* Univ. of Arizona Press, Tucson.

Carleton, W.P., and W.A. Traub 1972. Detection of molecular oxygen on Mars. *Science* **177**, 988-992.

Chamberlain, J.W., and D.M. Hunten 1987. *Theory of Planetary Atmospheres,* 2nd Ed., Academic Press, San Diego.

Coustenis, A. 2000. Titan. In *The Encyclopedia of Astronomy and Astrophysics* (D. Emerson, Ed.). Institute of Physics, Angleterre.

Coustenis, A., and F. Taylor, Eds. 1999. *Titan, the Earth-like Moon.* World Scientific Publishing, Singapore.

Cruikshank, D.P., Ed. 1995. *Neptune and Triton.* Univ. of Arizona Press, Tucson.

Drossart, P., J.-P. Maillard, J. Caldwell, S.J. Kim, J.K.G. Watson,, W.A. Majewski, J. Tennyson, S. Miller, S.K. Atreya, J.T. Clarke, J.H. Waite, Jr., and R. Wagener 1989. Detection of H_3^+ on Jupiter. *Nature* **340**, 539-541.

Gautier, D., F. Hersant, O. Mousis, J.I. Lunine 2001. Enrichment in volatiles in Jupiter: A new interpretation of the Galileo measurements. *Astrophys. J.* **550**, L227-L230.

Gehrels, T., Ed. 1976. *Jupiter.* Univ. of Arizona Press, Tucson.

Gehrels, T., and M.S. Matthews, Eds. 1984. *Saturn.* Univ. of Arizona Press, Tucson.

Houghton, J.T. 1986. *The Physics of Atmospheres*, 2nd Ed. Cambridge Univ. Press, Cambridge.

Huestis, D.L., and T.G. Slanger 1993. New perspectives on the Venus nightglow. *J. Geophys. Res.* **98**, 10839-10847.

Hunten, D.M., L. Colin, T.M. Donahue, and V.I. Moroz, Eds. 1989. *Venus.* Univ. of Arizona Press, Tucson.

JGR Special Issue, 1998. The Galileo Probe Mission to Jupiter. *J. Geophys. Res.* **103**(E10), 22671-23070.

Keiffer, H.H., B.M. Jakosky, C.W. Snyder, and M.S. Matthews, Eds. 1992. *Mars.* Univ. of Arizona Press, Tucson.

Krasnopolsky, V.A. 1986. Oxygen emissions in the night airglow of the Earth, Venus, and Mars. *Planet. Space Sci.* **34**, 511-518.

Lewis, J.S., and R.G. Prinn 1984. *Planets and their Atmospheres, Origin and Evolution.* Academic Press, Orlando.

Lorenz, R.D. 2000. Post-Cassini exploration of Titan: Science rationale and mission concepts. *J. Brit. Interplanet. Soc.* **53**, 218-234.

Luhmann, J.G., M. Tatrallyay, and R. O. Pepin, Eds. 1992. *Venus and Mars: Atmospheres, Ionospheres, and Solar Wind Interactions*, Geophysical Monographs **66**, Am. Geophys. Union, Washington, DC.

Lunine, J.I. 1999. *Earth: Evolution of a Habitable World.* Cambridge Univ. Press, Cambridge.

Morrison, D., Ed. 1982 *Satellites of Jupiter.* Univ. of Arizona Press, Tucson.

Nair, H., M. Allen, A.D. Anbar, Y.L. Yung, and R.T. Clancy 1994. A photochemical model of the Martian atmosphere. *Icarus* **111**, 124-150.

National Optical Astronomy Observatory 2001. *The Telescope System Instrumentation Program (TSIP).* (http://www.noao.edu/system/tsip/).

National Research Council 1994. *An Integrated Strategy for Planetary Sciences 1995-2010.* National Academy Press, Washington DC.

National Research Council 2001a. *Astronomy and Astrophysics in the New Millennium.* National Academy Press, Washington DC.

National Research Council 2001b. *Astronomy and Astrophysics in the New Millennium: Panel Reports.* National Academy Press, Washington DC.

Nature Special Supplement 1979. Jupiter and Io. *Nature* **280**, 725-804.

Nier, A.O., and M.B. McElroy 1977. Composition and structure of Mars' upper atmosphere: Results from the neutral mass spectrometers on Viking 1 and Viking 2. *J. Geophys. Res.* **82**(28), 4341-4349.

Owen, T.C., P.R. Mahaffy, H.B. Niemann, S.K. Atreya, T. Donahue, A. Bar-Nun, and I. de Pater 1999. A low-temperature origin for the planetesimals that formed Jupiter. *Nature* **402**, 269-270.

Science Reprint Collection 1981. Voyager 1 Mission to Saturn. *Science* **212**, 159-239.

Slanger, T.G., P.C. Cosby, D.L. Huestis, and T.A. Bida 2001. Discovery of the atomic oxygen green line in the Venus night airglow. *Science* **291**, 463-465.

Traub, W.A., N.P. Carleton, P. Connes, and J.F. Noxon 1979. The latitude variation of O_2 dayglow and O_3 abundance on Mars. *Astrophys. J.* **229**, 846-850.

Trauger, J.T., and J.I. Lunine 1983. Spectroscopy of molecular oxygen in the atmospheres of Venus and Mars. *Icarus* **55**, 272-281.

Visconti, G. 2001. *Fundamentals of Physics and Chemistry of Atmospheres.* Springer-Verlag, Berlin.

Wayne, R.P. 2000. *Chemistry of Atmospheres*, 3rd Ed. Oxford, New York.

Yung, Y.L., and W.B. DeMore 1999. *Photochemistry of Planetary Atmospheres.* Oxford, New York.

The Future of Solar System Exploration, 2003-2013
ASP Conference Series, Vol. 272, 2002
Mark V. Sykes, ed.

The Future of Io Exploration

J.R. Spencer[1], F. Bagenal[2], A.G. Davies[3], I. de Pater[4], F. Herbert[5], R.R. Howell[6], L.P. Keszthelyi[5], R.M.C. Lopes[3], M.A. McGrath[7], M.P. Milazzo[5], J. Moses[8], J. Perry[5], J. Radebaugh[5], J.A. Rathbun[1], N.M. Schneider[2], G. Schubert[9], W. Smythe[3], R.J. Terrile[3], E.P. Turtle[5], & D.A. Williams[10],

[1] *Lowell Observatory, Flagstaff, AZ*
[2] *Laboratory for Atmospheric and Space Physics, University of Colorado, Boulder, CO*
[3] *Jet Propulsion Laboratory, California Institute of Technology, Pasadena, CA*
[4] *Astronomy Department, University of California, Berkeley, CA*
[5] *Lunar and Planetary Laboratory, University of Arizona, Tucson, AZ*
[6] *Department of Geology, University of Wyoming, Laramie, WY*
[7] *Space Telescope Science Institute, Baltimore, MD*
[8] *Lunar and Planetary Institute, Houston, TX*
[9] *Department Earth and Space Science, University of California, Los Angeles, CA*
[10] *Department of Geology, Arizona State University, Tempe, AZ*

Abstract. Io's dynamism makes it a uniquely exciting target for future exploration. It has much to teach us about the fundamental process of tidal heating, about large-scale volcanic processes, and about basic planetary physics. Many questions remain, and most of these can be addressed with future missions to the Jovian system, and improved UV instrumentation on large near-earth telescopes. Large groundbased telescopes (for good spatial resolution) and medium to small groundbased telescopes (for good temporal resolution), also have important contributions to make.

EXECUTIVE SUMMARY

Io is an important target for future exploration both for its own fascination and for what it can tell us about wider issues in planetary science.

As the most dynamic solid body in the solar system, not excluding Earth, Io occupies a unique place in planetary science. It is the only place beyond Earth where we can watch geology happen on a large scale- we are not restricted, as on other planetary bodies, to reconstructing geological processes from their long-dead remains. The interconnections between its orbital dynamics, interior, tectonics, volcanism, surface chemistry, atmosphere, and its associated magnetospheric phenomena illuminate all areas of planetary science. As one of the most spectacular places in the solar system, Io is a perfect focus for E/PO efforts, with the ability to inspire children and adults to an appreciation of astronomy and science in general.

Io is also the best place to study tidal heating, a process of fundamental importance to the evolution of planetary satellite systems, and one which may greatly expand the habitability zone for extraterrestrial life. Specifically, Io's tidal heating is intimately connected to that of Europa, where the tidal heating that maintains its probable ocean is much less easily observed. Io can also illuminate the Earth's early history, providing a living example of how a planet responds to high heat flow similar to the Earth's heat flow when life began, and displaying extreme volcanism similar to large eruptions in our own past. Io also provides an ideal testbed for understanding fundamental atmospheric, planetary escape, and magnetospheric processes.

Much about Io is still not understood. Outstanding issues include:

- Interior composition, the size of its core, and why it apparently has no magnetic field.
- Mechanisms of tidal dissipation, and the magnitude of its heat flow, though heat flow appears to be greater than is sustainable in steady state, suggesting that Europa may also have time-variable tidal heating.
- The composition of its lavas, which appear different from terrestrial basalts, and the extent of magmatic differentiation on Io.
- The source of the sodium, potassium, and chlorine that Io supplies to the Jovian magnetosphere.
- The origin of its spectacular mountains, or the mysterious scarps that lace its surface and appear to have been eroded despite the lack of a dense atmosphere.
- The full range of its eruption mechanisms, as most of its spectacular eruptions have been seen only during fleeting snapshots.
- The spatial distribution, dynamics, and composition of its atmosphere, and how the atmosphere responds to volcanism.
- How the 1 ton/second that Io supplies to the Jovian magnetosphere escapes Io.
- How Jupiter's rotation powers the terawatts of EUV energy radiated by the torus.
- How plasma is transported radially to fill the Jovian magnetosphere.

As a general recommendation, we suggest that NASA avoid becoming so focused on missions narrowly dedicated to the search for life that we will lose the chance for the broad new perspectives and serendipitous discoveries that are likely to come from a more balanced exploration program. Many of the major breakthroughs in our understanding of life's place in the universe have come from missions that were not primarily looking for life or organic chemistry.

Radiation and Δv considerations make an Io orbiter unlikely in the next decade or so, but much can be learned from an eccentric Jupiter orbit with repeated close Io flybys. A high-inclination orbiter could combine Io studies with studies of the Jovian polar regions, aurorae, and high-latitude magnetospheric processes. Alternately, a low-inclination orbiter could study the other Galilean satellites in addition to Io. Multiple flybys with similar geometry could watch the evolution of selected volcanic centers, with distant monitoring during apojove periods, supplying the vital temporal coverage that Galileo's few snapshots could not provide. Radiation doses experienced by Galileo suggest that 50 Io flybys could be accomplished for half the 4-Mrad radiation dose expected for the Europa Orbiter.

Instrumentation for such a mission should include high-resolution multicolor imaging, near-infrared imaging spectroscopy for compositional and volcanic emission studies, 1000-3000Å UV spectroscopy for spatial mapping of the atmosphere, and mid-infrared imaging for heat flow mapping, as well as plasma instruments and a neutral mass spec. If practical, a pair of penetrators with seismometers and in-situ compositional capabilities could tell us much about interior structure and surface composition in a very short time, using Io's probable violent and frequent natural seismicity as an interior probe. Such a mission is likely to be in the half-billion-dollar "New Frontiers" class, and could carry much more capable instrumentation than the Europa Orbiter because it would not need the propellant mass used by EO for satellite orbit insertion.

Space-based telescopes, in particular HST, have contributed greatly to Io studies in recent years, particularly in the study of its atmosphere, which is most easily seen at ultraviolet wavelengths inaccessible from the ground. However, HST is limited by its spatial resolution, which is not diffraction limited in the UV, and by the limited sensitivity of its UV instrumentation which makes most studies of Io's atmosphere S/N limited. Most seriously, HST is due to retire in about 2010, and no space telescope with even comparable capabilities is currently planned to replace it. NGST will have no UV capability.

We recommend that a general-purpose ultraviolet orbital telescope, optimized for planetary work, should be a priority for NASA in the next decade. Such an instrument would have many applications beyond Io science, and if diffraction-limited in the UV, and equipped with high quantum efficiency UV detectors, could greatly exceed the capabilities of HST. The cost of such a telescope is likely to be in the billion-dollar range, but it would benefit many areas of planetary science, and wider astronomy, beyond Io studies.

Io-specific UV space telescopes, such as the proposed "JIST" and "JMEX" missions, can also tell us much about Io on Discovery-class budgets or less ($100-300 M), because they can devote long periods to studying Io's temporal variability that is key to understanding physical processes.

Many Io phenomena, such as thermal emission from its volcanos, emissions from its atmosphere and torus, its reflectance spectrum, and even its large-scale albedo patterns, can be studied from the Earth's surface. Recent advances in adaptive optics (AO) with large telescopes provide 20 or more pixels across Io's disk, opening up a new and exciting field of ground-based disk-resolved studies. Io's dynamism requires frequent observations to capture and understand the full range of phenomena that it exhibits. Such monitoring is most easily done on smaller telescopes like the IRTF that have limited spatial resolution, but queue scheduling can allow frequent snapshots even on heavily subscribed large 8-meter-class telescopes. One exciting possibility is high-resolution 7-8 micron spectroscopy of Io, coupled with AO on an 8-meter-class telescope, which could directly map the atmospheric distribution and characterize its temporal variability.

We recommend that NASA expand the time available for general planetary science on 8-meter class telescopes, by purchasing more time on existing facilities or by construction of a dedicated large planetary telescope with AO capabilities. Creative scheduling, including queue-scheduled, remote, service, and daytime IR observing, can maximize the efficiency of these expensive facilities. Continued support for smaller facilities that can do crucial temporal studies is also important.

REPORT

1. Why Io Should be a Priority for Future Exploration

It is obvious that Io is a fascinating body, and while its intrinsic interest is summarized here, we spend more time discussing its "extrinsic" interest: why Io is also important for understanding wider issues in planetary science.

1.1. Io's intrinsic interest

Io is the most dynamic solid body in the solar system, not excluding Earth, thanks to intense tidal heating by Jupiter. Global heat flow is estimated at 2 W/m^2, compared to 0.06 W/m^2 for the average Earth. From this one remarkable number flows a huge variety of interconnected phenomena seen nowhere else to the same degree in the modern solar system. Io thus greatly expands the parameter space of observable planetary bodies.

Io presents an extremely rich array of geological, geophysical, geochemical, atmospheric, and plasma phenomena. Diurnal tidal flexing amplitudes may reach 100 meters, dissipating huge amounts of heat in the interior, though the sites and mechanisms of the dissipation are still unknown. Lavas of unknown composition inundate the surface, while volcanic plumes feed a tenuous, inhomogeneous atmosphere controlled by a combination of volcanic sources and surface frost sinks. In addition to the pervasive volcanic landforms, Io's surface is studded with some of the solar system's highest and most dramatic mountains, and by scarps of both tectonic and erosional origin.

Most remarkably, we can watch geology happen on Io, and are not restricted, as on other planetary bodies, to reconstructing geological processes from their long-dead remains. We can watch plumes cover the surface with pyroclastic debris, observe lava flows advance across the surface, follow chemical changes as surface materials anneal over time, and with better temporal and spatial coverage we may also be able to watch tectonic and erosional processes at work.

Io is also the dominant influence on the Jovian magnetosphere, filling it with a dense plasma of heavy ions, neutral atoms, and dust grains. The processes by which these materials leave Io, and subsequently evolve, are complex and still poorly understood.

Io is so energetic that many Io-related processes are observable from Earth, including thermal emission from its volcanos, large-scale albedo changes and plume activity, temporal changes in its atmosphere, and the ionized and neutral clouds that fill the magnetosphere. Much can thus be learned relatively inexpensively from the ground or Earth orbit.

Finally, as one of the most spectacular places in the solar system, Io has unique public appeal, and Io exploration offers many opportunities to attract and engage public interest in planetary science.

1.2. History and mechanics of tidal heating

Io is the prototype body for tidal heating, a process that is fundamental to the evolution of giant planet satellite systems, and one which may greatly expand the habitability zone for extraterrestrial life. By understanding Io, where this process is displayed to extremes and is easily observed, we can better understand the importance of tidal heating in a wide range of circumstances, from the history of satellites in our own solar system to the possibility of life-sustaining tidal heating elsewhere in the universe.

Io is very directly involved in the evolution of the currently most promising site for extraterrestrial life, Europa, via the coupled orbital evolution of the two bodies. Jovian tidal forces act more strongly on Io than on Europa, but because of the 2:1 Laplace resonance between the bodies, whatever affects Io affects Europa too. Io's current heat flow appears to be at least a factor of two higher than sustainable by steady-state orbital evolution, suggesting that time-variable orbital and thermal evolution of Io is likely. If true, this would imply time-variable tidal heating of Europa as well, with obvious implications for the sustainability of its possible subsurface ocean.

1.3. What Io can tell us about Earth history, and volcanism elsewhere?

Volcanism has been a fundamental process in the Earth's evolution, providing communication between the planet's interior and exterior. Volcanos probably brought much of the Earth's volatile inventory to the surface, supplying the modern atmosphere and ocean. Much of the Earth's crust is made of volcanic and other igneous rock. Studies of Io's volcanism can thus help us to understand the Earth.

Io's extreme heat flow is similar to that experienced by the Earth shortly after its formation, at the time life began, so Io provides a window into the Earth's formative years. This analogy has already proved useful, in that at least some Io's lavas appear to have an unusually high eruption temperature (Figure 1) and may be similar to the high-temperature komatiitic lavas that were common in the Archaean but have been extremely rare in the past billion years.

The present-day Earth also experiences several types of large-scale volcanism such as flood basalts, which (fortunately) occur too rarely to have been observed by modern humans. Such eruptions can have devastating regional and even global effects, and have been implicated in mass extinctions. Eruptions on this scale occur much more frequently on Io than on Earth, and are thus available for direct study. For instance, Io appears to show examples of large lava flows that grow slowly by inflation (at Prometheus) as well as rapidly-emplaced flows of comparable size (at Pillan).

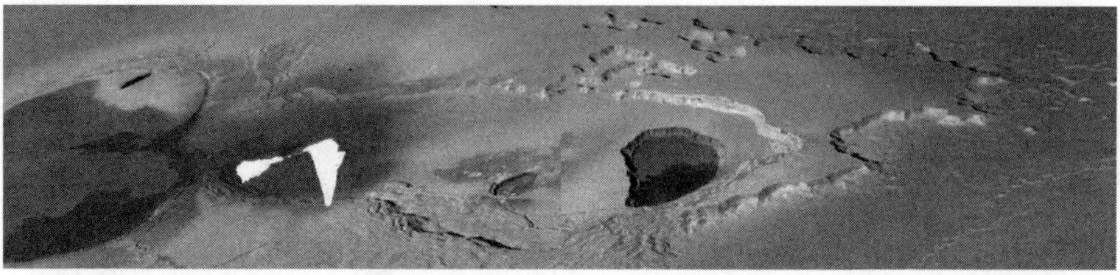

Figure 1. A high-temperature silicate eruption, probably a line of fire fountains, saturates Galileo's CCD at Tvashtar Catena, Io. This single image was Galileo's only high-resolution view of a major eruption in progress. The image also shows good examples of Io's fretted scarps produced by mysterious erosion processes. The image is 300 km across, with a resolution of 185 m/pixel, and was taken by Galileo on November 25, 1999 (NASA).

Other planetary bodies, particularly Venus, Mars, and the Moon, have undergone massive volcanic eruptions in the past, so Io's modern-day examples will teach us much about the history of these bodies. Analogies to the Moon in particular, because it is also airless, may be particularly instructive.

1.4. Atmosphere

Io's atmosphere is unlike any other in the Solar System. It plays a central role in controlling surface properties (through frost condensation) and the magnetosphere (as the buffer of escaping material). It provides a fundamentally contrasting case study for several reasons. First, because the atmosphere is so closely coupled to the surface temperature via vapor pressure equilibrium (VPE) of the SO_2 gas, and because the VPE of SO_2 varies by at least 6 orders of magnitude for the various temperatures appropriate for Io's surface (i.e., day, night, volcanic hot spots), Io's atmosphere provides access to a temperature-pressure regime unlike any other atmosphere known. Strong horizontal winds that reach hundreds of meters per second carry SO_2 towards regions of lower pressure. Vertical transport is also rapid, so that volcanic material reaches the exobase almost immediately. Many volcanic plumes exceed the scale height of the atmosphere. Second, unlike other atmospheres, Io's position deep within the magnetosphere of Jupiter, surrounded by a dense corotating plasma of heavy ions, leads to a wide variety of nonthermal processes involving charged particles which affect the surface and atmosphere, and ultimately nearly every other aspect of the Jovian magnetosphere including the other satellites. These conditions challenge, and therefore ultimately extend, our ability to deal with basic plasma and atmospheric physics.

1.5. Magnetospheric interaction

Io's magnetospheric interaction provides an excellent opportunity to study basic physical problems of general interest in space plasma physics. Io's torus is a prominent example of a mass loaded plasma. Mass loading occurs when ionized material is introduced into a fast-flowing magnetized plasma. Pickup ions created from Io's atmosphere and extended neutral clouds will generate field-aligned currents which transfer momentum from the ambient medium to the pickup ions via an Alfvenic disturbance. Other examples of mass loaded plasmas include solar wind flow by comets, planets, moons and other solar system objects. Io also represents a dramatic example of electrodynamic interaction. The plasma flow around Io and its conducting atmosphere generates a MegaAmp current across Io which closes along the Jovian magnetic field lines. These field-aligned currents are propagated toward Jupiter via Alfven waves and couple Io to Jupiter's ionosphere.

The coupling between distinct plasma populations (i.e., the Io plasma torus and Jupiter's ionosphere) is a fundamental and unresolved problem in space physics. In many cases, coupling is imperfect due to an electric field component parallel to the ambient magnetic field. Parallel electric fields inhibit the propagation of the momentum-transferring Alfven wave and a magnetic decoupling (or violation of the frozen-in condition) results. Perhaps the most familiar example of a decoupled system is the Earth's magnetosphere-ionosphere system where the decoupling is manifested by electron acceleration and the formation of discrete aurora. Likewise, the Io-Jupiter system is at least partly decoupled by parallel electric fields which are manifested by auroral emissions in Jupiter's atmosphere at the base of Io's flux tube. The persistence of auroral emissions wakeward of Io's flux tube suggests that decoupling persists for a significant period of time following the initial interaction of a given flux tube with Io. Io's magnetospheric interaction is therefore highly appealing for studying the general problem of coupling and electron acceleration processes.

More parochially, Iogenic plasma completely dominates the Jovian magnetosphere, and via sputtering and implantation, has a major influence on the surfaces of the other Galilean satellites. It's even possible that Iogenic plasma provides a source of chemical energy for possible Europan life.

Finally, the unique spectral signatures and large spatial scale of emission from a plasma-rich planetary magnetosphere provides a possible means of detecting and characterizing extrasolar planetary systems, increasing the importance of understanding the physics of such systems.

2. What We Need To Know

2.1. Interior composition and structure

Measurements of the spherical harmonic degree 2 gravity coefficients of Io leave no doubt that Io has a metallic core. The size of Io's metallic core is between about 550 km and 900 km for an Fe-FeS core or between 350 and 650 km for an Fe core.

The size of the core is uncertain mainly because we do not know its chemical composition. Io's core contains an unknown amount of sulfur; the larger the sulfur content is, the larger Io's core is. A determination of the size of Io's core would fix its density and by inference lead to a determination of core composition, i.e., core sulfur content. Another basic unknown about Io's core is its physical state, i.e., liquid or solid or partially frozen. Apparently, Io does not have an intrinsic magnetic field so knowledge of the physical state of the core would enable us to understand why Io has no magnetic field. Further, an understanding of Io's lack of a magnetic field will contribute importantly to our general understanding of dynamo action in the cores of Earth and other planetary bodies.

Additional gravity data, together with topography data, would tell us about density anomalies in Io's mantle and the thickness and variations in thickness of Io's crust, and would constrain crustal density and the degree of isostatic compensation, and thus give a handle on crustal chemistry and the degree of magmatic differentiation, if any. We could also learn about anomalous mass concentrations at or near Io's surface associated with lava flows and magma chambers, for example.

2.2. Heat flow

Io's heat flow is so prodigious that it can be measured remotely, by inventorying the infrared radiation that Io emits. Recent estimates, using a combination of ground-based and spacecraft observations, have been in the range 2 - 3 W/m^2, while steady-state tidal heating rates, as constrained by historical observations of Io's orbital evolution, should be no higher than about 1 W/m^2. Possible solutions to this problem include coupled variations in Io's heat flow, internal rheology, and orbital elements (which would necessarily also involve Europa), or episodic release of heat despite constant tidal input, due to convective instabilities. It is still possible, however, that the heat flow estimates are incorrect, and more precise estimates are sorely needed in order to address these questions. In addition, spatial variations in Io's heat flow provide important clues to its internal state and the site of tidal dissipation: dissipation in the deep mantle will produce excess heat flow at the poles, while shallow asthenospheric dissipation will concentrate heat flow at the equator.

To improve our understanding of the heat flow, we require accurate measurements of the spatial distribution of Io's total thermal radiation at all wavelengths and in all directions, and accurate measurements of its bolometric albedo (which controls the input of solar heat).

2.3. Surface chemistry

Io's surface composition remains a mystery despite several spacecraft visits and more than 70 years of photometric, calorimetric, and spectral observations. Sulfur dioxide frost, with its many diagnostic infrared absorption features, is the only constituent that has been definitively identified on Io's surface. Indirect hints as to the identity of the other surface constituents are provided by (1) the satellite's overall high (and bland) reflectance in the visible and infrared, (2) its strong absorption in the ultraviolet, (3) the identification of atomic and ionized S, O, Na, Cl, and K in the Io torus and neutral clouds, (4) active volcanism across the satellite, (5) the variety of colors (yellow, red, orange, black, green) observed in different regions across the surface, (6) the very high temperatures (indicative of ultramafic silicates) in many active volcanic centers on Io, and (7) the high bulk density of the satellite (suggesting a silicate composition for the satellite as a whole). Elemental sulfur (S_8) has long been a contender as a dominant surface constituent based on points (1), (2), (3), (4), and (5) above (as have polysulfur oxides). Surprisingly, Io's surface exhibits few spectral features indicative of silicates, other than a 0.9-m absorption feature tentatively identified as a Mg-rich silicate (e.g., pyroxene) observed in some of the dark spots across Io. The only other infrared spectral features not linked to SO_2 include an absorption band at 3.915 μm that has been tentatively identified as pure solid H_2S or Cl_2SO_2 diluted in solid SO_2, and an enigmatic broad 1-1.5 μm band. Candidates for the red plume deposits and other red materials include S_4, S_2O, Cl_2S, and elemental sulfur contaminated by As, Se, or Te. The dominant Na^-, Cl^-, and K-bearing surface constituents are still uncertain, although NaCl, KCl, and perhaps alkali sulfides are favored by models of volcanic outgassing. Many outstanding questions about Io's surface composition remain. What causes Io's latitudinal albedo variations (e.g., darker and redder poles)? What is responsible for Io's complex surface color patterns (e.g., multicolored flows and plume deposits)? Answers to these questions and many more must await future Io missions.

2.4. Tectonic and surface processes

Io provides a laboratory for understanding surface processes on bodies without atmospheres, and tectonic processes on bodies without plate tectonics, e.g., the Moon and perhaps the early Earth.

Our few very-high-resolution images of Io show a myriad of baffling features, including "ripples" in plains material, gullies, and fretted terrain suggestive of sapping. To understand these features, we need global imaging in at least 4 colors at ~100 m/pixel to inventory landforms and understand their distribution, and very high-resolution (1 m/pixel or better, with 1-10 m/pixel context) imaging of select regions.

Questions important to tectonic processes:

- What is the thickness and composition of the crust?
- What creates and destroys Io's mountains? A plausible recent theory for mountain formation, that they result from crustal compression caused by resurfacing, needs to be tested (e.g., by looking for thrust faults), and alternatives explored.
- How do tectonism and volcanism interact? E.g., are some or all of the ~500 large calderas tectonic in origin?

To answer these questions, we need: global imaging at ~100 m/pixel, and topography, gravity, and seismic data.

The various types of observations need to be obtained in conjuction, over an extended period of time, so the processes and the timescales over which they occur can be watched.

2.5. Volcanism

Io is the only place beyond the Earth where active volcanism can be studied. A wide range of Earth-based instruments and spacecraft (e.g., Voyager, Galileo, Hubble Space Telescope) have been used to study the nature of Io's volcanic processes. Much has been learned, but after more than two decades of study, new questions with broad implications have arisen, including:

- How does volcanism operate in extreme environments (i.e., no plate tectonics, no atmosphere, low gravity)? Io is the only place in the solar system where we can observationally test physical models of silicate volcanism that can be applied to Venus, the Moon, Mars, and asteroids. The key to answering this question is to separate the influences of magma composition and volatiles, tectono-physical processes (e.g., tidally-induced superheating, mantle plume ascent), and local environmental conditions. Such studies will improve not only our understanding of Ionian volcanism but also of planetary volcanism in general.

- What is the compositional range of Io's magmas? High magma temperatures suggest widespread ultramafic eruptions, but basaltic and sulfurous eruptions may also be common. We need to determine magma and gas compositions via remote sensing and in-situ studies, in order to create more realistic physical and chemical models of eruption processes. Magma chemistry also provides a vital window on Io's interior composition and structure.

- How do very large volcanos behave? The key to answering this question is to frequently monitor active Ionian volcanoes, which enables identification of eruption size and style from visible imaging and thermal emission analysis, and to compare resulting volcanic eruptions, their products, and morphologies with those on Earth and the other planets.

2.6. Surface age and cratering timescales

While it's obvious that Io is being resurfaced rapidly, the actual resurfacing rate is not known. Many of the large plume deposits are ephemeral and large areas of the surface are unchanged since Voyager. Heat flow considerations give mean resurfacing rates of about 1 cm/year, but much of this may be repeated resurfacing of caldera floors.

It's possible that much of Io is resurfaced at a rate of less than 1 mm/year, allowing the possible survival of a few impact craters that could be detected with sufficient high-resolution imaging coverage. If these were found, comparison of crater densities with independent estimates of resurfacing rates derived from techniques such as plume particle size analyses and lava flow thickness measurements would provide a valuable independent estimate of cratering timescales on the Galilean satellites, improving age estimates for the surfaces of the other satellites.

2.7. Atmosphere

Observations made so far only weakly constrain Io's atmosphere, but suggest that it is dominated by a combination of volcanic and sublimation processes. Models suggest that Io's is the only solar system atmosphere with orders of magnitude spatial variation in surface pressure, producing transsonic winds and bulk atmospheric composition that may be completely different between high and low pressure regions. SO_2 in the sublimation component of the atmosphere will condense at night, while the volcanic component will be diurnally constant. Ly-alpha images of the atmosphere also indicate that it condenses out at the poles. Atmospheric vertical structure is unusual, being probably dominated by electrical and plasma impact heating.

In many ways the atmosphere is the "missing link" between Io's surface and magnetosphere. Because the observational constraints are so weak, many major unsolved questions persist, whose answers require new observations. Such questions include:

- What is the main immediate atmospheric source, volcanism or recycled surface frost, and what is the main volcanic effluent gas?
- What gases other than SO_2, SO, and S_2 are present in the high and low pressure regions, and how do they interact?
- By what mechanisms are the various visible and UV emissions excited?
- How is the ionosphere maintained and is it global or local?
- Do particulate aerosols, visible in the plumes, interact with the atmosphere and provide less volatile constituents such as Na and Cl to the atmosphere and magnetosphere? Are the plumes the main source of Io's dust streams?
- Is the upper atmosphere greatly expanded by the electrodynamic interaction between the magnetosphere and atmosphere, as predicted by models?
- Are there local high-current electrodynamical interactions at the plumes, and if so, do they affect the plumes or atmospheric mass loss?
- How does SO_2 that freezes out at the poles get recycled back into the atmospheric circulation? Why are there no extensive polar caps?

2.8. Mass loss and the Torus

While the torus has been observed from the ground, Earth orbit and spacecraft, there remain several critical issues relating to the production, transport and loss of plasma mass and energy.

Different measurements are consistent with a production rate on the order of 1 ton per second of Iogenic plasma, but we do not understand the detailed mechanism of the near-Io source (e.g., ionization by the impacting plasma, stripping of an ionosphere, ionization of a sputtered corona, etc;) nor the relative importance of localized ionization vs. ionization of an extended neutral cloud. Moreover, one naturally expects that the highly variable volcanism would lead to dramatic changes in Io's atmosphere, neutral corona and, hence, ion production rate.

The torus is emitting several terrawatts of energy in EUV emissions. The energy ultimately comes from Jupiter's rotation, coupled by the magnetic field to torus plasma but the mechanisms which transfer energy within and between species are poorly understood. Possible heating mechanisms are ionization, charge exchange, wave-particle interactions and acceleration by global electric fields.

To first order the plasma is confined and coupled to the planet's 10-hour rotation by Jupiter's strong magnetic field. But we also know that over time scales of days the plasma is transported radially outwards (either by global or microscale processes) and the plasma lags a few percent behind corotation. While these dynamical processes are telling us that the magnetospheric plasma gets decoupled from the Jupiter fly-wheel, we cannot describe quantitatively the causes or mechanism(s).

To address issues of plasma production, energization and transport we need to monitor variations in Io's atmosphere at the same time as monitoring changes in torus properties. The mechanisms allowing energy to flow within and between species requires in situ measurements of the ion and electron 3-d velocity distributions (particularly separation of the dominant ion species O^+ and S^{++} which have the same M/Q=16). To understand the coupling between the torus plasma and Jupiter's rotation we need to measure the heretofore unexplored high latitude regions between the torus and the planet with polar orbiting spacecraft.

3. How We Can Make Progress

There is a danger in NASA becoming so focused on missions narrowly dedicated to the search for life that we will lose the chance for the broad new perspectives and serendipitous discoveries that are likely to come from a more balanced exploration program. Many of the major breakthroughs in our understanding of life's place in the universe have come from missions that were not primarily looking for life or organic chemistry - e.g., our understanding of the history and importance of planetary impacts derived from studies of our dead moon, the discovery of outflow channels on Mars by Mariner 9, the discovery of the importance of tidal heating by Voyager 1 at Io, the discovery of circumstellar dusk disks by IRAS, and so on. We thus urge a continued balance in solar system exploration between life-focused and general exploration programs.

3.1. Missions

There is much we will never learn about Io until we return there for close-up observations. From Earth we are unlikely to achieve spatial resolution better than many tens of km, and we cannot make observations of the poles or night hemisphere, or obtain the detailed compositional analysis possible with in situ observations.

Though Io orbiters have been proposed in the past, e.g., in the 1996 and 1999 Roadmap studies, the radiation hardness and delta-V issues that have emerged from studies of the Europa Orbiter concept make an Io orbiter seem very difficult with technology available in the next decade. Both the delta-V and radiation issues are even more severe at Io than at Europa. At the same time, Io's dynamism is not well suited to study from a one-shot flyby. The next mission to study Io in detail is thus likely to do so from Jovicentric orbit, perhaps assisted by penetrators. Of course, a Jovicentric orbit also allows studies of the other Galilean satellites. While such a mission would superficially resemble Galileo, they could accomplish much more for less cost due to advances in instrument capabilities and miniaturization over the 20 years since Galileo was designed, and the likely 100-fold improvement in data return possible with a functioning high-gain antenna. The following mission concept is for an Io-dedicated mission, but could be readily adapted to include other targets as described above.

3.2. A strawman mission concept

Orbit: Jovicentric, eccentric, period roughly 1 month, perijove near Io.

Duration: Galileo has survived seven Io flybys, with a radiation dose of roughly 40 krad each. A four-year mission with 50 monthly Io flybys would accumulate only half the 4 Mrad radiation dose expected for the Europa orbiter.

Payload: 1000-3000Å UV spectrometer for atmospheric studies, with solar occultation capability; high-resolution multicolor visible imager (1-10 m/pix samples, 100 m/pix global); 1-5 μm spectrometer with 1 km resolution; 10, 20 μm imager with 10 km resolution; laser altimeter; mass spectrometer; plasma package. Possibly, a pair of penetrators with 20-hour lifetimes (comparable to DS2), each including a seismometer, atmospheric mass spectrometer, surface composition package, and possible ranging capability.

Orbiter Operations: Repeated flybys of the same hemisphere of Io, with similar lighting geometry, emphasizing studies of time variability. Active regions found in early in the mission would become the focus of intensive repeated study in later flybys. The remainder of the orbit would be used for both data playback and distant studies of Io, using either a scan platform, simple mirror system, or frequent attitude maneuvers.

Penetrator Operations: Penetrators would be released after impact sites had been chosen on early orbits, and will require retro-rockets to reduce impact speed. Mass specs. would determine atmospheric composition during entry, and surface composition could also be determined. Io is likely to be so seismically active that 20 hours of simultaneous coverage from three stations may be sufficient to map internal structure, though studies are needed to confirm this. Low-frequency seismometers would allow tidal flexing to be measured directly.

Such a mission could go a long way towards answering some of the questions posed above. The penetrator seismometers would reveal Io's interior structure, and by measuring the core size could distinguish between Fe and FeS core composition. Laser altimetry would provide regional topographic information and lava flow thicknesses, and conceivably could measure tidal flexing even from Jovicentric orbit. Spectroscopy of fresh lavas, and direct sampling and UV spectroscopy of atmospheric gases would constrain interior composition and differentiation. Repeated wide-area imaging, multicolor photometry of plume dust, and laser altimetry would constrain resurfacing rates, and repeated detailed observations of active volcanic centers would reveal eruption mechanisms. Mid-IR mapping would determine global heat flow and its spatial distribution, thus constraining dissipation mechanisms. UV and mass spectroscopic and solar occultation mapping would reveal the spatial distribution and dynamics of the atmosphere, and its sources and sinks.

3.3. Space-based telescopes

Space-based telescopes provide a critical capability that is largely complimentary to both ground-based and in situ spacecraft observations. Fundamental contributions to our understanding of both the Io atmosphere, and its interaction with the plasma torus, have been made by several space-based facilities in the past 20 years. For example, discovery of the atomic sulfur and oxygen emissions at Io, which are a basic diagnostic of the plasma interaction with the satellite, was made using the IUE satellite in 1986. More recently, observations with the Hubble Space Telescope (HST) have provided the only quantitative spatially resolved information about Io's SO_2 atmosphere, and determined to first order its basic spatial distribution. HST has also provided the only spectroscopic observations of volcanic plumes, in which molecular sulfur (S_2) was discovered in the Pele plume. The UV imaging capability of HST has provided detailed information about the plasma interaction with the satellite via its images of the atomic emission morphology, which provides basic ground truth for the detailed MHD models of the interaction. EUVE observations of the Io torus discovered Na^+ emission. None of the spacecraft sent to the Jovian system (Pioneer, Voyager, Galileo) has had the ability to make these measurements.

Generally speaking the space-based capability that has proven most useful for studying Io and the torus is access to the ultraviolet portion of the electromagnetic spectrum for two reasons: SO_2 and other atmospheric molecules have strong absorption bands throughout the UV spectral region; and the electronic ground state transitions of both sulfur and oxygen occur in the ultraviolet. Additionally, the Io torus radiates primarily at UV wavelengths. A UV capability combined with the superior spatial resolution and small spectroscopic slits on the HST instruments have provided the most comprehensive results compared with other more specialized facilities such as EUVE or FUSE. It is therefore of very high priority for continued success in unraveling Io's secrets to strongly endorse a widely accessible successor to HST with UV capability. An obvious next step would be to have diffraction-limited capability in the UV, which HST lacks, and UV detectors with higher quantum efficiency than HST's. NGST's lack of UV capability, its tailoring to astrophysical, and particularly cosmological, problems, and its lack of planetary capability (specifically, the ability to track moving targets) will give it only limited usefulness for Io observations.

A second important capability that is dearly needed for progress in Io studies is a long-term synoptic monitoring capability in space-based facilities. Observations such as the ones described above have given us only the shortest glimpse (e.g., the science time available in one HST orbit is only about 40 minutes) at the true nature of Io's atmosphere and torus interaction. While the first order spatial structure of Io's SO_2 atmosphere is now known, these observations also make it obvious that, as expected, the atmosphere is highly variable. What they are unable to untangle, due to the general scarcity of the observations, is how much of the variability is due to actual physical causes (e.g., the volcanos) and how much is due to factors due only to the viewing geometry (such as correlation with specific features on Io's surface). Space-based facilities such as HST that provide only a brief, in-depth brief snapshot will never be capable of telling the full story. A facility that provides long term synoptic monitoring capability, such as the proposed JMEX or JIST missions, or the more general-purpose SCOPE proposal, provides the best hope of allowing us to explain the complex, and intricately interconnected, Io-torus-magnetosphere system.

3.4. Ground-based telescopes

Many Io phenomena, such as thermal emission from its volcanos, emissions from its atmosphere and torus, its reflectance spectrum, and even its large-scale albedo patterns, can be studied from the Earth's surface, providing the cheapest way to study Io's time variability. Io's dynamism requires frequent observations to capture and understand the full range of phenomena that it exhibits- for instance, the largest infrared volcanic "outbursts" are seen only a few percent of the time. Such monitoring is most easily done on smaller telescopes like the IRTF that have limited spatial resolution, but queue scheduling can allow frequent snapshots even on heavily subscribed large 8-meter-class telescopes. Multi-wavelength infrared observations of the volcanic thermal emission can constrain magma temperatures, eruption mechanisms, spatial distributions, and the time evolution of these quantities. Recent advances in AO with large telescopes provide 20 or more pixels across Io's disk, opening up a new and exciting field of ground-based disk-resolved studies. New instrumentation also opens up new possibilities: one possibility is high-spectral-resolution 7-8 micron spectroscopy of Io, coupled with AO on an 8-meter-class telescope, which could directly map the SO_2 atmospheric distribution and characterize its temporal variability. The atmosphere has also recently been detected at 19 μm from the IRTF. We recommend that NASA expand the time available for general planetary science on 8-meter class telescopes, by purchasing more time on existing facilities or by construction of a dedicated large planetary telescope with AO capabilities. Creative scheduling, including queue-scheduled, remote, service, and daytime IR observing, can maximize the efficiency of these expensive facilities, particularly for bright objects like Io where integration times are often short. Continued support for smaller facilities that can do crucial temporal studies is also important. Telescope time on "smaller" facilities may become available as these are replaced by the next generation of giant telescopes: for instance Caltech time on the Palomar 5-meter might be available for purchase by NASA if the CELT 30-meter telescope is built. Future Io space missions should include in their budgets support for ground-based monitoring programs (including construction of new instruments of telescopes if appropriate) that can enhance the spacecraft science return, e.g., by providing better temporal coverage of volcanic eruptions, for a small fraction of the mission cost.

The Future of Solar System Exploration, 2003-2013
ASP Conference Series, Vol. 272, 2002
Mark V. Sykes, ed.

Europa Exploration: Science and Mission Priorities

J.F. Cooper[1], C.B. Phillips[2], J.R. Green[3], X. Wu [3], R.W. Carlson[3], L.K. Tamppari[3], R.J. Terrile[3], R.E. Johnson[4], J.H. Eraker[5], & N.C. Makris[6]

[1] *NASA Goddard Space Flight Center, Greenbelt, MD*
[2] *SETI Institute, Mountain View, CA*
[3] *Jet Propulsion Laboratory, California Institute of Technology, Pasadena, CA*
[4] *Department of Engineering Physics and Astronomy, University of Virginia, Charlottesville, VA*
[5] *Ball Aerospace Technology Corporation, Boulder, CO*
[6] *Department of Ocean Engineering, Massachusetts Institute of Technology, Cambridge, MA*

Abstract. This report surveys the present state of knowledge for Europa and recommends key science objectives and spacecraft missions for the next decade of 2003 - 2013 and beyond. The highest priority initial objective is to confirm the presence of a subsurface water ocean, for which the evidence from multiple observations is already very compelling but not yet definitive. This should be addressed by a coupled set of gravity, altimetry, and magnetic field measurements in Europa orbit. Other related objectives include remote surveys for recent ocean-associated activity on the surface, the determination of chemical composition for Europa's surface, atmosphere, and dust ring, global mapping of energy flux and composition for magnetospheric particles incident at the surface, and the search for biochemical markers of any organisms originating from habitats below the visible surface. These objectives can partly be addressed by the same initial orbiter with addition of a moderate suite of remote-sensing and in-situ analysis instruments. Alternatively, a flyby mission with return to Earth of atmosphere and dust ring samples for chemical and isotopic analysis could achieve some of the compositional objectives. Science and engineering development are needed to support planning for lander missions in future decades.

EXECUTIVE SUMMARY

In the context of the present decadal survey we find strongly compelling, but not yet definitive, evidence from multiple types of observations that there may be an ocean of liquid water below the frozen surface of Europa. Europa's fractured surface shows relatively few craters and appears only tens of millions of years old. The visible ice crust shows strong evidence of tidal stresses from gravitational coupling to Jupiter and other Galilean satellites. Linear and cycloidal fractures extend far across the surface with layer upon layer of such features indicating a long sequential history of surface deformation from tidal, radiogenic, and other heating of subsurface ices. Abundance of chaos regions, domes and pits possibly associated with diapiric plumes, localized regions of emergent flooding, and apparent extrusion of sulfate hydrates further point to the continuing existence of a subsurface ocean to recent geologic times. Magnetometer data from many Europa flybys by the Galileo Orbiter are consistent with expected levels of electric current induction in a modern salty ocean, but sustained measurements from Europa orbit will be needed to filter out ionospheric effects and to determine the global extent of the ocean. Despite the strong geologic and magnetic evidence for an ocean, image comparisons from Voyager and Galileo have yielded no visible sign of surface changes over the intervening twenty years, and continuing future observations by various types of missions to Europa and the Jovian system should search for such activity.

Various aspects of the surface chemical composition inferred from remote sensing are alternatively attributed to internal and exogenic sources. The ocean could provide a mixture of magnesium and sodium sulfate salts suggested by features in spectroscopic data from the Near-Infrared Mapping Spectrometer (NIMS) on Galileo Orbiter. Models for chemical evolution of Europa's interior allow for large abundance of such salts, which could also act to help maintain the ocean in a liquid state. A neutral cloud of sodium is observed to be emanating from Europa, as expected from magnetospheric sputtering if this element is abundant on the surface. However, there is an alternate interpretation of the NIMS surface data in terms of sulfuric acid hydrate with sulfur coming either from the ocean or from implantation of magnetospheric ions. Higher spectroscopic and spatial resolution are needed for future observations to resolve this debate by remote sensing (flyby and orbital) and in-situ surface measurements. Return of atmospheric gas and dust ring samples to Earth for elemental, molecular, and isotopic analysis could also increase our knowledge of the surface chemistry.

The extremely high (0.1 $mW\text{-}m^{-2}$) energy inputs from magnetospheric and extreme ultraviolet irradiation make the Europan surface a chemical hotbed for rapid processing of ice and non-ice materials. The observed chemical composition (also including H_2O_2, O_2, and CO_2) in the millimeter-thick optical layer of the outer surface must be strongly affected by the irradiation. Regolith gardening by high-speed impacts of interplanetary meteoroids mixes immediate products of irradiation to meter depths over tens of millions of years. Instabilities of the kilometers-thick ice crust, and chemical leaching by any emergent liquid water, may also convey these products to the putative ocean on similar time scales. There are many uncertainties regarding what fraction of these products might actually reach the ocean, but upper limits for oxygen production at the surface could allow full oxygenation of the entire ocean in 10^7 years.

Since photosynthetic processes within the putative subsurface ocean would be inhibited by the overlying ice crust, an alternative source of chemical energy for evolution of life might come from reactions of surface oxidants with reduced compounds within the ocean from primordial, meteoritic, hydrothermal, and radiolytic sources. The surface environment is lethal to known terrestrial extremophiles but dormant spores or biochemical markers of oceanic organisms emerging to the surface might be detected by remote sensing and in-situ measurements. There is some current speculation that spectroscopic signatures at near-IR wavelengths of the observed sulfate hydrates on Europa's surface might actually be associated with such organisms. A flyby or orbiter mission could provide higher resolution measurements at appropriate wavelengths to further investigate this possibility, preparatory to detailed in-situ measurements at surface sites where such material appears to be concentrated.

We recommend that the primary focus of planning and implementation for Europa-related missions in the next decade should be the confirmation of the subsurface ocean. Only then would there be sufficient justification to pursue the more costly objectives of compositional and biological in-situ surveys, at and below the surface, that may tell us whether the ocean is truly a habitat for life. The presently proposed Europa Orbiter would be the most effective technical approach towards fulfillment of the prime objective but may be too costly in its present form to meet present budget requirements. The minimum science objectives for a first orbiter mission should include gravity modeling and laser altimetry to set limits on the tidal flexure of the global ice surface by an underlying ocean. These two techniques could resolve an expected equatorial bulge of a few tens of meters in vertical amplitude. A magnetometer is also clearly needed to globally map spatial and temporal variations of magnetic fields near Europa, so that the induced magnetic field components can be confirmed as direct evidence for a modern ocean. Mapping for non-dipolar moments of the induced field is required to test the global extent of liquid water which might otherwise be confined to ocean basins or pools of brine within soft subsurface ice or denser mantle material.

Other important objectives of a moderate-scale mission could include measurements of surface composition and search for geologic change with remote sensing instruments, global mapping of magnetospheric radiation inputs to the surface, and in-situ orbital measurements of Europa's atmosphere and dust ring. Composition of the latter may be directly related to surface composition. If overall mission cost could be sufficiently reduced, the radar sounding experiment of the present Europa Orbiter design might also be included to survey the ice crust to several kilometers in depth. This would be particularly important in the choice of sites for future lander missions, but a series of orbiters may be advisable (as for Mars) before the more ambitious objective of landing on Europa's inhospitable surface is attempted.

A complete range of missions for Europa would include occasional flybys to search for signs of recent surface activity, probe the surface composition for inorganic and possible biological components with advanced spectrometers, and collect samples of ring dust and atmospheric constituents for return to Earth. Isotopic analysis of even small amounts (e.g., micrograms) for such samples under laboratory conditions would be extremely interesting. Initial missions to the surface might include the Europa Pathfinder concept under study at NASA's Jet Propulsion Laboratory with balloon landing systems inherited from the highly successful Mars Pathfinder. Aside from the obviously unique opportunity for in-situ compositional analyses, such a mission could also implant a sensitive seismic instrument for deep sounding of the ice crust, the ocean, and Europa's deeper interior. In the more distant future, well beyond the present decade, we envisage the placement of networks of seismic and electromagnetic sounding instruments on Europa and the other Galilean satellites to probe for magma (Io) or water oceans already suggested by magnetometer and other results for Ganymede and Callisto as well as for Europa.

In support of such missions we also discuss pressing needs for development of spacecraft power, navigation, and science instrument systems that would enable collection and return of data from Europa's challenging orbital and surface environments. Realistic laboratory simulations are also needed for spacecraft and instrument operations at 80-130 K in a highly irradiated and potentially toxic chemical environment at the surface which is likely to be strongly oxidized by surface interactions with the Jovian magnetosphere. Interpretation of remote sensing and in-situ compositional data will also be greatly aided by simulations of such interactions and of resultant chemical evolution of the icy regolith to meter depths. The multi-mission scope recommended for future Europa exploration also calls for long-term data analysis and archiving strategies to facilitate the search for surface activity and to make data from these missions available to, and usable by, future generations of Europa scientists.

REPORT

1. Key Science Questions

1.1. Europa surface geology: Evidence for subsurface ocean?

Europa, an ice-covered body with a radius of about 1560 km (about the size of Earth's moon), is locked in a tidal resonance with its neighboring satellites, Io and Ganymede. This resonance keeps the satellites from perfect circular orbits, inducing a forced eccentricity that varies their distances from Jupiter and results in tidal flexing and internal dissipation of heat and energy. This heating is most intense at Io, but a substantial amount of heat is dissipated at Europa's distance as well (Cassen et al. 1979, 1980).

Gravity measurements taken by the Galileo spacecraft (Anderson et al. 1998) show that Europa is an internally differentiated rocky body with about 100 km of water and/or ice at its surface. The center of the current debate about Europa is the past and current state of this water, which if predominantly liquid would exceed the volume of all of Earth's oceans combined. The first study of the habitability of Europa was by Reynolds et al. (1983).

A number of lines of evidence have been suggested for the presence of liquid water beneath the icy surface of Europa (Pappalardo et al. 1999, Stevenson 2000). These include theoretical studies, spectroscopic measurements, geologic imaging, gravity measurements, and magnetic field measurements. Before the Voyager flybys of Europa in the late 1970's, little was known about Europa's surface except for its anomalous high albedo and the spectroscopic suggestion of the presence of water there. Images of Europa taken by Voyager revealed a surface covered with crack-like features, and very few impact craters (Malin and Pieri 1986).

The relative lack of craters on Europa means that the surface is young, perhaps as young as a few million years (Zahnle et al. 1998), though more likely at least a few tens of millions of years. A young surface requires geologic activity, which is more likely in the presence of liquid water. However, a search for

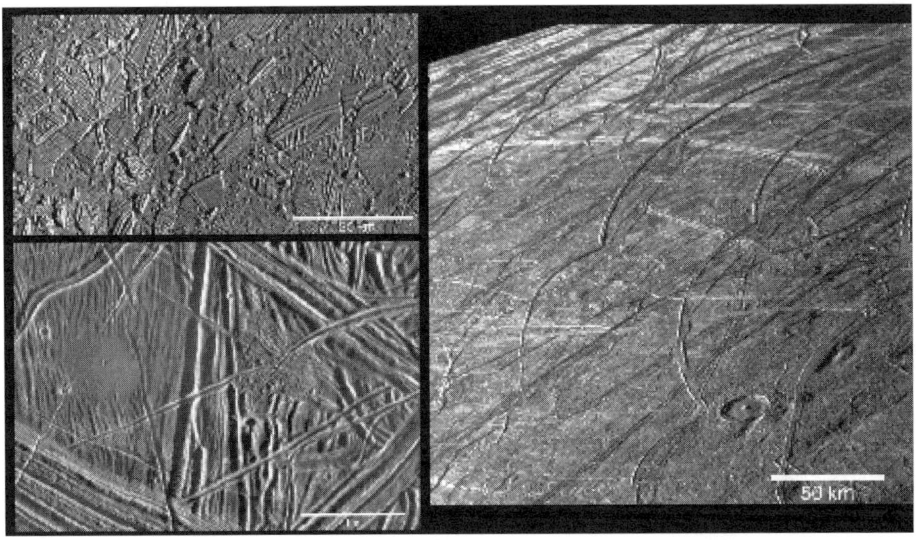

Figure 1. Galileo images of geological features on Europa (NASA). Chaotic terrain (upper left; scalebar 25 km), ridges and a possible cryovolcanic flow (lower left; scalebar 4 km), arc-shaped cycloidal ridges (right; scalebar 50 km) and a shallow impact crater (right).

signs of geologic activity on Europa's surface between the Voyager and Galileo observations, over a 20-year period, has shown no detectable changes (Phillips et al. 2000).

Images of Europa's surface taken by the Galileo spacecraft have shown surface features which could be consistent with the presence of liquid water beneath Europa's surface, but again do not prove it (Figure 1). Europa's surface is primarily covered by a vast set of interconnecting cracks and ridges. Also present are areas of disrupted 'chaotic terrain', where the surface appears to have been broken up into coherent iceberg-like blocks that seem to have 'rafted' into new positions (Carr et al. 1998). Such areas can be reconstructed by fitting the preexisting features on the blocks back together (Spaun et al. 1998). Other features of interest on Europa's surface include regions that could possibly be low viscosity surface flows, and impact craters that are anomalously shallow.

A key question for Europa is whether liquid water exists at all beneath its surface, and, if so, how thick the solid ice layer is at the surface. Many models attempting to address Europa's interior structure have been proposed for the formation of the variety of features visible in Galileo images of Europa's surface. There is currently no consensus among these often contradictory views of Europa's geophysics. For example, the morphology of Europa's impact craters suggests that they formed within a solid target, but their shallow depths are suggestive of post-formation viscoelastic relaxation (Moore et al. 1998). Such models suggest that most craters on Europa formed in a 5 - 15 km thick brittle surface layer, overlying a lower-viscosity subsurface layer. This subsurface material, however, could either be liquid water or warm, low-viscosity ice. A new model (Turtle and Pierazzo 2001) suggests that the presence of central peaks in some of the larger craters on Europa requires a surface layer at least three kilometers thick.

Cryovolcanic surface flows would be intriguing, but there are very few regions where the morphology is suggestive of flows (Fagents et al. 2000; see Figure 1 for one example). There is also a substantial buoyancy problem in their formation, as it is difficult to get liquid water to the surface of Europa since it is denser than ice.

Models of ridge formation range from cryovolcanism to tidal squeezing to linear diapirs to compression and plastic deformation (Pappalardo et al. 1999). These models range in requirements from a very thin crust overlying liquid water (the tidal squeezing model) to completely solid-state models with a thin brittle crust on top of a lower-viscosity, warm ice layer (diapirism or compression). For instance, cycloidal ridges have been shown (Hoppa et al. 1999) to correspond in orientation and location to cracking of the surface in response to the changing diurnal tidal stresses. Hoppa et al. argue that this model requires the existence of a global ocean near the surface to obtain sufficient tidal stresses to crack the ice. Clearly, current ridge formation models in the literature are contradictory and have very different implications for Europa's subsurface structure.

Chaotic terrain was first seen as a 'smoking gun' for the presence of liquid water beneath Europa's surface, but solid-state models are also possible (Carr et al. 1998, Greenberg et al. 1998, Pappalardo et al. 1998). At the liquid end of the spectrum, regions of chaotic terrain are seen as areas of localized heat flow where the ice layer melted all the way to the surface. In this model, the blocks are buoyant remnants of the preexisting icy crust that move about in a slushy matrix, both translating and tilting. Eventually the matrix freezes solid, ending the blocks' motion and preserving their final positions. This model requires localized heating of the crust, but it may be difficult to concentrate the heating in both space and time. The solid-state formation model suggests that ice rises to the surface in a diapir, eventually disrupting the brittle surface. These convective upwellings make it difficult to tilt the blocks as observed, however. A third, intermediate model (Head and Pappalardo 1999) suggests that runaway melting within rising diapers produces chaos. Thus, models of chaos formation do seem to favor the existence of either water or an ice-water slurry at shallow depths near the surface, but such patches could be localized and not require the existence of a global liquid ocean layer.

The major question relates to what surface geology tells us about what may lie beneath for a few hundred kilometers, a liquid water ocean or layers of soft ice? Is the hard ice crust a few kilometers in thickness, or tens of kilometers? How accessible is the putative ocean to direct exploration by future cryobotic probes? Do material and chemical (see next section) exchanges between the surface and ocean occur and on what time scales? Does this mean that full planetary protection protocols must be invoked for future landings?

1.2. Europa surface composition: Relative contributions from magnetospheric and endogenic sources? Astrobiological implications?

Various aspects of Europa's surface chemical composition inferred from remote sensing are alternatively attributed to endogenic and exogenic sources. A major limitation on such measurements at the surface is that molecular composition is inferred looking 'through a glass, darkly' with respect to a heavily irradiated optical layer of only millimeters in thickness. Times for full radiolytic processing of exposed material are orders of magnitude shorter than the probable geologic ages of all visible surface features (Cooper et al. 2001). Relatively 'pristine' material indicative of endogenic composition may lie only at depths of meters or greater, below the ice regolith continually stirred by meteoritic impact gardening. An unresolved debate is whether the observed sulfate hydrates in near infrared observations are associated with acidic H_2SO_4 (Carlson et al. 1999b) or the $MgSO_4$ (epsom) salt hydrate (McCord et al. 1998a, 1998b, 1999, 2001). Far more speculative are similarities of the associated near-infrared absorption features to those of a few extremophile bacteria on Earth (Dalton 2002). What does this difference bode for evolution of life in the putative ocean? Even the question of whether the sulfur source is magnetospheric or internal appears unresolved - do dissolved sulfates come up from the ocean through breaks in the crust or do ice sublimation and brine mobilization, produced by diapiric heating, concentrate implanted sulfur from the magnetosphere in these same regions? Hemispheric leading/trailing asymmetries suggest a magnetospheric source but higher concentrations are usually found in disrupted regions even away from the center of the trailing hemisphere.

The known energy inputs from solar UV and magnetospheric particle irradiation (Cooper et al. 2001, Paranicas et al. 2001) make the surface a chemical hotbed for rapid processing of ice and non-ice materials into astrobiologically interesting substances such as oxygen and hydrocarbons that could be conveyed to the ocean over millions of years. Detection of H_2O_2 (Carlson et al. 1999a) suggests a highly oxidized environment consistent with high levels of radiation processing and preferential escape of hydrogen as from the earth's atmosphere, but do these oxidants and other irradiation products persist at deeper levels? H_2O_2 is known to persist for thousands of years in terrestrial polar ice cores (Neftel and Fuhrer 1993) and is highly soluble in water, thus potentially providing an efficient medium for oxygen transfer from the surface to the ocean (Chyba 2000, Cooper et al. 2001, Chyba and Phillips 2001, Chyba and Hand 2001), even in the likely absence of abundant photosynthetic production. Gaidos et al. (1999) had presumed not long before that lack of effective solar light penetration through the ice would render photosynthesis ineffective and the ocean oxygen free. Various other authors, including Arthur C. Clarke in his Space Odyssey (2001 to 3001) novels, have made the same assumption of an oxygen-poor ocean, which may no longer be tenable.

Two different interpretations have been proposed for Europa's hydrated material, hydrated salt minerals (McCord et al. 1999) and sulfuric acid (Carlson et al. 1999b). The salt hypothesis suggests an internal briny ocean source whereas the sulfuric acid is a radiolysis product of sulfurous material in water ice. In this case, the original source is hidden, but candidates include (1) the implantation of Iogenic sulfur ions, (2) an acidic sulfuric acid ocean (Kargel et al. 2000, 2001; Marion 2002), or (3) a briny ocean containing sulfate salts (Kargel et al. 2000). The magnetic signature of Europa is consistent with both acidic or salty oceans, however the extent to which oceanic materials are introduced to the surface is unknown. NIMS spectral mapping could be consistent with exogenic deposition of magnetospheric sulfur ions, since bright central bands in Europa's linear ridges, presumably consisting of upwelling subsurface material, are brighter with less non-ice material than the surrounding regions.

The case for a salty ocean source would be enhanced if evidence for salt cations were found on the surface. Magnesium and sodium are the most likely cations, but Na can be introduced from the Io neutral and ionized tori so its presence is not a definitive indicator of an oceanic source. However, Johnson (2000) argued that Europa is a net magnetospheric source of Na for the observed Na neutral cloud component of Europa's extended atmosphere. Brown's (2001) observations revealed a significant increase in the Na/K ratio from the cloud of Io to that of Europa, so there is substantial support for the endogenic source argument. The Na cloud results from preferential sputtering of neutral Na, so Mg abundance may be relatively enhanced on the surface. Radiolysis predominately drives the surface chemical composition so that Mg may be found as a salt cation or as an oxide or hydroxide. An upper limits of three percent (by number) has been obtained for $Mg(OH)_2$ (Shirley et al. 1999).

Key composition questions to be resolved by ongoing (Galileo Orbiter) and future (orbiter, lander) measurements include the elemental, isotopic, and molecular composition of the sensible surface in different geologic regions. Sputtering by magnetospheric ions preferentially removes more volatile species (e.g., H_2O, O_2, Na), and lighter isotopes of these elements, from the surface. 'Fresh' material from below might therefore be distinctive in terms of having more volatiles of non-radiolytic origin and relatively lighter isotopes. The latter is also a potential biological signature. Are distinctive non-ice compositions near surface fractures, ridges, chaos regions, apparent flood zones, pits, and domes representative of deep subsurface composition or more so of liquid water emergence and subsequent effects of brine mobilization? Are determinations of composition on the exposed upper surface meaningful in view of heavy processing by magnetospheric irradiation, solar UV, and meteoritic impacts?

What are the depth profiles within meters of the surface for abundance of oxidants, other volatile gases, and hydrocarbon products of irradiation, and could these materials serve as potential resource reservoirs for life in liquid water regions at kilometers-deep levels? Is there significant chemical exchange between the surface and these deeper regions? Is there an accessible depth at meters or deeper in which more 'pristine' material can be sampled by lander instruments?

Cooper et al. (2001) and Chyba and Hand (2001) computed radiolytic energy inputs in ice and liquid water from radioisotope decays. Models for thermal evolution of the satellite interior over billions of years rely on expected radioisotope abundance, assumed to be chondritic (Kargel et al. 2000), within the core, mantle, and lower density water/ice layers. The radioisotope K^{40} may be found in sea salts on Europa in greater abundances than on the Earth, due to scaling from meteoritic abundances, and could provide significant radiolytic energy input for evolution of life within Europa (Chyba and Hand 2001), in addition to that from magnetospheric irradiation of the exposed ice surface. Can interior abundances of these radioisotopes be inferred from isotopic measurements at the more accessible surface?

What is the chemistry of the ocean, if it exists, and of the underlying core and mantle? How do hydrothermal sources of chemicals and heat from the mantle affect the ocean? Is the ocean oxidizing like the irradiated surface, and the modern terrestrial ocean, or more reducing like the early Archean ocean of Earth due to outgassing (H_2, CO, CH_4) and outflow (H_2S, NH_3) of other reductants from the interior? Kargel et al. (2000) noted that one possible outcome of early interior evolution within Europa could be formation of a large bed of sulfur below the ocean, conceivably consistent with findings of sulfuric acid on the surface and perhaps indicative of a highly acidic ocean. Another evolutionary scenario produces a high abundance of salt in the ocean, consistent with surface measurements of mineral salt hydrates and an ocean environment more like the Dead Sea on Earth where only halophilic organisms prosper. Different interpretations of small spectroscopic features in the near-infrared data lead to vastly different outcomes for Europa and its putative ocean. Which interpretation is correct? Are spores or remnants of biological material already visible on the surface, needing only improved spectroscopic measurements to be confirmed?

As in the previous section, does the surface chemical and irradiation environment warrant in-situ biochemical and biological investigations while also motivating full implementation of planetary protection protocols to protect the putative ocean from contamination by spacecraft-borne bacteria from Earth? E.g., what are the chances that a radiation-resistant species like Deinococcus radiodurans could survive and be carried through the crust to the ocean? Would this hardy terrestrial bacterium also survive a highly oxidizing environment? Inversely, what would be the likelihood that subsurface organisms of Europa origin could reach the surface from the ocean and leave recognizable biochemical signatures or even living spores?

1.3. Other geophysical evidence for the ocean and data on the deeper interior?

Gravity surveys with the Galileo Orbiter spacecraft during many Europa flybys (Anderson et al. 1998) indicate a differentiated satellite interior with a 10^2-km thick outer layer of unit-density material, presumably water, although these data do not distinguish between water in the liquid and ice phases. Tidal gravity interactions with Jupiter and the other Galilean satellites, mostly understood from gravitational resonance models, are of course reputedly responsible for internal heating of Europa and maintenance of the liquid ocean, but limits on heat dissipation within Europa's ice from this source seem marginal compared to the heat required to keep the ocean liquid. However, the accompanying magnetometer surveys (Khurana et al. 1998; Kivelson et al. 1999, 2000) are consistent with expected levels of electric current induction in a modern salty ocean, not just one in the recent but lately lamented geologic past millions of years ago. The currents are induced by time variation of Jupiter's magnetospheric magnetic field at Europa as this satellite moves in its orbit relative to the Jovian dipole equator during Jupiter's 10-hour rotation period and Europa's 3.5-day (one Eurosol) orbital period.

Similar inferences from magnetometer and other (geology and composition) data for potential oceans within Ganymede and Callisto, more likely buried deeper at hundreds of kilometers, raise the question of whether Europa is unique. Is it merely representative of a large new class of such objects within the Jupiter system, the greater solar system, and the universe beyond? A major question must remain whether the magnetometer findings on induced fields can be accounted for by any other means? E.g., an explanation arising somehow from magnetospheric interactions without need for subsurface oceans might qualitatively account for all three satellites, although an effect induced by magnetospheric magnetic field or trapped radiation would necessarily fall off quantitatively from Europa to Callisto.

There is also the unexplored question of the seismo-acoustic environment which might reveal much about the interior by obvious analogy to the long history of seismic studies of the terrestrial interior of Earth. The constantly shifting crust under effects of tidal stresses may produce a veritable cacophony of sounds that could be exploited to probe the interior ice, ocean, mantle, and core layers and constrain the compositional and thermal states of each layer through detectable effects on sound speed and propagation times. Buried within the ambient noise of this environment may also be some sonic signature of life in the ocean. Future missions to Europa's surface would be well advised to include monitors for this environment, as we discuss later.

1.4. Do potentially habitable zones exist anywhere from near Europa's surface to within the ocean and below? What would be the accessible signatures of such life for investigation by future missions?

In traditional debates about life in the universe, Europa has not been favored until relatively recently. Its position in the Jovian system places it far outside the 'habitable zone' defined by orbits of terrestrial planets such as Earth (and early Mars?) which could receive energy for biological evolution from direct sunlight via photosynthetic production of oxygen and carbohydrates, while being neither too hot (Mercury, Venus) nor too cold (Jupiter and beyond). Small pockets of liquid water providing media for photosynthetic organisms might exist near Europa's ice-encrusted surface below meter-thick ice covers over tiny fractions of the surface (Reynolds et al. 1983). There was some geologic and chemical evidence for a subsurface ocean (e.g., Squyres et al. 1983), but it seemed unlikely that photosynthetic processes provided sufficient energy for any abundant life. As reflected in Arthur C. Clarke's novels of the pre-Galileo era, hypothetical life within the ocean of Europa evolved over billions of years in lonely struggles for rather pathetic existences within small isolated niches around hydrothermal vents. McCollom (1999) showed that such vents would supply limited chemical energy for life via methanogens, known bacteria which combine hydrothermal CO_2 and H_2 to form CH_4. Although the presence of ionizing radiation and resultant oxidants on the surface of Mars was widely (if not universally) recognized as being counterproductive to life on the surface of that planet, there was no real speculation (that we are aware of) concerning potentially positive aspects of such an environment for evolution of life on either Mars or Europa. Gaidos et al. (1999) gave the most recent negative assessment regarding Europan life from arguments on scarcity of chemical redox energy within the subsurface ocean, inputs at the surface being considered to be negligible in the absence of global sunlight access for photosynthesis.

Clearly the increasing evidence during the Galileo Orbiter mission for a Europan liquid ocean have spurred new interest in the possibility of Europa life and resultant discussions of planetary protection issues (Esposito et al. 2000). On the assumption of a relatively thin ice crust, Greenberg et al. (2000) have argued for habitable niches within cracks near the surface that are bathed with oceanic nutrients due to tidal pumping during each Eurosol. Presence of cycloidal cracks may argue for the thin crust (Hoppa et al. 1999) but there are alternative models (Pappalardo et al. 1998, 1999) for much of the surface geology which invoke convection of warm soft ice in a layer extending tens of kilometers below a thin and stagnant upper crust, whether or not a liquid ocean exists at lower depths. In either case the apparent 10^7 - 10^8 year age of the Europan surface strongly suggests the possibility of material exchange over such times between the visible surface and a putative ocean.

After Carlson et al.'s (1999a) detection of hydrogen peroxide on Europa, Chyba (2000) noted that this and other products of surface radiolysis from magnetospheric irradiation could provide at least minimal chemical resources (O_2, H_2CO) for biological evolution within the ocean after downward transport through the crust and conversion of radiolytic H_2O_2 to O_2 within the ocean. Although the detected H_2O_2 at 10^2-μm depths had a short lifetime due to photolytic dissociation by solar UV, Cooper et al. (2001) later showed that more long-lived radiolytic products could be produced by penetrating electrons and protons at millimeter depths below the UV photolysis zone. Furthermore, they argued that continual overturn of the meter-thick regolith on Europa by meteoritic impact gardening could protect significant fractions of these products over millions of years until downward transport to the ocean might conceivably occur. Subsequent works (Chyba and Phillips 2001, Chyba and Hand 2001) now take most of these processes into account. However, remaining uncertainties about chemical composition of the ice crust, losses to the atmosphere, chemical reactions during transport between the surface and the ocean, and the oceanic composition leave a wide range of values for biological resource production and consumption rates.

Upper limits given by Cooper et al. (2001) allow for large (i.e., terrestrial) concentrations of oxygen $\sim$20 mM (millimoles of O_2 per liter of H_2O) within the entire 10^2-km thick ocean, while lower concentration limits $\sim$0.2 mM (Chyba and Hand 2001) could arise now from continuous O_2 production by decay of the radioisotope K^{40} within salts in crustal ice and ocean water. The peak dissolved oxygen density in the photic zone (depth < 200 m) of Earth's oceans is $\sim$0.3 mM, while macrofauna in the deep oceans survive at $\sim$0.03 mM (Chyba and Hand 2001). Therefore even the lower limits on O_2 production leave much room for substantial O_2 losses in transit from the radiolytic source at the irradiated surface to a living ocean.

Habitability of the icy surface of Europa will clearly be limited in part by the extremely high radiation dosages from magnetospheric particles. The radiation-resistant bacterium Deinococcus radiodurans thrives happily at dosages up to 0.5 Mrads (Battista 1997), which would allow lifetimes greater than one month at

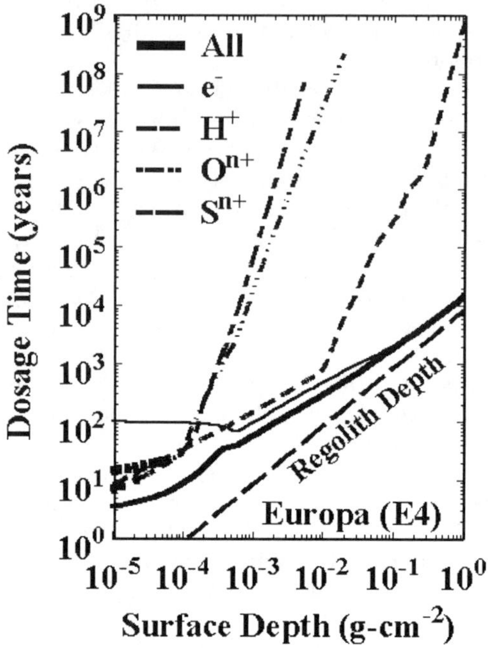

Figure 2. Depth profiles in water ice for times in years to accumulate 100-eV-(16-amu)$^{-1}$ radiation dosages from magnetospheric particles, collectively and individually shown in curves for protons, heavy ions, and electrons (Cooper et al. 2001). The O^{n+} and S^{n+} ions are of undetermined positive charge n. The ice is at unity density. Times are also shown in the lower dashed curve to overturn the surface regolith to given depths by micrometeoroid impacts.

depths greater than 0.4 cm. Less tolerant organisms (like humans) with lifetime dosage limits of 100 rads could get along for a month at the base of the meter-thick (Cooper et al. 2001, Chyba and Phillips 2001) Europan regolith. This also the thickness of ice that would quickly form over any fresh liquid water emerging at the surface and potentially carrying small organisms.

At dosage levels many orders of magnitude above those for survival limits, even the biochemical remnants of oceanic organisms carried to the 'hot' surface environment may not last long enough for detection by remote sensing or in-situ instruments. Figure 2 from Cooper et al. (2001) shows depth profiles of the relevant time scales for complete chemical modification of carbonaceous molecules with a standard molecular mass of 16 amu (i.e., CH_4), as also compared to regolith gardening times. Protons and heavy ions (O, S) dominate the dosage rates at μm (10^{-4} g-cm^{-2}) depths, while electrons and more penetrating protons are the main radiolytic components at greater depths. Molecules exposed at remotely sensible depths down to one centimeter are fully processed in less than 10^4 years, while this happens in only $\sim$10 years at μm depths. Sampling of biochemical markers of potential Europa life must then be done at meter depths to avoid prior decomposition by radiation effects over ice crust global resurfacing times ~10^7 to 10^8 years.

2. How Can the Key Science Questions be Addressed?

2.1. Geology and the search for signs of Europa's ocean

Looking for evolutionary change: The exciting prospect of a subsurface ocean is tempered by the lack of any recognizable morphological change in surface geology over the twenty-year gap span of observations from Voyager to Galileo Orbiter (Pappalardo et al. 1999, Phillips et al. 2000). The extension of an existing cycloidal crack, evolution of terrain around a linear ridge feature, formation of new pits and domes, emergence of a new flood zone, outgassing from a cryogenic plume through a surface fracture, growth of local hot spots, or any related change in spectroscopic reflectance would give some sign of recent geologic activity possibly associated with an ocean. If magnetometer data tell us that the ocean truly exists today, then one part of the daunting task ahead is to search for any signs of such activity through image mapping on consecutive missions in the years (and perhaps centuries or even millennia) ahead. We should take advantage of any opportunity to regenerate maps with new observations for at least portions of the visible surface geology, UV to IR reflectance, and thermal emission which can be compared to maps from earlier missions. Such observations, however, must be matched in filter, resolution, and viewing geometry as closely as possible to those of existing data sets. Also, long term monitoring with earth-based spectroscopic telescopes of the extremely thin atmosphere, normally produced by magnetospheric sputtering, may reveal localized enhancements from outgassing events.

Future Jovian system missions focusing on the planet, its atmosphere and rings, the Io plasma torus and the Jovian magnetosphere, and satellites including Europa will provide potential opportunities for brief snapshots or more detailed surveys of Europa's surface. Europa orbiter and lander missions will provide high resolution global or landing site data but may be constrained by available science payload mass and power, while targeted or opportunistic flybys by other missions may not be so limited. An important advantage of post-Galileo missions making such flybys is the inevitable improvement in remote sensing and in-situ instrument technologies.

Missions to the outer solar system beyond Jupiter and local interstellar space will likely continue to rely on Jovian Gravitational Assist (JGA) for the foreseeable future. One highly successful precedent for such opportunistic observations is the Cassini Orbiter's flyby of Jupiter in December 2000 enroute to Saturn. Cassini provided simultaneous and complementary observations of the Jovian system during the ongoing Galileo Orbiter mission, although it remained too far away from the Galilean satellites for high-resolution surface imaging.

Future missions flying closer to Europa should be encouraged, no matter what their final destinations, and could enormously expand our imaging databases for the Europan surface and provide the necessary extension forward in time, while also making more direct searches for the ocean by other means described below. Needless to say, comparisons of maps from different missions will not be feasible if the imaging and spectroscopic data are not preserved indefinitely in appropriate data formats within secure archives funded for operations into the indefinite future, and if there is no continuity in science support for review of the archival data in comparison to new observations. These points are further discussed later in the report.

In the meantime, it should be recognized that an enormous amount of crucial information has already been collected from the Voyager and Galileo Orbiter missions, and that long-term data analysis and modeling efforts should be considered part of a separate 'mission' to fully exploit this information. Understanding processes for evolutionary change on the visible surface of Europa, and within the overlying atmosphere, is critical to future confirmation of the ocean's existence. The geology appears very alien but there may be terrestrial analogs (e.g., Lake Vostok in Antarctica) that can be applied with models originally developed for earth systems science, particularly for polar regions (see Farr et al., this volume, pg. 35). The relation of thermal map features to geology and underlying processes such as ice convection and diapirism needs further modeling, as do the chemical thermodynamics of variants on the possible successive hard ice, soft ice, liquid ocean, mantle, and core layers plus phase changes in local pockets (e.g., brine inclusions) within each layer. If the chemistry of the trailing hemisphere in particular, and the global surface overall, has been significantly affected by magnetospheric irradiation, perhaps by preferential sputtering of volatile elements and radiolytic formation of oxidants and sulfate compounds, then modeling is needed for determination of how this affects evolutionary change and thermal profiles of material in the visible crust and within deeper layers. For astrobiology and planetary protection we want to know from observations and models whether one or more 'conveyor belt' processes exists for exchange of material between the irradiated surface and the ocean. If so, one what time scales could these processes act? Chemical reaction rate modeling under crust chemistry and thermal conditions is needed for determination of chemical abundances of products delivered on each such time scale. Laboratory simulations of such reactions under Europan cryogenic conditions are needed to validate the chemical rate models.

Tidal gravity: One possible way to detect an ocean is to measure tides and tidal effects on the trajectory of a low-altitude Europan orbiter, since a body with internal fluid layers deforms more than an otherwise similar but solid body. Greenberg's (2002) review notes that the peak tidal height at Europa's surface due to Jupiter's pull on a sub-surface ocean for a circular orbit around Jupiter would be $\sim$500 m at Europa's sub-jovian and anti-jovian nodes. Since Europa's rotation during each Eurosol is phase-locked to its orbital motion of the same period around Jupiter, this primary tidal component is constant. However, there is a smaller, but more critical, time-varying component of approximate height $\sim$30 m, also dependent on respective thickness of the ice crust and ocean, that is induced by eccentricity of Europa's orbit. This diurnal tide peaks when Europa is at perijove and may alternatively be only a few meters in vertical amplitude without a liquid ocean. Time-dependent global measurements of both components during a month-long orbiter mission would provide critical new information, the time-varying component being more sensitive to presence of a low-viscosity ocean layer.

For a nearly spherically-symmetric Europa, the time-variable gravity potential perturbed by tidal deformation at the spacecraft position is

$$\Delta V_2(R,\theta,\lambda) = k_2 R^3 r^{-3}\, V_2^J(R,\theta,\lambda)$$

and the ground uplift is

$$u_2(R,\theta,\lambda) \;=\; h_2 g^{-1}\, V_2^J(R,\theta,\lambda)$$

V_2^J is the 2nd-degree tide-generating potential raised by Jupiter, which is fairly well known. R and g are the surface radius and gravity of Europa. k_2 and h_2 are tidal Love numbers that describe the response of Europa under tidal forcing.

For a four-layer Europa model (ice shell, liquid ocean, rocky mantle and an iron rich core (Anderson et al. 1998), k_2 falls from a peak value of about 0.26 for a 10-km ice shell to 0.17-0.22 (dependent on other model parameters) for 150 km, before dropping precipitously to 0.02 at 160 km (the presently assumed total thickness of ice and ocean) with no ocean in the model. k_2 (and h_2) also depends significantly on the thickness and rigidity of the ice shell, the rigidity of the mantle and core, and the density of the salty ocean water (not shown). On the other hand, the Love numbers are relatively insensitive to the thickness of the ocean and to the densities of core, mantle, and the ice shell. The sensitivity of the Love numbers to multiple interior parameters requires a systematic measurement system to determine both Love numbers, the tidally induced rotational variation of the ice shell, and high resolution static topography and gravity, for successful resolution of all relevant interior parameters (Yoder 2002).

Such measurement objectives can be met by the planned Europa Orbiter with a preferably high inclination, near circular orbits, and a science mapping duration of a month or even less. The spacecraft would carry an X-band radio Doppler tracking facility, an altimeter, an imaging instrument, an ice penetrating radar sounder, and/or other instruments. For example, it has been demonstrated that k_2 can be determined to an astonishing accuracy of 0.0004 along with the static gravity field (Figure 3) using only 15 days of Doppler tracking from two Deep Space Network stations on Earth (Wu et al. 2001). This should be compared with the nominal value of 0.28 for this Love number. The uplift Love number h_2, the amplitude of ice shell rotational variation, and the static topography can be determined by the altimeter and/or imager. If sufficient accuracy can be achieved for the other measurements, the high k_2 accuracy implies not only that a global ocean can be detected, but also that an ice shell thickness can be determined with accuracy between 200 m and one km.

The high accuracy for k_2 also raises the hope to detect possible near-resonant excitation of internal oscillation modes that may be used to further constrain the interior structure such as the thickness of the ocean and the existence of a molten outer core. In addition, with a global liquid ocean in between, the ice shell and the rocky mantle will most likely have different amplitudes of rotational oscillations. Such amplitude of the mantle can be measured by Doppler orbit tracking with an accuracy of about one arc-second with three stations on Earth, through rotational effects on the gravity field. Differential amplitude of rotation would be further unambiguous evidence of the global subsurface ocean.

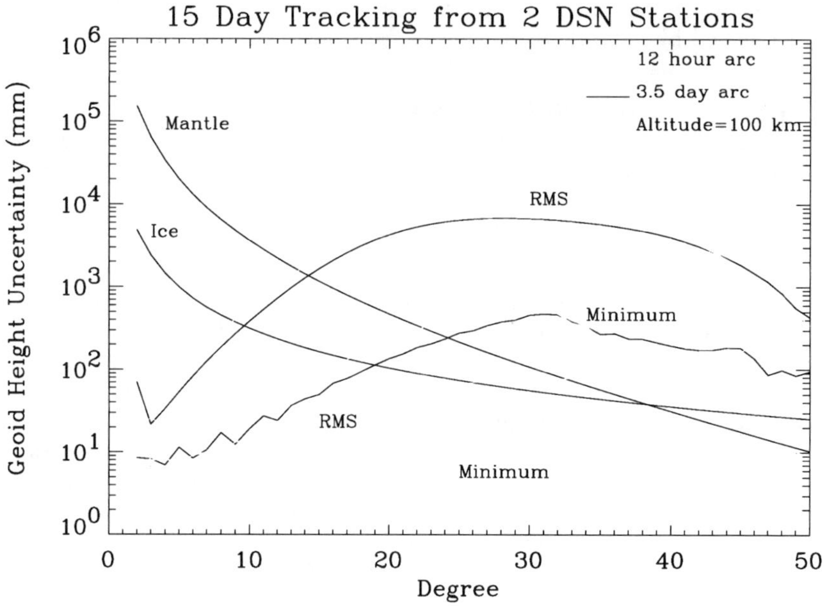

Figure 3. Anticipated static geoid anomaly (equipotential surface, proportional to gravitational potential) spherical harmonic coefficient accuracy expressed as Root Mean Square (RMS) and minimum per degree for three orbit scenarios (Wu et al. 2001). The nominal orbit inclination is 83° for an altitude of 200 km.

Another scenario that should draw our attention is the possibility of separate subsurface ocean basins instead of a global ocean. Obviously, this possibility significantly strengthens the case for global coverage and systematic mapping from a robust orbiter mission to determine the distribution and dimensions of these ocean basins. In this case, the tidal response will not be spherical-symmetric and will contain a richer spherical harmonic spectrum than the simple 2nd degree tidal-forcing potential. With the help of a non-spherical-symmetric tidal deformation model, a low-flying orbiter currently remains the best candidate mission for this challenge. An orbiter could resolve a large number of deformational response harmonics with high accuracy. Any substantial non-spherical-symmetric tidal deformation (greater than five percent of normal tides) would be a clear indication of non-global ocean distribution.

Laser altimetry: Another method of measuring Europa's surface gravity tides utilizes laser altimetry. A well-known example of the former is the Mars Orbiter Laser Altimeter (MOLA) in orbit around Mars on the Mars Global Surveyor (MGS) spacecraft. MOLA transmitted infrared laser pulses towards Mars at a rate of 10 Hz and measured the time of flight to determine the range of the MGS spacecraft to the Martian surface. These range measurements have been used to construct precise topographic maps of Mars accurate to about one meter in topographic altitude and 100 meters in horizontal location after detailed reconstruction of the spacecraft orbit (Smith et al. 2001). A MOLA-type instrument on the Europa Orbiter could therefore be very sensitive to tidal topographic changes of tens of meters on the global surface of Europa during each Eurosol for a liquid ocean.

Radar sounding: The proposed radar sounder experiment on Europa orbiter would probe the Europan ice crust to maximum depth ~10 km (20 km for very clean, non-convecting ice) in search of the ice-liquid boundary. Effectiveness of this instrument will be severely constrained by the physical state of the ice, including fractures, brine inclusions, and other variations. This facility-class instrument would use an array of three standard 3-element Yagi antennas forming a narrow downward 50-MHz beam from a baseline perpendicular to the ground track of the orbiter. Measurements at an orbital altitude of 100 - 200 km would have a surface horizontal resolution of one to two km and depth resolution of 100 m for boundary layers down to 20 km in depth. Mean altimetric elevation and surface roughness scale resolutions would be 15 m and 30 m, respectively, possibly not sufficient to resolve the time-varying diurnal tidal component for a liquid ocean.

Induced magnetic field: Notable by its absence from the strawman payload for Europa Orbiter is a magnetometer capable of high-resolution mapping for global time variations of induced magnetic fields (Figure 4) in an apparently salty modern ocean during the month-long orbital mission phase. Although the putative ocean was first discovered using low-order, dipole components of the induced magnetic fields during multiple flybys of Europa (Khurana et al. 1998; Kivelson et al. 1999, 2000), a protracted magnetic survey in orbit could allow resolution of higher order moments and more precisely model induced current systems at depths far below those accessible to surface altimetry and radar sounding. This would be one of only two techniques (the other is seismo-acoustic sounding) capable of detecting and surveying oceans at depths of hundreds of kilometers as appear to be present on Ganymede and Callisto, again primarily from magnetometer results.

Magnetospheric ion shadow of Europa: Orbital monitoring of interactions of high energy (MeV) magnetospheric heavy (O, S) ions with the surface would be highly complementary to a magnetic survey. These ions globally sample the magnetic environment along incident trajectories from the magnetospheric environment to the spacecraft and the surface, not just along the spacecraft orbit. A spinning spacecraft or instrument platform would be required to measure directional distributions of the ions and image the 'shadow' produced by ion absorption at the satellite surface as seen at the orbiting spacecraft. This type of shadow is unlike that seen for electrons and low energy ions in the satellite's magnetospheric wake, since the level of ion depletion is strongly dependent on incidence direction at the sensor. The spatial shape and extent of an energetic ion shadow is related to the gyroradii of the ions in the local magnetic field. For iogenic ions with low charge states (e.g., O^{2+}, S^{3+}) these gyroradii are of the order of the Europan radius (Cooper et al. 2001) at MeV energies. Such shadows were observed (Cooper 2001) during most of the Io and Europa flybys

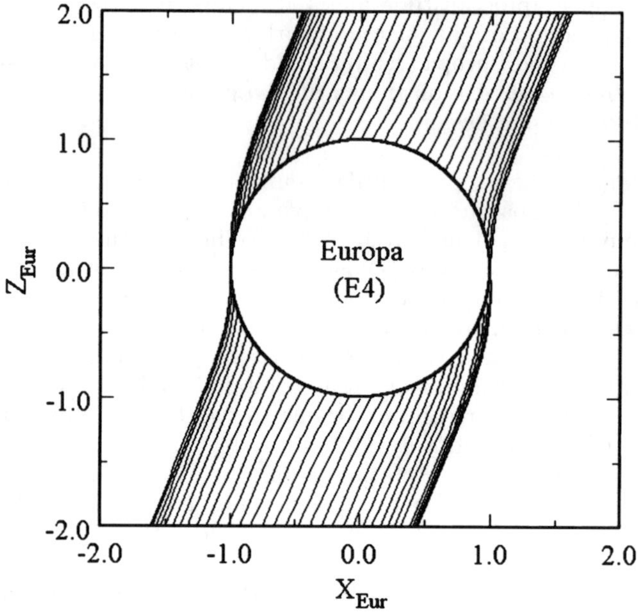

Figure 4. Deflection of Jovian magnetospheric field lines near Europa by an induced magnetic field from a subsurface ocean (Cooper et al. 2001).

by the Galileo Orbiter in anisotropic intensity data for > 40 MeV ions detected by the Heavy Ion Counter (HIC). Radiation dosage monitoring for less energetic particles, particularly electrons and protons, would also be complementary to surface composition mapping in terms of models for radiolytic processing of surface ices.

Seismo-acoustic sounding: A second deep sounding technique could utilize seismo-acoustic waves measured at one or more monitoring stations (Figure 5) on the surface to probe both for the Europa ocean and into the deeper interior down to the satellite core. Robust measurements can be made of ice crust thickness and ocean depth even with a single seismometer by measuring multiple arrivals and returns of waves from various sub-surface layers, as is routinely done on Earth to measure ocean depth and sub-bottom layering. Makris et al. (2001) predicted for such measurements that waves in the 0.1 to 100 Hz range with wavelengths of about one kilometer would be emitted at thirty-second intervals from individual cycloidal crack features and at monthly intervals from meteor impacts. While radar sounding would be susceptible to interference by small-scale ice fractures and voids, possibly including brine channels associated with the reported surface sulfates, these density inhomogeneities would be relatively transparent to the long-wavelength seismo-acoustic waves.

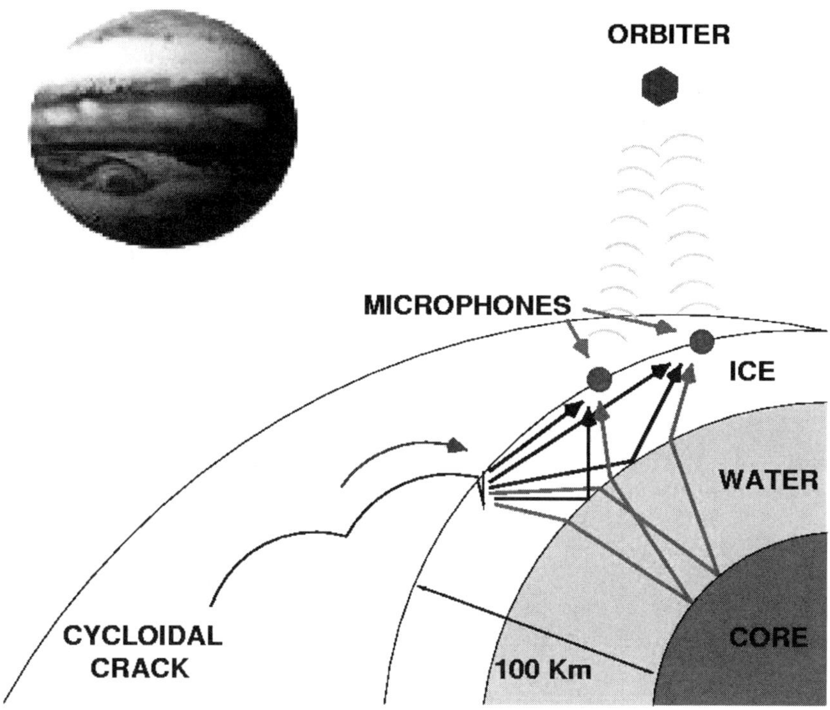

Figure 5. Seismo-acoustic deep sounding for the ocean and interior core within Europa.

An isolated sound source could produce numerous echoes from multiple reflections of compressional, shear, and compressional-shear waves off various sub-surface layers. Makris et al. (2001) found that spacing of active surface cracks would need to be about that of a typical cycloidal crack length for optimal signal-to-noise ratios to allow resolution of isolated wave sources from ambient background with a single seismometer. Time correlation of background noise level to Europa's orbital motion would also set useful limits on global gravity tides. The time series and spectrum of ambient noise (Urick 1986) will also be of astrobiological interest if living sound sources are present.

Kovach and Chyba (2001) noted that the ice crust thickness above a Europan ocean could be deduced by measurements of Crary waves, a type of vertically (perpendicular to surface) polarized shear wave produced by critical angle reflections at upper and lower ice surface boundaries. Constructive interference between multiply reflected waves occurs for waves of a characteristic frequency dependent on vertical separation of these boundaries. Crary waves from discrete distant sources within the crust would arrive at the seismometer after the first compressional waves but before the horizontally polarized (parallel to surface) shear waves. For compressional and shear wave velocities of 4.0 and 2.0 km-s^{-1}, respectively, the Crary wave frequencies vary from 0.48 Hz for a 5-km crust to 0.24 Hz for 10 km.

Another diagnostic for presence of an underlying liquid boundary would come from the frequency dispersion of initial arrival time at the seismometer for horizontally polarized shear waves, in this case called Love waves. In traveling through the ice 'waveguide' above a liquid ocean, the first arrivals would be the higher frequency Love waves, which is inverse to the result for ice overlying a denser (e.g., rocky) substrate, for which faster shear wave velocities in the substrate produce the first arrivals at lower frequencies. The predominant group velocity of the arriving wave train also provides a measure of the ice crust thickness. Kovach and Chyba (2001) concluded that monitoring of waves in the 0.08 to 10 Hz range would be useful for limits on ice shell thickness up to tens of kilometers.

2.2. Chemical composition of Europa's surface

Solar reflectance at near-infrared to ultraviolet wavelengths: Targeted and opportunistic observations of Europa's surface and atmospheric composition are needed into the foreseeable future. The principal findings to date on water ice, sulfate hydrates, H_2O_2, CO_2, and SO_2 are derived from absorption features in sunlight reflected off the surface at near-IR and UV wavelengths as observed at limited spectral and spatial resolutions with the Galileo Orbiter Near-Infrared Mapping Spectrometer (NIMS) and Ultraviolet Spectrometer (UVS) instruments. Hubble Space Telescope (HST) observations with high spectral (but little spatial) resolution have been used at visible to UV wavelengths to measure or set limits on oxygen, sodium, and potassium in Europa's near surface and extended atmospheres. HST observations also indicate that O_2 is also present in condensed phase at the surface.

A significant limitation to varying degrees of many present observations is that derived compositional information could not be well correlated to geologic features due to limited spatial resolution. Improved spatial mapping at UV and IR wavelengths of H_2O_2 and OH would be of great interest for correlation to composition affected by radiolytic processes. Higher IR spectral resolution would allow measurement of structural OH absorption bands (e.g., NaOH) at 1.48 μm and perhaps other bands not yet detected on Europa. High resolution mapping of SO_2 at UV to IR wavelengths, and correlation to dark material and hydrates, might resolve current questions about sources of these materials. Improved IR spectra near 1.5 μm could detect band structures produced by hydrated salts as compared to hydrated H_2SO_4. Spatially coordinated visible, UV, and X-ray spectra would be best of all. A major question to answer is whether Europa's SO_2 correlates spatially to H_2SO_4? These molecules do correlate on coarse scales, and this also be true on fine scales.

Extension of near-IR measurements beyond the 5.2 μm wavelength limit of NIMS would be highly desirable, but the fundamental bands at longer wavelengths generally involve H (e.g., -OH, -CH, -NH). There are some fundamentals (e.g., CO_2 and CO) in the 4 - 5 μm region, but most are beyond 5 μm. However, at NIMS wavelengths the solar spectrum intensity diminishes quickly and thermal emission becomes the dominant source of photons. Thermal emission spectroscopy of surfaces hasn't proven to be very diagnostic of the icy satellites. Extending the measurement region from 5 to 8 μm would allow measurement of carbonates, nitrates, chlorates, and many sulfites and sulfates.

Spectral resolution of future observations should be better than NIMS but not excessively so. The goal would be to see narrow absorption features such as structural OH. These bands occur at ~3 μm, 2.2 - 2.4 μm, and ~1.4 μm. Ten times better spectral resolution would enable these lines to be resolved and their positions to be measured accurately. The profile of the 1.65-μm H_2O feature is indicative of temperature, but NIMS's resolution was insufficient to resolve this feature. An O_2 band at ~1.27 μm might be discerned from an H_2O band at higher resolution than available from NIMS. The shape of the CO_2 bands on the icy satellites is indicative of the local environment and could be addressed with resolution ten times better than NIMS.

Advanced technology imaging instruments on future missions going to or through the Jovian system will likely offer improved spectral and spatial resolution as opportunities to target the Europa surface arise. Obtaining UV, visible, and IR spectra at similar spatial resolutions of a few kilometers on future remote sensing missions would allow high resolution delineation of compositional gradients, and temporal changes thereof, across linear ridges, chaos zones, lenticulae (e.g., pits and domes), potential flood zones, and other disrupted terrain. Such observations may reveal the presence of chemical species originating either directly from the subsurface ocean, as perhaps might be found in bright central bands of linear ridges, or indirectly from radiolytic processing of ocean ejecta in darker regions near the brighter source material.

Relatively reduced molecular compounds of some elements (e.g., H_2S as compared to H_2SO_4 for sulfur, CH_4 and CO vs. CO_2 for carbon, NH_3 vs. HNO_3 for nitrogen) could be signatures of fresh injection to the oxidizing surface from below, if the ocean was relatively reducing as in the case of the Archean (2.5 - 3.8 billion years ago) ocean of Earth before the rise of biotic oxygen to modern levels. Measurement of the elemental ratio Fe/O could be partly diagnostic of an Archean versus a modern Earth-like ocean, since this ratio decreased by several orders of magnitude in Earth's ocean as rising oxygen production precipitated out the dissolved Fe from hydrothermal sources. This ratio is therefore inversely correlated to the oxidation state of the ocean, which might in turn be affected by the radiolytic oxygen source at the surface and efficiency of downward transport through the ice crust.

Even higher spectral resolution would allow measurements of isotopic ratios - these are significant because magnetospheric sputtering preferentially removes the light isotopes over time. Thus the ratios of heavy to light isotopes may be correlated to surface age in observed regions, while local concentrations of light isotopes may be indicative of fresh material ejected from below. Isotopic abundances of C, S, O, N, etc. within material from the ocean are also of interest due to isotopic fractionation in biogeochemical processes (Schlesinger 1997) such as methanogenesis, sulfate reduction, photosynthesis, and nitrogen cycling. 'Hotspots' of biological materials might then be manifested through relatively high ratios of C^{12} to C^{13}.

X-ray, Gamma ray, neutron, and optical emissions: It may be possible to use emissions generated by charged particle impact to determine the composition of the surface. Oxygen Kα X-rays have been observed from Europa (Howell 2001). The excitation mechanism may be impact of high-energy oxygen ions (Howell 2001) or electron excitation. Magnesium Kα X-rays occur at higher energies and are absorbed by O, but this fluorescence may be observable if Mg is abundant. Other elements (S, Al, Fe, etc.) are discernable through X-ray spectroscopy. Magnetospheric proton and heavier ion bombardment would also produce gamma-ray line and neutron emissions through excitation and spallation of target nuclei in Europan surface material. Energies of these nuclear emissions can be related to elemental composition of the irradiated material, as successfully done by the Lunar Prospector mission for the Moon and more recently by Mars Odyssey for Mars. Separation of the compositional signatures from general background emissions will, however, be nontrivial in the intense radiation environment of the Jovian magnetosphere.

Protons impacting magnesium sulfates sputter excited Mg atoms, and fluorescence of these atoms is seen in ultraviolet and visible emission lines (Nash and Fanale 1977). These lines are superposed on a luminescence continuum and molecular emission bands may be produced. The energy efficiency for producing optical emissions from proton impact is $\sim 10^{-3}$ to 10^{-6} (Nelson and Nash 1979). The efficiency for electron impact needs to be measured and further work is needed for proton impact on Mg salts and mixtures. Molecular emissions such as the ultraviolet OH band produced in electron impact of ice may also be used to determine the spatial distribution of the incident electron flux. Cerenkov radiation can also be used as a probe of electron impact while absorption features in the Cerenkov spectrum provide compositional information.

Atmospheric mass spectroscopy: Magnetospheric sputtering ejects more volatile neutral atoms and molecules (O_2, CO_2, CO, SO_2, H_2S, Na, K) from the surface into Europa's neutral atmosphere at altitudes up to a few hundred kilometers. Ionization of the ejected neutrals by solar UV photons and interactions with magnetospheric plasma forms a thin ionosphere. Some of the neutrals fly far out from the surface before ionization and magnetospheric pickup occurs, such that an extended atmosphere forms, as observed around Europa for Na (Brown and Hill 1996) and K (Brown 2001). Johnson et al. (1998), and Johnson and Sittler (1990), noted that direct atmospheric sampling of abundance for these sputtering products with neutral and ion mass spectrometers could provide information on the composition of satellite surfaces.

It would be highly desirable to include neutral and ion mass spectrometers on future spacecraft making Europa close flybys (< 200 km) or going into low altitude orbit. Another valuable experiment would be to directly image energetic neutrals being ejected from the surface using instruments of the type now carried on the Cassini Orbiter mission to Saturn and the IMAGE spacecraft in earth orbit. Since the neutrals are unaffected by magnetic fields around Europa, compositional information on specific source sites imaged by such instruments could be derived.

Dust ring composition: Dust grains are ejected from the Galilean satellite surfaces by hypervelocity micrometeoroid impacts and form a ring around each satellite, as already observed and modeled for Ganymede (Krüger et al. 1999, 2000). The dust rings around Europa and Callisto have similar densities to that around Ganymede (Krüger et al. 2002). After impacts at the source site, the ejected dust particles with sizes up to μm's come direct off the surface on ballistic trajectories and mostly fall back to the surface within seconds to minutes. This means no significant accumulation of radiation or atmospheric interaction, so the dust ring composition reflects the source site composition.

However, there is a problem of dust speed relative to the spacecraft, since the spacecraft speed at 200 km in circular orbit is only 1.5 km-s^{-1}. At this speed only the dust grains at the tail end of the ejecta size distribution can be measured with present instruments, and only a few grains would be analyzed during each orbit. Krüger et al. (H. Krüger, private communication, 2002) counted eight ring grains on average during each of eight Europa flybys by Galileo Orbiter. Since there would be about 330 orbits in a month-long mission, and efficiency for compositional analysis would be less than for mere counting, some elemental abundance information with limited statistics might be collected from several hundred total events during an orbiter mission. Future improvements in measurement techniques might increase the orbital sampling rate.

A large-area collection system (e.g., aerogel) on a single flyby mission returning to Earth for recovery and ground laboratory analysis would be another possibility and could also include a high resolution search for spores of microscopic life forms or their chemical signatures within the sample grains. There is now significant heritage for such systems from the ongoing Stardust and Genesis missions, which will respectively return samples of cometary dust and solar wind plasma ions to Earth. An earlier Discovery mission proposal called Europa Ice Clipper would have flown once by Europa and collected plume material produced from high speed impact of a pre-released metal ball on Europa's surface. It would be far simpler, however, to collect dust, neutral gas, and plasma ions from Europa's natural environment for return to Earth.

Raman spectroscopy: For in-situ chemical analyses on the Europan surface it will be necessary to use techniques which are not susceptible to radiation and magnetic field backgrounds from the satellite's interaction with the magnetosphere. One such technique may be Raman spectroscopy, which uses inelastic scattering of monochromatic light photons from lasers at visible to infrared wavelengths for derivation of information on composition of targeted material. Energy changes of the incident photons are related to various rotational and vibrational energy states of the target molecule and give information on its energy levels. Since the energy of the laser can be highly focused on small (e.g., μg) samples of target material, natural backgrounds from magnetospheric irradiation can be overcome and information can be obtained from trace samples filtered out of bulk material. Ability to change the laser wavelength allows avoidance of specific fluorescence backgrounds. Applications of modern CCD arrays, interferometry, and fast fourier transforms for data processing provide high spectral resolution and excellent background rejection.

Gas chromatography: Recent meetings of the NASA Astrobiology Institute's Europa Focus Group have defined mass spectrometers employing gas chromatography as being highly desirable instruments in the detection and analysis of inorganic, organic, and biochemical materials at the Europan surface and within the local atmosphere. These instruments operate at specific temperatures to vaporize selected samples and inject the resultant vapor into the head of a chromatographic column. Rates of vapor flow through the column are used to derive the molecular mass distribution of the sample material for masses ranging up to thousands of atomic mass units. Stepwise heating allows analysis of a wide range of molecular species with different vaporization temperatures.

2.3. Habitability and biochemical signatures of life

The search for habitable zones and associated life within Europa is first of all a search for liquid water. It will be an extreme risk to send an expensive, primarily astrobiological mission like Mars Viking to the surface of Europa before more definitive evidence is obtained on the presence of liquid water below the surface. The previous discussion indicates that the two most sensitive techniques available from Europa orbit to remotely sense tidal flexure from a liquid subsurface ocean include tidal gravity and laser altimetry mapping. These techniques provide vertical surface position resolution of a few meters or better which is not available from a radar mapping experiment. An orbital magnetic survey would also be highly desirable and cost effective, since this technique has given the only direct evidence for a modern ocean.

Observations at maximal spatial and spectral resolution of central bright bands within linear ridges would provide upper limits on fractional area of the global surface covered by thin ice lids over freshly emergent liquid water. This fractional area is likely to be extremely small as noted by Reynolds et al. (1983), so that photosynthetic activity could take place only in small niches widely separated on the global surface. This situation has motivated searches for other energy sources such as magnetospheric irradiation and electric current induction, frictional heating in cracks within the ice crust, diapiric plumes, radioisotope decay, and chemical energy supplied by hydrothermal fluids. Continuing analysis and modeling of geologic and compositional observations relevant to such energy sources are needed. Details of chemical and thermal gradients for temporally evolving structures such as diapiric plumes need further understanding.

The visible surface of Europa to at least centimeter depths is highly inhospitable to most terrestrial life forms. Worse, the rate of radiolytic reactions from magnetospheric irradiation is so high that even the remnants and biochemical markers of micro- and macro-faunal corpses could be highly degraded unless these are either freshly emerged within a few years or less from subsurface reservoirs, shielded from direct irradiation by overhanging geologic structures, or buried more deeply in the regolith. If remote monitoring from Europa orbit or elsewhere (Jupiter orbit, Earth) could identify a zone of fresh emergence (e.g., an Io-like plume) on such time scales before a final landing site is targeted, direct in-situ search for relatively fresh biological material might be possible. A remotely operated surface rover could presumably search for protected niches, within which relatively pristine material might be found and analyzed.

A lander platform with astrobiological objectives must otherwise have the capability to probe the underlying surface chemistry to several meters in depth. Cooper et al. (2001) define one meter as the average growth depth of the ice regolith under the effect of meteoritic impact gardening over a surface age of 10^7 years. To probe beneath the regolith one could land at the center of the trailing hemisphere where the meteoritic impact rate is reduced by an order of magnitude from that at the center of the leading side, but then the irradiation effects would be proportionately higher than at meter depths. Since it is commonly assumed that living organisms could only exist on Europa in liquid environments at temperatures far above that ~100 K at the surface, these would also likely only be found by deep surface probes, e.g., for brine pockets perhaps associated with diapiric plumes or with hot spots of chemical oxidation from reactions of buried oxidants (i.e., from radiolytic sources) with carbonaceous materials. Orbital surveys of hot spots, oxidant concentrations, and related geologic features would be needed to define where these pockets might most likely be located.

3. Recommendations for 2003 - 2013 Science and Mission Priorities

3.1. Find the water now, look for bugs later

From the perspectives of both planetary geology and astrobiology the most important question raised initially by Voyager observations, and now in far greater detail by those of Galileo Orbiter, is the possible and perhaps even likely presence of liquid water below the icy surface of Europa. There is much visual evidence that water may been responsible for disruption of the surface topography in recent geologic times, measured in millions of years, but there is no consensus that this evidence is totally definitive. Current models of the Galileo Orbiter magnetometer data near Europa do strongly indicate, in the absence of any other viable explanation for the induced magnetic fields, that a salty ocean could be present today. Evolutionary models for Galilean satellite interiors do not appear to rule out such an ocean, even at Ganymede and Callisto where the evidence from induced magnetic fields is also compelling. Evidence for bear-infrared detection of 'sea salts' ejected to the surface from below is extremely interesting but also not yet definitive. The alternative model for H_2SO_4 hydrate, the sulfur perhaps or not being of magnetospheric origin, has not been ruled out, although this may still require brine mobilization from subsurface liquid water to account for the observations.

The central theme of mission planning and operational missions in the next decade should be to determine whether Europa's ocean really exists and to what degree. Are we talking about slushy ice, some scattered pockets of brine, or a liquid ocean greater in volume than that of Earth? The search must be done in stages reasonable in technical feasibility, likely science return, and cost with much emphasis on the latter. Although liquid water inevitably suggests the possibility of life under the present astrobiological paradigms, an all-out mission like Mars Viking straight to the surface to look first for living organisms or their biochemical markers would be very inadvisable. Chemical evidence for life in Europa's harsh surface environment may turn out to be just as controversial as for Mars until we first fully understand the nature of that environment and its coupling to the magnetosphere and the subsurface.

3.2. Keep looking for Old Faithful

Mining the present data: We must first complete full exploitation of the available observations from Voyager and Galileo Orbiter over a time span of more than twenty-one years to look for any evidence of recent geologic or chemical activity on the surface, and to understand how the present geology and composition has evolved. We see volcanic plumes all the time on hyperactive Io, but nothing at all like that on Europa. Where are the cryogenic plumes and the fresh ejecta liquid water reservoirs below the surface? Is the pace of action on Europa just more glacial, perhaps by thousands to millions of years, as compared to the daily fireworks on Io, but still far faster than on the eons-old surfaces of Ganymede and Callisto. Perhaps this just argues that a lithosphere of soft ice insulates the visible surface from an ocean still potentially accessible at tens of kilometers below the surface? Do the relatively fast time scales of radiolytic processing on Europa's surface require that any fresh material be rapidly changed (in so far as we can see it) to 'old' stuff, so it's unlikely we'll see a compositional change? Are bright central bands within linear ridges very young in this sense?

Continuing work on observational interpretation and modeling of geologic and compositional data might determine answers to these questions and provide firmer science foundations for future missions. A key element of this work may also be increasingly realistic simulations of Europan surface chemistry in an environment constantly battered by particle and solar UV irradiation, meteoritic bombardment, day/night thermal cycling at 80 - 130 K temperatures, probable seismic disturbances (shaken, not stirred) during the 3.5-day tidal period, and emergence of subsurface material.

Keeping watch into the future: The time scale of recognizable activity and change on Europa currently appears to be longer than twenty years. Thus there should be a long term program of repeated observations for selected surface regions and features by future spacecraft with imaging capabilities targeted for Europa, other objectives in the Jovian system, or for the outer solar system via Jovian Gravitational Assist (JGA). Also, even small ejections of fresh material from cryogenic plumes could have significant and visible impacts on Europa's extremely thin atmosphere. Long term support is therefore needed for occasional monitoring of Europan atmosphere density and composition from ground (e.g., Keck) and orbital (HST) telescopes at Earth with high spectral resolution, and planetary observation capabilities should be maintained and improved for future facilities such as the Next Generation Space Telescope and the Terrestrial Planet Finder. Extension of Europa observations to X-ray energies, as now being done with the orbital Chandra X-Ray Observatory, could provide valuable new information on composition of the surface and interactions with the magnetosphere. Atmospheric variability induced by magnetospheric interactions will need to be distinguished by such means from that due to subsurface contributions.

At Jupiter a Europa orbiter mission would offer the most global coverage at high spatial resolution but will likely be limited to a one-month window of observation time by cumulative radiation damage to the spacecraft. Spectroscopic imaging instruments at infrared to X-ray (and even gamma-ray) wavelengths, perhaps also complemented by energetic neutral imaging sensors as on the presently operational Cassini Orbiter and IMAGE spacecraft, could be directed for Europa flybys as part of potential flyby or orbital missions to Jupiter. Other Jovian missions targeted for spectroscopic imaging of the planet, its rings and other satellites, the Io torus, and the magnetosphere will offer additional opportunities for occasional snapshots of Europa. A follow-on Galileo Orbiter II mission touring all the Galilean satellites again with multiple flybys over many years with modern remote sensing technology would of course be extremely welcome but would somehow have to be significantly scaled down in cost from the present Galileo Orbiter mission.

3.3. Europa mission concepts

Europa flyby - We come in peace: A quick flyby mission to Europa would probably not provide direct sensing of the ocean but could address many science goals without the cost burdens of long-term radiation exposure for an orbiter and of biological sterilization requirements for orbiters and landers. Earlier, the Europa Ice Clipper (otherwise named here as Earth Attacks) was proposed as a Discovery-class mission to fly through an impact plume that would be generated by a metal ball released prior to arrival and targeted to the surface to produce an observable crater. Collected material would then be returned to Earth for laboratory analysis, presumably with strong protection against biohazards from any surviving Europan bacteria, while in-situ and remote sensing data would be collected during the flyby.

Since Europa has its own dust cloud and atmosphere, it would be simpler, cheaper, safer, and much friendlier just to fly through the natural environment at low altitude and then return to Earth with collected samples of gas and dust. Molecular and isotopic composition of the collected samples, even if these amounted only to micrograms or less, could be analyzed in enormous detail back on Earth. An advanced suite of remote sensing and in-situ measurement instruments could greatly improve our knowledge about the surface and atmospheric composition, search for any signs of recent change in potentially active regions, and look for organic signatures in high resolution spectroscopic surveys of these regions. E.g., prevalence of light carbon isotopes in the dust grains might be indicative of biological processes, even if bacterial spores were absent.

Europa Orbiter redux: In the present absence of discernable surface activity a Europa orbiter mission is needed in the present decade to perform global mapping of tidal gravity and surface flexure by the most accurate and compatible techniques of orbital geodynamics and laser altimetry. A Sputnik-like satellite consisting only of a radio transmitter might support the first technique and provide a 'minimal' Europa mission at acceptable cost, but the tremendous science return and cost effectiveness of the Mars Orbital Laser Altimetry (MOLA) experiment argues for addition of such an instrument on a Europa orbiter. Since induced magnetic field data from Galileo Orbiter flybys have provided the only direct evidence for a modern Europa ocean, not just one in 'recent' geologic time, a magnetometer experiment also seems called for. In pursuit of ocean-induced geologic and chemical activity on the visible surface, a suite of IR to UV mapping spectrometers at comparable spatial resolutions of a few kilometers is clearly also needed.

Additional and valuable new instruments could include neutral and ion mass spectrometers for atmospheric sampling and an energetic neutral imager for correlation of directly sputtered particles to composition at surface sputtering sites. Energetic particle sensors are needed for global mapping of magnetospheric irradiation dosages and verification of magnetic field models derived from magnetometer data. X-ray and neutron mapping spectrometers could provide further data on surface composition from orbit. Trajectory and composition sensors could determine points of origin, and chemical abundance distributions, of dust cloud grains ejected from the surface by high-speed interplanetary meteoroid impacts. However, the low orbital speed (1.5 km-s^{-1}) is not optimal for dust measurements of this type with present instruments, so sample statistics may be very limited.

Scientific utility, payload mass and power requirements, and cost limitations must dictate the breaking point in the instrument ensemble between a full and a minimal mission. The radar mapping facility originally planned for Europa Orbiter would generate enormously valuable data on the ice crust to about ten kilometers in depth but may have marginal capability for ocean detection, particularly for an ice crust tens of kilometers in thickness. Lack of detectable activity at present argues for a thick crust. Inclusion of the radar experiment, and some above in the 'new' category, might not constitute a minimal mission but may be possible if overall mission costs could be significantly reduced, in the event that such reductions become necessary. If low cost can be achieved for 'minimum' missions to Europa, a series of such missions could be launched over an extended period of time to achieve incremental objectives, as with the Mars program, and also to reduce overall science loss from failure of any one mission. If Europa has no liquid ocean, however, the question of astrobiology becomes moot, so it is critical to carefully choose the right instruments to answer this question via the initial orbital mission(s) at minimum cost.

Europa Pathfinder - Our first step onto the surface: The Europa Pathfinder (EPF) concept under study at JPL relies on successful design and operational heritage from Mars Pathfinder's landing system and may provide the best present approach to a minimal lander mission following global surveys by an initial Europa orbiter. The lander vehicle would be carried into Europa orbit at altitude ~200 km by a carrier spacecraft, descend via retrofire propulsion stages to a couple of kilometers altitude, and then be dropped within a protective envelope of air bags to the surface. The targeted landing site would likely be that showing evidence from orbital surveys of freshly emergent material, e.g., in an apparent flood zone. Present design mass for the landed payload of Europa Pathfinder is just over thirty kilograms with a total dry mass of over two hundred kilograms for the delivery (Europa orbit to surface) vehicle including propulsion, thermal, mechanical, landing airbag, and lander systems. No Sojourner-type rover is included.

Some strawman science instruments include a seismometer, imagers, Raman spectrometer, and radiation sensors. It would be important to do complementary measurements in orbit as well as the surface, e.g., by having imagers, magnetometers, and radiation sensors on both platforms. A single seismometer on EPF could make critical measurements on the thickness of the ice crust which may not have been well constrained from tidal gravity and altimetry data on an initial orbiter mission. This measurement would be a necessary prelude to any future attempt to penetrate the crust to the ocean with a cryobotic system.

The likely follow-on missions to EPF could be impact probes penetrating to meters in depth to reach 'pristine' material below the regolith, and surface rover missions capable of seeking out protected niches of emergent subsurface material for chemical analysis of potential ejecta from a Europan ocean. In niches protected from direct irradiation by magnetospheric particles and solar UV, it may be possible to find biochemical markers or even dormant spores of living organisms from the ocean.

Galilean Satellite Ocean Explorer - The ultimate mission: A highly ambitious method of investigating deeply buried oceans on all the Galilean satellites, including a potential magma ocean on Io, would be implantation of distributed networks of geophysical sounding instruments on each satellite. We propose the concept of a Galilean Satellite Ocean Explorer mission which carries a large set of twenty light ($\sim$20 kg) impact probe systems for implantation on the surfaces of all these satellites to support in-situ chemical analysis and interior sounding. This mission would cover all four satellites with five probes targeted to each one. The descent probes would be dropped during low altitude passes at 5 - 10 km-s^{-1} ground speed by the mother spacecraft as the latter conducts a Galileo-type orbital tour of the Jovian system. We assume that accumulated orbital gravity and navigational information from earlier missions would allow reasonably precise targeting of surface impact sites from a flyby platform, although current navigational margins require descent from a Europa orbiter.

Definitive location of seismic epicenters associated with ice fracture events would require survival of at least three monitoring sites in each network. The probes would accumulate and store accumulated data on seismic activity, gravitational tides, magnetic induction, diurnal heat flow, chemical composition to meter depths, and radiation dosages. These probes would then need to remain operational for months between successive flybys of the mother spacecraft, so that data could be recovered for uplink either through the mother spacecraft or direct to Earth.

3.4. Technology development issues

Power systems: Future exploration of Europa in orbit and at the surface will require the development of next generation power sources. A strong case should be made for development of new nuclear power sources like advanced radioisotope thermal generators (RTGs) but reduced in size, mass, and cost. After the two Cassini Orbiter RTG spares are used, there are no available power sources of this type. Operation in the high radiation environment at Europa requires high margins of performance and allows little time for correction. The cost, mass, stability and reliability of RTGs provides that necessary margin. At present there is no indication that planetary protection protocols will limit the use of nuclear power sources at Europa in the future - the intense radiation environment really makes this a moot point. There may be innovative alternative methods of obtaining power in the Europa environment that could be investigated in coming years. For example, the $\sim$0.1 W-m^{-2} (Cooper et al. 2001) energy flux from magnetospheric irradiation, and the likely high oxidation potential of the irradiated surface, are two energy sources that have both astrobiological as well as engineering potential.

Radiation hardening of components: Any future exploration of Europa and Io will require radiation hardened or radiation tolerant electronics and sensors. Some of this work is underway for Europa Orbiter and needs to be completed. Additionally, work is needed in sensor technology and in computer memory. The volatile computer memory on the present Europa Orbiter design is a relatively soft component requiring a massive vault to shield it. This shielding is not only expensive (each kg of shielding requires more than two kg of fuel), but also adds an additional burden for planetary protection. Microbes can survive a long time within the shielded vault, and years of bombardment by the magnetospheric radiation environment in jovicentric orbit are required for sterilization prior to initiation of operations at Europa. This is a strong requirement for the Europa orbiter and will be a large problem for future landed missions. We strongly endorse the continued development of radiation tolerant components for spacecraft systems and science sensors.

Surface operation requirements: Operations to and on the Europa surface will require advances in many other systems. Lander-probe mechanical and electronic systems will not only have to be radiation hardened but will also need to function under cryogenic conditions of the ambient 80 - 130 K temperatures. Heaters can raise the spacecraft temperature but at the risk of melting or subliming away the underlying ice and making possible an unintended exploration of the subsurface. Exposed components may be degraded by chemical reactions with strong oxidants from the highly irradiated ice surface. Seismic activity will be of great scientific interest but strong local shaking could be catastrophic. All systems must meet decontamination requirements of planetary protection before being allowed to reach Europa orbit or the surface, which may require prior exposure of components to high heat or radiation dosages, or to highly active chemical disinfectants.

Implementation of all surface missions discussed above will require increasingly sophisticated developments in descent propulsion, attitude control, navigation, landing, autonomous control, and uplink telemetry systems. Miniaturization of descent probe and science instrument systems will clearly be required to support network implantation for geophysical sounding objectives. Power system developments for small probes will clearly be needed to allow surface network operations over many months.

Surface environment simulation: Accurate interpretation of remote sensing observations from orbit, planning for surface landing operations, and direct measurements on the surface all call for development of engineering and science system simulations matching environmental conditions on Europa. Computer models from Galileo Orbiter and other observations can help define the environmental parameters but physical facilities contained within cryogenic chambers are needed for full simulations. Size scales of simulated components within such chambers would vary from those of the micron- to millimeter-thick optical layers accessible for remote sensing to that of the meters-deep regolith containing the accumulated products of millions of years of radiolytic processing, meteoritic bombardment, subsurface material emergence. The ongoing debate on whether the observed sulfate hydrates on Europa are Mg-Na mineral salts or sulfuric acid hydrates is an example of the difficulty in interpretation of compositional measurements in a highly irradiated and chemically hostile environment. This debate might be resolved with our recommended initial focus on an orbiter mission, but only if the needed remote sensing instruments are appropriately calibrated in simulation facilities for a variety of possible surface conditions. Materials testing will clearly be required for lander components directly exposed to the surface environment. Buildup of salts, if present in significant abundance, could inhibit mobility of surface rovers and subsurface probe systems. Survival of potential Europan microorganisms or their biomarkers in this environment should be tested with terrestrial extremophiles such as D. radiodurans and instruments needed to detect such organisms. Planetary protection requirements for sterilization of orbiter and lander systems will be major cost drivers for Europa missions and should be evaluated in the context of realistic environmental simulations.

3.5. Data analysis and archiving issues

Inproving our understanding of Europa's present geologic and chemical states, continuing to search for evidence of surface activity in present and future observations, and modeling Europa's complex interactions with the Jovian magnetosphere all require a robust program of support for analysis, modeling, and archiving of multi-mission data from missions to the Jovian system and Europa. Facilitation of comparisons for imaging and composition data from consecutive missions to or by Europa requires definition of data format standards in cooperation between data providers, archive facilities, and data user communities. To search for surface change it will be required to have observations at comparable spatial and spectral resolutions, already a difficult task for comparisons of Voyager and Galileo Orbiter data.

Since the study of Europa in particular, and the Jovian system in general, is likely to span multiple missions into the foreseeable future, a research funding program supporting long-term data analysis and modeling investigations for up to five years is called for, independent of individual mission programs. The Jovian System Data Analysis Program (JSDAP) was created primarily for Galileo Orbiter investigations but should be extended to form the needed multi-mission program. There may not be billions of dollars in funding available now for new Galileo-class missions, but there should certainly be some millions of dollars per year to exploit the data we already have and allow development of firm scientific foundations for future missions. Timely processing of grant or contract awards for selected investigations is an absolute necessity, so that critical human and other resources are not lost before such investigations are actually begun.

References

Anderson, J.D., G. Schubert, R.A., Jacobson, E.L. Lau, W.B. Moore, and W.L. Sjogren 1998. Europa's differentiated internal structure: Inferences from four Galileo encounters. *Science* **281**, 2019-2022.

Battista, J.R. 1997. AGAINST ALL ODDS: The survival strategies of Deinococcus radiodurans. *Annu. Rev. Microbiol.* **51**, 203-224.

Brown, M.E. 2001. Potassium in Europa's atmosphere. *Icarus* **151**, 190-195.

Brown, M.E. and R.E. Hill 1996. Discovery of an extended sodium atmosphere around Europa. *Nature* **380**, 229-231.

Carlson, R.W., M.S. Anderson, R.E. Johnson, W.D. Smythe, A.R. Hendrix, C.A. Barth, L.A. Soderblom, G.B. Hansen, T.B. McCord, J.B. Dalton, R.N. Clark, J.H. Shirley, A.C. Campo, and D.L. Matson 1999a. Hydrogen peroxide on the surface of Europa. *Science* **283**, 2062-2064.

Carlson, R.W., R.E. Johnson, and M.S. Anderson 1999b. Sulfuric acid on Europa and the radiolytic sulfur cycle. *Science* **286**, 97-99.

Carr, M.H., M.J.S. Belton, C.R. Chapman, M.E. Davies, P. Geissler, R. Greenberg, A.S. McEwen, B.R. Tufts, R. Greeley, and R. Sullivan 1998. Evidence for a subsurface ocean on Europa. *Nature* **391**, 363-365.

Cassen, P., R.T. Reynolds, and S.J. Peale 1979. Is there liquid water on Europa? *Geophys. Res. Lett.* **6**, 731-734.

Cassen, P., S.J. Peale, and R.T. Reynolds 1980. Tidal dissipation in Europa: A correction. *Geophys. Res. Lett.* **7**, 987-988.

Chyba, C.F. 2000. Energy for microbial life on Europa. *Nature* **403**, 381-382.

Chyba, C.F. and K.P. Hand 2001. Planetary science - Life without photosynthesis. *Science* **292**, 2026-2027.

Chyba, C.F. and C.B. Phillips 2001. Possible ecosystems and the search for life on Europa. *Proc. Natl. Acad. Sci. USA* **98**, 801-804.

Cooper, J.F. 2001. High energy magnetospheric ion interactions with Io and Europa. *Bull. Am. Astron. Soc.* **33**, 1084.

Cooper, J.F., R.E. Johnson, B.H. Mauk, H.B. Garrett, and N. Gehrels 2001. Energetic ion and electron irradiation of the icy Galilean satellites. *Icarus* **149**, 133-159.

Esposito, L.W., A.F. Cheng, B.C. Clark, M.J. Daly, E.I. Friedmann, B.M. Jakosky, R.Y. Morita, A.-L. Reysenbach, and D.A. Stahl 2000. *Preventing the Forward Contamination of Europa.* National Academy Press, Washington, D.C.

Dalton, J.B. 2002. Detectability of potentially entrained microorganisms at the surface of Europa. *Lunar Planet. Sci. Conf. XXXIII*, pp. 1555. Lunar Planet. Inst., Houston.

Fagents, S.A., R. Greeley, R.J. Sullivan, R.T. Pappalardo, L.M. Prockter, and the Galileo SSI Team 2000. Cryomagmatic mechanisms for the formation of Rhadamanthys Linea, triple band margins, and other low albedo features on Europa. *Icarus* **144**, 54-88.

Gaidos, E.J., K.H. Nealson, and J.L. Kirschvink 1999. Life in ice-covered oceans. *Science* **284**, 1631-1633.

Greenberg, R. 2002. Tides and the biosphere of Europa. *Amer. Scientist* **90**, 48-55.

Greenberg, R., G. Hoppa, B.R. Tufts, and P.E. Geissler 1998. Chaos regions: Widespread melt-through to the surface of Europa? *Bull. Am. Astron. Soc.* **30**, 1086.

Greenberg R., P. Geissler, B.R. Tufts, and G.V. Hoppa 2000. Habitability of Europa's crust: The role of tidal-tectonic processes. *J. Geophys. Res.* **105**, 17551-17562.

Head, J.W. and R.T. Pappalardo 1999. Brine mobilization during lithospheric heating on Europa. *J. Geophys. Res.* **104**, 27143-27156.

Hoffman, P.F., A.J. Kaufman, G.P. Halverson, and D.P. Schrag 1998. A neoproterozoic snowball Earth. *Science* **281**, 1342-1346.

Hoppa, G.V., B.R. Tufts, R. Greenberg, and P.E. Geissler 1999. Formation of cycloidal features on Europa. *Science* **285**, 1899-1902.

Howell, R.R. 2001. X-ray emission from the Galilean satellites. *Bull. Am. Astron. Soc.* **33**, 1083.

Johnson, R.E. 2000. Sodium at Europa. *Icarus* **143**, 429-433.

Johnson, R.E. and E.C. Sittler, Jr. 1990. Sputter-produced plasma as a measure of satellite surface composition: The Cassini mission. *Geophys. Res. Lett.* **17**, 1629-1632.

Johnson, R.E., R.M. Killen, J.H. Waite, Jr., and W.S. Lewis 1998. Europa's surface composition and sputter-produced ionosphere. *Geophys. Res. Lett.* **25**, 3257-3260.

Kargel, J.S., J.Z. Kaye, J.W. Head, G.M. Marion, R. Sassen, J.K. Crowley, O.P. Ballesteros, S.A. Grant, and D.L. Hogenboom 2000. Europa's crust and ocean: Origin, composition, and the prospects for life. *Icarus* **148**, 226-265.

Kargel, J.S., J.W. Head III, D.L. Hogenboom, K.K. Khurana, and G. Marion 2001. The system sulfuric acid - magnesium sulfate - water: Europa's ocean properties related to thermal state. *Lunar Planet. Sci. Conference XXXII*, pp. 2138. Lunar Planet. Inst., Houston.

Khurana, K.K., M.G. Kivelson, D.J. Stevenson, G. Schubert, C.T. Russell, R.J. Walker, and C. Polanskey 1998. Induced magnetic fields as evidence for subsurface oceans in Europa and Callisto. *Nature* **395**, 777-780.

Kivelson, M.G., K.K. Khurana, D.J. Stevenson, L. Bennett, S. Joy, C.T. Russell, R.J. Walker, C. Zimmer, and C. Polanskey 1999. Europa and Callisto: induced or intrinsic fields in a periodically varying plasma environment. *J. Geophys. Res.* **104**, 4609-4625.

Kivelson, M.G., K.K. Khurana, C.T. Russell, M. Volwerk, R.J. Walker, and C. Zimmer 2000. Galileo magnetometer measurements: A stronger case for a subsurface ocean at Europa. *Science* **289**, 1340-1343.

Kovach, R.L. and C.F. Chyba 2001. Seismic detectability of a subsurface ocean on Europa. *Icarus* **150**, 279-287.

Krüger, H., A.V. Krivov, and E. Grün 2000. A dust cloud of Ganymede maintained by hypervelocity impacts of interplanetary micrometeoroids. *Planet. Space Sci.* **48**, 1457-1471.

Makris, N.C., S. Lee, A. Thode, J.D. Wilson, M. Zanolin, and R.T. Pappalardo 2001. Probing Europa's interior structure with natural ambient noise. *Eos (Fall Supplement)* **82**(47), P22B-0552 (abstract).

Malin, M.C. and D.C. Pieri 1986. Europa. In *Satellites* (J.A. Burns and M.S. Matthews, Eds.), pp. 689-717. Univ. Ariz. Press, Tucson.

Marion, G. 2002. A molal-based model for strong acid chemistry at low temperatures (187 to 298 K). *Geochim. Cosmochim. Acta*, submitted.

McCollom, T.M. 1999. Methanogenesis as a potential source of chemical energy for primary biomass production by autotrophic organisms in hydrothermal systems on Europa. *J. Geophys. Res.* **104**, 30729-30742.

McCord, T.B., G.B. Hansen, F.P. Fanale, R.W. Carlson, D.L. Matson, T.V. Johnson,, W.D. Smythe, J.K. Crowley, P.D. Martin, A. Ocampo, C.A. Hibbitts, and J.C. Granahan 1998a. Salts on Europa's surface detected by Galileo's Near Infrared Mapping Spectrometer. *Science* **280**, 1242-1245.

McCord, T.B., G.B. Hansen, R.N. Clark, P.D. Martin, C.A. Hibbitts, F.P. Fanale, J.C. Granahan, M. Segura, D.L. Matson, T.V. Johnson, R.W. Carlson, W.D. Smythe, and G.E. Danielson 1998b. Non-water-ice constituents in the surface material of the icy Galilean satellites from the Galileo near-infrared mapping spectrometer investigation. *J. Geophys. Res.* **103**, 8603-8626.

McCord, T.B., G.B. Hansen, D.L. Matson, T.V. Johnson, J.K. Crowley, F.P. Fanale, R.W. Carlson, W.D. Smythe, P.D. Martin, C.A. Hibbitts, J.C. Granahan, and A. Ocampo 1999. Hydrated salt minerals on Europa's surface from the Galileo near-infrared mapping spectrometer (NIMS) investigation. *J. Geophys. Res.* **104**, 11827-11851.

McCord, T.B., T.M. Orlando, G. Teeter, G.B. Hansen, M.T. Sieger, N.G. Petrik, and L. Van Keulen 2001. Thermal and radiation stability of the hydrated salt minerals epsomite, mirabilite, and natron under Europa environmental conditions. *J. Geophys. Res.* **106**, 3311-3319.

Moore, J.M., E. Asphaug, R.J. Sullivan, J.E. Klemaszewski, K.C. Bender, R. Greeley, P.E. Geissler, A.S. McEwen, E.P. Turtle, C.B. Phillips, B.R. Tufts, J.W. Head, R.T. Pappalardo, K.B. Jones, C.R. Chapman, M.J.S. Belton, R.L. Kirk, and D. Morrison 1998. Large impact features on Europa: Results of the Galileo nominal mission. *Icarus* **135**, 127-145.

Nash, D.B. and F.P. Fanale 1977. Io's surface composition based on reflectance spectra of sulfur/salt mixtures and proton irradiation experiments. *Icarus* **31**, 40-80.

Nelson, R.M. and D.B. Nash 1979. Spectral reflectance change and luminescence of selected salts during 2-10 keV proton bombardment: Implications for Io. *Icarus* **39**, 277-285.

Neftel, A. and K. Fuhrer 1993. A record of atmospheric oxidant from polar ice cores over the past 100,000 years: dream or real possibility? In *The Tropospheric Chemistry of Ozone in the Polar Regions*, NATO ASI Series, Vol. 17, (H. Niki and K.H. Becker, Eds.), pp. 219-233. Springer-Verlag, New York.

Pappalardo, R.T., J.W. Head, R. Greeley, R.J. Sullivan, C. Pilcher, G. Schubert, W.B. Moore, M.H. Carr, J.M. Moore, and M.J.S. Belton 1998. Geological evidence for solid-state convection in Europa's ice shell. *Nature* **391**, 365-368.

Pappalardo, R.T., M.J.S. Belton, H.H. Breneman, M.H. Carr, C.R. Chapman, G.C. Collins, T. Denk, S. Fagents, P.E. Geissler, B. Giese, R. Greeley, R. Greenberg, J.W. Head, P. Helfenstein, G. Hoppa, S.D. Kadel, K.P. Klaasen, J.E. Klemaszewski, K. Magee, A.S. McEwen, J.M. Moore, W.B. Moore, G. Neukum, C.B. Phillips, L.M. Prockter, G. Schubert, D.A. Senske, R.J. Sullivan, B.R. Tufts, E.P. Turtle, R. Wagner, and K.K. Williams 1999. Does Europa have a subsurface ocean? Evaluation of the geological evidence. *J. Geophys. Res.* **104**, 24015-24056.

Paranicas, C., R.W. Carlson, and R.E. Johnson 2001. Electron bombardment of Europa. *Geophys. Res. Lett.* **28**, 673-676.

Phillips, C.B., A.S. McEwen, G.V. Hoppa, S.A. Fagents, R. Greeley, J.E. Klemaszewski, R.T. Pappalardo, K.P. Klaasen, and H.H. Breneman 2000. The Search for current geologic activity on Europa. *J. Geophys. Res.* **105**, 22579-22598.

Reynolds, R.T., S.W. Squyres, D.S. Colburn, and C.P. McKay 1983. On the habitability of Europa. *Icarus* **56**, 246-254.

Schlesinger, W.H. 1997. *Biogeochemistry: An Analysis of Global Change.* Academic Press, San Diego.

Shirley, J.H., R.W. Carlson, and M.S. Anderson 1999. Upper limits for magnesium and sodium hydroxides on Europa. *Eos (Fall Supplement)* **80**(46), P21A-09. (abstract)

Smith, D.E., M.T. Zuber, H.V. Frey, J.B. Garvin, J.W. Head, D.O. Muhleman, G.H. Pettengill, R.J. Phillips, S.C. Solomon, H.J. Zwally, W.B. Banerdt, T.C. Duxbury, M.P. Golombek, F.G. Lemoine, G.A. Neumann, D.D. Rowlands, O. Aharonson, P.G. Ford, A.B. Ivanov, C.L. Johnson, P.J. McGovern, J.B. Abshire, R.S. Afzal, and X. Sun 2001. Mars Orbiter Laser Altimeter (MOLA): Experiment summary after the first year of global mapping of Mars. *J. Geophys. Res.* **106**, 23689-23722.

Spaun, N.A., J.W. Head, G.C. Collins, L.M. Prockter, and R.T. Pappalardo 1998. Conamara Chaos Region, Europa: Reconstruction of mobile polygonal ice blocks. *Geophys. Res. Lett.* **25**, 4277-4280.

Squyres, S.W., R.T. Reynolds, P.M. Cassen, and S.J. Peale 1983. Liquid water and active resurfacing on Europa. *Nature* **301**, 225-226.

Stevenson, D. 2000. Europa's ocean - the case strengthens. *Science* **289**, 1305-1307.

Turtle, E.P. and E. Pierazzo 2001. Thickness of a Europan ice shell from impact crater simulations. *Science* **294**, 1326-1328.

Urick, R.J. 1986. *Ambient Noise in the Sea.* Peninsula Publishing, Los Altos.

Wu, X., Y.E. Bar-Sever, W.M. Folkner, J.G. Williams, and J.F. Zumberge 2001. Probing Europa's hidden ocean from tidal effects on orbital dynamics. *Geophys. Res. Lett.* **28**, 2245-2248.

Yoder, C.F. 2002. Detection of a Europan ocean from tides, static fields and rotation. *Icarus*, in press.

Zahnle, K., L. Dones, and H.F. Levison 1998. Cratering rates on the Galilean satellites. *Icarus* **136**, 202-222.

The Future of Solar System Exploration, 2003-2013
ASP Conference Series, Vol. 272, 2002
Mark V. Sykes, ed.

Titan

R.D. Lorenz[1], J.R. Green[2], C.A. Wood[3], A. Coustenis[4], G.G. Ori[5], P. Paillou[6], R. Yelle[1], S.J. Bolton[2], K.H. Baines[2], R.J. Terrile[2], M.A. Gurwell[7], C.A. Griffith[1], A.-S. Wong[8], & K.A. Rages[9]

[1] *Lunar and Planetary Laboratory, University of Arizona, Tucson, AZ*
[2] *Jet Propulsion Laboratory, California Institute of Technology, Pasadena, CA*
[3] *Columbia University Biosphere 2 Center, Oracle, AZ*
[4] *Observatoire de Paris-Meudon, Meudon, France*
[5] *International Research School of Planetary Sciences, Universita' d'Annunzio, Pescara, Italy*
[6] *Observatoire de Bordeaux, Floirac, France*
[7] *Harvard-Smithsonian Center for Astrophysics. Cambridge, MA*
[8] *Division of Geological and Planetary Sciences, Calfornia Institute of Technology, Pasadena, CA*
[9] *SETI Institute, Mountain View, CA*

Abstract. Titan offers outstanding appeal as a planetary exploration target. Titan is both a planet-scale prebiotic chemistry laboratory, and a physical laboratory in which new flavors of oceanography and meteorology can be explored, as well as the traditional planetary science disciplines. Future exploration should involve the extension of the Cassini mission, and a mobile airborne platform to conduct wide-ranging, but close-up, exploration of Titan's surface and lower atmosphere.

EXECUTIVE SUMMARY

Titan is the Mars of the Outer Solar System. Everything that makes Mars interesting (landscape, meteorology, connections to the origin of life) applies in equal measure to Titan. As a body with a thick 'Earth-like' atmosphere, 'moon-like' gravity and the possibility of large expanses of surface liquid, Titan - already known to be a giant organic chemistry laboratory - promises to be an excellent physical laboratory for the study of terrestrial processes such as oceanography and meteorology and thus may engage a wider scientific community than has been involved in planetary science to date. Titan's importance as a scientific target may be considered in four areas:

As yet, we have little data on Titan's surface, although both radar reflectivity and the crude maps made from groundbased telescopes and HST point to strongly heterogenous regions on the surface. The deposition of photochemical materials, including liquid ethane, suggest that Titan's surface will have widespread lakes and seas of liquid hydrocarbons, making a uniquely evocative landscape. In addition to the usual panoply of volcanic, tectonic and impact-induced landforms, Titan is likely therefore to have many exotic variations on familiar landforms such as tidal flats and perhaps fluvial features like river beds.

Methane in Titan's atmosphere behaves much as water does on Earth. It is a condensible greenhouse gas that may participate in important climate feedbacks, as well as forming clouds that grow and dissipate as rainfall on short (hours) timescales. This methane weather appears more similar to - and may shed new insights on - Earth's weather than does the activity in any other solar system atmosphere. Titan study may assist important problems such as the treatment of moist convection and ocean-atmosphere fluxes. Photochemistry of methane leads to a thick organic haze which accumulates on the surface. The haze shows seasonal effects, notably a dark polar hood that builds up during polar winter and may be analogous to the polar stratospheric clouds associated with the 'ozone hole' on Earth.

The internal structure of the Galilean satellites suggests that Titan, helped with ammonia antifreeze, will have an internal water ocean like that of Europa. Unlike Europa, however, Titan has abundant organics - and in this respect Titan may be better endowed with the ingredients for life than is Europa. Titan, despite its distance, is energetically an easier object on which to land advanced instrumentation, since its thick atmosphere allows entry and descent via heat-shield and parachute, rather than the heavy retropropulsion required for Europa. Titan also does not suffer from the challenging radiation environment at Europa. Access to the internal ocean itself is not necessary for a meaningful astrobiological investigation - surface deposits will be present where transient exposures of liquid water from cryovolcanic activity or impact melt will have interacted over thousands to millions of years with the organics deposited from the atmosphere to form a frozen primordial soup, an experiment that cannot be adequately reproduced in terrestrial laboratories.

Titan, even stripped of its unique atmosphere, also deserves study simply as an important icy satellite. It is clearly a unique member of the Saturnian system, being considerably larger than any other Saturnian satellite. It is large enough to possess significant intrinsic activity (volcanism, tectonics) and the differences in its evolution and constitution from the Galilean satellites beg to be understood. Furthermore, its orbital eccentricity presents a challenging puzzle - even without dissipation from possible oceans of surface hydrocarbon liquids, the eccentricity (unforced by resonances) points to an anomalous orbital history, perhaps indicating a recent collision or near-miss. In sum, Titan even without its atmosphere is a more important an icy satellite than any other than perhaps Europa.

NASA and the scientific community must be ready to respond to the explosion of public interest when Cassini arrives in 2004/5. While designs should maintain flexibility until the early Cassini/Huygens results are analyzed, early technology development efforts should be undertaken so that technology readiness is not on the critical path to a follow-on mission.

Early discussion with ESA and its member states should be undertaken to exploit the resource-sharing opportunities with these international partners who will themselves have considerable Titan scientific activities.

An extension to the nominal 4-year Cassini mission will offer significant scientific return at modest cost. At present the limiting factor on Cassini Titan science is the number of Titan flybys possible in the tour. At the end of the present nominal mission, Cassini is in a short period orbit that would permit additional Titan flybys more often than the 'average' for the 4-year tour and is therefore poised to add considerably to the Titan science return.

The most important element in a post-Cassini mission will be an airborne exploration platform exploiting Titan's thick atmosphere and low gravity to give lander-like close-up investigation capability at many locations on Titan's diverse surface. This platform could be an airship with buoyancy control or conceivably some sort of heavier-than-air vehicle. Such a platform offers tremendous scientific capability as well as unparallelled scope for public engagement. Since a Titan explorer will require a radioisotope power source in any case, its duration need not be tightly constrained - a mission of 1-2 years should be easily attainable.

It is possible that an orbiter may be required to support the data return from such a platform (although direct-to-Earth options have yet to be fully explored); a limited scientific payload on the orbiter should complement the capabilities of Cassini and exploit the extended period of proximity to Titan.

Critical technologies that require early development are the adaptation of a radioisotope power source to operation in an atmosphere, on-board autonomy, instrumentation and aerobraking technologies.

The interaction of water-rich liquids with Titan organics ('tholins') are not well-understood, and further laboratory study should be undertaken, not only as a scientific investigation, but to anticipate and shape the design of compact instrumentation for future missions to look for prebiotic compounds such as amino acids and sugars in surface deposits. Detailed interpretation of Cassini result will require laboratory work in other fields, such as photochemistry, haze formation, rheology of volatile-rich ices etc.

Titan has shown itself to be a dynamic object, with meteorological changes on hourly timescales (but yet of unknown frequency) as well as a seasonal cycle. The difficulties in securing adequate observing opportunities for monitoring this activity (i.e. fractions of a night for tens to hundreds of nights a year) can be substantially mitigated by permitting more creative scheduling on groundbased telescopes via queued/service/remote observing. A program to make available modest instrumentation (1-2 micron photometers and/or CCD spectrometers) to amateurs or small colleges would provide valuable additional data and engage a wider community in Titan study.

REPORT

1. Current State of Knowledge

Titan's radius is 2575 km, fractionally smaller than Ganymede, and intermediate between that of Mars and the Moon. Its density has been determined at 1880 kg/m^3 - suggesting a roughly 50:50 mix of rock and ice. Titan is large enough that the energy of accretion should have softened and melted the outer layers of ice, allowing the rock component to settle into the interior forming a rocky core, although there is at present no direct measure of Titan's internal structure.

The surface gravitational acceleration is roughly 1/7 of the Earth's. The surface atmospheric pressure is 1.5 bar, and the temperature of the atmosphere just above the surface is 94K. The composition is predominantly molecular nitrogen, with a few per cent methane, an undetermined amount (less than a few per cent) of argon, and traces of many (over 20 at the last count) organic compounds.

One particularly intriguing aspect of Titan is that methane, known since 1944 to be present in its atmosphere, is destroyed on short ($\sim 10^7$ year) timescales by solar ultraviolet radiation. This implies that its presence in the atmosphere is buffered by resupply and/or a surface reservoir. Both methane and ethane, which is the dominant photochemical product of methane photolysis are liquids at Titan's surface conditions - making a surface at least partially covered with hydrocarbon liquids a likely scenario. Thus Titan's landscape may have been modified by marine, lacustrine or fluvial processes, like the Earth.

Titan's atmosphere is a powerful analog of the Earth's - a condensible greenhouse gas (methane/water) is photolyzed in the stratosphere yielding a UV-absorbing product (haze/ozone) which keeps the stratosphere warm. Heterogenous chemistry and/or condensation during the polar night participates in the removal of this absorber. The condensible greenhouse gas participates in a hydrological cycle and has important surface-atmosphere interactions such as climate feedback and erosion.

Some recent findings (since the development of the Cassini/Huygens mission, designed largely on the photochemistry-dominated post-Voyager view of Titan) deserve mention.

Titan has a strong seasonal cycle. Its equator is tilted to the ecliptic to a similar degree to the Earth and Mars - the haze in Titan's atmosphere is thicker in one hemisphere that the other, and 'sloshes' over to the winter hemisphere in response to the seasonal forcing. A dark hood forms over the winter pole, probably the result of accumulation and condensation of organics, and may be an important control on meteorology beneath.

Titan has weather. While methane has been long suspected to be a participant in an active hydrological cycle, observations since 1995 show evidence of transient clouds in Titan's troposphere - including a large 'storm' event which covered 10% of Titan's disk, and events which suggest clouds come and go on timescales of hours.

Titan's surface has been imaged by near-infrared telescope cameras which can penetrate the haze. These maps show a variegated surface with some pitch-black regions suggestive of organic lakes and a large, bright (perhaps mountainous) region on Titan's leading face. Lakes and seas will experience tides due to Saturn's strong gravity, and wind-driven waves - they may therefore have cliffs, beaches and tidal flats.

Millimeter-wave measurements suggest Titan's atmosphere is highly enriched in heavy nitrogen, suggesting that the thick atmosphere we see now (10x more mass per unit area than the Earth's) is only a weak remnant of a much (5-200x) thicker early atmosphere. Understanding the events associated with this profound atmospheric loss (apparently not fractionating methane, unless it is delivered by volcanism to the surface today?) is a crucial problem.

Another outstanding mystery is Titan's orbital eccentricity, which is not forced by resonances. Unless the eccentricity is due to a recent near-miss or giant impact, it is difficult to understand why it has not been damped by tidal dissipation in Titan's interior and/or seas.

The presence (and surprising ubiquity) of internal water oceans on icy satellites has only been realized in the past couple of years. Titan seems all but certain to have such an ocean, although probably somewhat deeper under an ice crust than is that of Europa. This opens the possibility of Titan sharing characteristics of Europa in terms of being an incubator for prebiotic compounds.

2. Key Science Questions

Many outstanding Titan science questions exist, and are addressed, at least in principle, by Cassini. If we anticipate a more-or-less successful Cassini/Huygens mission, the following set of science questions will then (probably) assume prominence

- What is the composition and distribution of organic materials on and beneath Titan's surface?

  Rationale: Cassini should determine the gas-phase atmospheric composition and its variations with latitude quite well. It will also make pyrolysis-GCMS measurements of the haze material and perhaps constrain gross surface composition at moderate spatial scales. No detailed surface composition measurements will be made, nor are there measurements of the depth of deposits to permit a good organic inventory to be determined.

  Measurement: In-situ measurement of surface material from a variety of locations - requiring a mobile platform with surface sample acquisition as well as sophisticated chemical instrumentation. Sounding radar to determine depth of surface deposits.

- What controls Titan's weather and how does it change with time?

  Rationale: Cassini's Titan coverage during the 4-year mission will allow the identification and gross characterization of weather systems, but not detailed monitoring of their evolution. Winds will only be determined indirectly - difficult if cloud features evolve rapidly.

  Measurement: In-situ meteorological measurements from a near-surface platform, augmented perhaps by global monitoring from orbit (microwave spectroscopy for winds and trace chemistry; IR spectral imaging for synoptic views)

- What unique small-scale geological features does Titan possess?

  Rationale: With the exception of the small region around the Huygens landing site, Cassini's optical and radar resolution is inadequate to characterize small-scale and possibly dynamic geological features (methane geysers, stream deposits, dunes etc.) To fully capture Titan's range of geology requires meter-scale or better imaging that needs to be performed from below the obscuring haze layer (the discovery of apparently recent gullies on Mars is an analogous situation.)

Measurement: High-resolution imaging from a near-surface platform (which will be required in any case to select appropriate sampling sites for surface chemistry)

- What is Titan's internal structure and how does it affect the surface?

  Rationale: Cassini will determine the gross structure of Titan from gravity measurements. Variations in crustal thickness and the presence of small-scale geothermal anomalies like near-surface melt may not be detectable from Cassini remote measurements. Titan is excited tidally and thus seismic measurements will be easy.

  Measurement: Sounding radar from a near-surface platform. Seismic measurements during extended surface contact periods of the platform mission.

3. Recommendations

3.1. Missions

An extension to the nominal 4-year Cassini mission will offer significant scientific return at modest cost. At present the limiting factor on Cassini Titan science is the number of Titan flybys possible in the tour. At the end of the present nominal mission, Cassini is in a short period orbit that would permit additional Titan flybys more often than the 'average' for the 4-year tour. Funding and Deep Space Network support capability for an extended mission of at least 2 years should be factored into NASA planning.

The post-Cassini mission that satisfies the broadest range of anticipated post-Cassini objectives and promises the greatest scope for public engagment will be an airborne exploration platform exploiting Titan's thick atmosphere and low gravity to give lander-like close-up investigation capability at many locations on Titan's diverse surface. This platform could be an airship with buoyancy control or conceivably some sort of helicopter. Early studies have been made of these concepts, and while significant development effort is required, there are no obvious technological barriers. Since a Titan explorer will require a radioisotope power source in any case, its duration need not be tightly constrained - 1-2 years.

It is *possible* that an orbiter may be required to support the data return from such a platform; a limited scientific payload on the orbiter should complement the capabilities of Cassini and exploit the extended period of proximity to Titan.

Critical technologies that require early development are the adaptation of a radioisotope power source to operation in an atmosphere, on-board autonomy, instrumentation and aerobraking/capture technologies. Solar-electric propulsion opens up mission delivery opportunities, albeit at higher cost than conventional propulsion.

It is clear that such a mission will not fall under a Discovery budget: An early JPL study has already been performed of a mission with an airship and heavy orbiter, together costing about \$1 billion: further studies may identify significant cost-saving options (in particular using direct-to-Earth communication). There are no significant technological barriers. Furthermore, development of autonomous airborne vehicles may attract investment from other US agencies, and the participation of experienced international partners offers considerable scope for cost-sharing.

3.2. Laboratory Studies

The interaction of water-rich liquids with Titan organics ('tholins') are not well-understood, and further laboratory study should be undertaken, not only as a scientific investigation, but to anticipate and shape the design of compact instrumentation for future missions to look for prebiotic compounds such as amino acids and sugars in surface deposits.

Detailed interpretation of Cassini result will require laboratory work in other fields, such as photochemistry, haze formation, rheology of volatile-rich ices etc.

3.3. Telescope Facilities

Titan has shown itself to be a dynamic object, with meteorological changes on hourly timescales (but yet of unknown frequency) as well as a seasonal cycle. The difficulties in securing adequate observing opportunities for monitoring this activity (i.e. fractions of a night for tens to hundreds of nights a year) can be substantially mitigated by permitting more creative scheduling on groundbased telescopes via queued/service/remote observing.

Targeted studies of particular questions can be largely addressed by existing and anticipated ground- and space-based capabilities.

An additional augmentation to such a monitoring program would be the provision, perhaps via the Planetary Astronomy program, of suitable near-IR instrumentation since the crudest (photometry, spectroscopy) measurements can be performed with modest telescopes available to amateurs or small colleges. 1-2 micron photometers and CCD spectrometers are becoming available off the shelf for only a few $K). The former instrumentation would be also useful for Io monitoring.

4. Closing Remarks

It is vital that the prospects for a future Titan mission be sufficiently advanced that the expertise gained from the Cassini/Huygens development and recent groundbased observations not be lost. The long lead times associated with outer solar system exploration make this a particular problem.

Although Titan is unlikely to play a significant role as a target for human space activities in the near term, various space environments may be useful in studying Titan (as an example, haze particle formation and coagulation could be usefully studied in microgravity since settling and wall effects are a significant perturbation to terrestrial laboratory experiments.)

As a scientific target, Titan promises to reach a more diverse range of disciplines than perhaps any other target. Not only conventional planetary geology, atmospheres and space physics topics, but also hydrology, oceanography, protein chemistry, meteorology and so on may adopt Titan as a case study. Titan - and in-situ exploration by airborne platform - can capture the imagination in ways few places can) and thus offers tremendous outreach opportunities.

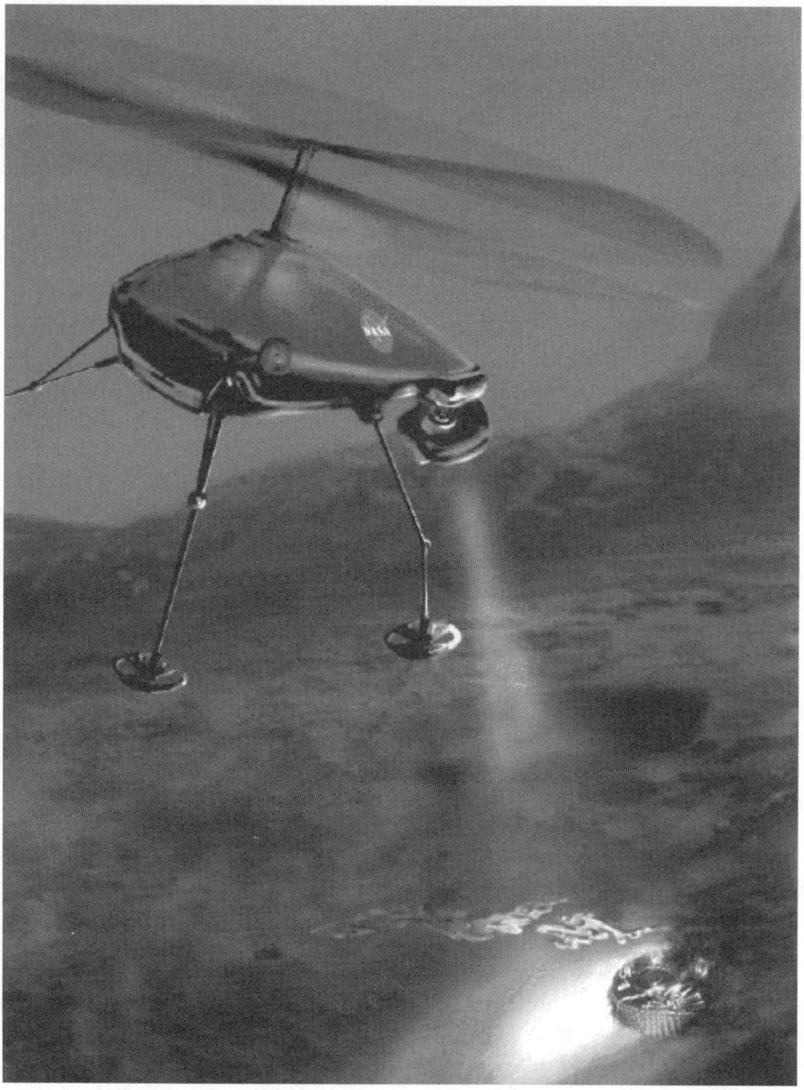

Figure 1. Artists Impression (see www.fastlight.demon.co.uk) of a Titan helicopter circa 2015 inspecting the Huygens probe landing site to detect surface changes (courtesy, James Garry). A helicopter on Titan would use 38 times less power to fly in the low gravity and thick atmosphere than on Earth. A blimp would be a less technologically-aggressive, but still scientifically-capable, option.

References

Coustenis, A. and R.D. Lorenz 1999. Titan. In *Encyclopaedia of the Solar System* (P. Weissman, L.-A. McFadden, and T.V. Johnson, Eds.), pp. 377-404. Academic Press, San Diego.

Coustenis, A. and F. Taylor 1999. *Titan: The Earth-Like Moon*, World Scientific, Singapore.

Griffith, C.A., J.L. Hall, and T.R. Geballe 2000. Detection of daily clouds on Titan. *Science* **290**, 509-513.

Lorenz, R.D. 2000. Post-Cassini exploration of Titan: Science rationale and mission concepts. *J. Brit. Interplanet. Soc.* **53**, 218-234.

Lorenz, R.D. and J.M. Mitton 2002. *Lifting Titan's Veil*, Cambridge Univ. Press, Cambridge.

The Future of Solar System Exploration, 2003-2013
ASP Conference Series, Vol. 272, 2002
Mark V. Sykes, ed.

Planetary Rings

M.K. Gordon[1], S. Araki[2], G.J. Black[3], A.S. Bosh[4], A. Brahic[5], S.M. Brooks[6], S. Charnoz[5], J.E. Colwell[6], J.N. Cuzzi[7], L. Dones[8], R.H. Durisen[9], L.W. Esposito[6], C. Ferrari[5], M. Festou[10], R.G. French[11], S.M. Giuliatti-Winter[12], A.L. Graps[13], D.P. Hamilton[14], M. Horányi[6], R.M. Karjalainen[15], A.V. Krivov[16], H. Krüger[13], S.M. Larson[17], H.F. Levison[8], M.C. Lewis[18], J.J. Lissauer[7], C.D. Murray[19], F. Namouni[8], P.D. Nicholson[20], C.B. Olkin[4], F. Poulet[7], N.J. Rappaport[21], H.J. Salo[15], J. Schmidt[16], M.R. Showalter[22], F. Spahn[16], L.J. Spilker[21], R. Srama[21], G.R. Stewart[6], & P. Yanamandra-Fisher[21]

[1] *San Jose State University Foundation, San Jose, CA*
[2] *Department of Information Science, Tohoku Fukushi University, Sendai, Japan*
[3] *National Radio Astronomy Observatory, Charlottesville, VA*
[4] *Lowell Observatory, Flagstaff, AZ*
[5] *CEA Saclay, Service D'Astrophysique, Gif-Sur-Yvette, France*
[6] *Laboratory for Atmospheric and Space Physics, University of Colorado, Boulder, CO*
[7] *NASA Ames Research Center, Moffett Field, CA*
[8] *Southwest Research Institute, Boulder, CO*
[9] *Department of Astronomy, Indiana University, Bloomington, IN*
[10] *Observatoire Midi-Pyrénées, Toulouse, France*
[11] *Department of Astronomy, Wellesley College, Wellesley, MA*
[12] *Departamento de Matematica, Campus de Guaratingueta-UNESP, Guaratingueta, Brazil*
[13] *Max-Plank-Institut für Kernphysik, Heidelberg, Germany*
[14] *Department of Astronomy, University of Maryland, College Park, MD*
[15] *Department of Physical Sciences, Astronomy Division, University of Oulu, Oulu, Finland*
[16] *Institut für Physik, University of Potsdam, Potsdam, Germany*
[17] *Lunar and Planetary Laboratory, University of Arizona, Tucson, AZ*
[18] *Department of Computer Science, Trinity University, San Antonio, TX*
[19] *Astronomy Unit, School of Mathematical Sciences, Queen Mary College, University of London, London, England*
[20] *Astronomy Department, Cornell University, Ithaca, NY*
[21] *Jet Propulsion Laboratory, California Institute of Technology, Pasadena, CA*
[22] *Center for Radar Astronomy, Stanford University, Palo Alto, CA*

Abstract. The past two decades have witnessed dramatic changes in our view and understanding of planetary rings. We now know that each of the giant planets in the Solar System possesses a complex and unique ring system. Recent studies have identified complex gravitational interactions between the rings and their retinues of attendant satellites. Among the four known ring systems, we see elegant examples of Lindblad and corotation resonances (first invoked in the context of galactic disks), electromagnetic resonances, spiral density waves and bending waves, narrow ringlets which exhibit internal modes due to collective instabilities, sharp-edged gaps maintained via tidal torques from embedded moonlets, and tenuous dust belts created by meteoroid impact onto, or collisions between, parent bodies. Yet, as far as we have come, our understanding is far from complete. The fundamental questions confronting ring scientists at the beginning of the twenty-first century are those regarding the origin, age and evolution of the various ring systems, in the broadest context. Understanding the origin and age requires us to know the current ring properties, and to understand the dominant evolutionary processes and how they influence ring properties. Here we discuss a prioritized list of the key questions, the answers to which would provide the greatest improvement in our understanding of planetary rings. We then outline the initiatives, missions, and other supporting activities needed to address those questions, and recommend priorities for the coming decade in planetary ring science.

EXECUTIVE SUMMARY

The discovery of the narrow, dark rings of Uranus (1977) was followed by the first Voyager images of the tenuous Jovian ring system (1979), and the spectacular data of many kinds returned during the twin Voyager flybys of Saturn (1980, 1981). In subsequent years, ground-based stellar occultations, the Voyager flybys of Uranus (1986) and Neptune (1989), a handful of Galileo images, coupled with ground-based and HST observations continuing to the present, have defined planetary rings as a class of objects which broadened our perspective and stretched our theories to operate in different environments and regimes. The focus remains on the physics responsible for the phenomena we observe, as well as the basic need to understand the origin and evolution of planetary rings in the general context.

Perhaps the greatest surprise was the complexity and variety of structure in the various ring systems revealed in the Voyager images and occultations. Recent studies have identified complex gravitational interactions between the rings and their retinues of attendant satellites. Among the four known ring systems, we see elegant examples of Lindblad and corotation resonances (first invoked in the context of galactic disks), electromagnetic resonances, spiral density waves and bending waves, narrow ringlets which exhibit internal modes due to collective instabilities, sharp-edged gaps maintained via tidal torques from embedded moonlets, and tenuous dust belts created by meteoroid impact onto, or collisions between, parent bodies. We are starting to realize that bombardment of rings by interplanetary meteoroids can have very significant effects on both ring structure and ring composition. Studies of planetary rings have increased our understanding of the behavior of astronomical disks in general and have been fundamental in improving our understanding of nebular dynamics and extrasolar planets.

Nonetheless there remain many unresolved issues in our attempt to establish the origins and ages of the various ring systems. One glaring unresolved problem is identifying the mechanism(s) that produce the "record-groove" structure pervading Saturn's dense rings. Another major unsolved puzzle is Saturn's multi-stranded, clumpy F ring, which continues to defy a simple explanation more than twenty years after it was first glimpsed in grainy images taken by Pioneer 11. The composition of planetary rings remains a major unknown, beyond some zero-order understanding. While only Saturn has a substantial system of dense rings, all four giant planets maintain rings of fine dust, yet none of the inner planets appears to support an equivalent ring system. As we continue to study planetary rings we will continue to identify similarities and differences that may ultimately provide insight into the processes that produce, modify, and maintain ring systems.

Finally, one unanticipated realization has been that all ring systems may be much younger than their parent planets. That is, the processes by which they evolve are so vigorous that they are hard to reconcile with primordial ages. This raises the so called "short timescale" problem, most serious for Saturn's rings. Are we just lucky to be living in the "age of rings"? How is it possible to create a ring system as massive as Saturn's in the last few times 10^8 years?

The fundamental questions confronting ring scientists at the beginning of the twenty-first century are those regarding the origin, age and evolution of the various ring systems, in the broadest context. Understanding the origin and age requires us to know the current ring properties, and to understand the dominant evolutionary processes and how they influence ring properties. Here we discuss a prioritized list of the key questions, the answers to which would provide the greatest improvement in our understanding of planetary rings. The answers to these questions may be different for each giant planet, since the rings and their environments are sufficiently dissimilar that they may not have common origins or ages.

Not surprisingly, many of our current key questions will be addressed either directly or indirectly by the Cassini mission. We have considered that mission in our deliberations and have attempted also to look beyond Cassini. Once Cassini data has been analyzed, priorities may change and unanticipated questions, missions and activities may need to be added to the lists below.

With the exception of two key questions raised in our discussions, the detailed questions which, if answered, will significantly improve our understanding of the origin and age of the rings can be grouped under a few prioritized broad tasks. Embedded within these tasks are key questions which will enable us to focus our search for better understanding of planetary rings. Based on a survey to which more than sixty percent of the participants responded, the key questions have been prioritized. The survey produced clear first and second choices (the question in section A and the first in section B below), followed by a large middle group of questions which were in turn rated as more important than the final two (the questions in sections E and F).

A. Characterize the physical properties of the various rings.

- What are the current physical properties (particle size distributions, particle shapes, the nature of particle aggregations, local kinematics, intrinsic strengths of particles and particle aggregations) of the various rings and of distinct regions within the rings?

B. Identify the most significant evolutionary mechanisms.

- What are the most important mechanisms for ring evolution on relatively long time scales?
- What are the most important mechanisms for evolution on relatively short time scales?

C. Determine the underlying kinematics and dynamics of the various ring systems.

- How fast are angular momentum and energy being transferred between rings and ringmoons?
- How do self gravity, viscosity, ballistic transport and collisions interact?

D. Determine the chemical compositions of the various ring systems.

- What is the current composition of the various rings and of distinct regions within the rings?

E. Characterize the mass flux into the various ring systems.

- What is the current mass flux into the various ring systems? What are the current size, mass, velocity, and composition distributions of the influx population? How did those distributions change with time?

F. Determine the influences of the magnetospheric and plasma environments of the various rings.

- What is the influence of the magnetosphere and plasma environment of the rings?

There were two additional questions, different in nature to the others, which were included in the survey and fall into the mid group of key questions. The first is a question we must continually ask as our understanding of individual ring systems improves. The other "different" question must be answered to enable us to better understand the data we gather from Earth-based activities.

G. What do the differences among ring systems tell us about differences in ring progenitors and/or differences in initial and subsequent processes?

H. What is the relationship between ring local properties and those properties observable via remote sensing?

We divide our recommendations into two broad categories: interplanetary spacecraft and activities that can be performed on or near Earth. In establishing priorities, we have produced ordered lists with the most important first. Within each list items tended to cluster naturally; consequently, the recommendations are divided into three groups: essential, critical, and important.

Among all of the possible activities that we considered for the decade, our overall number one priority is for robust support and augmentation of the Cassini nominal mission.

The current Cassini teams are excellent, but they are thinly staffed and funded. Most of the team members have been committing time far above the levels for which they have been funded. To capitalize on NASA's investment in Cassini, and to optimize the treasure house of data it will return, calls for an influx of new funding for the current teams, and for new blood to work alongside the current teams. At current funding levels it is not clear that the teams will have sufficient time to do the early data analysis required to refine the later mission sequences, not to mention to archive the incoming data in the PDS on the planned timescale. PDS archival is required before a long term data analysis program can be initiated, so the data archive pipeline must be fully functional before the wider community can become involved in Cassini data analysis. Early augmentation of the teams can ensure that both the teams' necessary internal data analysis, and prompt data archiving, can be accomplished in a timely way.

In the category of interplanetary space craft we recommned:

- ESSENTIAL. Increased funding for Cassini nominal mission: a participating scientist program (competitive, but probably limited selection), support for additional team associates (selected by teams), and a comprehensive data analysis program (competition open to the community). *Supports all key questions with respect to Saturn.*

- ESSENTIAL. Cassini extended mission (including associated support for data analysis and archiving). The initial extension should be for a minimum of approximately three years to encompass local observation of the next solar ring plane crossing and significant reopening of the rings. *Particularly important for key questions in sections A-D and the last additional question.*

- CRITICAL. Close-in orbiter to study Saturn ring microphysics at the level of individual particles, and the local ring environment in general. *Particularly important for key questions in sections A-C and F.*

- IMPORTANT. Neptune orbiter. *Supports all key questions with respect to Neptune.*

In the category of Earth-based or near-Earth activities, we recommend:

- ESSENTIAL. Long term observing campaign for the rings of Saturn. *Supports key questions in sections B-E.*
- CRITICAL. Simulations and laboratory experiments. *Supports all key questions.*
- CRITICAL. Periodic astrometric observations of Saturnian inner satellites. *Supports key questions in sections B, C and the last additional question.*
- CRITICAL. Develop a virtual theoretical laboratory to provide a vehicle for more coordination and integration of theoretical work. *Supports all key questions.*
- IMPORTANT. Long term observing campaign for the rings of Neptune. *Key questions in Sections B-D and the additional questions.*
- IMPORTANT. Long term observing campaign for the rings of Uranus. *Key questions in Sections B-D and the additional questions.*

REPORT

1. Current State of Knowledge

Galileo first recorded observations of the rings of Saturn in 1610. Subsequent observations produced more puzzlement than enlightenment as with time the features seemed to wax and wane. The fundamental question for seventeenth century scientists concerning these observations was "What are they?". The answer was provided in 1659 when Christian Huygens first demonstrated that the appearance was the result of a thin disk of material surrounding Saturn. The new fundamental question became "What is the structure of this disk, solid or many individual particles?". That question was not answered until 1859 when James Clerk Maxwell provided a mathematical proof that the rings could not be solid. The next fundamental question to be answered, which started to address the broader issues, was "Why is Saturn the only planet with rings"? This was drastically revised in the late 1970s and early 1980s with the discovery of the rings of Uranus, Jupiter, and Neptune, to "Why is Saturn the only planet with broad, bright rings?"

The discovery of the narrow, dark rings of Uranus (1977) was followed by the first Voyager images of the tenuous Jovian ring system (1979), and the spectacular data of many kinds returned during the twin Voyager flybys of Saturn (1980,1981). In subsequent years, ground-based stellar occultations, the Voyager flybys of Uranus (1986) and Neptune (1989), a handful of Galileo images, coupled with ground-based and HST observations continuing to the present, have defined planetary rings as a class of objects which broadened our perspective and stretched our theories to operate in different environments and regimes. The focus remains on the physics responsible for the phenomena we observe, as well as a newly compelling need to understand the origin and evolution of planetary rings in the general context.

Figure 1. Voyager data like this Voyager 2 image of Saturn's rings, demonstrated that rings are far more spectacular and complex than our earlier theories predicted. This image shows Saturn's rings at high phase angle so that the radial spokes which are comprised of fine dust appear as bright patches above the complex and for the most part poorly understood structure of the B ring. Beyond the dark Cassini Division, the A ring too shows fine structure including the dark narrow band of the Enke gap. Unlike the structure in the B ring, most of the structure in the A ring is understood. Visible beyond the edge of the main rings is the enigmatic F ring, a narrow, multistrand ring which possesses a braided, kinked structure unlike any other known ring. (Image courtesy of NASA, and the Voyager mission).

Perhaps the greatest surprise was the complexity and variety of structure in the various ring systems revealed in the Voyager images. Recent studies have identified complex gravitational interactions between the rings and their retinues of attendant satellites. Among the four known ring systems, we see elegant examples of Lindblad and corotation resonances (first invoked in the context of galactic disks), electromagnetic resonances, spiral density waves and bending waves, narrow ringlets which exhibit internal modes due to collective instabilities, sharp-edged gaps maintained via tidal torques from embedded moonlets, and tenuous dust belts created by meteoroid impact onto, or collisions between, parent bodies. We are starting to realize that bombardment of rings by interplanetary meteoroids can have very significant effects on both ring structure and ring composition.

One glaring unresolved problem is the "record-groove" structure pervading Saturn's dense rings. This so-called "irregular structure" - so opaque that no light has ever been seen through parts of it - has scales from 10-300 km, is sometimes nonaxisymmetric, probably varies radially in particle packing density. In spite of a few hypotheses which have been advanced, its cause remains a mystery. Another major unsolved puzzle is Saturn's multi-stranded, clumpy F ring, which continues to defy a simple explanation more than twenty years after it was first glimpsed in grainy images taken by Pioneer 11. Voyager, HST and ground based active optics images reveal a complex, probably chaotic, dynamical interaction between unseen parent bodies within and perhaps surrounding this ring and its two shepherd satellites, Pandora and Prometheus. The orbits of the shepherds themselves seem to exhibit irregular jumps every few years.

The composition of planetary rings remains a major unknown, beyond some zero-order understanding. Infrared spectroscopy and microwave radiometry show that Saturn's rings are dominated by water ice; however, their pinkish color - not dissimilar from Triton at visual wavelengths - calls for at least one important secondary component. Different regions of the rings exhibit noticeable color variations on all scales, from local to regional, indicating compositional variation on all those scales. The color of the rings differs from the color of any of Saturn's icy moons. The composition of the rings of Jupiter, Uranus, and Neptune is unknown, but markedly different. The Uranian rings are nearly, but not entirely, colorless. The Jovian ring shares the reddish color of Amalthea, but the brightness of the particles is unknown because of the coupling with the unknown cross section of particles. The Neptunian rings are fairly dark, but nothing is known about their color.

The study of planetary ring dynamics has been instrumental in developing our understanding of the processes at work in the nebula during the formation of the solar system. In particular, the interactions between embedded objects and disks (spiral wave generation, angular momentum transport, truncation of disks) has been illuminated through our study of rings. Migration of bodies within an untruncated disk, due to angular momentum exchange with surrounding material, was first advanced in the ring context, and is now a mainstay of planetary formation models. The discovery and study of bending waves in the Saturn system was important in understanding how nebulae can be warped. Furthermore, our understanding of the development of chaos where resonances overlap has been elucidated through ring studies; certainly, similar studies of asteroid dynamics and the Uranian satellites also played a role here. Thus studies of planetary rings have been fundamental in improving our understanding of nebular dynamics and extrasolar planets.

As we have made progress in understanding the processes that influence planetary ring structure, there has been a growing realization that these processes can act to influence ring properties on timescales that are short compared to the age of the solar system. Global angular momentum transfer (torques) between Saturn's outer main rings and the inner ringmoons should be acting to expand the orbits of the moons, and to cause the rings to fall inwards accordingly. Bombardment of Saturn's rings by interplanetary projectiles should darken the ring particles to a charcoal-like appearance if the rings were as old as the solar system. Gas drag should remove the inner rings of Uranus on a short timescale as well. Transfer of tiny charged grains from Saturn's inner rings to the planet should deplete those rings on comparably short timescales. There are uncertainties in the parameters of all these processes, but they all go the same way and they all cause us to wonder about the origin of all four ring systems. Are they as old as the planets and their retinues of moons, or are they geologically recent and transient features of the planetary landscape?

Not surprisingly, many of our current key questions will be addressed either directly or indirectly by the Cassini mission. We have considered that mission in our deliberations and have attempted also to look beyond Cassini. We are at this time more than thirty months before Cassini's Saturn orbit insertion. Forecasting needs by first anticipating discoveries is an unsatisfactory process at best. Thirty months prior to Galileo orbit insertion, no one predicted that one of highest priority follow-on missions to Jupiter would be a Europa orbiter. The need for that mission only became apparent once the Galileo observations had been analyzed. Once Cassini data has been analyzed, priorities may change and unanticipated questions, missions and activities may need to be added to the lists below.

2. Key Science Questions

The fundamental questions confronting ring scientists at the beginning of the twenty-first century are those regarding the origin, age and evolution of the various ring systems, in the broadest context. Understanding the origin and age requires us to know the current ring properties, and to understand the dominant evolutionary processes and how they influence ring properties. Here we discuss a prioritized list of the key questions, the answers to which would provide the greatest improvement in our understanding of planetary rings. The answers to these questions may be different for each giant planet, since the rings and their environments are sufficiently dissimilar that they may not have common origins or ages.

With the exception of two key questions raised in our discussions, the detailed questions which, if answered, will significantly improve our understanding of the origin and age of the rings can be grouped under a few prioritized broad tasks. Embedded within these tasks are key questions which will enable us to focus our search for better understanding of planetary rings. Based on a survey to which more than sixty percent of the participants responded, the key questions have been prioritized. The survey produced clear first and second choices (the question in section A and the first in section B below), followed by a large middle group of questions which were in turn rated as more important than the final two (the questions in sections E and F).

A. Characterize the physical properties of the various rings.

What are the current physical properties of the various rings and of distinct regions within the rings?

In order to determine the origins and ages of the various rings, we must first establish the current state of the rings with far greater certainty than is currently possible. We need to know the particle size distributions of the various regions within the rings. In order to interpret remote observations, we need to know the particle shapes and the nature of particle aggregations. In order to refine models of local dynamics, we need to know more about the local kinematics down to the level of establishing the local value for the coefficient of restitution, as well as the intrinsic strengths of both the particles and particle aggregations. For dusty rings, particle size is a key parameter defining how the grains respond to non-gravitational forces, including electromagnetic perturbations and solar radiation pressure. The size also determines particle lifetimes against sputtering and micrometeoroid erosion. Finally, each dusty ring requires an embedded population of larger source bodies. These populations are rarely seen but they play a critical role in defining the origin and evolution of dusty rings.

B. Identify the most significant evolutionary mechanisms.

What are the most important mechanisms for ring evolution on relatively long time scales?

While we have a fairly good idea that the main mechanisms are related to gravity, viscosity, and bombardment, the statement of this key question is misleadingly simple. Planetary rings are far from the relatively homogenous and static distributions of particles envisioned prior to the revelations of the Voyager spacecraft. The complex macro-structures, rings, ringlets, gaps, and waves vary with time. The composition of a ring changes as material is lost from the ring and replenished by new material from a variety of sources; individual particles change size and other characteristics due to a variety of processes. Different mechanisms will be more important in different regions and for different particle distributions. Since the regions are not static, the set of most important evolutionary mechanisms will itself evolve.

What are the most important mechanisms for evolution on relatively short time scales?

Saturn's F ring, and its surrounding regions, show clumps which come and go on timescales of days to weeks. There are several alternate hypotheses which explain this by a combination of external impacts and collisions between embedded moonlets too small to see directly. Is there a belt of as-yet unseen moonlets undergoing collisional evolution? Is the F ring itself a recent member of the system? The shepherds themselves undergo sporadic orbital jumps. Is dynamical chaos, produced by overlapping resonances in this tightly coupled region, responsible? Or, are there massive bodies buried within the F ring which perturb the orbits of these objects? The enigmatic spokes in Saturn's B ring form and disperse in half of a ring revolution. If meteoroid bombardment onto the rings is the trigger, can the current observed frequency of meteoroid impacts explain the frequency of the spokes? Are the Neptune ring arcs indeed confined in corotational resonances, or are they themselves transient? While gravity and collisions are likely to dominate in relatively massive rings, dusty rings have a rather different set of evolution mechanisms, including electric charging, electromagnetic forces, solar radiation pressure, and sputtering.

C. Determine the underlying kinematics and dynamics of the various ring systems.

How fast are angular momentum and energy being transferred between rings and ringmoons?

Two decades ago it was realized that current theories, now well supported by detailed analysis of spiral density waves and bending waves, predict that the orbits of Saturn's inner ringmoons should be expanding at a rate that could be observed by comparing Cassini and Voyager observations. This would allow the "age" of the rings to be determined. It has now been revealed by HST observations that these orbits display even larger, more sporadic, fluctuations. This greatly complicates the problem of unraveling the mean global rate of momentum transfer. Nevertheless, the problem is a central one and more sophisticated approaches of solving it must be sought.

How do self gravity, viscosity, ballistic transport and collisions interact?

Viscosity and gravity compete to form "overstabilities" - stable, nonaxisymmetric structures which are still poorly understood. Both viscosity and local self gravity probably play a role in forming "wakes" which trail embedded moonlets, but only one of these wakes has as yet been studied in the Voyager data. Viscosity depends on particle random velocities, and local "particles" influence particle random velocities. Because "particles" might be transient, gravitationally formed clumps themselves, the aspects of transport usually lumped into "viscosity" may have to be handled somewhat differently for different applications or different particle sizes. Although the time scale for ballistic transport is longer than collision or orbital time scales, its ability to produce ring structure is strongly influenced by these other transport mechanisms.

D. Determine the chemical compositions of the various ring systems.

What is the current composition of the various rings and of distinct regions within the rings?

In order to determine the origins and ages of the various rings, we need to know the chemical composition of the ring particles. Different material might be left over from ring formation, deposited from outside the system, or possibly modified in place. The spatial distribution of material of different composition will allow these possibilities to be separated.

E. Characterize the mass flux into the various ring systems.

What is the current mass flux into the various ring systems? What are the current size, mass, velocity, and composition distributions of the influx population?

New material is constantly being added to the rings as the result of meteoroid bombardment, and this mass flux is one of the mechanisms for ring evolution. The compositional and structural effects of this bombardment can provide constraints on ring age independent from angular momentum arguments. However, those effects remain undetermined because of the poorly known interplanetary mass flux in the outer solar system. Ideally, for each of the ring systems, it would be best to know the combined size, mass, velocity and composition distributions of incoming material as a function of ring radius and ring longitude. To fully constrain ballistic transport (the evolution of ejecta from impacts), we need the equivalent of a mass intensity for the projectiles (he mass flux of small projectiles into the rings as a funtion of both ring radius and direction). The direction matters because the relative amount of impacts on the forward and trailing ring particle hemispheres matters. We need to know mass and velocity separately because the ratio of impact ejecta mass to projectile mass depends on the impact speed.

F. Determine the influences of the magnetospheric and plasma environments of the various rings.

What is the influence of the magnetosphere and plasma environment of the rings?

The faint, dusty rings of Jupiter and Saturn illustrate a number of the possible links between ring dust and the planet's magnetosphere and plasma environment. The three dimensional structure of Jupiter's inner halo is defined by orbital resonances with the planet's rotating magnetic field. The structure of Saturn's broad E Ring is probably defined by a particular match between electromagnetic forces, solar radiation pressure and Saturn's gravity field, which enables some dust particles to be dispersed widely throughout the Saturn system. Unfortunately, the electric charge on grains, which is so critical to defining their motion, depends on properties of the plasma environment that are very poorly known. The "spokes" which flicker across the face of Saturn's rings might be triggered by meteoroid impacts or by electromagnetic instabilities. Their propagation and occurrence display unmistakable signatures of the planet's magnetic field, including connections to its longitudinal asymmetry. Orbital instabilities of tiny charged grains might be a significant loss mechanism of ring material to the planet. The properties of the local plasma and charging environment at the ring plane are unknown, and yet are critical to advancing our understanding of these important phenomena.

There were two additional questions, different in nature to the others, which were included in the survey and fall into the mid group of key questions. The first is a question we must continually ask as our understanding of individual ring systems improves. The other "different" question must be answered to enable us to better understand the data we gather from Earth-based activities.

G. What do the differences among ring systems tell us about differences in ring progenitors and/or differences in initial and subsequent processes?

   Much of our understanding of the geologic and atmospheric activities of the planets comes from comparative planetology. So too with planetary rings, comparisons between the various ring systems will provide insights beyond those achievable by viewing each ring system in isolation. This key question will be asked repeatedly as our understanding of individual ring systems improves.

H. What is the relationship between ring local properties and those properties observable via remote sensing?

   Much of our current knowledge is based on modeling local ring properties theoretically, predicting remotely observed properties from these models, and adjusting the various models to achieve the closest match to observations. We need to validate our understanding of the parameters within those models. Planetary scientists have gained insights into the geology of Mars by interpreting Martian observations within the knowledge context of Terrestrial field geology. Geologists have walked the shores and deserts, climbed the mountains, descended into the canyons of Earth, have studied the rocks of Earth in the field and in the laboratory, all of which establishes a foundation for remote studies of distant bodies. No one has yet been to a ring.

3. Recommendations

We divide our recommendations into two broad categories: interplanetary spacecraft and activities that can be performed on or near Earth. In establishing priorities, we have produced ordered lists with the most important first. Within each list items tended to cluster naturally; consequently, the recommendations are divided into three groups: essential, critical, and important.

Among all of the possible activities that we considered for the decade, our overall number one priority is for robust support and augmentation of the Cassini nominal mission.

In the area of interplanetary spacecraft, our recommendations are as follows:

- ESSENTIAL. Increased funding for Cassini nominal mission: support for a participating scientist program (competitive but probably limited selection), additional team associates (selected by team members) and a comprehensive data analysis program (competition open to the community).

  Cassini has a powerful array of instruments either new or improved, and a scheduled four year tour with numerous observing geometries. It's hard even to imagine the new things we'll discover and the new questions which will emerge as we get tens of thousands of images in many filters and several polarizations, hundreds of detailed UV and IR spectra at resolutions adequate to resolve most of the ring structure, approximately a hundred stellar and radio occultations, thermal emission in infrared and microwave spectral regions, and in-situ measurements of dust, gas, and plasma abundance and composition around the rings (and even right above them during Saturn Orbit Insertion). Cassini is NASA's single largest investment in planetary science.

  The current Cassini teams are excellent but they are thinly staffed and funded. Most of the team members have been committing time far above the levels for which they have been funded. To capitalize on NASA's investment in Cassini, and to optimize the treasure house of data it will return, calls for an influx of new funding for the current teams, and for new blood to work alongside the current teams.

  At current funding levels it is not clear that the teams will have sufficient time to do the early data analysis required to refine the later mission sequences, nor to archive the incoming data as rapidly as planned. Data archiving is the necessary intermediate step between data acquisition and data distribution to the community at large. The planned data archive pipeline must be fully functional before the wider community can become involved in Cassini data analysis.

  Early augmentation of the current teams can ensure the staffing for the teams' necessary internal planning, data analysis and publication, as well as prompt data archiving for the benefit of the broader community. Broader community involvement can be achieved through the implementation of a sustained data analysis program beginning approximately one year after Saturn Orbit insertion and continuing several years beyond the end of the mission.

- ESSENTIAL. Cassini extended mission (including associated support for data analysis and archiving). The initial extension should be for a minimum of three years to encompass local observation of the next solar ring plane crossing and subsequent significant reopening of the rings.

  Cassini's nominal tour ends in July 2008, well within the period covered by this decadal study. Hopefully, it will be in good health and have ample propulsion reserve to go several years longer. Based on the earliest look at science planning for the nominal four year tour, there will not be enough time or data volume to do any more than a "must have" subset of all the ring observations one could imagine doing. From the standpoint of addressing any of the more detailed level "key questions" we can identify now, we can't think of a single higher priority approach we could recommend than the full-up continuation of Cassini into an extended mission, bolstered by a budget line built right into NASA's planning wedge which would include robust participating scientist opportunities, continuation of the data analysis program and maintenance of the data archiving pipeline to allow that flood of data to be preserved, disseminated, analyzed and interpreted.

During such an extended mission, the rings will close up as seen from Earth, and then reopen again. This provides new opportunities for radio occultations to pick up great sensitivity in the optically thin regions. Close up observation of the rings with edge on illumination will allow high resolution studies of vertical corrugations and provide an opportunity to observe the illuminated side of the rings at high phase angles (not possible in Earth based observations). Spokes might become visible again for the first time in the mission (spokes have never been seen when the Sun/Earth elevation angle is larger than about 4-5 degrees). All instruments will continue to monitor the rings for changes, zoom in on interesting features, improve spectral resolution and coverage, and refine radial resolution and longitudinal coverage in new illumination and viewing geometries. Such an extension will provide the longer time needed to observe the several dynamical phenomena which are known to operate on 5-10 year timescales in the main rings and in the F ring. Perhaps most important of all will be the opportunity to follow up on the inevitable new discoveries with dedicated, multi-instrument campaigns, to distinguish between the competing hypotheses which will quickly arise.

- CRITICAL. Close-in orbiter to study Saturn ring microphysics at the level of individual particles, and the local ring environment.

Despite everything Cassini will accomplish, we will not learn much directly about ring microphysics, physics at the level of individual particles. Currently we infer such processes indirectly from remote sensing data, but those inferences have never been truly validated by detailed in-situ observations. In the main rings, we are dealing with a self-gravitating, collisional, particulate system inside its planet's Roche zone with disparate particle sizes, a wide range of densities, random velocities and packing densities, subject to a variety of influences. Even subtle influences can have important effects over time. We need direct information about the microphysics.

Second, minor constituents of the ring material, or complex organic molecules, might never be detectable in remote observations, even by Cassini. Certainly, isotopic compositions and the presence or absence of noble gases will never be detectable remotely. All of these contain important clues to ring origin.

Third, the effort to understand fully several important processes requires us to know what the local environment of the ring particles is. What is the electron density in the surrounding plasma, and how is it maintained? What is the charge on small dust grains and what are their motions? What sort of impact plasma or cloud forms when a meteoroid slams into the rings at 70 km/second, and how does it evolve? Are there electromagnetic instabilities? Can material from the planet's ionosphere modify ring particle properties?

Some of these questions might be addressed by a focused, relatively short term mission perhaps in the Discovery class. The spacecraft would operate just out of the ring plane, "hovering" above various regions within Saturn's main rings. It would need the capability of altering its radial distance from the planet in order to sample different regimes. This mission, depending on instrument packages and mobility, could address six to eight of the key questions at levels of detail that cannot be achieved by Cassini. Such a mission might be a suitable test bed for an ion drive engine.

The mission needs to arrive at Saturn near the time when the rings are the most open to the sun - both for optimum illumination, and for easy insertion into equatorial orbit. The next opportunity for such a mission would be with arrival in the interval between 2014 and 2016. The subsequent window would be approximately fifteen years later.

- IMPORTANT. Neptune orbiter.

The next most interesting ring system after Saturn, is probably that of Neptune. It has the additional scientific interest of being the most remote in the solar system, and the closest to the Kuiper Belt. There is a massive inner ringmoon belt with probably many smaller moons Voyager couldn't find. There are diffuse dusty rings. This is the only ring system known to possess relatively stable arcs, although the proposed confinement mechanism has recently come into question as a result of recent HST observations of the clumps. The vertical and radial structure of the Neptunian rings remain unresolved; we are completely ignorant of the ring and ringmoon composition or how it varies with location, except that the material seems to be dark. We are likely always to remain ignorant if we have to rely on Earth-based observations alone. Are the ring arcs indeed trapped or are they evolving? Are they "rubble belts" containing large boulders as well as dust? Are they the results of disruption of moons? How long ago? What is the meteoritic in fall rate and composition? Voyager data contains hints of other narrow, kinky ringlets in a few as-yet unstudied images.

New ground-based observations may provide more insights, but ground-based observations of the Neptunian rings have serious limitations. It is best to view dusty rings in forward-scattered light, which places the rings between the observer and the Sun. This requires spacecraft.

If more new answers and discoveries come from lifting the veil on areas where our ignorance is the deepest, the Neptunian system clearly offers the greatest potential for expanding our knowledge. Bringing our understanding of the Neptunian rings up to a similar level as the Saturnian and

Jovian rings would greatly enhance the ability of the rings community to do real comparative studies. We would be better able to address the key question "Why are the ring systems of the giant planets so different?" which necessarily requires that we get one or two of the outer most ring systems up to a level of understanding approaching the Saturn ring system (Voyager-Saturn level, if not Cassini-level).

In the area of Earth-based or near-Earth activities, we recommend:

- ESSENTIAL. Long term observing campaign for the rings of Saturn.
- CRITICAL. Simulations and laboratory experiments.
- CRITICAL. Periodic astrometric observations of Saturnian inner satellites.
- CRITICAL. Develop a virtual theoretical laboratory to provide a vehicle for more coordination and integration of theoretical work.

  We need to consider more coordinated theoretical efforts than are currently the norm. We have begun to develop "virtual observatories". A similar effort has proven beneficial in the study of stellar structure. Some coordination of theoretical efforts does occur in ring science, but not in a coherent and consistent fashion. Now is the time to develop a virtual theoretical laboratory to facilitate better coordination and more systematic integration of those efforts.

  Useful facilities can be envisioned at very different levels of sophistication. Goals could range from modest ones, such as coordinating activities among various ring theory groups, to ambitious ones, such as providing and maintaining easy-to-use, robust, rigorously tested on-line modular software laboratories which permit users to undertake complex simulations and to visualize their results without undo concern about where or how the computations are being done. In the latter case, the "laboratory" must be funded well enough that requisite computational and software facilities can be developed, maintained, and documented by teams of planetary scientists, computational scientists, software analysts, and system managers and technicians. The lab would be "virtual" in that the teams responsible for different software areas and the computational facilities utilized by the laboratory could have a wide geographical distribution. Although the concept originated in our Ring Panel discussions as a way to help provide more focus and cooperation among ring theorists, there is no reason to limit its scope to rings over the coming decade. It may perhaps be time to work toward establishment of a national if not global virtual laboratory for astrophysics and planetary science. The larger facility could evolve from virtual laboratories designed for specific problems deemed of central importance to various disciplines.

- IMPORTANT. Long term observing campaign for the rings of Neptune.
- IMPORTANT. Long term observing campaign for the rings of Uranus.

Uranian ring plane crossings occur approximately every forty-two years. The next such ring plane crossing will be in 2007. Thus during the decade of this study we will be afforded a rare opportunity to observe the edge-on rings. In order to maximize the return from this viewing opportunity we need to establish a high resolution viewing campaign that extends throughout the decade in order to encompass the entire cycle as the ring opening angle closes and then reopens to a significant amount.

4. Data Archive and Access Issues

Some processes within planetary rings operate on sufficiently short timescales to permit the detection of changes on timescales ranging from days to decades or tens of decades. In order to gain insights into the underlying processes, it must be possible to access and analyze historical data. Spacecraft data associated with planetary rings studies are archived and distributed by the NASA Planetary Data System. A limited amount of Earth based and HST data have also been archived. The volume of planetary data will increase dramatically during the coming decade. The PDS should expand its capability to provide online distribution of data sets, provide tools for fine grained searches of the data, and tools to return basic observational parameters. Efforts to incorporate Earth based data should continue and tools to simplify the submission of data to the PDS be developed.

5. Education and Public Outreach

Planetary rings are among the most spectacular and appealing objects in the solar system and possess the capability to capture the imagination of young people. Various aspects of rings studies can be organized to allow high school and undergraduate students to actively participate in the search for better understanding. Properly organized such efforts should leave them with a sense of the search for understanding rather than the all too frequent classroom approach that encapsulates the proposed question with the neatly presented correct answer.

6. Summary

The decade covered by this White Paper is critical to our quest to understand the age, origin and evolution of planetary rings. The Cassini spacecraft arrives at Saturn in the second year of the study decade for a nominal mission of four years and the potential to continue operations throughout the bulk of the decade. The unique viewing geometry and illumination of the rings associated with solar ring plane crossings for Saturn (every fifteen years) and Uranus (every forty two years) both occur during this decade.

The Cassini mission provides the cornerstone and focus for efforts to better understand the composition and processes of Saturn's rings, and must be done right. Cassini has received significant funding drawbacks from HQ over the last few years and the science teams are stretched extremely thin. To capitalize on

NASA's investment in Cassini, and to optimize the treasure house of data it will return, we call for an influx of new funding for the current teams, for new blood to work alongside the current teams, and for a comprehensive Cassini Data Analysis program to be established.

Beyond the Cassini nominal mission should be more Cassini. Even before Cassini arrives in orbit at Saturn, it is clear that if the spacecraft remains healthy, a Cassini extended mission will provide the best science for dollar return in the outer solar system. Such a mission extension should be anticipated and fully funded.

Following Cassini a dedicated mission to study the detailed interactions of Saturn's main rings at the individual ring particle level should be undertaken.

Throughout the decade comprehensive, systematic observation programs need to be supported for the all of the planetary ring systems. Theoretical work will be greatly enhanced by influx of new data. New approaches to the coordination and sharing of that work need to be explored taking advantage of the potential inherent in the high speed, reliable connectivity possible with advances in modern computing.

References

There is a wealth of information available in the literature. We give here a few of the review papers written in recent years that relate to the study of planetary rings. These papers and the references contained within them will provide the interested reader a detailed look at the past, present, and future focus of planetary ring science.

Burns, J.A. 1999. Planetary rings. In *The New Solar System* (J.K. Beatty, Ed.), pp. 221-241. Sky Publishing Co., Cambridge.

Cuzzi, J.N. 1995. Evolution of planetary ringmoon systems. In *Comparative Planetology* (M.T. Chahine, M.F. A'Hearn, J.H. Rahe, Eds.), pp. 179-208. Kluwer Press, Dordrecht.

Cuzzi, J.N. 1995. Evolution of planetary ringmoon systems. *Earth, Moon, Planet.* **67**, 179-208.

Cuzzi, J.N., J.E. Colwell, L.W. Esposito, C.C. Porco, C.D. Murray, P.D. Nicholson, L. Spilker, E.A. Marouf, R.C. French, N. Rappaport, and D. Muhleman 2002. Saturn's rings: Pre-Cassini status and mission goals. *Space Sci. Rev.*, in press.

Esposito, L.W. 1993. Understanding planetary rings. *Annu. Rev. Earth Planet. Sci.* **21**, 487-523.

Esposito, L.W. 2002. Planetary rings. *Rep. Prog. Phys.*, in press.

Murray, C.D. 1994. Planetary ring dynamics. *Phil. Trans. R. Soc. Lond. A* **349**, 335-344.

Nicholson, P.D. and L. Dones 1991. Planetary rings. *Rev. Geophys. Supp.* **29**(1), 313-327.

Porco, C.C. 1995. Highlights in planetary rings. *Rev. Geophys. Space Phys. Supp.* **33**, 497-504.

The Future of Solar System Exploration, 2003-2013
ASP Conference Series, Vol. 272, 2002
Mark V. Sykes, ed.

Dust Astronomy: New Venues in Interplanetary and Interstellar Dust Research

E. Grün[1,2], P.G. Brown[3], A.L. Graps[1], J.M. Hahn[4], D.P. Hamilton[5], W.M. Harris[6], M. Horányi[7], D.L. Huestis[8], A. Krivov[9], M.J. Kuchner[10], A.C. Levasseur-Regourd[11], D.J. Lien[12], J.-C. Liou[13], C.M. Lisse[5], D.D. Meisel[14], W.T. Reach[15], M.L. Sitko[16], T.P. Snow[7], R. Srama[1], J.A. Stansberry[17], M.V. Sykes[17], H. Yano[18], & M.E. Zolensky[19]

[1] *Max-Planck-Institut für Kernphysik, Heidelberg, Germany*
[2] *Hawaii Institute for Geolophysics and Planetology, University of Hawaii, Honolulu, HI*
[3] *Department of Physics and Astronomy, University of Western Ontario, London, Canada*
[4] *Lunar and Planetary Institute, Houston, TX*
[5] *Department of Astronomy, University of Maryland, College Park, MD*
[6] *Space Astronomy Laboratory, University of Wisconsin, Madison, WI*
[7] *Laboratory for Atmospheric and Space Physics, University of Colorado, Boulder, CO*
[8] *Molecular Physics Laboratory, SRI International, Menlo Park, CA*
[9] *Institut für Physik, University of Potsdam, Potsdam, Germany*
[10] *Harvard-Smithsonian, Center for Astrophysics, Cambridge, MA*
[11] *Service d'Aeronomie, University of Paris, Verrieres, France*
[12] *Department of Physics, Oklahoma State University, Stillwater, OK*
[13] *Lockheed-Martin, Houston, TX*
[14] *State University of New York at Geneseo, Geneseo, NY*
[15] *Infrared Processing and Analysis Center, California Institute of Technology, Pasadena, CA*
[16] *Department of Physics, University of Cincinnati, Cincinnati, OH*
[17] *Steward Observatory, University of Arizona, Tucson, AZ*
[18] *Planetary Science Division, Institute of Space and Astronautical Science, Kanagawa, Japan*
[19] *NASA Johnson Space Center, Houston, TX*

Abstract. Dust Astronomy is a new research area which investigates the properties of and the link between the interplanetary and interstellar dust. Both, in-situ methods as well as remote sensing methods like optical and infrared observations are used to study the physical and chemical properties of the dust. These measurements need to be complemented by ground-based meteor studies and the analysis of collected dust samples. Major goals for the next decade are to characterize the Kuiper belt dust, the zodiacal and interstellar dust complex. It is recommended that (1) all missions to the outer solar system carry an in-situ dust detector, (2) a Cosmic Dust Observatory should be launched into heliocentric orbit, and (3) a Heliocentric High-Inclination Explorer infrared mission should be realized.

EXECUTIVE SUMMARY

Dust particles, like photons, are born at remote sites in space and time. From knowledge of the dust particles' birthplace and the particles' bulk properties, we can learn about the remote environment out of which the particles were formed and how those particles have evolved physically and dynamically. Moreover, dust dominates the thermal emission of our planetary system and other planetary systems, therefore studying interplanetary dust teaches us how the solar system appears from space, and it helps us to interpret observations of other planetary systems.

The study of dust in the planetary system has been, and will be, a prime goal of many NASA missions. Missions to the dust rich environments of comets and planetary rings carried, and will carry, one or several dust instruments. In addition, interplanetary missions, like Pioneer 10 and 11, Helios, and Ulysses performed studies of the interplanetary dust environment among their primary objectives. Future missions to the unexplored regions on the solar system map, such as the Solar Probe missions, missions to study long-period comets, orbiter missions to Uranus, Neptune, and Pluto, and missions to the Kuiper belt and beyond, will carry dust instrumentation. Small dedicated missions to study the dust environment at 1 AU and in near-Earth space will also carry sophisticated dust detectors. In addition to in-situ methods, optical and infra red remote sensing methods of observing dust will allow us to make a global assessment of dust origin and production in our solar system and its context within the local interstellar environment, and will provide insights into basic processes within our solar system that are directly applicable to planetary systems around other stars. Because of the complex nature of dust particles, the study of space dust requires a multi-disciplinary approach: space measurements, both by in-situ and remote sensing methods need to be complemented by ground-based meteor studies, analyses of collected samples, simulation experiments and theory. Priority questions for the next decade are:

- Is there dust in the Kuiper belt and what is its distribution and density?
- What are the compositions of the components of the zodiacal dust complex and how do they vary spatially and temporally?

Recommendations for the next decade are, in order of importance:

- All missions to the outer solar system, in particular missions to Pluto and beyond, should include an in-situ dust experiment for detecting and determining the properties of local dust
- A "Cosmic Dust Observatory" should be launched into heliocentric orbit near 1 AU to measure compositional, physical, and dynamical properties of interplanetary dust particles (which should include dust from identified comets). There should be coordination between this observatory and collection experiments from the Earth's upper atmosphere.

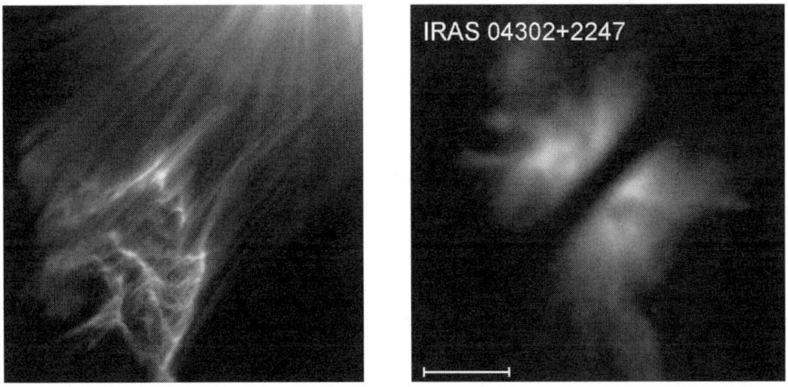

Figure 1. Reflection nebula in the Pleiades (left). An interstellar dust cloud is illuminated by a near-by star (above the top). Right: Young stellar disk around IRAS 4302-2247. The dark dust band obscures the central star and is probably the birth place of planets (HST images).

- A Heliocentric High-Inclination Explorer mission consisting of a cryogenically cooled thermal infrared telescope that 'looks over' the inner zodiacal cloud to study dust in the outer solar system and investigate the structure, sources, and sinks of the inner zodiacal cloud.

All recommendations require support for continuing theoretical studies of dust dynamics, modeling of past observations, and laboratory studies to characterize the composition and physical properties of dust.

REPORT

1. Current State of Knowledge

Cosmic dust particles evolve chemically, physically and dynamically. The evolution of dust traces out paths in which the universe recycles material: production, storage, processing, collection, consumption, and discarding. In stellar winds of evolved stars, new dust is formed and is injected into interstellar space. Young stardust is mixed with old heavily-processed diffuse interstellar dust, and is subject to passing supernova shocks and ultraviolet radiation. Dusty clouds form (Figure 1). Star formation in cool molecular clouds becomes both a sink of old dust, as well as a source of new dust. A typical dust grain anywhere in space will have undergone several cycles.

Dust in a planetary system is the most processed (being formed, destroyed, and/or locked into a near-pristine state) of the different populations of cosmic dust. Interplanetary dust is permanently replenished by dust ejected from cometary nuclei, the most pristine bodies in the solar system, and released from collisions in the asteroid and Kuiper belts. In our solar system, interplanetary dust exists alongside interstellar dust, which is flowing through the solar system, offering a tangible, physical link between our planetary system and the stars.

The planetary system hosts a variety of dusty phenomena, such as scattered light, thermal emission, impacts on solid bodies and on atmospheres. Cosmic dust can be found in cometary comae and tails, in planetary rings, on asteroidal regolith, in the Earth's stratosphere, in polar ices, and in the interplanetary dust cloud, which is permanently replenished by interstellar, cometary and asteroidal dust.

One of the source mechanisms for dust is hypervelocity impacts of micrometeoroids, interstellar grains, or ring particles onto the surfaces of solar system bodies (planetary satellites, asteroids, atmosphereless planets Mercury and Pluto, KBOs etc.). For example, Mercury, an atmosphereless planet orbiting the Sun in the densest part of the interplanetary dust complex, must experience intensive meteoroidal bombardment and should be surrounded by a pronounced ejecta cloud. Recent in-situ detections of ejecta clouds around Galilean satellites of Jupiter and future measurements of dust swarms around other bodies (e.g., outer Saturnian moons by Cassini) can be considered as a natural impact experiment. Such studies will allow us to get more insight into impact physics in space and to extend the lab impact experiments to a broader range of masses and speeds, astrophysically relevant materials and morphologies of targets and projectiles, and microgravity conditions.

Once produced, dust grains are moving through complex gravitational, radiation, and magnetic fields and plasma environments. Studying the dust dynamics and comparing the results with remote and direct observations allow one to indirectly probe the properties of these fields and environments and in such a way, to complement direct measurements.

Planets and extrasolar dust disks (Figure 2) are manifestations of the same phenomenon called a planetary system. Circumstellar dust is the material from which, and in which, planets are formed. Dust is an excellent indicator of physical processes and larger bodies in a circumstellar disk - comets and planetesimals (that continuously replenish the dust cloud) and planets (that perturb dust disks directly as well as indirectly, through dynamical sculpturing of the dust parent body populations). Dust, however, is much easier to observe than the planets. The masses of the observable disks in Vega-type systems are estimated to be less than the Earth's mass, whereas planets of such masses are far beyond the detection limits. The properties of dust and their change with time fully reflect the evolution of a planetary system and may serve as a measure of the evolutionary stage of an evolving planetary system.

An early motivation for the study of dust in space was the risk imposed by impacts of meteoroids onto man-made satellites. From that early motivation a large number of primitive dust detectors were launched into near-Earth space. After it was recognized that the natural meteoroid flux is low and that the risk can be mitigated by simple technical means, the early motivation ceased to exist. Subsequently, the motivation for the study of dust in space shifted to astrophysical questions about the physical and chemical properties of the grains, their contemporary sources and sinks, and their significance as probes of the conditions during the formation of the planetary system. This is documented by dust measurements in the inner planetary system (Helios), at comets (Giotto,

Figure 2. Dust disks. Top: Edge-on view of the dust disk around the star beta Pictoris (HST image). This disk is a model for the suspected Kuiper belt dust ring of our solar system. Middle: Inner zodiacal dust cloud, partially obscured by the moon (NASA Clementine image). Bottom: Jupiter dust ring. Outside the bright central ring is the faint gossamer ring that derives from the Jovian satellites Amalthea and Thebe. (NASA Galileo image)

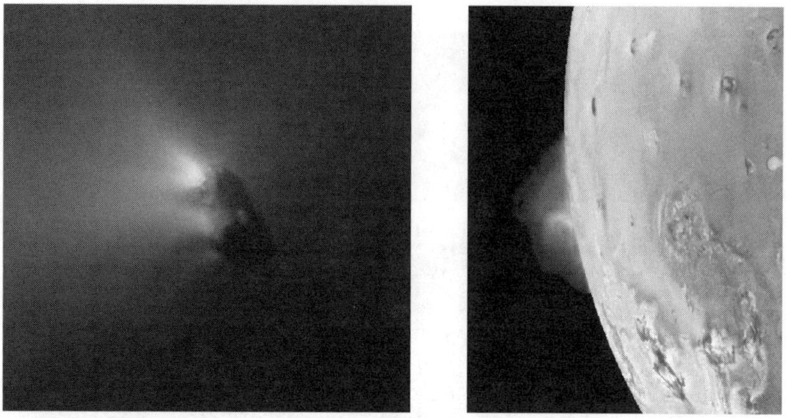

Figure 3. Dust formation. Dust emission from Comet Halley (Giotto image, courtesy H.U. Keller). Right: Volcanic plume on Jupiter's moon Io. Nanometer-sized dust escapes Io and is ejected by Jupiter's magnetic field (NASA Galileo image).

VeGa, Stardust), in the vicinity of Mars, Jupiter, and Saturn (Nozomi, Galileo, Cassini) and above the poles of the Sun (Ulysses) (Figure 3). The recognition that manned activity in near-Earth space left a growing amount of space debris lead to the deployment of an increasing number of small dust impact detectors in low Earth orbit, like Spadus and Debie. However, within the Earth debris belts, the natural meteoroid background is obscured.

The dust detectors on the Galileo, Ulysses and Cassini missions have a high sensitivity (0.1 μm particles) together with a large sensitive area (0.1 m^2) and enable the study of difficult-to-detect effects, such as the passage of interstellar grains through the planetary system (which cannot be identified by smaller detectors because of a too low number of recorded dust impacts). The combination of an impact ionization detector with a time-of-flight mass spectrometer provides the capability of a chemical analysis of the recorded dust grains. Such instruments have been used in the Halley missions and an upgraded version flies on the Stardust mission to comet Wild 2. The Cosmic Dust Analyzer, CDA on the Cassini mission, is the most versatile dust detector. It combines a large impact area with a mass spectrometer. In addition, it measures the electrical charge of dust particles, which provides a refined measurement of the dust speed and trajectory. A smaller impact ionization detector, the Mars Dust Counter, MDC on the Nozomi mission, is currently en route to Mars, where it will characterize the Martian dust environment.

There are other methods to collect and analyze cosmic dust at Earth: Interplanetary dust particles can be collected in the stratosphere by aircraft or micrometeoroids can be extracted from the Antarctic ice (Figure 4). From analyses of these particles, compositional, mineralogical, and structural information is obtained. The composition, shape, and density of the particle affects all physical processes acting on the particle, such as the Poynting-Robertson drag force and the charging currents. One result from particle density measurements is

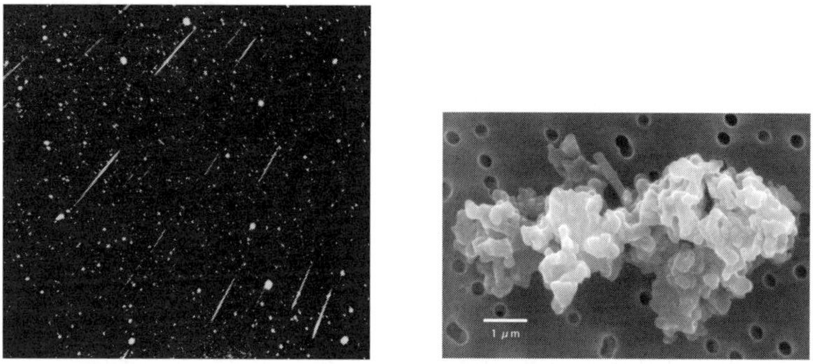

Figure 4. Dust reaching the Earth. Left: Leonid meteor storm as observed in 1966. Centimeter-sized particles enter the Earth's atmosphere. Right: Interplanetary Dust particle collected by high-flying airplane (NASA image).

that common IDPs are much more porous than meteorites, with a typical density of ca. 2 g/cm^3 corresponding to a porosity near 40%. Particles with a density below 1 g/cm^3 do exist, but they are rare. Microcraters on satellite surfaces (e.g., LDEF or on lunar rocks) provide knowledge of the overall dust flux over a wide size range from micrometer to millimeter sizes. On the basis of this information, an interplanetary dust flux model at 1 AU distance from the sun has been formulated. The model covers a mass range from 10^{-18} to 10^2 g.

Zodiacal light observations have been performed from the ground, from Earth satellites and from deep spaceprobes. With the photometers on board Helios, a radial brightness profile of zodiacal light was determined between 0.3 and 1 AU. Outside the Earth's orbit, zodiacal light was observed by photometers on board Pioneer 10 and 11. The zodiacal light brightness was found to exceed the background out to 3.3 AU distance from the Sun.

Once the technology for spacebased infrared observations was developed in 1984, the IRAS satellite and, later, the COBE satellite, obtained unprecedented information on the overall structure of the zodiacal cloud. Besides the large-scale structure of the zodiacal cloud, broad asteroidal bands, as well as narrow comet trails, were discovered that directly relate to asteroidal and cometary sources of the zodiacal cloud. Models have been derived that describe the infrared brightness of these phenomena as observed by COBE with an accuracy of better than 3%.

In the mass range $m > 10^{-7}$ g, radar and photographic meteor observations provide independent information on the orbits of these particles. The Arecibo radio telescope in radar mode can detect meteors from grains of masses $m > 10^{-11}$ g.

Dynamical models have been constructed on the basis of meteor data, zodiacal light observations, and spacecraft measurements that predict the distributions in mass and orbital elements of interplanetary dust grains. While these models describe the current state of the meteoroid population, the origin of these particles is generally unknown. No evolutionary models have been constructed that reliably describe the relative contributions from asteroidal, cometary, Kuiper belt and other sources to the solar system dust cloud.

2. Key Science Questions

The field of space dust research is far from complete. Despite the great advances made in the last years of the understanding of the interplanetary and planetary dust environments, there remain many important questions to be answered. Starting close to the Sun, nature separates meteoroid material according to its volatility. Analysis of the spatial distribution of matter close to the Sun will immediately give us information on the volatility of its constituents.

Closer to home, the dust environment of the Earth is of interest, because mankind is affecting this environment due to its space activities. However, the natural dust environment is also of technological and scientific interest. Hazards from meteor streams (like the Leonids) will require continuous attention. Meteoroids that pass the Earth are of scientific interest because they are messengers from distant worlds, like asteroids, comets and even interstellar space. Once we know from where dust grains originate, compositional analysis of grains can tell us many things about these worlds. Therefore, the goal of such dust studies is to identify the sources of dust particles together with their in-depth analysis. The future of dust measurements in Earth orbit lies in three areas: (1) environmental monitoring, (2) use of dust telescopes to separate and analyze dust populations of different origin, and (3) collection and sample return of dust from various sources for in-depth analysis in laboratories.

Analysis of particulates from comet Halley brought us new and important information that has relevance to the understanding of the formation of our planetary system. Currently, the Stardust mission is on its way to analyze, collect and return dust from comet Wild 2. This will give us a second example from the large variety of comets. Analyzing different comets will tell us about the spatial and compositional variations in the protoplanetary nebula through the comets, which may have sampled different regions of this nebula.

Enhanced dust densities in the Martian planetosphere have long been suspected. The comparison of future dust observations at Mars with those in the dusty rings of Saturn and the other giant planets will tell us the effects of solar radiation pressure (strongest at Mars) and planetary magnetospheres (negligible at Mars).

All giant planets have ring systems which display a variety of different "dusty" phenomena. While Saturn's ring system is of high complexity, the rings of Jupiter, Uranus and Neptune show other features that have not yet been found elsewhere. The Pluto-Charon system may have rings with yet other features. To understand the common characteristics of all these rings, and the reasons why they are so different, requires detailed measurements of the rings and their environments.

Detection of Kuiper belt objects (KBOs) of up to a few 100 km diameter confirmed the existence of objects beyond the orbits of the planets. Mutual collisions among KBOs, as well as impacts of interstellar grains, generate dust locally. The action of the Poynting-Robertson effect, together with resonances with the outer giant planets, interaction with the solar wind and neutral interstellar gas, may have lead to radial and azimuthal structure of the distribution of dust at the edge of the planetary system. The detection of infrared excesses in main sequence stars began a renewed interest in the outer extensions of our own solar system dust cloud. The observation of dust disks around beta Pictoris and epsilon Eridani stimulated this interest. Observations of the Kuiper dust belt in our solar system can, therefore, be used as a model for extra-solar dust clouds and can help to reveal information about other planetary systems.

Effects of the solar cycle-dependent heliosphere reach into interstellar space out to about 300 AU from the Sun, where it interacts with the small (<0.1 micron) particles entering the heliosphere and modulates their flow. The tiny particles' origin, however, may be different from that of bigger grains that are accessible in Earth orbit. We know that evolved stars continuously lose mass. This "stardust" provides the seeds for ISD grains, that grow in cool interstellar clouds by accretion of atoms and molecules and by agglomeration. An unbiased look into this interstellar dust factory will provide us with information on processes that are difficult to quantify by astronomical observations alone. Therefore, in-situ dust measurements will be an important analysis tool when automated probes will leave our solar system.

Dust studies may be broken down into the following key areas:

- Dust inventory in the solar system
    - What are the spatial, size, and temporal distributions of dust in the solar system?
    - How important are comets, asteroids, and Kuiper belt objects as sources of the solar system dust cloud?
- Structure of the solar system dust cloud
    - What are the radial and latitudinal distributions of the solar system dust cloud?
    - What is the configuration of the symmetry plane dust of solar system dust cloud?
    - What is the fine structure dust of solar system dust cloud (comet trails, asteroidal bands, planet shepherding rings, and other local phenomena)?
- Specific dusty environments in interplanetary space
    - How is dust of different volatilities separated in the F-corona?
    - What is the chemical, isotopic, and mineralogical composition of dust from short-period and long-period comets?
    - What is the dust environment in the asteroid belt?
    - What are the processes of dust generation and dispersion in the Kuiper belt?
    - What is the temporal and spatial evolution of meteor stream dust, e.g., in the Leonids?
- Dust in planetary environments

- What are the characteristics of the Mercury ejecta cloud?
- How do natural and man-made debris evolve in near-Earth space?
- What is the dust mobility on the lunar surface?
- What is the dust environment of Mars?
- How different are the rings and dust environments of Jupiter, Saturn, Uranus and Neptune?
- How much dust is in the Pluto-Charon system?
- What role does dust play as a medium for material transport in planetary environments?

- Interstellar dust inside and outside the heliosphere
  - What is the flow of ISD through the heliosphere?
  - How important is size-dependent filtering of ISD in the heliosphere and at the heliospheric boundary?
  - What is the chemical and isotopic composition of interstellar dust grains in the heliosphere?
  - How do dust grains evolve in interstellar space?
  - What is the interstellar dust size distribution and how significant are large interstellar dust grains contributing to the total dust content of the local interstellar cloud?
  - How significant are spatial and compositional variations of the local interstellar dust cloud?
  - How strongly is the chemical evolution of the local interstellar cloud coupled to the overall chemical evolution of the Galaxy?
  - What are the key ejection mechanisms for ejection of dust from a planetary system into interstellar space?

- What do dust observations tell us about the space environment and about basic planetary processes?
  - Dust ejecta clouds as natural impact experiments
  - Dust as probes of remote plasma and fields environments
  - Dust as monitors of volcanoes (Io) and geysers (Triton)
  - Dust processes in the Kuiper belt as a model for extra-solar dust clouds
  - Dust-magnetosphere interactions, dusty plasmas in space
  - Levitation and dust transport on planetary and asteroidal surfaces

3. How Do We Address Key Science Questions?

3.1. From space

- Determine the large- and fine-scale structure of interplanetary and planetary dust clouds.

  This requires remote sensing observations of the dust distribution using scattered sunlight and thermal radiation throughout the solar system. Dedicated instrumentation may be needed to provide successful measurements.

- Characterize the dust environment throughout the solar system

  Perform in-situ measurements of dust distribution and of dust properties, and fly dust instrumentation on missions to unknown territory, e.g., the F-corona, the Kuiper belt and the planetospheres of Uranus, Neptune, and Pluto. Appropriate dust instrumentation should determine the radial profile of dust density, the orbital distribution, physical, and chemical properties of dust.

- Analyze the orbital characteristics and chemical composition of various dust components in interplanetary space: interstellar dust and interplanetary dust of cometary and asteroidal origins.

  These types of measurements require a spacebased "Dust Telescope". Such a dust telescope is capable of providing mass, speed, physical and chemical information of dust grains in space. Particle origins are identified by their trajectories, which need to be determined with sufficient accuracy. Information on the physical and chemical composition is obtained from impact signals when dust particles hit versatile dust detectors. Targets for such telescopes are: dust from meteor streams, comets, asteroids, the Moon, and interstellar space (passing dust through the solar system). It provides a cost-effective means of geochemically sampling targets which are otherwise inaccessible by spacecraft. Even though a dust telescope should first be employed in near Earth space (outside the dense debris belts), it is foreseen that this kind of telescope will provide valuable information also in planetary environments and in the outer regions of the solar system.

- Collect and return samples of dust from various sources (comets, asteroids, interstellar dust) at 1 AU and beyond.

  Interplanetary grains collected by high-flying aircraft have been successfully analyzed in many laboratories. Stardust is the first mission planned to provide dust samples from a short-period comet, and, perhaps, a few interstellar grains. Future missions in interplanetary space at 1 AU and beyond will be able to return dust samples from asteroids, long-period comets and interstellar grains.

3.2. From the ground

- Extend the trajectory studies of meteoroids to larger masses by performing optical and radar meteor studies.

  Optical and radar meteor studies will expand the dynamic studies of meteoroid orbits to larger masses than those obtained by space measurements.

- Establish a "Dust Laboratory" which provides the infrastructure for the development and calibration of advanced dust instruments

  A Dust Laboratory should serve the scientific community with state-of-the-art simulation facilities for the study of dusty phenomena. Key facilities should be dust accelerators that extend the capabilities of the only operational dust accelerator facility in Germany. A dust accelerator serves for instrument development and calibration and for studies of effects of hypervelocity impacts on space systems. Besides studying impact processes, it provides a means for characterizing detrimental effects on sensitive space instrumentations, e.g., optical surfaces, x-ray detectors, and neutral particle detectors.

- Develop advanced in-situ and remote sensing space dust instrumentation

  A dust telescope requires new advanced analyzers for trajectory and composition measurements. These analyzers should have large (0.1 to 1 m^2) sensitive areas in order to provide statistically meaningful numbers of impacts in interplanetary space. Light weight detectors of even larger impact areas require novel design of sensors and signal processors. Dedicated infrared instruments that work in the 10 to 200 micron wavelength range without requiring active cooling could provide key information on the structure of and the dust production in the outer zodiacal cloud and Kuiper belt.

- Maintain and extend the laboratory facilities to analyze dust particles

  A wide range of micro-analytic tools is being used for elemental, isotopic, mineralogical, and morphological analysis of extraterrestrial dust particles: electron microscopy in conjunction with energy-dispersive X-ray analysis, X-ray fluorescence, neutron activation, and proton as well as synchrotron induced X-ray emission, and secondary ion mass spectrometry. These and even more powerful methods need to be developed in the future in order to make the best use out of samples returned from planetary bodies. Some methods require major facilities, such as the synchrotron light facility at the Brookhaven National Laboratory, which should be available for future cosmic dust studies.

- Perform theoretical studies of dust origin and evolution

  Further development of evolutionary models of the interplanetary dust cloud and of the dusty planetary environments. The current empirical dynamic models of the interplanetary dust cloud do not provide reliable extrapolations to regions were no previous measurements were taken.

4. Priority Issues for the Next Decade

A. Is there dust in the Kuiper belt and what is its distribution and density?

The dust population in the outer solar system is the largest single unknown in our understanding of the dust inventory of the solar system and its structure. Determination of dust densities in that region would provide an important constraint to the level of past and ongoing collisional activity.

B. What are the compositions of the components of the zodiacal dust complex and how do they vary spatially and temporally?

Through impact detection and remote observations, we have identified the general morphology of the interplanetary dust cloud within Jupiter's orbit, as well as identified significant sources of dust production in some asteroid families and comets. Knowledge of the relative and specific contributions from sources (which are presently very uncertain) would reveal structures (otherwise degenerate along our line of site) and allow us to better model the evolution of the cloud.

5. Recommendations

In order of importance:

- All missions to the outer solar system, in particular, missions to Pluto and beyond, should include an in-situ dust experiment for detecting and determining the properties of local dust

  This directly addresses priority issue A and is a low cost alternative to a stand alone mission.

- A "Cosmic Dust Observatory" should be launched into heliocentric orbit near 1 AU to measure compositional, physical, and dynamical properties of interplanetary dust particles (which should include dust from identified comets). Such a facility could be constructed, making use of the Cassini Composition Dust Analyser design. There should be coordination between this observatory and collection experiments from the Earth's upper atmosphere.

  This directly addresses priority issue B.

- A "High-Inclination Explorer" mission consisting of a cryogenically cooled thermal infrared telescope that "looks over" the inner zodiacal cloud to study dust in the outer solar system.

  This directly addresses both priority issues A and B.

NOTE: All recommendations require support for continuing theoretical studies of dust dynamics, modeling of past observations, and laboratory studies to characterize the composition and physical properties of dust.

The Future of Solar System Exploration, 2003-2013
ASP Conference Series, Vol. 272, 2002
Mark V. Sykes, ed.

Exploration of the Neptune System

H.B. Hammel[1], K.H. Baines[2], J.D. Cuzzi[3], I. de Pater[4], W.M. Grundy[5], G.W. Lockwood[5], J. Perry[6], K.A. Rages[7], T. Spilker[2], & J.A. Stansberry[8]

[1] *Space Science Institute, Boulder, CO*
[2] *Jet Propulsion Laboratory, Caifornia Institute of Technology, Pasadena, CA*
[3] *NASA Ames Research Center, Moffett Field, CA*
[4] *Department of Astronomy, University of California, Berkeley, CA*
[5] *Lowell Observatory, Flagstaff, AZ*
[6] *Lunar and Planetary Laboratory, University of Arizona, Tucson, AZ*
[7] *SETI Institute, Mountain View, CA*
[8] *Steward Observatory, University of Arizona, Tucson, AZ*

Abstract. We propose a mission to the Neptune system comprised of an orbiter with a Neptune atmospheric multi-probe. NASA's Solar System Exploration theme listed a Neptune mission as one of its top priorities for the mid-term (2008-2013). A recent NASA study also gave it top ranking for rich scientific return and connections to astrophysical problems outside the Solar System (atmospheric structure and dynamics, geology, ring systems/dynamics, magnetic fields/dynamos, pre-biotic chemistry on Triton, local extrasolar planet analog for Neptune- and Uranus-sized objects), calling it "almost Cassini-like in scope, near Discovery-like in cost" (though the costing was admittedly done in more optimistic times). The Neptune mission's unmatched diversity of science yield should place it at the top of the queue for outer planet exploration.

EXECUTIVE SUMMARY

Our knowledge of the Neptune system has been developed in three phases. The first, from 1846 to 1989, was based entirely on Earth-based observations. Then in 1989, the Voyager 2 flyby revolutionized our understanding of the Neptune system. During the third phase, from 1989 through to the present, HST, Keck, NASA IRTF, other ground-based observatories have continued to probe Neptune and its environs, revealing that much more remains to be learned about this system.

In spite of (perhaps due to) Voyager's success at Neptune and subsequent studies with HST, many questions about Neptune remain, and we identify what is needed to answer them. What powers the winds, and why are the winds and thermal structure similar to those of Uranus, though the internal heat sources differ? How deep does the zonal structure go? For these, we need visible imaging and thermal mapping at various phase angles with scales down to 10 km, and occultations of radio telemetry signals to probe atmosphere down to ~2 bar. What is the composition of discrete features (bright and dark), and of the atmosphere as a function of altitude? UV occultations to measure density, scale height, temperature and composition, and compositional mapping at near-IR wavelengths would address this. Why are the magnetic fields much more asymmetric in ice giants than in gas giants? This question requires measurements of magnetic field and magnetospheric particles at a variety of latitudes and longitudes.

Short of exploring Pluto, exploring Triton may provide our best opportunity to examine the surface and atmosphere of a Kuiper Belt Object analog. What is Triton's atmospheric composition and structure, and how has it changed since Voyager? To answer this, we need radio occultations to determine atmospheric size/structure, high phase and high-resolution (100-300 m) limb imaging for hazes/plumes, UV occultations (density, scale height, temperature, composition), and atmospheric sampling (fly-through). Is there evidence for "recent" solid-state convective activity in an icy mantle? How does composition vary between/within surface features? What causes geologic structures on Triton's surface? Has the geyser distribution changed since Voyager? Have atmospheric changes modified the surface? To answer these questions requires UV to near-IR global imaging (< 100 m), high-resolution imaging (10-30 m) of selected locales, thermal (50 and 100 μm) mapping, and global 1-km imaging spectroscopy at 1-5 μm with $\lambda/\Delta\lambda = 300$.

Is a resonant model for arc stability correct? If not, how do arcs remain stable? Do Neptune's inner satellites show the effect of extreme tidal stress? Are they a "major ring system waiting to happen"? To better understand the Neptune ring system, we need low-phase 100-m scale imaging of arcs to find embedded bodies, high-phase 1-km scale imaging to detect new rings/arcs and to characterize ring/arc morphology, and spectroscopic capability to determine composition.

We divide our recommendations into two categories: interplanetary spacecraft and activities that can be performed on or near Earth. Among all of the possible activities that we considered for the coming decade, our overall number one priority is for a robust program of technology development for deep space missions, of which a Neptune System mission should be a flagship. Of almost equal importance is continued support of the Earth-based and future near-Earth observational programs for the Neptune System (planet, rings, and Triton).

If new discoveries come from exploring those areas where our ignorance is deepest, then the Neptunian system clearly offers the greatest potential for expanding our knowledge. Bringing our understanding of the Neptune system up to a similar level as the (Galileo) Jupiter and (expected Cassini) Saturn systems would greatly enhance the ability of the planetary community to do meaningful comparative studies. We should also consider the fact that some day in the not-so-distant future, astronomers will discover Neptune-sized bodies around other stars. Having a more complete understanding of our own Solar System's ice giants will be invaluable for providing context for the study of extra-solar ice giant planets.

Our top priority among potential missions is a Neptune System Orbiter. Given the technological advances since the Voyager 2 Mission, a flyby would yield excellent scientific return. However, an orbiter could significantly enhance the science: long-term coverage of atmospheric dynamics on Neptune, the opportunity for multiple high-resolution mapping passes for Neptune and Triton to measure small-scale and short-term changes, coverage over a wide range of phase angles for multiple targets (atmospheres, surfaces, rings).

The orbiter is the core of the mission, providing a remote sensing platform, in-situ probes of the magnetic field and environs, and primary data links. Multi-probes are an essential part of an investigation of the deep (~100 bar) atmospheric structure and chemistry on Neptune. However, significant technology advances would be required to enable high S/N transmission from depth in a cost-effective manner.

Recent studies indicate a Neptune System mission is feasible given innovative technologies (Porco 1998): high-power lightweight SEP and solar sails, qualified aeroshells, aerocapture, autonomous spacecraft communications, advances in miniaturization, lightweight power generation systems, temperature-tolerant electronics (~50 K), and lightweight structures. These technology drivers are required for many outer planet missions.

After an orbiter mission, our next priority is a Neptune flyby mission. An orbiter is strongly preferred for reasons outlined above, but a flyby would produce excellent scientific return. Given the cost constraints of the current political situation, a flyby merits discussion.

The main spacecraft would carry a remote sensing platform, in-situ probes of the magnetic field and environs, and primary data links. As with an orbiter, multi-probes should still be considered for a flyby mission, and can be an essential part of an investigation of Neptune's deep atmospheric structure and chemistry.

In the category of Earth-based and Near-Earth activities, our top priority recommendations are moving target capability on future space-based telescopes, e.g., NGST) and access to existing and future large telescopes for planetary system observations (e.g., IRTF and NGST). High priority recommendations are long-term observing campaigns for the atmosphere of Neptune and the reflectivity of Triton, and laboratory experiments at temperatures and pressures relevant to the Neptune system. Priority recommendations are a long term observing campaign for the rings of Neptune, periodic astrometric observations of Neptune's inner satellites, and improvement in models and simulations pertinent to issues in the Neptune planetary system.

REPORT

Neptune serves as an extra-solar planet analog. Triton serves as an analog Kuiper Belt object (KBO) and provides ties to pre-biotic chemistry via surface organics (ring particle composition may also be of relevance). We have already revisited Jupiter and (soon) Saturn, which are (fraternal) twin gas giants, so a visit to an ice giant would be timely. Furthermore, smaller extrasolar planets will soon be detected. Sub-Saturn-sized are already detected, and it is only a matter of time before Neptune-sized bodies are detected. A comprehensive understanding of such bodies in our own solar system should therefore be of top importance to NASA, and this is in fact reflected in the ranking of such a mission in recent NASA reports (Table 1).

Table I - Importance of Neptune System Exploration to NASA

Date	Comment	Source
1998	Very high scientific priority	SSES Astrophysical Analogs Campaign Strategy Working Group Report
1999	Top priority	Exploration of the Solar System, Science and Mission Strategy (SSE Roadmap)
2000	Important component of post-2007 exploration	NASA Space Science Enterprise Strategic Plan
2000	High-level roadmap: contribution to national priorities (2012-2025)	NASA Strategic Plan 2000

During the preparation of this work, we assumed that a mission to Pluto is in the NASA queue. Pluto and Triton are an excellent pair for comparative planetological studies. A Pluto mission finishes the job of the Voyager era of initial exploration; a Neptune system mission is part of the Galileo and Cassini era of more in-depth studies. We strongly support a Pluto mission. We also ruled out a Uranus system mission because of timing. Uranus is approaching equinox right now (2007). By the time a spacecraft could be planned, launched, and had traveled to Uranus, the planet would probably be well past equinox and approaching northern solstice. Since this is the same season that Voyager observed, the atmospheric observations would likely be far less interesting that those of Neptune. A series of missions to Centaurs, KBOs, and the Uranian system could, taken together, perhaps provide extensive information about giant planet atmospheres, small outer solar system bodies, rings, and magnetospheres, but only in the Neptune system are all of these fascinating areas of investigation available in a single system.

1. Current State of Knowledge

Based on size and heliocentric distance, the four giant planets in our Solar System break neatly into two pairs. The Gas Giants, Jupiter and Saturn, consist of a deep, massive atmosphere of gases such as hydrogen and helium surrounding a small core of rock, water, and ammonia. Both have been the target of intensive study, most recently with Galileo (for Jupiter) and with Cassini (for Jupiter and soon Saturn). Further out lie the smaller giants Uranus and Neptune. The term Ice Giant is now being used more frequently to characterize these planets, which consist mostly of a core of "ices" – of water, methane, and ammonia – surrounded by a moderately thick atmosphere of hydrogen and helium.

Our knowledge of the Neptune system has been developed in three phases. The first, from 1846 to 1989, was based entirely on Earth-based observations. Then in 1989, the Voyager 2 flyby revolutionized our understanding of the Neptune system. During the third phase, from 1989 through to the present, HST, Keck, NASA IRTF, and other ground-based observatories have continued to probe Neptune and its environs, revealing that much more remains to be learned about this system.

1.1. Neptune

Atmospheric dynamics and structure: The major development in our understanding of Neptune's atmosphere since Voyager has been our increasing appreciation of just how active and variable that atmosphere is, though there were indications of this activity before and during Voyager (Joyce et al. 1977, Cruikshank 1985, Smith et al. 1989, Hammel et al. 1992, Ingersoll et al. 1995). During the Hubble era (which is now merging into the Adaptive Optics era), we have seen Neptune completely change character two or three times. The Voyager Great Dark Spot disappeared (Hammel et al. 1995), to be replaced by one or two northern hemisphere Great Dark Spots (Hammel et al. 1995, Hammel and Lockwood 1997, Sromovsky et al. 2001a), at least one of which has also subsequently disappeared (Sromovsky et al. 2001b). In the southern hemisphere as a whole, the Voyager encounter period now appears to have been rather quiescent. The latitudes between 45° and 55°S, which were featureless in 1989 except for a couple of discrete spots (Dark Spot 2 and Scooter) have erupted into nearly continuous bright features (Sromovsky et al. 2001b, Rages et al. 2001). Northern hemisphere cirrus-like features have come and gone, and the intermittent bright South Polar Feature at 70°S continues to form and dissipate within 1-2 Neptunian days while remaining fixed in latitude for more than a decade (Rages et al. 2001). What powers the winds, and why are the winds and thermal structure similar to those of Uranus (Figure 1), though the internal heat sources differ? How deep does the zonal structure go? What determines the many time scales of Neptune's variable weather?

NEED: Visible imaging and thermal mapping at many phase angles down to 10 km; occultations of radio telemetry signals to probe atmosphere down to ~2 bar.

Atmospheric composition: Our understanding of Neptune's atmospheric composition (Figure 2) is still largely derived from Voyager (Bishop et al. 1995, Baines et al. 1995, Gautier et al. 1995). The atmosphere is mostly hydrogen with an ortho-para ratio near the thermal equilibrium value, and a mole fraction of 0.190±0.032 helium (Conrath et al. 1991, Gautier et al. 1995). Methane is also present in significant amounts, with a mixing ratio of 0.02-0.03 in the troposphere (Baines and Smith 1990, Lindal 1992) and 10^{-4} to 10^{-3} in the stratosphere (Orton et al. 1992, Baines and Hammel 1994). Figure 15 of Baines et al. (1995) gives the nominal atmospheric aerosol structure, a hydrocarbon haze layer at a few tens of mbar in the stratosphere, the methane cloud with optical depth of order unity and a base near 1.5 bars, and an optically thick cloud (most likely hydrogen sulfide) with a top at 3.5-4.5 bars. What is the composition of discrete features (bright and dark), and of the atmosphere as a function of altitude?

NEED: UV occultations to measure density, scale height, temperature and composition; compositional mapping at near-IR wavelengths.

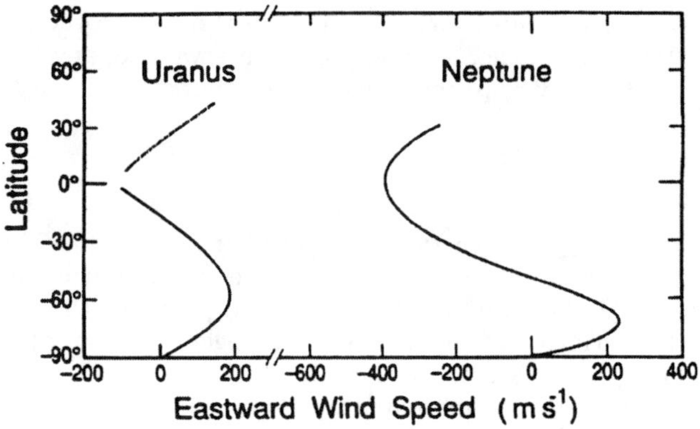

Figure 1. Comparison of zonal wind profiles of Uranus and Neptune. Adapted from Ingersoll et al. (1995) with Uranus data from Hammel et al. (2001).

Planetary interior and magnetic field environs: Neptune's field (Figure 3), like that of Uranus, is highly tilted and offset from the planet's center (Ness et al. 1989, 1995). Outside several planetary radii, Neptune's magnetic field can be represented by a dipole moment offset by at least 0.55 radii (about 8,500 miles) from the physical center of Neptune and tilted by 47 degrees with respect to the spin axis of the planet. The field strength at the surface varies from a maximum of more than 1 gauss in the southern hemisphere to a minimum of less than 0.1 gauss in the northern. (Earth's equatorial magnetic field at the surface is 0.32 gauss.) Because of its unusual orientation and the tilt of the planet's rotation axis, Neptune's magnetic field goes through dramatic changes as the planet rotates in the solar wind. For example, at the time of the Voyager 2 encounter Neptune's spin axis tipped directly away from the sun ~25 degrees from ecliptic north and Neptune's equatorial plane was inclined some 29 degrees to its orbital plane. The combination of the spin axis orientation and the large tilt angle of the magnetic dipole moment caused the angle between the magnetic axis and solar wind flow to vary from ~20 degrees ("Earth-like" magnetosphere) to ~114 degrees ("pole-on" magnetosphere) and back every planetary rotation of 16.11 hours during the Voyager flyby. Why are the magnetic fields much more asymmetric in ice giants than in gas giants?

NEED: Measurements of magnetic field and magnetospheric particles at a variety of latitudes and longitudes.

1.2. Triton

Short of exploring Pluto, exploring Triton may provide our best opportunity to examine the surface and atmosphere of a Kuiper Belt Object analog. Furthermore, some aspects of its terrain suggest that Triton may be "a cousin to Europa in unexpected ways" (Schenk 1999).

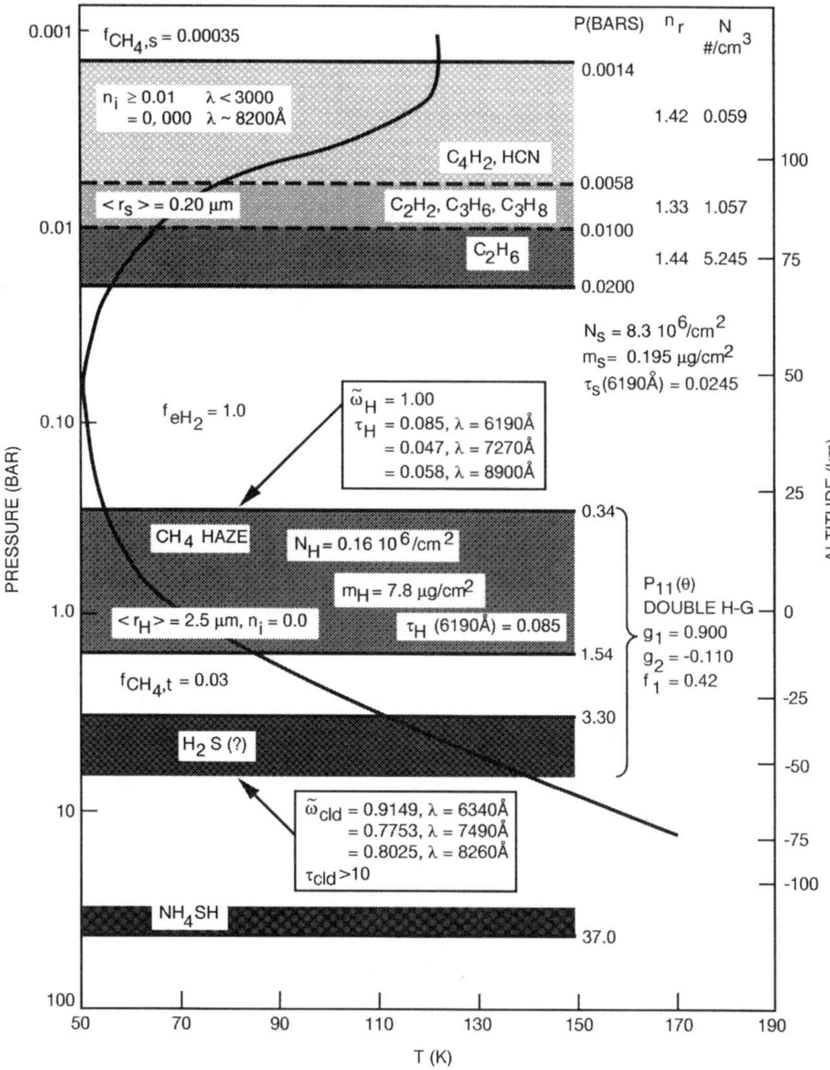

Figure 2. Neptune's vertical aerosol distribution. From Baines et al. (1995).

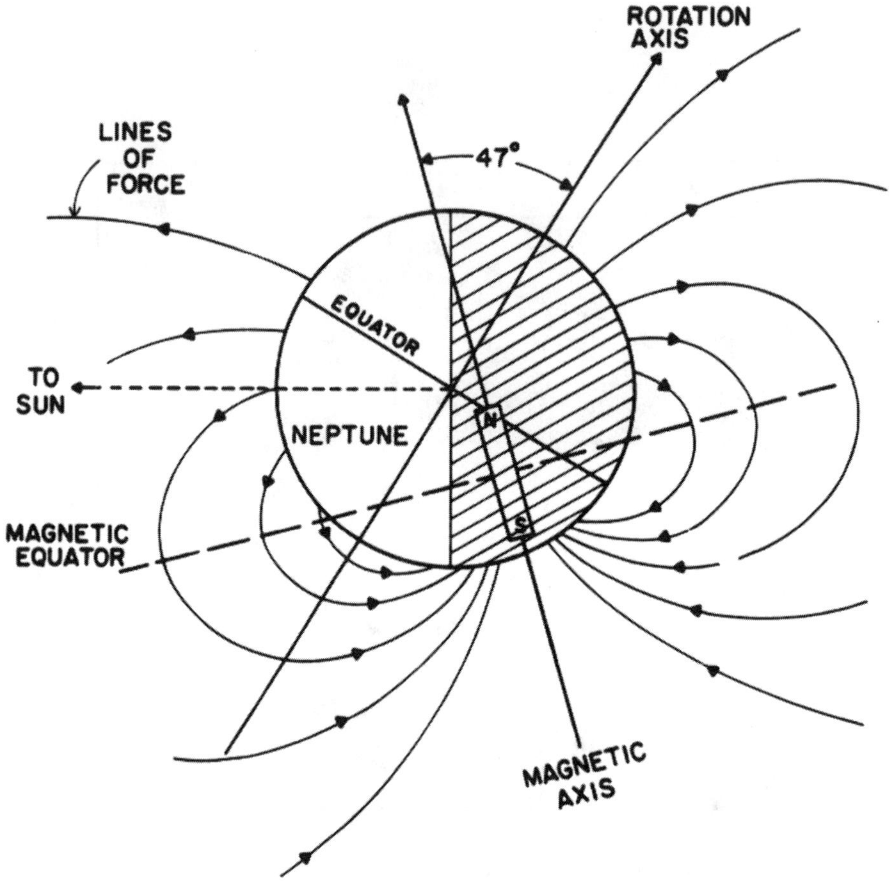

Figure 3. Neptune's magnetic field: an off-set tilted dipole. From Ness et al. (1989).

Surface geology: Triton's surface (e.g., Smith et al. 1989, McEwen 1990, Stern and McKinnon 1999) is widely thought to be younger, and more active, than almost any other planetary satellite. Croft (1990) and subsequently others (e.g., Schenk 1999) have provided overviews of the varied geology, but a key note is that less than half of Triton's surface was mapped by Voyager. What secrets does this "Triton Incognita" hold? Of the known terrain, the weirdest terrain on Triton may be the so-called cantaloupe terrain (Figure 4), characterized by closed topographic "cells" 30-40 km across. These cells may be formed by diapirs (Schenk and Jackson 1993), blobs of material that rise from depth and penetrate through a surface layer. This suggests that Triton's crust is layered.

The southern hemisphere consists of at least two distinct units: spotted and ridged terrains. The spotted terrain (Figure 5) consists of large irregular regions and is also the site of "wind streaks" and plumes (see next section). These

Figure 4. Triton's cantaloupe terrain. Image by P. Schenk.

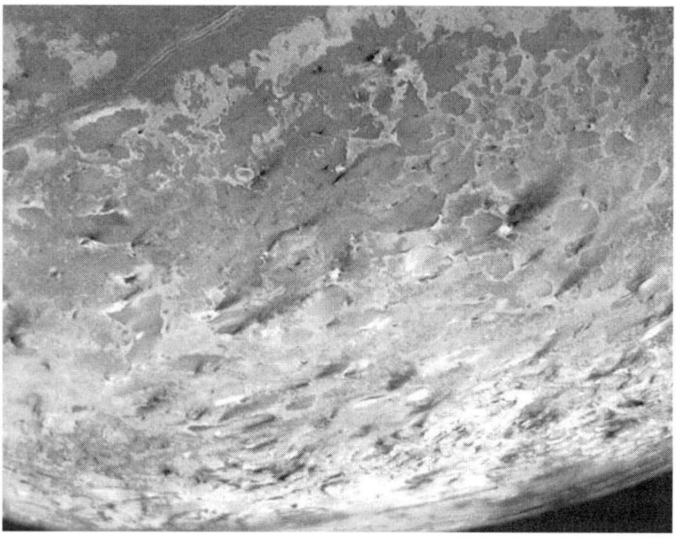

Figure 5. Wind streaks on Triton southern hemisphere. Voyager revealed the source of these dark streaks to be active cryovolcanism ("geysers"). Courtesy, NASA.

terrains show craters and cross-cutting linear features, suggesting that they do not constitute a seasonally-deposited "polar cap" (Schenk and Moore 1993).

Stratigraphic mapping suggests that cantaloupe terrain may be the oldest unit on Triton. However, recent cratercounts (Schenk 1999) reveal unambiguously that there are no fresh impact craters larger than 4 km on cantaloupe terrain or within the transition zone between cantaloupe and volcanic terrains (which corresponds to Triton's leading-trailing hemisphere boundary). The absence of any significant impactor flux from the Kuiper Belt suggests that the cratering record on Triton cannot be used to characterize the Kuiper Belt number density. It also suggests that Triton's surface age is extremely young, perhaps effectively zero, and may well prove to be younger than Europa's! This lends support to ideas that Triton's plumes may be related to ongoing geologic, thermal, and/or seasonal activity (next section).

Is there evidence for "recent" solid-state convective activity in an icy mantle? How does composition vary between/within surface features? What causes geologic structures on Triton's surface? Has the plume distribution changed since Voyager? Have atmospheric changes modified the surface?

NEED: UV to near-IR global imaging (< 100 m); high-resolution imaging (10-30 m) of selected locales; thermal (50 and 100 μm) mapping; global 1-km imaging spectroscopy at 1-5 μm with $\lambda/\Delta\lambda = 300$.

Atmosphere-surface interactions: The atmosphere of Triton, like that of Mars and Pluto, is in vapor-pressure equilibrium with the frost on its surface. Triton's icy surface is thought to contain the volatile frosts N_2, CO, and CH_4, as well as less volatile CO_2, coloring agents, and absorbers (see Brown and Cruikshank 1997). Changes in Triton's subsolar latitude (and the resulting changes in the insolation patterns) are predicted to lead to changes in the surface pressure of factors of 10 or more because of the exponentially sensitive temperature dependence of the volatility of the frosts (e.g., Trafton 1984, Spencer 1990, Spencer and Moore 1992; see Yelle et al. 1995 for a review). The seasonal change may be driving the sublimation of ices into the atmosphere, resulting in observable (perhaps dramatic) changes in Triton's atmospheric structure, photometric properties, and surface composition and microphysical structure, including widescale frost migration. Both ground-based and Hubble Space Telescope (HST) data indicate that Triton appears to have begun its long-awaited epoch of global seasonal change.

Occultations from 1995 to 1998 show Triton's pressure at 50 km altitude increased by 40% between mid-1995 and late-1997 (Elliot et al. 1998, 2000a, 2000b). Visible photometry and spectroscopy show changes in Triton's visible colors that last roughly one year (Buratti et al. 1999). Ultraviolet spectra obtained with the HST Space Telescope Imaging Spectrograph show that Triton's albedo and spectral slope in the UV appears extremely variable longward of 2400Å (Young and Stern 2001). Finally, near-infrared spectra (1.4 to 2.5 μm) show pronounced changes in the strength of CH_4 absorption features from 1980 to 1992, but little change between 1995 and 1998 (Brown et al. 1995, Hilbert et al. 1998).

It is fascinating to speculate about how these effects are changing the surface appearance, and about whether they are related to the active cryovolcanism detected on Triton by Voyager (Figure 5). However, Triton's size and distance preclude much more than speculation. A spacecraft to the Neptune system is required to understand the details of the surface-atmosphere interaction.

What is Triton's atmospheric composition and structure? How has it changed since Voyager? Can interactions with specific surface regions be tied to seasonal change?

NEED: Radio occultations for atmospheric size/structure; high phase and high-resolution (100-300 m) limb imaging for hazes/plumes; UV occultations (density, scale height, temperature, composition); atmospheric sampling (fly-through); global mapping of surficial units.

1.3. Rings and small satellites

Rings are addressed more fully by Gordon et al. (this volume, p. 263). We reiterate here a few of their points that pertain to the Neptune ring system. Voyager images revealed the complexity and variety of structure in the various ring systems. The processes that influence planetary ring structure can influence ring properties on timescales that are short compared to the age of the solar system (such processes include global angular momentum transfer, darkening caused by bombardment of rings by interplanetary projectiles, gas drag, and depletion by transfer of tiny charged grains from rings to their planet). There are uncertainties in the parameters of all these processes, but they nevertheless cause us to wonder about the origin of all four ring systems. Are they as old as the planets and their retinues of moons, or are they geologically recent and transient features of the planetary landscape?

Recent studies have identified complex gravitational interactions between the rings and their retinues of attendant satellites. The study of planetary ring dynamics has been instrumental in developing our understanding of the processes at work in the nebula during the formation of the solar system. In particular, the interactions between embedded objects and disks (spiral wave generation, angular momentum transport, truncation of disks) has been illuminated through our study of rings. Thus studies of planetary rings have been fundamental in improving our understanding of nebular dynamics and extrasolar planets.

The composition of planetary rings remains a major unknown, beyond some zero-order understanding. Saturn's rings are dominated by water ice, but their pinkish color - not dissimilar from Triton at visual wavelengths - calls for at least one important secondary component. The composition of the rings of Jupiter and Uranus is unknown: the Uranian rings are nearly, but not entirely, colorless; the Jovian ring shares the reddish color of Amalthea. Neptune's rings are fairly dark, but nothing is known about their color.

The next most interesting ring system, after Saturn, is probably that of Neptune (Figure 6). It has the additional scientific interest of being the most remote in the solar system, and the closest to the Kuiper Belt. There is a massive inner ringmoon belt with probably many smaller moons Voyager couldn't find. There are diffuse dusty rings. This is the only ring system known to possess relatively stable arcs, although the proposed confinement mechanism has recently come into question as result of recent HST observations of the clumps. The vertical and radial structure of the Neptunian rings remain unresolved. We are completely ignorant of the ring and ringmoon composition or how it varies

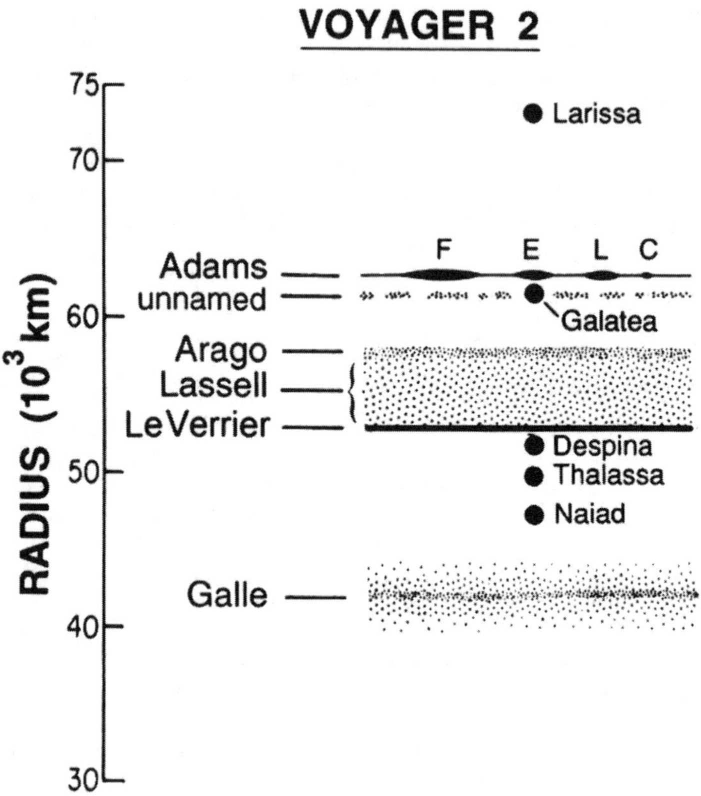

Figure 6. Diagram of Neptune's rings and satellites within the ring region. The arcs are labeled C for Courage, F for Fraternité, E for Egalité, and L for Liberté. From Porco et al. (1995).

with location, except that the material seems to be dark. We are likely always to remain ignorant if we have to rely on Earth-based observations alone.

New ground-based observations may provide more insights, but ground-based observations of the Neptunian rings have serious limitations. It is best to view dusty rings in forward-scattered light, which places the rings between the observer and the Sun. This requires a spacecraft beyond Neptune.

Is a resonant model for arc stability correct? If not, how do arcs remain stable? Do Neptune's inner satellites show the effect of extreme tidal stress? Are they a "major ring system waiting to happen"? Are the ring arcs indeed trapped or are they evolving? Are there "rubble belts" containing large boulders as well as dust? Are they the results of disruption of moons? How long ago? What is the meteoritic in fall rate and composition? Voyager data contains hints of other narrow, kinky ringlets in a few as-yet unstudied images. Are they real?

NEED: Low-phase 100-m scale imaging of arcs to find embedded bodies; high-phase 1-km scale imaging to detect new rings/arcs and to characterize ring/arc morphology; spectroscopic capability to determine composition.

2. Key Science Questions

The science questions naturally drive measurement objectives and, in turn, instrumentation requirements.

2.1. Neptune

- What is the three-dimensional atmospheric circulation pattern as a function of depth?
- What is the composition of deep atmosphere/interior, including the H/He ratio, isotopic compositions, etc.?
- How does the atmosphere vary with time at smaller scales (indicated by HST-based observations for large-scale changes)?
- What are the atmospheric convection patterns and, if present, zonal circulation patterns at depth in thermal emission?
- What is the aerosol composition and particle size in the stratosphere and upper troposphere?
- What are the populations and energies of magnetospheric constituents, which are needed to understand the energy inputs to both the upper atmosphere of Neptune and the surface/atmosphere of Triton?
- What is the three-dimensional compositional structure of the atmosphere, including the upper atmosphere where aerosols are likely produced?
- What are the higher moments of the magnetic field?
- What are the higher moments of the gravity field, to determine internal mass distribution?
- What is Neptune's temperature field; how does it affect Neptune's internal heat flux?

Measurement Objectives

- Thermal mapping of the entire planet at various phase angles.
- Compositional mapping of the entire planet.
- Repeated, simultaneous narrow- and wide-angle imaging to capture atmospheric motions and to measure winds. Narrow-angle imaging scales can be 10 km.
- Imaging of limb at high phase to detect hazes.
- UV-to-IR spectral mapping and broad phase angle coverage of the atmosphere.
- Measurement of the magnetic field and magnetospheric particles at a variety of latitudes and longitudes in the rotating (magnetic) coordinate system.

- Multiple occultations of radio telemetry signals for atmospheric sounding down to 2-bar level.
- Multiple UV (stellar or solar) occultations of Neptune's atmosphere to determine density, scale height, temperature and composition.

2.2. Triton

- What are the physical properties and structure of Triton's atmosphere (temperature, pressure, scale height), including any variation with latitude?
- What is the composition of the atmosphere, including local variations arising from erupting geysers, and the abundance and distribution of aerosols?
- What is the global distribution of volatiles, including the presence of a north polar cap?
- What are the relevant surface geologic processes, including tectonic activity, volcanic/geyser activity, glacial activity, subsidence, processes of erosion and degradation, and processes affecting crater morphology (mass wasting, relaxation)?
- What is Triton's internal structure and what is the magnitude of tidal distortion?
- Have there been changes in orbital eccentricity?
- Does Triton have an induced or intrinsic magnetic field?

Measurement Objectives

- Map surface compositions, searching for correlations between particular geological and compositional units.
- Map surface temperatures to identify the sources of volatiles in the atmosphere.
- Global imaging at resolutions of 100-m from UV to near IR, especially the polar regions. (Dark side may be visible in Neptune-shine.)
- Very high resolution imaging (tens of meters/pixel) of specific locations at different latitudes, including the south polar cap region.
- Global-imaging spectroscopy from 1 to 5 μm at resolutions of 1 km, with spectral resolution of $\lambda/\Delta\lambda = 300$ in the 1-5 μm range.
- Multiple occultations of radio telemetry signals for atmospheric structure.
- Multiple UV (stellar or solar) occultations of Triton's atmosphere to determine density, scale height, temperature and composition.
- Imaging of limb at high phase to detect hazes and plumes.
- Thermal mapping of Triton's surface at 50 and 100 μm.
- Direct sampling of the Triton's upper atmosphere.

2.3. Rings and small satellites:

- What are the orbital elements of the rings (particularly the ring arcs) and the inner satellites, and do they exhibit unexpected or predicted changes over time?

- What are the details of the interactions between rings and satellites, and are there new satellites and new rings that cannot be detected from ground-based observations or HST/NICMOS?

- What are the morphologies of the ring arcs/Adam's rings, in particular is there evidence for embedded bodies and Lagrange satellites?

- Can we characterize surface processes of ring-region satellites, e.g., through examination of crater ejecta patterns and fracture patterns (perhaps similar to those seen on Phobos and Deimos)?

- What are the compositions of the rings and, if possible (at low phase), the compositions of embedded bodies/Lagrangian satellites?

- What are the compositions of ring-region satellites, looking for methane, water, carbon dioxide, other hydrocarbons, organics and silicates (particularly, the 0.9- and 1-μm bands of the silicates olivine and pyroxene)?

- What are the densities (masses), compositions, and gross geologic histories of the satellites Larissa, Proteus, and Nereid?

Measurement Objectives

- Imaging resolutions of at least 1-km on rings and satellites for orbital determination and monitoring and for mapping satellites.

- Low-phase high-resolution (100-m) imaging of ring arcs to search for arc-embedded bodies and for compositional determination of such bodies.

- High-phase imaging searches at 1-km scale for detection of new rings and/or arcs and for characterizing ring/arc morphology.

- Compositional determination of larger ring bodies and of both known and new satellites, with the ability to distinguish CH_4, CO_2, silicates, C-H bonds, etc.

- Multiple occultations of radio telemetry signals for ring-structure and particle-size characterization.

- Global mapping at resolutions of 100-m of Proteus and Larissa; imaging at resolutions of 100-m of Nereid, if possible.

- Composition and mass of Proteus, Larissa, and, if possible, Nereid.

- Orbital refinement of all of the small satellites.

Table 2. Neptune & Triton Missions: Architecture Trade Space

Mission Type	Science Return	Cost Class
Simple Flyby	Voyager-like; remote sensing, in-situ fields and particles	Discovery Plus?
Flyby With Probe(s)	Add in-situ atmosphere; Galileo-like, minus satellites & time variability	Double Discovery?
Multi-Element Flyby With Probe(s)	Multiple viewing aspects; more time variability	Flagship
Simple Orbiter	Remote sensing, in-situ fields and particles, interior (via field maps), time variability	Double Discovery?
Complex Orbiter With Probe(s)	Cassini-like; "Flagship science"	Flagship, ~$1B
Multi-Element Orbiter With Probe(s)	More detailed investigations in every aspect; synergistic	Large Flagship
NEP Orbiter With Probes	Triton orbiter; Ring observer; Nereid rendezvous	Large Flagship

3. Recommendations

We divide our recommendations into two categories: interplanetary spacecraft and activities that can be performed on or near Earth. Among all of the possible activities that we considered for the coming decade, our overall number one priority is for a robust program of technology development for deep space missions (Table 2), of which a Neptune System mission should be a flagship. Of almost equal importance is continued support of the Earth-based and future near-Earth observational programs for the Neptune System: planet, rings, and Triton.

3.1. Neptune System Mission

Given the advances in technology since the Voyager 2 Mission, a flyby would produce excellent scientific return. However, an orbiter could contribute significantly enhanced science return: long-term coverage of atmospheric dynamics on Neptune, the opportunity for multiple high-resolution mapping passes for Neptune and Triton to measure small-scale and short-term changes, and coverage over a wide varieties of phase angles for multiple targets (atmospheres, surfaces, rings), to name just some of the advantages.

The overall science goals of a Neptune System Mission are to examine the composition, structure, and dynamics of Neptune's atmosphere; to map Triton's surface features, examine its geologic history, surface composition, and internal structure, and monitor its atmosphere and seasonal cycles; to probe Neptune's magnetosphere with extended temporal and spatial sampling; to study the rings, ring arcs, and shepherd satellites over at least 2 years; and to image and determine the densities of the satellites Larissa, Proteus, and Nereid (if possible).

Top Priority - Neptune system orbiter: The orbiter is the core of the mission, providing a remote sensing platform, in-situ probes of the magnetic field and environs, and primary data links. An integrated imaging package would include: visible imager, IR imaging spectrometer, and UV imaging spectrometer. Other remote sensing devices might include a thermal IR spectrometer and a microwave radiometer. Space physics detectors might include a magnetometer, an ion-and-neutral-mass spectrometer, and perhaps other instruments. Radio science instruments would also be necessary.

Multi-probes are an essential part of an investigation of Neptune's deep ($\sim$100 bar) atmospheric structure and chemistry. However, significant technology advances are required to enable high S/N transmission from depth in a cost-effective manner. An optimal probe package would include a main probe (GCMS; sensors for temperature, pressure, and acceleration; solar and IR radiometers; nephelometer) and at least three mini-probes (species-specific sniffers; temperature, pressure, and acceleration sensors) to sample diverse atmospheric regions.

A stretch goal would be a miniature surface lander to make in-situ studies of Triton's lower atmosphere and surface geology/composition.

Sample mission profiles have been discussed in depth elsewhere (Oberto 1999) but we mention two here just to give a flavor of the possibilities. These are merely illustrative, and a specific mission profile would have to be developed in the context of a more robust mission proposal. The first (Figure 7a) is a traditional ballistic trajectory, with launch in 2007 (probably unrealistic in the current fiscal climate, the next window for a Jupiter Gravity Assist is 2017 or later). Such a trajectory would have a 12-year flight time when launched from a Delta III/Star 48 (a more capable launch vehicle, e.g., an Atlas IIAS/Star48, could reduce the flight time by several years). The primary advantage of this type of trajectory is the lower cost inherent in the older technology. An alternative (Figure 7b) is based on Solar Electric Propulsion (SEP) technology, which has the advantage that the launch window is not constrained by Jupiter. The limiting factor for a launch window is the development of SEP technology. The flight times are similar (of order a decade). This particular orbit shows an Earth Gravity Assist, but others were considered which omit this.

An orbital mission sequence, like the specific flight profile, would be determined in detail at later planning stages, but an example follows for an optimal 2-4 year mission. The sequence has roughly four stages (Figure 8). The aerocap-

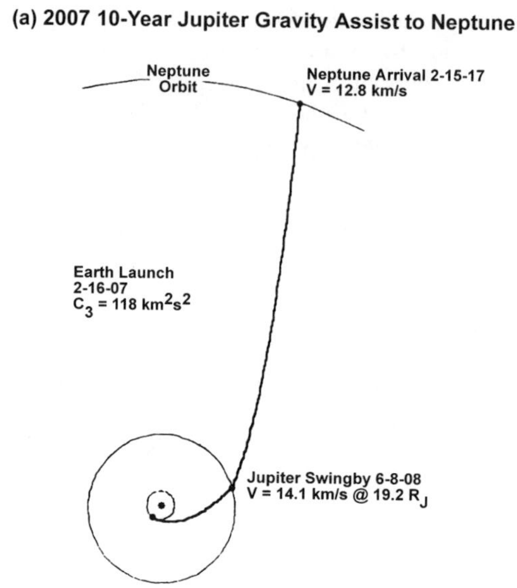

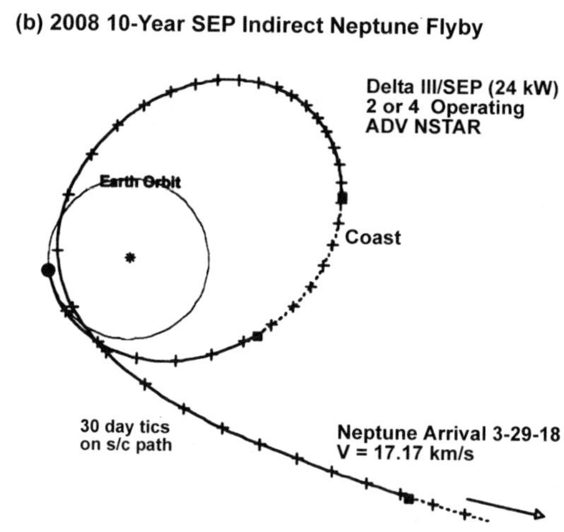

Figure 7. Sample flight profiles. (a) Traditional Jupiter gravity assist. (b) Solar Electric Propulsion. Adapted from Porco (1998).

(a)

(b)

(c)

(d)

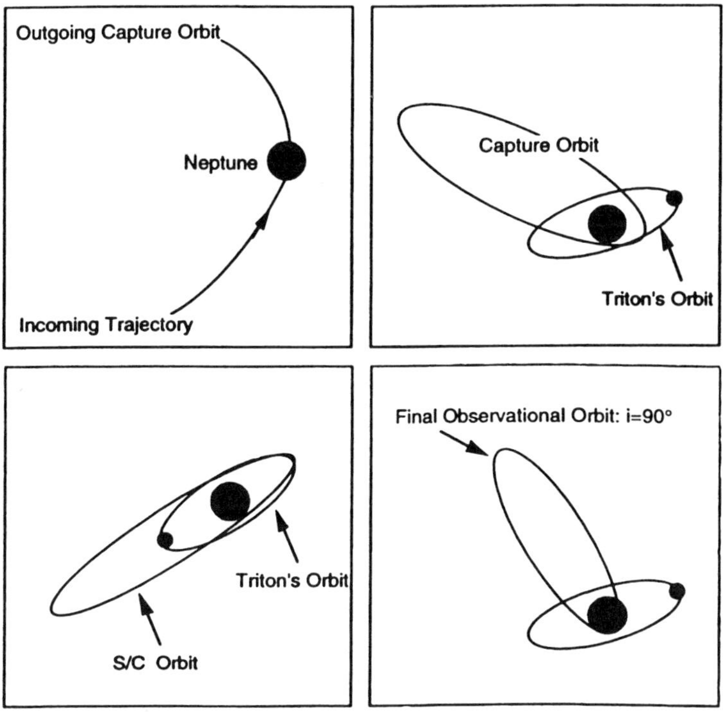

Figure 8. Sample orbital mission sequence. (a) Aerocapture. (b) Initial orbit. (c) Triton orbit. (d) Polar orbit. Adapted from Porco (1998).

ture phase (Figure 8a) employs a ballute entry into a retrograde orbit. During this phase, the spacecraft periapsis is raised above Neptune's atmosphere post-capture. In the initial orbit phase (Figure 8b), a Triton gravity assist is used to crank the spacecraft down into the Triton orbital plane. During the Triton orbit phase (Figure 8c), the spacecraft flies in a Triton coplanar orbit with a 10- to 30-day period. Triton is encountered over a range of phase angles and orbital positions. There are, of course, concomitant Neptune encounters during this phase also. During the final polar orbit phase (Figure 8d), the spacecraft orbit is cranked out of the plane by 60 to 90 degrees. Observations of Neptune and the ring plane would be the focus of this phase. Again, this example is illustrative. Data collection strategy for a Neptune System Orbiter could include the following:

- Approach coverage: atmospheric monitoring of Neptune and a possible Nereid flyby (these data can be stored and sent back later).
- Capture phase: accelerometer and fields and particles data collected for later return.

- Orbital science phase: repeated flybys and observations of Neptune, Triton, the rings, and, possibly, Proteus and Larissa; fields and particles measurements.

  This phase should last at least 2 years (3 years desired) in order to monitor changes on Triton and within the atmosphere and magnetosphere of Neptune. During this phase, the following geometries are required:

  - Closest-approach flybys of Triton at altitudes of 1000 km.
  - Closest-approach flybys of Triton at different phases in its orbit (i.e., different longitudes) and at different latitudes, including both poles.
  - Very-high-phase (>165°) viewing of Triton and the rings.
  - Closest-approach distances to the innermost satellites of ~100,000 km at various phases in their orbits.
  - Very-close-approach, low-phase encounters with the ring arcs at ~10,000 km.
  - Closest-approach flybys of Neptune at a variety of latitudes.
  - Repeated low-phase and contiguous Neptune observing from distances of 38 Neptune radii (~940,000 km).

Recent studies indicate a Neptune System mission is feasible given innovative technologies (Porco 1998): high-power lightweight SEP and solar sails, qualified aeroshells, aerocapture, autonomous spacecraft communications, advances in miniaturization, lightweight power generation systems, temperature-tolerant electronics (~50 K), and lightweight structures. These technology drivers are required for many outer planet missions. Their solutions will be broadly applicable.

High Priority - Neptune system flyby: An orbiter is strongly preferred for reasons outlined above, but a flyby would produce excellent scientific return. Given the cost constraints of the current political situation, a flyby merits discussion.

The main spacecraft would carry a remote sensing platform, in-situ probes of the magnetic field and environs, and primary data links. An integrated imaging package would include: visible imager, IR imaging spectrometer, and UV imaging spectrometer. Other remote sensing devices might include a thermal IR spectrometer and a microwave radiometer. Space physics detectors might include a magnetometer, an ion-and-neutral-mass spectrometer, and perhaps other instruments. Radio science instruments would also be necessary.

Multi-probes should still be considered for a flyby mission, and can be an essential part of an investigation of Neptune's deep (~100 bar) atmospheric structure and chemistry. However, significant technology advances would be required to enable high S/N transmission from depth in a cost-effective manner. An optimal probe package would include a main probe (GCMS; sensors for temperature, pressure, and acceleration; solar and IR radiometers; nephelometer) and at least three mini-probes (species-specific sniffers; temperature, pressure, and acceleration sensors) to sample diverse atmospheric regions.

Data collection strategy for a Neptune System Flyby could include the following:

- Approach coverage: atmospheric monitoring of Neptune and a possible Nereid flyby.
- Closest-approach flyby of Triton at altitudes of 1000 km.
- Very-high-phase (>165°) viewing of Triton and the rings.
- Closest-approach distances to the innermost satellites of ~100,000 km at various phases in their orbits.
- Low-phase and contiguous Neptune observing from distances of 38 Neptune radii (~940,000 km).

3.2. Earth-based and Near-Earth Activities

Remarkable advances have been made in understanding the Neptune System from Earth-based observations, and given the rate of technological advancement, much more remains to be discovered this way. However, the inherent challenges in studying this distant planetary system are many (e.g., Neptune's small apparent disk - only 2.3 arcseconds, and even worse for Triton). Future requirements include: access to large telescopes; ensuring that future telescopes have the capability to track moving targets and have relevant filter selections; continued access to existing planetary telescopes; and adequate funding for observations and analyses.

Top Priority:

- Moving target capability on future space-based telescopes, e.g., NGST.
- Access to existing and future large telescopes for planetary system observations, e.g., IRTF and NGST.

High Priority:

- Long-term observing campaign for the atmosphere of Neptune.
- Long-term observing campaign for the reflectivity of Triton.
- Laboratory experiments at temperatures and pressures relevant to the Neptune system.

Priority:

- Long-term observing campaign for the rings of Neptune.
- Periodic astrometric observations of Neptune's inner satellites.
- Improvement in models and simulations pertinent to issues in the Neptune planetary system.

3.3. Education and public outreach

The Neptune System offers a wide range of phenomena to capture the public interest and to use as a springboard for a range of Education and Public Outreach activities. Following after the Galileo and Cassini mission exploration of our solar system's gas giants, a focus on an "Ice Giant" system would be timely and complementary. Given Neptune's complex rings, its unusual large moon Triton, and the fact that Neptune is the "other" blue planet, a variety of activities and educational projects can be envisioned.

3.4. Data archive and access issues

Many processes within the Neptune System operate on sufficiently short timescales to permit the detection of changes on timescales ranging from days to years, and certainly from decades or tens of decades. In order to gain insights into the underlying processes, it must be possible to access and analyze historical data. Voyager spacecraft data are archived and distributed by the NASA Planetary Data System. HST data have also been archived. However, other earth-based observations are seriously underrepresented in the archival state. The volume of planetary data will increase dramatically during the coming decade. Efforts to incorporate Earth based data should continue and tools to simplify the submission of data to the NASA Planetary Data System be developed. Funding for archiving of ground-based data would also enhance the utility of the database.

4. Summary

We are only beginning to realize and appreciate the subtle (and perhaps not so subtle) dynamics and temporal variability taking place in the Neptune system. From the planet's disappearing dark spots, to the evolving thin atmosphere of Triton, to the clumpy ring system, to the unusual magnetic field, this system provides a cornucopia of planetary system phenomena. The anticipated science return of a Neptune System Mission is limited only by the scope of the imagination of those designing the tools of exploration. The inherent difficulty of deep space exploration should not dissuade us from pursuing exploration of the Neptune System. The challenge should be met head on. The results will be worth it. Bringing our understanding of the Neptune system up to a similar level as the (Galileo) Jupiter and (expected Cassini) Saturn systems would greatly enhance the ability of the planetary community to do meaningful comparative studies. In the mean time, vigorous support of Earth-based and near-Earth facilities and programs is essential for obtaining high quality observations of this distant yet dynamic system.

References

Baines, K.H., and H.B. Hammel 1994. Clouds, hazes, and the stratospheric methane abundance in Neptune. *Icarus* **109**, 20-39.

Baines, K.H., and W.H. Smith 1990. The atmospheric structure and dynamical properties of Neptune derived from ground-based and IUE spectrophotometry. *Icarus* **85**, 65-108.

Baines, K.H., H.B. Hammel, K.A. Rages, P.N. Romani, and R.E. Samuelson 1995. Clouds and hazes in the atmosphere of Neptune. In *Neptune and Triton* (D.P. Cruikshank, Ed.), pp. 489-546. Univ. of Arizona Press, Tucson.

Bishop, J., S.K. Atreya, P.N. Romani, G.S. Orton, B.R. Sandel, and R.V. Yelle 1995. The middle and upper atmosphere of Neptune. In *Neptune and Triton* (D. P. Cruikshank, Ed.), 427-488. Univ. of Arizona Press, Tucson.

Brown, R.H., and D.P. Cruikshank 1997. Determination of the composition and state of icy surfaces in the outer solar system. *Annu. Rev. Earth Planet. Sci.* **25**, 243-277.

Brown, R.H., D.P. Cruikshank, J. Veverka, P. Helfenstein, and J. Eluszkiewicz 1995. Surface composition and photometric properties of Triton. In *Neptune and Triton* (D.P. Cruikshank, Ed.), 991-1029. Univ. of Arizona Press, Tucson.

Buratti, B., M.D. Hicks, and R.L. Newburn 1999. Does global warming make Triton blush? *Nature* **397**, 219.

Conrath, B.J., D. Gautier, G.F. Lindal, R.E. Samuelson, and W.A. Shaffer 1991. The helium abundance of Neptune from Voyager measurements. *J. Geophys. Res.* **96**, 18907-18919.

Croft, S.K. 1990. Triton: Geology and Geologic History. *Lunar Planet. Sci.* **XXI**, 248. (abstract)

Cruikshank, D.P. 1985. Variability of Neptune. *Icarus* **64**, 107-111.

Elliot, J.L., H.B. Hammel, L.H. Wasserman, O.G. Franz, S.W. McDonald, M.J. Person, C.B. Olkin, E.W. Dunham, J.R. Spencer, J.A. Stansberry, M.W. Buie, J.M. Pasachoff, B.A. Babcock, and T.H. McConnochie 1998. Global warming on Triton. *Nature* **393**, 765-767.

Elliot, J.L., D. F. Strobel, X. Zhu, J.A. Stansberry, L.H. Wasserman, and O.G. Franz 2000a. The thermal structure of Triton's middle atmosphere. *Icarus* **143**, 425-428.

Elliot, J.L. M. J. Person, S.W. McDonald, M.W. Buie, E.W. Dunham, R.L. Millis, R.A. Nye, C.B. Olkin, L.H. Wasserman, L.A. Young, W.B. Hubbard, R. Hill, H.J. Reitsema, J.M. Pasachoff, T.H. McConnochie, B.A. Babcock, R.C. Stone, and P. Francis 2000b. The prediction and observation of the 1997 July 18 stellar occultation by Triton: More evidence for distortion and increasing pressure in Triton's atmosphere. *Icarus* **148**, 347-369.

Gautier, D., B.J. Conrath, T. Owen, I. de Pater, and S.K. Atreya 1995. The troposphere of Neptune. In *Neptune and Triton* (D.P. Cruikshank, Ed.), pp. 547-612. Univ. of Arizona Press, Tucson.

Hammel, H.B., and G.W. Lockwood 1997. Atmospheric structure of Neptune in 1994, 1995, and 1996: HST imaging at multiple wavelengths. *Icarus* bf 129, 466-481.

Hammel, H.B., S.L. Lawson, J. Harrington, G.W. Lockwood, D.T. Thompson, and C.W. Swift 1992. An atmospheric outburst on Neptune from 1986 through 1989. *Icarus* **99**, 363-367.

Hammel, H.B., G.W. Lockwood, J.R. Mills, and C. Barnet 1995. Hubble Space Telescope imaging of Neptune's cloud structure in 1994. *Science* **268**, 1740-1742.

Hammel, H.B., K. Rages, G.W. Lockwood, E. Karkoshka, and I. de Pater 2001. New measurements of the winds of Uranus. *Icarus* **153**, 229-235.

Hilbert, B.N., J.A. Stansberry, W.M. Grundy, R.V. Yelle, M.W. Buie, and C. dalla Piazza 1998. Search for secular changes in the near-infrared spectrum of Triton. *Bull. Am. Astron. Soc.* **30**, 4904.

Ingersoll, A.P., C.D. Barnet, R.F. Beebe, F.M. Flasar, D.P. Hinson, S.S. Limaye, L.A. Sromovsky, and V.E. Suomi 1995. Dynamic meteorology of Neptune. In *Neptune and Triton* (D.P. Cruikshank, Ed.), pp. 613-684. Univ. of Arizona Press, Tucson.

Joyce, R.R., C.B. Pilcher, D.B. Cruikshank, and D. Morrison 1977. Evidence for weather on Neptune. I. *Astrophys. J.* **214**, 657-662.

Lindal, G.F. 1992. The atmosphere of Neptune: An analysis of radio occultation data acquired with Voyager 2. *Astron. J.* **103**, 967-982.

McEwen, A.S. 1990. Global color and Albedo Variations on Triton. *Geophys. Res. Lett.* **17**, 1765-1768.

NASA 1999. *Exploration of the Solar System: Science and Mission Strategy.* Prepared by the NASA Solar System Exploration Subcommittee, Jet Propulsion Laboratory, Pasadena, California.

Ness, N.F., M.H. Acuña, L.F. Burlaga, J.E.P. Connerney, R.P. Lepping, and F.M. Neubauer 1989. Magnetic fields at Neptune. *Science* **246**, 1473-1478.

Ness, N.F., M.H. Acuña, and J.E.P. Connerney 1995. Neptune's magnetic field and field-geometric properties. In *Neptune and Triton* (D.P. Cruikshank, Ed.), pp. 141-168. Univ. of Arizona Press, Tucson.

Oberto, B. 1999. Neptune Orbiter 11-99. Advanced Project Design Team (JPL Team X) Report. Jet Propulsion Laboratory, Pasadena, California.

Orton, G.S., J.H. Lacy, J.M. Achtermann, P. Parmar, and W.E. Blass 1992. Thermal spectroscopy of Neptune: The stratospheric temperature, hydrocarbon abundances, and isotopic ratios. *Icarus* **100**, 541-555.

Porco, C.C., P.D. Nicholson, J.N. Cuzzi, J.J. Lissauer, and L.W. Esposito 1995. Neptune's Ring System. In *Neptune and Triton* (D.P. Cruikshank, Ed.), pp. 703-804. Univ. of Arizona Press, Tucson.

Porco, C.C. 1998. Report of the NASA SSES Astrophysical Analogs Campaign Strategy Working Group. Prepared by the NASA Solar System Exploration Subcommittee, Jet Propulsion Laboratory, Pasadena, California.

Rages, K.A., H.B. Hammel, G.W. Lockwood, and M. Marley 2001. Variability on Uranus and Neptune during HST Cycle 9. *Bull. Am. Astron. Soc.* **33**, 1079.

Schenk, P., and M. Jackson 1993. Diapirism on Triton - A record of crustal layering and instability. *Geology* **21**, 299-302.

Schenk, P., and J. Moore 1993. Geology of the southern hemisphere of Triton: No polar cap. *Lunar Planet. Sci.* **XXIV**, 1247-1248. (abstract)

Schenk, P.M. 1999. Vigorous dynamic Triton. Abstract for Pluto and Triton: Comparisons and Evolution Over Time Workshop, 23-24 September 1999, Lowell Observatory, Flagstaff, Arizona.

Smith, B.A., L.A. Soderblom, D. Banfield, C. Barnet, A.T. Basilevsky, R.F. Beebe, K. Bollinger, J.M. Boyce, A. Brahic, G.A. Briggs, R.H. Brown, C. Chyba, S.A. Collins, T. Colvin, A.F. Cook II, D. Crisp, S.K. Croft, D. Cruikshank, J.N. Cuzzi, G.E. Danielson, M.E. Davies, E. de Jong, L. Dones, D. Godfrey, J. Gougen, J. Grenier, V.R. Haemmerle, H. Hammel, C.J. Hansen, C.P. Helfenstein, C. Howell, G.E. Hunt, A.P. Ingersoll, T.V. Johnson, J. Kargel, R. Kirk, D.I. Kuehn, S. Limaye, H. Masursky, A. McEwen, D. Morrison, T. Owen, W. Owen, J.B. Pollack, C.C. Porco, K. Rages, P. Rogers, D. Rudy, C. Sagan, J. Schwartz, E.M. Shoemaker, M. Showalter, B. Sicardy, D. Simonelli, J. Spencer, L.A. Sromovsky, C. Stoker, R.G. Strom, V.E. Suomi, S.P. Synnot, R.J. Terrile, P. Thomas, W.R. Thompson, A. Verbischer, and J. Veverka. 1989. Voyager 2 at Neptune: Imaging science results. *Science* **246**, 1422-1449.

Spencer, J.R. 1990. Nitrogen frost migration on Triton - A historical model. *Geophys. Res. Lett.* **17**, 1769-1772.

Spencer, J.R., and J.M. Moore 1992. The influence of thermal inertia on temperatures and frost stability on Triton. *Icarus* **99**, 261-272.

Sromovsky, L.A., P.M. Fry, T.E. Dowling, K.H. Baines, and S.S. Limaye 2001a. Coordinated 1996 HST and IRTF imaging of Neptune and Triton III. Neptune's atmospheric circulation and cloud structure. *Icarus* **149**, 459-488.

Sromovsky, L.A., P.M. Fry, T.E. Dowling, K.H. Baines, and S.S. Limaye 2001b. Neptune's atmospheric circulation and cloud morphology: Changes revealed by 1998 HST imaging. *Icarus* **150**, 244-260.

Stern, S.A., and W.B. McKinnon 2000. Triton's surface age and impactor population revisited in light of Kuiper Belt fluxes: Evidence for small Kuiper Belt objects and recent geological activity. *Astro. J.* **119**, 945-952.

Trafton, L.M. 1984. Large seasonal variations in Triton's atmosphere. *Icarus* **58**, 312-324.

Yelle, R.V., J.I. Lunine, J.B. Pollack, and R.H. Brown 1995. Lower atmospheric structure and surface-atmosphere interactions on Triton. In *Neptune and Triton* (D. P. Cruikshank, Ed.), pp. 1031-1106. Univ. of Arizona Press, Tucson.

Young, L.A., and S.A. Stern 2001. UV observations of Triton in 1999 with HST/STIS. 2150-3180Å spectroscopy and disk-integrated photometry. *Astron. J.* **122**, 449-456.

The Future of Solar System Exploration, 2003-2013
ASP Conference Series, Vol. 272, 2002
Mark V. Sykes, ed.

The Study of Comets

M.R. Combi[1], M.T. Capria[2], G. Cremonese[3], M.C. De Sanctis[2], T.L. Farnham[4], Y.R. Fernandez[5], M.C. Festou[6], U. Fink[7], J.R. Green[8], W.M. Harris[9], C.W. Hergenrother[7], P.L. Lamy[10], S.M. Larson[7], H.F. Levison[11], D.J. Lien[12], C.M. Lisse[13], D.D. Meisel[14], D.T.F. Möhlmann[15], B.E.A. Mueller[16], N.H. Samarasinha[16], M.L. Sitko[17], H.A. Weaver[18], & P.R. Weissman[8]

[1] *Department of Atmospheric, Oceanic and Space Sciences, University of Michigan, Ann Arbor, MI*
[2] *Instituto di Astrofisica Spaziale e Fisica Cosmica, CNR, Rome, Italy*
[3] *Osservatorio Astronomico, Padova, Italy*
[4] *Department of Astronomy, University of Texas, Austin, TX*
[5] *Institute for Astronomy, University of Hawaii, Honolulu, HI*
[6] *Observatoire Midi–Pyrenees, Toulouse, France*
[7] *Lunar and Planetary Laboratory, University of Arizona, Tucson, AZ*
[8] *Jet Propulsion Laboratory, California Institute of Technology, Pasadena, CA*
[9] *Space Astronomy Lab, University of Wisconsin, Madison, WI*
[10] *Labaratoire d'Astronomie Spatiale, Marseille, France*
[11] *Southwest Research Institute, Boulder, CO*
[12] *Department of Physics, Oklahoma State University, Stillwater, OK*
[13] *Department of Astronomy, University of Maryland, College Park, MD*
[14] *Department of Physics and Astronomy, State University of New York, Geneseo, NY*
[15] *Deutschen Zentrum für Luft- und Raumfahrt, Köln, Germany*
[16] *National Optical Astronomical Observatories, Tucson, AZ*
[17] *Department of Physics, University of Cincinnati, Cincinnati, OH*
[18] *Applied Physics Laboratory, Johns Hopkins University, Baltimore, MD*

Abstract. The study of comets is critical to our understanding the origin, structure and evolution of the solar system in general and of life itself. We present here a brief summary of our current understanding of comets and their important place in the solar system, as well as our recommendations regarding priorities for their study over the next decade in the context of a coordinated program of spacecraft missions and ground-based and theoretical investigations.

EXECUTIVE SUMMARY

Comets are understood to represent the most fundamental building blocks of the Solar System. Their high volatile content is preserved by existing in low temperature and low gravity environments, and they are now our most accessible samples of the primitive material from which the Solar System formed. Knowledge of the composition and evolution of comets and the processes that affect their current state can provide the clearest picture of our origins. Key issues in our current understanding of comets include:

- What are the physical characteristics of cometary nuclei? Little is known of the size distribution, the mass, density and other similar fundamental characteristics.
- What is the composition of the comet nucleus? How do variations reflect different formation environments and/or evolution? What are the composition variations among comets and classes of comets? What is the role of comets as sources of water and pre-biotic compounds to the earth and other planets, and how might this affect the origin and evolution of life.
- What other aspects of comets can be used to constrain models of Solar System formation/evolution? How do KBOs evolve into comets?

Over the next decade, we recommend:

- In-situ measurements via orbiters, landers and sample return missions are key to validating remote sensing data and providing the next level of detail in answering the key issues. Such missions are highly likely to provide paradigm-altering data.
- Groundbased and near-earth observations of a large sample of comets to identify mission targets, support mission operations and most importantly to provide both the context in which to interpret spacecraft data and to elucidate the big-picture of the population of objects.
- Encourage and facilitate the archiving of all available comet data from observational programs and missions.

REPORT

1. Current State of Knowledge

A scientific consensus has developed over the past 50 years that comets are the most pristine remnants of the material that existed in the outer part of the solar system when it formed 4.5 billion years ago. The last half of the 20th century began with a seminal series of works, all around 1950, that revolutionized our understanding of comets and the uniquely important role they play in the formation and evolution of the solar system. The works of Kuiper and Edgeworth and Oort and Öpik explained how the distribution of cometary orbits was related to their origin and to the origin of the solar system. Basically, there are two broad

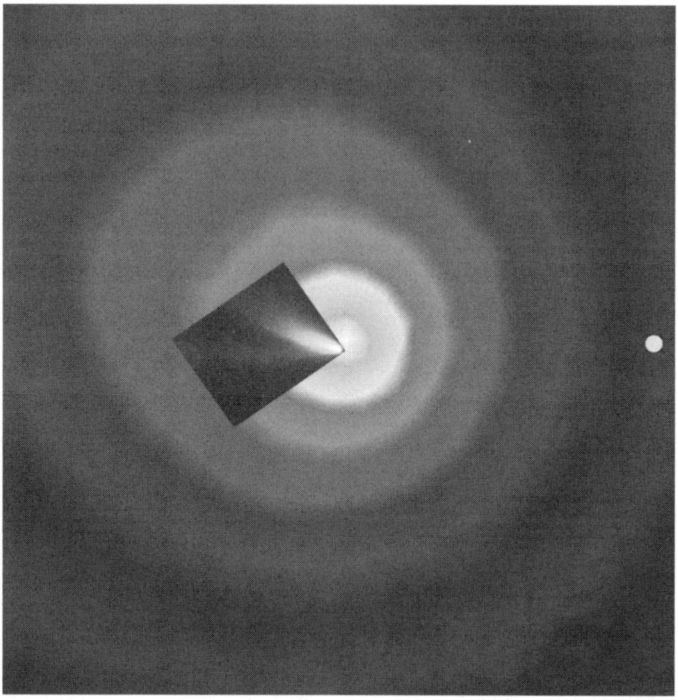

Figure 1. The visible dust and ion tails of comet Hale-Bopp (1995 O1), shown inserted to scale on an isophote map of the huge hydrogen Lyman-alpha coma obtained with the SWAN all-sky camera on the SOHO spacecraft. The white circle to the right shows the Sun's size to scale. (Combi et al., *Icarus* **144**, 191, 2000)

populations of comets: the Oort/Öpik cloud, which is comprised of comets in generally random inclination orbits having aphelia in the outer part of the solar gravitational potential, and the short-period Kuiper/Edgeworth Belt, which is comprised of comets in generally low-inclination prograde orbits and which get passed into the inner solar system by random gravitational assists by the giant planets. Cometesimals were the first objects to coalesce out of the outer protosolar nebula and became the building blocks, not only of the comets, but of the planets themselves. Comets that remained outside of the orbit of Neptune formed the so-called Kuiper Belt. Those which formed in the vicinity of the giant planets had three ultimate fates: accretion onto a giant planet, ejection on a hyperbolic orbit from the solar system, or entrapment in the gravitational "no-man's-land" at the edge of the solar system, the Oort/Öpik cloud. Thus, comets then have been preserved in the deep-freeze of space far from the Sun for most of the last 4.5 billion years, only to be rarely and randomly scattered into the inner solar system to provide a spectacular visual display with bright comae and long tails (Figure 1).

The work of Whipple showed that the source of the spectacular cometary displays was a small 'dirty-snowball' nucleus. Sublimation of the volatile ices in the nucleus feed gas into the coma, and the outflowing gases dragged small, refractory dust particles with them. Near the nucleus more collimated dust jets can follow spiral patterns in response to nucleus rotation. Farther out the action of solar radiation pressure on the dust produces long, curved dust tails, while the ionization of the gases and their subsequent interaction with the solar wind produces the long, straight ion tails. The jet force of sublimation on the nucleus produces the so-called Non-Gravitational Force (NGF), which perturbs the motions of comets from their otherwise predictable Newtonian orbits. Remarkably, it was the analysis of observations of the long, blue ion tails of comets, which are mostly produced by the fluorescence of CO^+, that allowed Biermann to demonstrate the existence of the global "solar wind" which permeates the entire solar system. Delsemme and Swings completed the icy conglomerate picture of Whipple by showing that water ice controlled the sublimation of cometary nuclei, rather than ices of ammonia and methane, and that this explained the variation in brightness along a comet's orbit and its rather dramatic increase inside two astronomical units from the Sun.

The next 50 years included ground-based optical, infrared, and radio observations of comets and space-based ultraviolet, extreme UV and X-ray observations. An armada of spacecraft flew by comets Giacobini-Zinner and Halley in 1985 and 1986 and then Grigg-Skjellerup in 1992, providing a wealth of new detailed measurements of the cometary gas, dust and plasma, and providing the first tantalizing glimpses of a small, dark nucleus. The recent flyby of periodic comet 19P/Borrelly by the Deep Space 1 spacecraft provided our most detailed images of a nucleus and from which the first cometary geology is able to be performed (Figure 2) However, the basic qualitative ideas, and some of the quantitative predictions made over 50 years ago, have largely been verified. Based on the evidence accumulated over the past 50 years, we are finding that comets have very likely retained some of the original unprocessed interstellar material from which the solar system formed. In addition, the thermochemical record detailing the formation and evolution of cometary matter is encoded in the molecular and mineralogical composition of cometary nuclei. That early cometary bombardment of the Earth is responsible for both the water and the prebiotic chemicals out of which life formed is still a very viable theory, although the specific measurements needed to provide irrefutable evidence have not yet been obtained. The results of many-year, multiple-comet surveys, and the preponderance of all comet observations, have provided important new insights into the nature of comets, but the objective of producing a general cometary classification scheme remains unfulfilled. Often the bright 'new' Oort cloud comets are rich in supervolatiles, with 10-20% CO, 5% CO_2 , a few percent of CH_3OH and H_2CO, and 0.1 to 1% of NH_3, HCN, CH_4, C_2H_2, and C_2H_6, but the number of comets having detailed abundance measurements is woefully small (roughly a dozen) and there are several counter-examples (e.g., long-period comets having CO abundances of order 1% or less). Ground-based surveys of 100 comets have characterized the production rates of the common carbon-bearing radicals C_2, C_3, and CN and sometimes NH or NH_2, which have then been compared to the production of OH (a proxy for water) and the observed amount of dust (actually the observed total cross section of dust particles of optical-wavelength size).

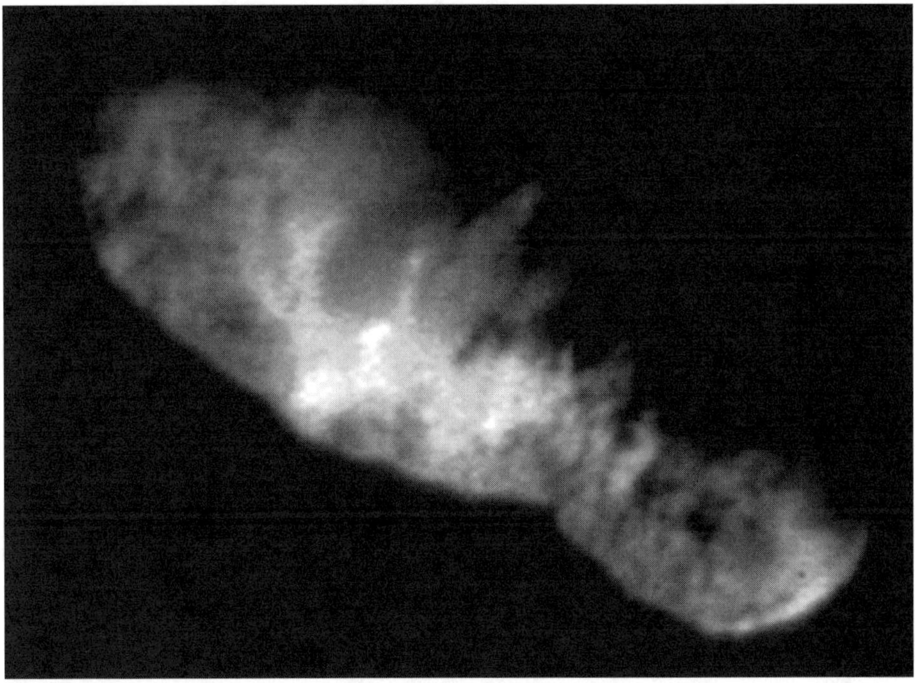

Figure 2. An image of the nucleus of comet 19P/Borrelly taken by the MICAS camera during the Deep Space 1 flyby in September, 2001 (NASA).

Typically, the 'new' Oort cloud comets, as well as one half of the short period comets, are characterized as having the 'normal' composition of these radicals compared to water. About half of the short period comets are depleted of the carbon-chain species (C_2, C_3), and this may be a primordial effect. However, the observed C_2 and C_3 are certainly not parent molecules (i.e., present in that form in the nucleus), and their production in the coma is still poorly understood. Basing a cosmogony on these optical surveys is still questionable at best. There also seems to be a difference in the observed spectral characteristics of the dust in the two major populations of comets. Small dust grains are present in the comae of the older Oort cloud comets, as evidenced by the presence of the silicate emission band in the mid-infrared. However, in the majority of short-period comets and in the dynamically new Oort cloud comets this feature is weak or absent. More direct studies of the dust particle size distribution seem to support the same general classification. Whether this is the result of differences the size distribution of refractory particles present in the Uranus-Neptune region compared to those outside the orbit of Neptune, or of differences in the subsequent processing of the bodies formed there is unknown. The grain fragmentation and sublimation in the production of the dusty-gas coma and their roles in producing the dust size distribution are also poorly understood.

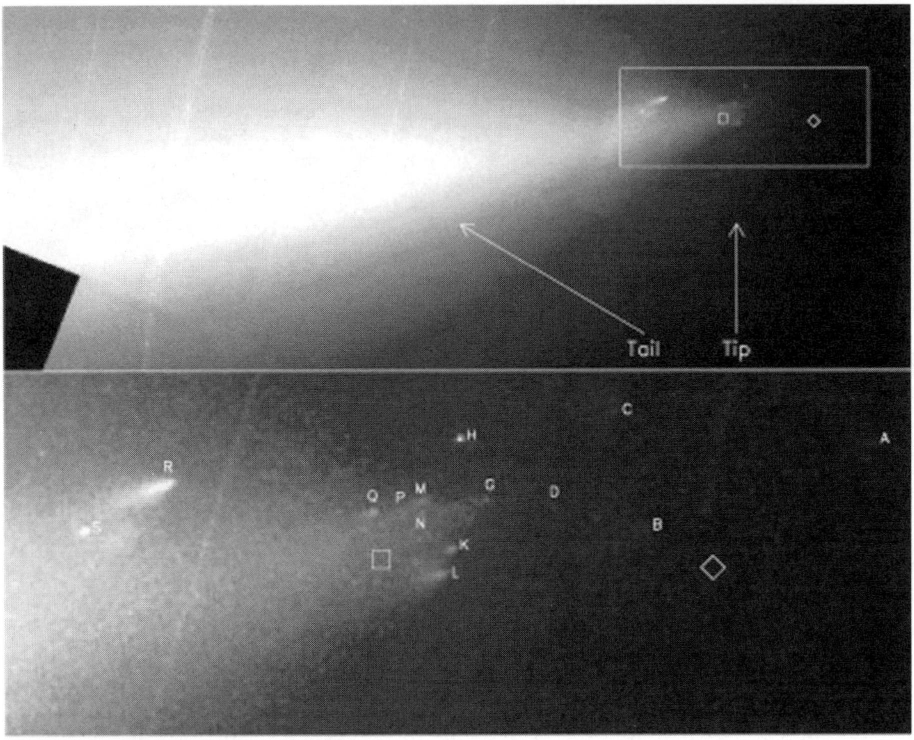

Figure 3. Two views of the head of comet LINEAR (1999 S4) taken with WFPC2 on the Hubble Space Telescope several days after the breakup of the nucleus. Letters in the image below show the locations of many of the individual remaining nuclei (Weaver et al., *Science* **292** 1249, 2001).

The recent example of comet C/1999 S4 (LINEAR) is a perfect illustration of the classification problem. It seems to violate the established schemes. C/1999 S4 was a long-period comet on a highly inclined orbit, and so presumably came originally from the Oort cloud. However, it was extremely depleted in CO (~0.4-1%) and evidently, based on the complete disruption of its nucleus at 0.85 AU from the Sun, had little or no internal strength (Figure 3). Both of these pieces of evidence suggest that this comet may be one of the minority of Oort cloud comets that formed in the vicinity of Jupiter, as opposed to the majority which formed near the orbits of Uranus and Neptune. However, most carbonaceous compounds including both C_2 and C_3 were depleted in C/1999 S4, which, by itself, might suggest that the comet formed in the Kuiper Belt. These apparently conflicting results highlight both the need to re-examine the meaning of the optical surveys of comets and the danger in basing conclusions on sample sizes that are too small. Regarding the latter, we must recognize that while space missions will undoubtedly provide huge leaps in our basic knowledge about a few cometary nuclei, and these data can provide "ground-truth" for remote observations, we also need remote observations of many comets to understand them as a population of varied objects.

Over the last century, many bits of evidence from meteor stream studies seemed to indicate that cometary debris (including that from KBOs) should be a major, if not the main, source of the present Zodiacal cloud through their associated debris streams, but how this happens in detail is not well understood. It is expected that the Stardust mission and other planned dust missions will elucidate the state of "pristine" cometary debris (at least from short-period comets) and possibly define the end of particle evolution as IDPs. But the degradation process whereby large cometary meteors (which are described as "fluffy") are converted into small zodiacal dust (which is granular) is not presently well understood enough even to determine if cometary debris is indeed a major source of zodiacal particles.

The study of comets, both individually in detail and more broadly as a diverse population of solar system objects, will provide unique and important insight into our origins: the origin of the solar system and the development of life on Earth. Their study is enabled by the serendipitous random gravitational interactions with passing stars through the Oort cloud and collisions in the Kuiper belt, processes which transport comets from the remote and cold outer solar system into the neighborhood of the terrestrial planets. In a very real sense, comets provide us with an easily accessible tool for studying the outer solar system, as they can be remotely observed continuously and they can be reached by spacecraft within periods of a few years, or even months under some circumstances, rather than the decades-long journey that is required for in-situ sampling of the solar system at Neptune and beyond. Because of their relative accessibility, and because of their importance in understanding the formation and evolution of the solar system, comets have been at or near the top of the priority lists of nearly every important scientific advisory committee during the past two decades. We have yet to follow the recommendations of these committees, and it is now time to devote the necessary resources to serious, in-depth, and long-term investigations of cometary nuclei.

2. Key Science Questions

What are the basic physical characteristics of cometary nuclei (size, shape, color, rotational period and axis orientation, mass, density, tensile strength, porosity, ice-to-dust ratio)?

Despite all that we have learned about comets in the last half century, we do not yet know the mass of a single cometary nucleus. None of the planned comet missions, except for the Rosetta orbiter in late 2011, will be able to determine the mass of a nucleus. While there is abundant evidence that cometary nuclei are extremely fragile and shed substantial fragments every century or so, the hypothesis that all cometary nuclei are strengthless rubble piles is still largely untested. As mentioned above, cometary nuclei are frequently described as being "dirty snowballs", but there is increasing evidence that some nuclei, especially those of the short-period comets, may be dominated by refractories rather than ices. The question of the size distribution of cometary nuclei, and whether it is consistent with a collisionally evolved distribution, is only starting to be addressed in a systematic way. Understanding of the current nuclear size distribution is important not only for understanding the mechanisms of assimilation and ultimate end state of accretion from ISM material intro planetisimals in the proto-solar nebula, but also for the impact hazard to Earth from cometary bodies.

What is the composition of cometary nuclei and how is that related to the formation of the solar system and to the origin and evolution of life? What are the ice-to-dust ratio, the molecular composition of ices, mineralogical composition of the dust, elemental and isotopic ratios in different classes of comets?

As discussed above, the history of cometary matter is encoded in the composition of the nucleus. The most volatile ices were mostly lost for nuclei that formed in relatively warm environments, whereas nuclei that formed outside of the planetary region may have retained a substantial fraction of the volatiles present in the interstellar cloud that became the solar nebula. Thus, we might expect to see a wide range of composition among the cometary nuclei, and, indeed, we are starting to see some evidence of this. However, we need to accumulate abundance information on many more comets, hundreds if possible and of both dynamical classes, before we are likely to recognize trends that correlate with formation locations. The D/H ratio is particularly important for judging the role of comets in bringing water to the Earth, but we currently have accurate data for only 3 long-period comets. Although we have the technical ability (e.g., through D/H and OH/OD measurements) to investigate the D/H ratio in many comets, this task is challenging and would require a substantial amount of time on an Earth-orbiting telescopic facility. Cometary nuclei are also rich in organic molecules, so cometary bombardment may have delivered most of the raw materials for life to the Earth. If so, then comets may be responsible for seeding life throughout the solar system. However, these ideas are speculative and controversial and need to be thoroughly investigated during the next decade. Furthermore, if comets can trigger life then they may also terminate it under certain conditions, and we need further elucidation of the cometary role in terrestrial impact events.

How does the composition of the gas and dust in the coma and its variation along a comets orbit relate to the bulk composition and evolution of the cometary nucleus?

Because spacecraft missions of all types will only sample a handful of comets it is necessary to coordinate the ground truth of these missions with systematic earth-based and near-earth observations of comets in order to better understand cometary processes and to enable the general population of comets to be characterized appropriately. We must recognize that while space missions will undoubtedly provide huge leaps in our basic knowledge about a few cometary nuclei, and these data can provide "ground-truth" for remote observations, we also need remote observations of many comets to understand them as a population of varied objects.

Is the interior of a cometary nucleus substantially different from its surface layers, and what are the physical processes of cometary activity?

The question of the distribution of active regions on the surface, sublimation from depth if there is very high porosity, and the resulting evolutionary changes are not well known. Our capabilities to perform complicated model calculations (porous lattice gas flows, dusty-gas dynamics flows) far outpaces our experimental and measurement knowledge. With the recent images of comet 19P/Borrelly there is a suggestion that the somewhat darker regions (albedo of 0.06 vs. 0.03) are the sources of the observed dust jets, and that the dust emission seems well-collimated just above more distributed (round) regions at the surface. This could imply significant dusty-gas dynamical collimation and gas cooling in the surface layer, in contrast to most models of coma activity. Such questions will not be answered by a few nucleus flyby missions, or possibly even the Rosetta lander, which has a relatively short lifetime on the surface at initial encounter when the comet is fairly inactive. Perhaps monitoring by the remote sensing instruments by the Rosetta orbiter will provide some clues, but this is not certain.

What other aspects of comets can be used to constrain and/or select different solar system formation and evolution models?

As already mentioned above, the composition of cometary nuclei provides clues to their processing history. But are there other ways for the study of comets to answer some of the "big" questions regarding the formation and evolution of the solar system? For example, if cometary nuclei are rubble piles of cometesimals, maybe the size distribution of those cometesimals can be used to constrain the formation scenario. Does the absolute size distribution of cometary nuclei shed any light on this subject? We need to identify specific observational tests that can serve as important links between comets and planetary system formation scenarios.

How do KBOs evolve into comets, and what is the relationship between KBOs, Centaurs and comets?

There is now good dynamical evidence that KBOs can be transported from orbits in the trans-neptunian region to orbits that resemble those of typical short-period comets, but the question of whether short-period comets really resemble the KBOs physically and chemically remains unanswered.

How pristine are cometary nuclei and what do they tell us of the fundamental building-blocks of the solar system?

Although comets are in a deep freeze for most of their lifetimes, various processes (such as collisions, surface sputtering, cosmic ray processing, multiple perihelion passages, etc.) may have altered their original composition and/or physical structure. These effects are undoubtedly important for certain individual cases, but we need better investigations of how the overall properties of comets as a population may be affected.

What is behavior of cometary ion tails as a function of time generally, and in particular orbital position, heliocentric distance, and solar latitude?

Comets can serve as probes of the solar wind, interacting with it not like the planets which are on fixed orbits in or near the ecliptic plane, but on high inclination orbits enabling them to be affected by both the turbulent, and generally slower, low latitude solar wind near the ecliptic and the steadier high speed streaming solar wind on open field lines at high solar latitudes. Continuing uncertainty over the cause of tail disconnection events (sector boundary crossings vs. solar wind impulses) as well as the recent exciting discovery of cometary X-ray emission provide important information on the interaction between comets and their environments.

What is the real distribution of cometary orbits in aphelia and orbital inclination?

Current surveys still suffer from search-based selection effects that do not significantly sample high inclination orbits. As NEO surveys go fainter and cover more sky at a higher frequency, the observational bias will move towards larger heliocentric distance and provide better statistics of the orbital distribution of comets. This has both scientific consequences for understanding the origin and distribution of comets and human consequences if we are seriously undersampling a significant population of potent Earth impact hazards.

How does cometary and KBO debris degrade to become a component of the zodiacal dust, particularly in the outer solar system ? How does this process compare in importance with other sources of zodiacal material ?

These questions are not well addressed by spacecraft experiments alone. Coherent spacecraft support studies using a combination of UHF radar (Arecibo, Kiruna, etc.) observations, which bridge the gap between dust studies and ordinary meteors, laboratory investigations of cometary debris degradation processes (leaching out of volatiles and solar wind effects), and a program of advanced cometary debris evolution simulation using massively parallel calculations can address this important question.

3. Recommendations

The next decade will see the fruition of long-planned spacecraft missions in several comet flybys, the return of cometary dust particles to Earth, and the impact of a comet nucleus by a fast projectile to release material from deep inside the nucleus. At the very end of the decade a rendezvous by the Rosetta spacecraft will perform a landing on and direct sampling of a mostly dormant nucleus followed by a three-year study by its orbiter to examine a comet during its transition from the dormant to a fully active state.

The recent disintegration of comet 1999 S4 (LINEAR) dramatically demonstrated that all comets are not alike. In order to understand comets as a population of objects, it is necessarly to use the rare spacecraft opportunities to provide the "ground truth" with which the potentially hundreds of comets, which can be observed remotely, can be understood. A balanced program consisting of remote ground-based and space-based observations, theoretical modeling of the nucleus and gas, dust and plasma environment, and relevant laboratory measurements is necessary to fully exploit the coming decade of in situ spacecraft encounters. In addition the next decade must specify the following decade of new spacecraft missions which must include orbiter/lander and sample return missions as the natural follow-on to the flyby missions that are now largely underway. These are detailed below.

The overarching issue is cometary public policy. For the past 20 years and more, scientific advisory committees have given nearly the highest priority to the importance of the studies of small solar bodies in general and to comets in particular, because of their unique and certain connection to the origin of the solar system and to the early surface processing of the Earth and even the origin of life itself. Despite these recommendations science policy has not been kind to comet missions. There were a series of cancellations of Halley missions at various stages of development for a once-in-a-lifetime opportunity. There was the choice to cancel the Comet Rendezvous Asteroid Flyby mission in favor of yet another giant planet orbiter mission, Cassini.

Our concern as members of the comet community is that the results of this decadal study will turn out the same way as in the past. Either cometary science will fall in priority because it is outweighed by other planetary science interests, or a clear recommendation in the decadal report for the importance of cometary science will again be ignored when NASA sets its mission priorities.

3.1. What missions, groundbased facilities, laboratory facilities, or other supportive initiatives would be needed to address or answer the above questions?

Missions: A basic problem throughout planetary science seems to be the $300M budget level for Discovery missions, making it difficult to go far from Earth's orbit. For this reason, comets have seemed to fare reasonably well with Discovery missions (Stardust, CONTOUR, and Deep Impact); but all are flyby missions. It is clear from discussions above that orbiter/lander and sample return missions are necessary to address the most important questions of cometary science, and that such missions are not possible within the Discovery budget cap. Modest orbiter/small-lander missions could possibly be built for roughly double the Discovery cap, however a significant sample return mission would likely be in the major-mission budget class of Galileo and Cassini. Given the importance of comets to the fundamental questions about our origins, we believe that serious consideration should be given to both orbiter/lander missions, possibly through a new class of larger openly competed planetary missions, and to a major Cassini-class planetary mission for a cold sample return. Finally, NASA should fully support the MO&DA phase of the ROSETTA mission. For roughly half the cost of a single Discovery mission, which is spread out over the next decade, the science yield and cost benefit to the US is quite high.

Earth and Near Earth-based Observatories: There are many problems for comets in particular, and for planetary science in general, which require long-term consistent remote monitoring from first-class ground-based and space-based facilities. The current system of sharing 6% of general astrophysics facilities does not address these scientific needs. In addition, standard astrophysics instrumentation is often not optimized for planetary observations. Studies of comets require observations at all time scales so that effects from irregular short-term variability can be separated from diurnal (rotation-driven) effects, and long-term effects involving seasonal variations caused by rotation axis geometry as well as heliocentric distance effects. Dedicated planetary facilities would be useful across sub-disciplines in solar system science.

- We recommend consideration of a relatively low cost orbiting UV observatory optimized for observations at small solar elongation angles and so that a sufficient time base of comet observations can be made. The small percentage of time granted for planetary observations from astrophysics facilities, such as HST and FUSE, cannot answer the important questions about comets, their variation and evolution.

- We recommend consideration of a relatively low cost orbiting thermal IR (2 – 100 microns) observatory optimized for surveying the parent molecule and refractory dust emissions as well as the nuclear size/albedo/rotation rate for a large number of comets, unbiased by solar elongation and unaffected by the terrestrial atmosphere. With a spacecraft that can track moving solar system objects, and with a limiting sensitivity 10 times worse than SIRTF's, a 1 meter class telescope can study all the currently known short period and long period comets, and the brightest 1/3 of the KBO's. The small percentage of time granted for planetary observations on SIRTF will not be enough to provide the answers to the important questions about comets, their variation and evolution.

- Modernize or replace the current IRTF with a current state-of-the-art large planetary-dedicated observatory. Again, a small percentage of time on such major facilities, such as Keck and Keck II, cannot alone answer the important questions about comets, their variation and evolution.

In the areas of groundbased observation, data storage and laboratory programs, we recommend:

- Expand/complete the inventory of the solar system small bodies to include better sampling of high inclination orbit comets.

- Maintain/create/add to data bases in all wavelengths to get physical parameters of the nucleus (size, albedo, rotation, strengths, etc.) and molecular abundances.

- Support theoretical/numerical studies to understand the spacecraft and ground-based data and those which make predictions that are testable by observation.

- Laboratory measurements and ab initio calculations regarding the rates, branching ratios and energetics for the photochemical breakup (dissociation and ionization) of all the common parent and daughter species are still lacking and limit our ability to fully interpret remote-sensing observations.

- More consistent electronic archiving of cometary data independent of planetary missions is needed. One answer would be an approach similar to the International Halley Watch effort.

Funding for grants for ground based support, spacecraft analysis, and theoretical studies is essential. Ground-based observations not only supply support for observations of spacecraft targets, but supply the big picture for understanding the context in which to interpret spacecraft data in particular and more generally to understand comets and their unique role in the origin of the solar system and life on Earth.

Laboratory and theoretical atomic and molecular physics and chemistry are nominally supported in a number of Research and Analysis programs. Unfortunately, these are competed against traditional planetary science programs. It is suggested that all the planetary programs pool a reasonable fraction of their current budgets to support a separate Planetary Atomic and Molecular Properties Program to support both laboratory and theoretical work. This would include photochemical, charge impact chemistry, gas phase chemistry and surface chemistry.

NASA's astrophysics and space physics/solar programs have special "Theory Programs" for long-term support of development of theoretical simulation models. NASA should consider implementing a comparable planetary program. Such a program would play a similar role to PIDDP program for space experiments.

The study of comets should remain at the highest priority level because of their unique importance in understanding our origins. Such a recommendation would be consistent with most of the important scientific advisory panels over the past few decades.

3.2. Data archive and access issues

NASA's Planetary Data System should be fundamentally reevaluated. The large databases of the astrophysics community (e.g., HST and IUE) are easily accessible, user-friendly, have user tools and use FITS standards with which everyone in the community can use. These should be used as a model for the planetary community.

With an improved system observers should be encouraged to archive data to send the data to the small bodies node (data from surveys, e.g., at NOAO telescopes will be made public after a short proprietary period) so that everyone will have access to the data. If the necessary support is provided to investigators they could then be required to archive data.

3.3. Education and public outreach

The appearance of a bright comet every few years captures the public's awareness of planetary science. Comets are a very good tool to get the public/schools interested in science especially when a bright comet is visible. Support (financial and materials/ideas) for such activities would be very helpful.

The Future of Solar System Exploration, 2003-2013
ASP Conference Series, Vol. 272, 2002
Mark V. Sykes, ed.

Probing The Solar System's Outermost Frontier: The Future of Kuiper Belt Studies

W.M. Grundy[1], H.F. Levison[2], J.W. Parker[2], R.L. Allen[3], L.C. Ball[4], J.F. Cooper[5], M.C. De Sanctis[6], T.L. Farnham[7], B. Gladman[8], J.M. Hahn[9], C.W. Hergenrother[10], J.J. Kavelaars[11], H. Krüger[12], D.J. Lien[13], R. Malhotra[10], R.M.E. Mastrapa[10], A. Quillen[14], R. Srama[12], J.A. Stansberry[14], G. Strazzulla[15], R.J. Terrile[16], & C.A. Trujillo[17]

[1] *Lowell Observatory, Flagstaff, AZ*
[2] *Southwest Research Institute, Boulder, CO*
[3] *Department of Astronomy, University of Michigan, Ann Arbor, MI*
[4] *Emerald Lane Observatory, Decatur, AL*
[5] *NASA Goddard Space Flight Center, Greenbelt, MD*
[6] *Istituto di Astrofisica Spaziale - CNR, Rome, Italy*
[7] *Department of Astronomy, University of Texas, Austin, TX*
[8] *Observatoire de Nice, Nice, France*
[9] *Lunar and Planetary Institute, Houston, TX*
[10] *Lunar and Planetary Laboratory, University of Arizona, Tucson, AZ*
[11] *Department of Physics and Astronomy, McMaster University, Hamilton, Canada*
[12] *Max-Planck-Institut für Kernphysik, Heidelberg, Germany*
[13] *Department of Physics, Oklahoma State University, Stillwater, OK*
[14] *Steward Observatory, University of Arizona, Tucson, AZ*
[15] *Osservatorio Astrofisico di Catania, Catania, Italy*
[16] *Jet Propulsion Laboratory, California Institute of Technology, Pasadena, CA*
[17] *Division of Geological and Planetary Sciences, California Institute of Technology, Pasadena, CA*

Abstract. Priorities are described for research on trans-Neptunian Objects (TNOs), a recently-discovered (thereby excluding Pluto and Charon) population of small, outer solar system objects with important implications for the early history of the solar system. We conclude that we must discover and obtain accurate orbits for an unbiased sample of ~5000 TNOs over the next decade and a significant fraction of these objects must be targeted for more detailed physical observations. A broad range of complementary theoretical and laboratory studies are also needed, and we must work towards direct spacecraft exploration of TNOs as soon as possible.

EXECUTIVE SUMMARY

Since the discovery of the first trans-Neptunian object (TNO) in 1992, the trans-Neptunian region has surprised researchers time and time again. For example, many TNOs have been found in mean motion resonances with Neptune while others are in highly inclined and/or eccentric orbits (eccentricities as large as 0.9). Colors of TNOs span a remarkably wide range (B−V from 0.5 to 1.3), and 6 objects are now known to be binaries. Unexpected properties such as these have led to a re-evaluation of the history of the solar system.

The trans-Neptunian region represents the best known source for the study of the formation and evolution of the outer solar system because TNOs are relatively unaltered remnants from that early epoch. The orbital element distribution of TNOs is expected to bear scars from the late stages of the formation of Uranus and Neptune. By understanding this population, we can obtain information about the masses and locations of planets during earlier times, as well as external influences which could have impacted the solar system. Similar arguments apply to physical studies of TNOs, since their small sizes and low temperatures should have prevented their bulk chemistry from substantially changing since their formation. The TNOs thus far discovered have supplied us with tantalizing clues about these issues. However, a complete understanding of this important region cannot be achieved without considerably more effort.

The observational challenges presented by TNOs are immense, owing to their extreme distance and faintness. A balanced approach involving observations, spacecraft exploration, and Earth-based investigations is needed to maximize our understanding of the trans-Neptunian population and to realize the greatest scientific benefit from its study. This panel recommends a research program including these components:

- Obtaining a large, unbiased sample of some 5000 TNOs with accurately determined orbits is a top priority. To achieve this goal will require carefully coordinated search and recovery efforts for well over 100 dark nights per year for a decade on 4 m class telescopes.

- A diverse array of observations is needed to measure physical properties of as many TNOs as possible, and deep surveys are needed to probe fainter and more distant objects. These observational projects will require access to state-of-the-art ground-based, airborne, and space-based observatories.

- Close up, direct exploration of a few TNOs and in-situ measurements of their environment are needed. These tasks can only be accomplished by spacecraft. Prompt development and launch of NASA's Pluto Kuiper-Belt (PKB) mission is an essential first step.

- A wide variety of theoretical and laboratory studies are urgently needed to enable us to properly interpret observational data from each of the above efforts.

These components complement one another in important ways. If any part of this balanced program is neglected at the expense of another, overall progress toward understanding the outer solar system will be significantly retarded.

This panel believes that a top priority of the community is obtaining an observationally unbiased sample of approximately 5000 TNOs brighter than $\sim 24^{\text{th}}$ mag, as well as a smaller sample of fainter objects. The orbits of this sample must be determined to sufficient precision that we can (*i*) discern the dynamical fine-structure of the belt and (*ii*) recover these objects for physical studies. To achieve these goals, we must discover new objects at rates exceeding current discovery rates, coupled with a greatly-expanded campaign of astrometric follow-up. The need for consistent recoveries cannot be emphasized strongly enough. Five years ago, a primary impediment was the paucity of known objects. Now, the major stumbling block is lack of dedicated recovery efforts. Recovery failures have produced a sample that is burdened with complicated and unknown biases. It would be far more scientifically valuable to have a thousand TNOs with well-determined orbits and minimal observational biases than to have ten thousand with poorly determined orbits and/or subject to unknown observational biases.

To cost-effectively obtain a suitable sample of TNOs will require both deep and wide surveys. The primary effort of accumulating 5000 good orbits can be accomplished with wide-field surveys carried out at telescopes in the 4 m aperture class. Two dedicated 4 m telescopes could do an excellent job of discovering and following this sample over the coming decade. Deep surveys at larger telescopes are also needed, for studies of the spatial and size distributions of fainter objects.

In addition to search and recovery efforts, a great variety of complementary observational studies are needed to determine rotation rates, radii, albedos, colors, masses, and compositions of a significant and minimally-biased sample of the known TNOs, enabling us to draw conclusions about the ensemble properties of distinct sub-populations. These observations require access to state-of-the-art astronomical facilities including HST, Keck, and eventually SIRTF, ALMA, and NGST, as well as possible future facilities such as LSST or SNAP. Additionally, Kuiper belt research needs must influence the design of future instrumentation for these observatories.

The community places a high priority on direct spacecraft exploration of a few objects. Close flyby investigations offer unique opportunities to apply the tools of planetary geology to learn about the surfaces and interiors of TNOs. Spacecraft are also the only way to obtain in-situ measurements of the trans-Neptunian environment which are essential to understanding the evolution of TNO surfaces.

All of these investigations must be supported by theoretical research. Comprehensive numerical studies of the formation and the dynamical and collisional evolution of the outer solar system are needed to interpret the observed distributions of the thousands of objects to be discovered in the trans-Neptunian region. Studies of the radiative transfer and thermal and compositional evolution of TNO surfaces and interiors are also essential to interpreting observational data.

Laboratory studies are required for interpreting diverse types of observational data and to provide inputs to theoretical and numerical models. Studies of optical constants, photolytic and radiolytic pathways, low temperature chemistry, gas-solid interactions, and thermal and mechanical properties of outer solar system materials are all urgently needed.

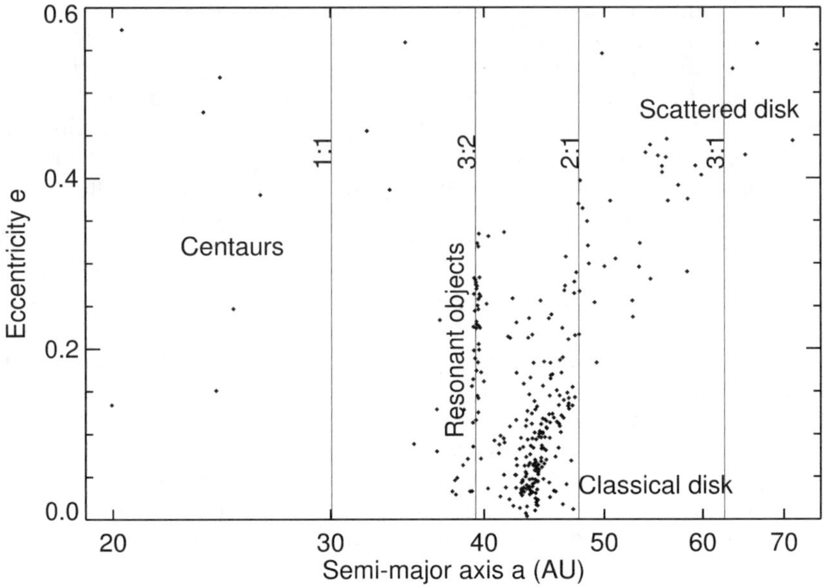

Figure 1. A still-incomplete picture of the dynamical structure of the Kuiper belt begins to emerge when objects with well-known orbits are shown on a plot of semi-major axis versus eccentricity. Mean motion resonances with Neptune's orbit are indicated by vertical lines and major dynamical classes are labeled.

REPORT

1. Current State of Knowledge

Since the discovery of the first trans-Neptunian object in 1992, there has been an explosion of activity directed towards learning about the contents of this fascinating part of our solar system. A decade later, we have discovered hundreds of objects, but our ignorance remains profound about many of their most basic characteristics (Figure 1). While we have determined the visual-wavelength colors and lightcurves for a small subset of the known TNOs, we know almost nothing about the sizes, shapes, compositions, and geology of the population as a whole. Infrared spectroscopy of a few TNOs points to a remarkable diversity of surface compositions, but a taxonomy of TNOs does not yet exist. We have found that the dynamical structure of this region is surprisingly rich, with some objects in Neptune mean motion resonances, others in near-circular orbits (classical Kuiper belt objects, or KBOs), and still others in highly inclined and eccentric orbits (Scattered Disk objects, or SDOs). The origin of this diversity remains mysterious. We do not even know how far away from the sun the trans-

Neptunian population extends, and have probably not yet discovered many of its largest members.

The potential value of the trans-Neptunian region as a source of information about the formation of the outer solar system is immense and thus our current state of ignorance is especially troubling. The orbital distribution of TNOs bears scars from the late stages of the formation of Uranus and Neptune and is indicative of a very violent past. We need to understand, in detail, the structure of this distribution in order to obtain vital clues to the masses and locations of planets at early times in the solar system. We may also be able to glean information about the Sun's stellar nursery and/or the more recent interstellar environment by studying these structures.

Furthermore, TNOs are believed to be among the most pristine objects in the solar system, with interior compositions little changed since the time of their accretion in the outer parts of the protoplanetary nebula. By studying TNO compositions along with the processes which have acted on their surfaces and interiors over the age of the solar system, we can gain a potentially unique window into the chemical evolution of the protoplanetary nebula and the source material from which our solar system formed.

So little is known about the Kuiper belt that it will be essential to attack our key scientific questions from many different directions simultaneously, if we are to make significant progress during the coming decade. The remainder of this paper will catalog our key scientific questions and then will describe a number of complementary investigations which this panel believes must all be pursued to achieve the kind of balanced approach that is needed.

2. Key Science Questions

As emphasized above, very little is known about the trans-Neptunian region and the objects which occupy it. Over the next decade we need to address the following four key science questions:

- What is the dynamical structure of the Kuiper belt and what processes sculpted it?
- What different classes of TNOs exist, and what is their origin?
- What are the compositions and structures of TNO interiors and surfaces and what processes act on them?
- What does the Kuiper belt tell us about formation of the solar system?

To effectively tackle these very broad and general questions, they can be broken into smaller, more readily addressable questions such as:

- How did TNO orbits get so dynamically excited and is there a dynamically cold population?
- What is the radial distribution of TNOs, and is the apparently sharp edge of the classical belt real? (if so, what caused it?)

- What are the compositions of TNO surfaces?
- How does composition relate to color and albedo, and what is the reason for the observed diversity of colors?
- What is the overall size-frequency distribution of the trans-Neptunian population, and what are the size distributions of different classes of TNOs?
- What relationships exist between dynamical and physical classes of TNOs?
- Are apparent correlations between absolute magnitude, inclination, and color in the main belt real, and what do they mean?
- What other large objects remain to be discovered? Are there more Plutos and Charons?
- What are the inclination distributions of TNO sub-populations and what caused them?
- What changes occurred in TNOs during and soon-after accretion?
- What does the distribution of TNO rotation periods look like?
- Where is the transition from irregular to spherical shapes, and what does this tell us about TNO interiors?
- Which TNOs have satellites, and how do they differ from moon-less TNOs?
- How do TNOs get into orbits like that of 2000 CR_{105}, with huge eccentricities and perihelia well beyond Neptune's?
- What caused the resonant structure of the Kuiper belt and how did it form?
- What are the connections between TNOs, centaurs, trojans, comets, and icy satellites, and how are their dynamical histories and compositions related?
- What can we learn about TNOs by extrapolation from comet and centaur studies?
- How does our Kuiper belt compare with disks around other stars?

It is readily apparent that a huge number of questions awaits our pursuit.

3. Recommendations

During the coming decade, this panel recommends:

- A pair of dedicated 4 m class telescopes (or their equivalent in dedicated dark time) be made available for wide-field search and recovery efforts.

- A LARGE amount of time be made available on various very large (6 to 10+ m and space-based) telescopes with adequate instrumentation for physical studies and deep surveys.
- NASA's "Pluto-Kuiper Belt" (PKB) mission, the only means of of doing crucial in-situ measurements of TNOs, proceed expeditiously to launch.
- A vigorous program of complementary laboratory, theoretical, and numerical investigations be supported.

A balanced approach will be essential to maximize our scientific understanding of the trans-Neptunian population. Indispensable research components include discovery of a large, unbiased sample of ~5000 TNOs with adequate astrometric follow-up, deep surveys to establish bounds on the Kuiper belt, observational efforts to determine TNO physical properties, in-situ spacecraft investigations of TNOs and their environment, and a wide variety of theoretical and laboratory studies. In the remainder of this section, we describe specific investigations which must be part of a balanced program to advance understanding of this important region of the solar system. Each component of this balanced program depends on progress in the others. If one area is neglected at the expense of another, overall scientific progress will suffer.

3.1. TNO recoveries and determination of orbital elements

The discovery rate of TNOs has dramatically increased over the last few years, making the need for a dedicated recovery program particularly urgent. 146 TNOs were discovered in 1999, more than in all previous years combined. 169 were discovered in 2000, and the rate of discovery continues to increase. Relative to discovery, it typically requires 5 times more telescope time for recovery efforts in order to sufficiently refine the orbital elements of a new TNO. This factor arises because recoveries must be made at least once during the year of discovery and each of the following 2 to 3 years (oppositions). As of mid-November 2001, ~540 TNOs have been discovered, but more than half of them have had no second-opposition recovery observations, and about half of those have had no measurements for more than a year (Figure 2); nearly 20% of all objects are irrecoverably lost and another ~10% are in immanent danger of being lost. This distressing record shows that current programs are not sufficiently following up their discovery observations.

In general, discovering objects only to lose them is of little use and wastes resources. Objects must be observed frequently enough that their orbits can be accurately determined. Accurate orbital elements are essential to ensure objects will be recoverable for physical observations. To provide useful constraints on dynamical models, it is necessary but not sufficient to acquire a large catalog of well-observed TNOs. It is even more important to avoid, or at least understand quantitatively, observational biases which favor recovering objects in particular types of orbits. Ephemerides used by observers for first recoveries are based on educated guesses, which in turn are based on the current understanding of Kuiper belt orbits. If an object is found at a heliocentric distance less than 40 AU, it is typically assumed to be in the 2:3 mean motion resonance with Neptune. If it is found beyond 40 AU, it is often assumed to be in a circular

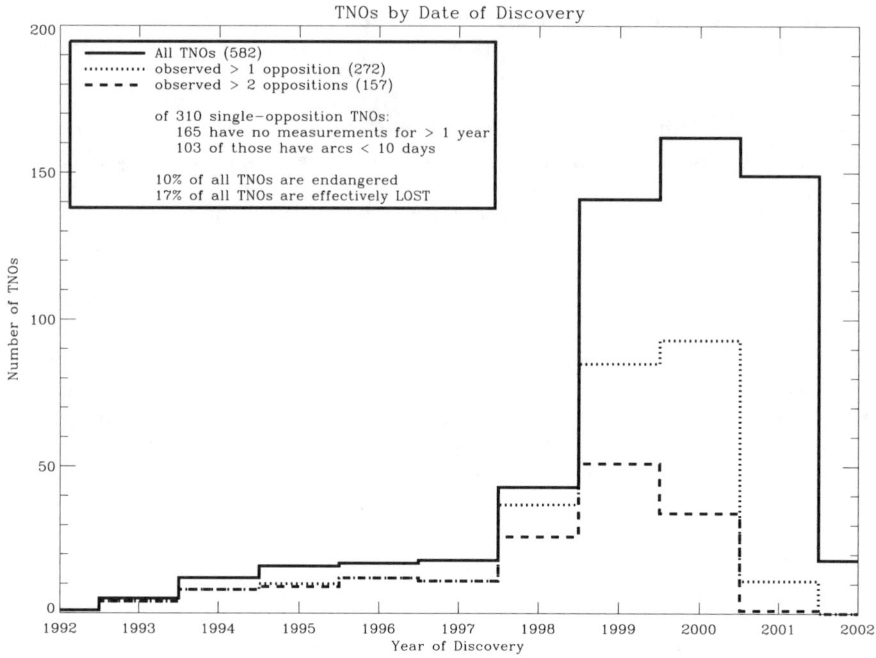

Figure 2. Discovery and recovery statistics (as of March 2002) of the currently known TNOs. Note that the recovery rate has significantly fallen behind the discovery rate (solid line) over the last several years. Without dedicated recovery efforts, the sample of objects in well-determined orbits will become increasing biased as the fraction of lost objects grows.

orbit. After enough time has passed since discovery, only those objects having nearly correct assumed orbits can be easily recovered. Thus, the Kuiper belt has probably taken on an appearance similar to our preconceptions and those objects that may be most unusual and interesting are the ones that have been systematically lost. The following examples illustrate this point and show why dedicated follow-up observations are essential:

- Recovery of the Scattered Disk object (SDO) 1996 TL_{66} was made possible only by early follow-up observations, which were done because it was bright enough to be observed by amateurs.

- Aggressive follow-ups were necessary to recover 2000 CR_{105}. It was far from its predicted position and was only recovered by virtue of a concerted effort. 2000 CR_{105}'s unusual orbit has a 44 AU perihelion and 415 AU aphelion, challenging current dynamical models.

- Theoretical models show that the 1:2 mean motion resonance should be populated. But it took second and third opposition observations to demonstrate that 1997 SZ_{10} and 1996 TR_{66} occupy this resonance that otherwise appeared to be empty.

If follow-up recoveries had not been made early and often enough, all these objects could very well have joined the ranks of "lost TNOs", and we would have been ignorant of their significance to solar system structure! Indeed, it is likely that many of the lost TNOs were on similarly unanticipated orbits so their initially assumed orbital elements were sufficiently incorrect that casual recovery efforts failed to find them. As the pace of Kuiper belt discoveries increases, the problem of recovery is exacerbated.

To assure recovery, the recovery observations need to equal, or preferably surpass the discovery observations in sensitivity to ensure that lack of recovery is due to ephemeris uncertainties and not due to worse limiting magnitude. Moderate aperture telescopes in the 4 m class are sufficient for both tasks (down to ~24 mag), and their cost effectiveness makes them preferable to larger instruments for discovering and following up the bulk of the ~5000 object sample needed to address the scientific questions posed in Section 3. Very large telescopes are useful for deep studies of size and spatial distribution, but it is unlikely that enough time can be allocated on these instruments to follow more than a small sub-sample of objects.

Establishing a reliable orbit for a newly-discovered TNO generally requires imaging it again a few days after discovery, then again a few months later, and then at least once a year thereafter for a few years until uncertainties in the predicted positions are sufficiently minimized. The recovery rate of TNOs must keep up with the discovery rate in order for discovery surveys to be most effective, and, as described above, the recovery observing plan must be designed to minimize orbital element biases by persevering in searching for objects, even when they are initially not found where they were expected to be. Because total recovery cost is about 5 times the telescope time used to initially discover each TNO, recoveries require a large time commitment on the part of the astronomer (who must quickly analyze the discovery data in order to prepare for the next recovery observations) and on the part of the observatory (which needs to allocate sufficient telescope time for discoveries and subsequent recoveries).

3.2. TNO discoveries

To answer many of the key scientific questions from Section 3, reliable orbits are required for a substantial and unbiased sample of TNOs, which in turn calls for extensive, coordinated search/discovery and recovery/astrometry observations that (*i*) survey a wider fraction of the sky and (*ii*) probe the solar system to greater depths. Two distinct surveys (which we will call "wide" and "deep") are needed. Wide surveys should cover a broad swath of the night sky to a limiting magnitude of ~24 in order to obtain some 5000 TNOs with well-determined orbits and subject to minimal observational biases (other than the obvious magnitude bias). This sample will be used to answer the many question of dynamics, structure, and classification as well as providing a reliable basis set of objects for physical studies. The wide survey effort must be complemented with deep surveys which probe a much smaller portion of the sky to greater depths (e.g., down to ~27th magnitude) in order to discover and recover a much smaller sample of especially faint TNOs. Deep surveys will illuminate the small-size end of the TNO size distribution and will also enlarge the "known" solar system by pushing its observable frontier outwards from the Sun.

Wide surveys: Wide surveys will play the primary role as we attempt to discern the dynamical structure of the Kuiper belt. The goal should be to discover and get reliable orbits for some 5000 TNOs in the next decade. To meet this objective will require a discovery rate of approximately 700 TNOs per year for the first seven years (the remaining three years are needed to adequately measure the orbits for the last set of discoveries). Objects brighter than R=24th mag can be detected with moderate-size (~4 m class) telescopes. The ecliptic surface density distribution at this magnitude is approximately 4 TNOs per square degree, and given an effective full width at half maximum of the ecliptic of ~6 °, there are expected to be some 15,000 such TNOs (including objects far from the ecliptic at present), of which the proposed survey would discover about one-third. Although some care must be taken in the planning of the survey to assure that observational biases do not affect the distribution of orbits in the discovery sample, the more critical point where observational biases appear is in the recovery observations, as discussed in Section 3.1.

To estimate the time commitment and facilities needed, we use the following illustrative example (the actual format of the observing plan could differ). If the search were performed at a single telescope during all dark-grey times (±5 nights of the new moon), or 120 nights per year, then the discovery rate would have to be about 6 objects per night (or more, since the galactic plane season may need to be avoided). With an average of ~4 TNOs per square degree, and assuming a survey instrument with a 1 square degree FOV, that would require 2 new search fields per night. Each field would need to be observed 4 to 6 times (2 to 3 times per night on two consecutive nights to verify the object's motion). The final tally would be about 10 to 15 pointings per night on average. Recoveries require a factor of ~5 more observing time, resulting in ~50 to 75 pointings per night during dark-grey times.

Two dedicated 4 m class telescopes, ideally one in each hemisphere, would be able to perform the Wide TNO Surveys efficiently and cost effectively. We strongly recommend that wide-field survey instruments of this type be built and used to discover and follow-up the proposed 5000 TNO sample.

The Large-aperture Synoptic Survey Telescope (LSST) as recommended by the Astronomy and Astrophysics Survey Committee (AASC) in *The Decade of Discovery* report has the potential for making a tremendous impact in Kuiper belt studies. The AASC-endorsed design features a 6.5 m mirror and an estimated $170M price for construction. It would be capable of surveying the entire night sky every week, enabling all ~15,000 TNOs brighter than 24th magnitude to be cataloged and tracked, resulting in unprecedented characterization of the extent and population of the TNOs and their dynamical sub-populations. Many TNOs with unique dynamical and physical properties (wide binaries, extreme colors, and unusual photometric variability) could be identified for further study. However, significant resources must be allocated for the detection and tracking of low apparent motion objects ($\leq$ 3 arcsec/hr) for the LSST to be useful for TNO studies. The Kuiper Belt community recognizes that such a project can be of great benefit to TNO research, but there are major concerns over the endorsement of the LSST: (*i*) If the LSST is primarily designed and implemented by the extragalactic astronomy and/or near-Earth hazard mitigation communities, it is possible that only a small fraction of the LSST time and resources would be available for TNO research or that observing sequences would be poorly optimized for TNO work. (*ii*) Since LSST would not address several essential types of physical observations (e.g., collection of high signal-to-noise broadband colors and lightcurves, near infrared spectrophotometry, and deep "pencil-beam" probes), substantial access to 8 m or larger telescopes will still be required; these two needs could end up competing instead of working together. (*iii*) LSST's construction cost, development time scale, and the fact that it is only a single telescope are difficult to justify when compared with the high cost effectiveness and straightforward engineering of the proposed dedicated pair of 4 m class telescopes. In summary, while the Kuiper Belt community is excited by the prospect of the LSST, the concerns outlined above do not permit the unfettered endorsement of such a project.

Deep surveys: Historically, faint TNOs discovered by deep surveys (sometimes called "pencil beam" surveys) have been lost soon after discovery, so these surveys have yet to achieve their full potential scientific yield. This need not be the case. Deep surveys of the next decade can and should perform the necessary follow-up observations to establish good orbits for at least some fraction of the faint TNOs they discover. Reliable orbits are crucial if faint TNOs are to most usefully constrain the outer solar system's dynamical structure, and thus early history.

While TNO surveys piggy-backing on future space-based wide-field imaging instruments might eventually offer an economical alternative approach to the task of determining orbits for a sample of fainter TNOs (if they do not systematically avoid targeting the ecliptic plane), deep surveys at present require access to the largest ground-based telescopes in the world, and thus are quite expensive. This approach would not be cost-effective for finding the ~5000 orbits needed for dynamical and physical studies.

However, the value of deep surveys for constraining the distribution of smaller and more distant objects is sufficient that they have an important role to play in our exploration of the Kuiper belt. A deep TNO survey probing 3 degree^2 of the sky to a limiting magnitude of $\sim 27^{\text{th}}$ magnitude could be expected to find several hundred objects, but following them up as they diverge from the discovery fields would consume more time than is likely to be available on 8 to 10 m telescopes unless a dedicated telescope were built for the project, something which would be an extravagant and inefficient use of the limited resources available for Kuiper belt research. Instead, we must be judicious in applying our very limited large telescope, wide-field imaging resources in the most scientifically productive way we can. Individual search fields should sample various ecliptic longitudes and should not be purely confined to the ecliptic plane. Astrometric follow-up should be primarily directed at particularly interesting objects, such as those discovered at the greatest heliocentric distances. Most importantly, telescope time allocations and funding for deep surveys need to take into account the necessity of doing adequate follow-up observations.

3.3. Observational studies of physical properties

In addition to learning about the dynamics of TNO orbits and their evolution, we have much to learn about the physical and chemical properties of individual objects and about how these properties are related to objects' dynamical histories. At present, we know remarkably little about even the most fundamental characteristics of most known TNOs. We are ignorant of their masses, shapes, albedos, rotation rates, compositions, and internal structures. The observational challenge of answering these questions for a significant sample of TNOs is immense, owing to their extreme distance and faintness. Progress in physical studies of TNOs over the coming decade will depend critically on access to the largest Earth-based, airborne, and space-based telescopes and on the availability of suitable instrumentation at those telescopes. Initially, only the brightest fringe of the TNO population is accessible to many types of observational investigations, but it is essential to survey a less-biased sample, if we are to understand the range of variability among the TNO population, and the differences between different sub-populations. To be effective, these observational efforts must be concurrently supported by a wide variety of laboratory and theoretical studies. Finally, spacecraft exploration of TNOs is needed since many crucial types of data can be obtained no other way.

Visible: It is already possible to do accurate broad-band, standard filter (B, V, R, and I) CCD photometry of TNOs from many larger telescopes. Numerous measurements have been reported, revealing many surprising things, such as a remarkable diversity of colors and large-amplitude lightcurves from objects thought to be much too large to sustain highly irregular shapes. Many, many more observations are needed, in order to reveal how colors, lightcurve amplitudes, etc., correlate with dynamical characteristics for different sub-populations of TNOs. Over the next decade there must be repeated observations of lightcurves and colors for a much larger sample of objects than has been observed to date. Access to CCD cameras mounted on 3 m and larger telescopes in both Northern and Southern hemispheres is essential for the timely acquisition of these data.

Near-infrared: With its plethora of distinctive, strong vibrational absorption bands, the near-infrared wavelength region has the potential to reveal far more compositional information than the visible region. Unfortunately, declining solar flux and increasing telluric sky brightness limit spectroscopic observations to the probably-unrepresentative, brightest fringe of the TNO population at the largest telescopes in the world, at least with the current generation of instruments. Standard broad-band J, H, and K filters permit observations deeper into the TNO population and/or from smaller telescopes, but they are too broad to distinguish distinct molecular absorption features, so they drastically limit the compositional information which can be extracted from this rich wavelength region.

High-throughput, low spectral resolution spectrophotometry is urgently needed. Although technically easy to realize by current instrumentation engineering standards, instruments with the required characteristics are unfortunately not available on any of the world's 6+ m aperture telescopes. Filters which divide each of the J, H, and K transmission windows into several bins would be useful, and instruments that could record all of these wavelengths simultaneously would be ideal, if mounted on 8 to 10 meter (or future, larger) aperture telescopes, with adequate time available to observe a significant sample of TNOs.

Thermal emission: Thermal emission observations are needed to constrain the sizes, albedos, and surface temperatures of TNOs. Their enormous distances from the sun and low surface temperatures make this task an extremely difficult one, but new facilities coming on line over the next decade, including SIRTF, SOFIA, ALMA, and NGST should provide the required capabilities to measure TNO thermal emissions, if we can obtain adequate access to them. Coordinated multi-wavelength visible and thermal observations are preferable for constraining sizes and albedos, and once again, a large, unbiased sample of objects needs to be observed so that patterns can be detected in their ensemble behavior.

In addition to providing information to constrain sizes, albedos, and temperatures, the thermal emission part of the far-infrared spectrum has the potential to provide valuable information on surface compositions. Exploiting this information will require sophisticated space-based instrumentation, as well as more information about the infrared optical behavior of cryogenic materials than is currently available.

Interiors/binaries: TNOs are thought to be primitive bodies, perhaps among the most primitive objects of the solar system. They could still contain ices and organic compounds in proportions little changed since their formation, and thus provide an incredibly valuable window into the chemistry of the outer protoplanetary nebula. Unfortunately, observations of TNO surface reflectance and emission probe only a very shallow surface layer, which is subject to relatively rapid alteration by processes such as photolysis, radiolysis, and volatile loss and thus is unlikely to be representative of TNO interiors.

We want to learn about TNO interiors, but for the next decade at least, we are probably limited to a few spacecraft encounters and to determining masses for the sub-set of TNOs having natural satellites. Discoveries of six binary TNOs have been reported as of early 2002. Combining the masses computed from satellite orbits with thermal determinations of albedos and radii, it is possible to calculate densities for these objects. The possibility of mutual eclipses also offers opportunities to determine individual sizes, densities, and albedos.

Studies of binary TNOs are valuable in other ways as well. By analyzing the ensemble properties of TNO binary orbits, we can compare formation scenarios (e.g., catastrophic disruption from collisions, rotational fission, mutual formation) that would allow us to estimate the number density, size distribution, collisional history, and dynamical evolution of objects in the Kuiper belt region. This in turn would enable us to peer into the formation history of the giant planets and the outer solar system.

Any discovery/recovery observing strategy should include procedures to identify and characterize binary objects. High spatial resolution (sub-arcsecond) imagery is generally needed, and would also be valuable for detecting possible outbursts, perhaps indicating on-going loss of volatile species or impact disruption of volatile-rich subsurface pockets. In addition to continuing the search for

TNOs with satellites, objects found to have satellites need to be aggressively targeted via all available observational techniques.

Comet and Centaur studies: Because many of them originated in the Kuiper belt, and they pass much closer to Earth than TNOs, observational studies of comets and centaurs offer special opportunities to study the compositions and structures of TNOs. Unfortunately, comets and centaurs do not preserve a unique dynamical record of their region of origin, so they cannot be used to directly probe specific sub-populations of the Kuiper belt. Nevertheless, observations of these more nearby objects can reveal unparalleled compositional details, particularly for gas species in cometary comae. Measurements of specific isotopic ratios are also possible, as are sample return missions. A future compositional taxonomy of comets would provide a valuable starting point for efforts to classify TNOs, and it may eventually become possible to link certain types of comets with specific elements of the trans-Neptunian population, by means of compositional markers yet to be discovered.

3.4. Numerical/theoretical models

Much of our current understanding of the Kuiper belt is the result of intense efforts to model their interiors, surfaces, and orbital evolution. Modeling efforts will continue to play a crucial role in unraveling the questions outlined in Section 3, and must be a part of a balanced program of Kuiper belt studies during the next decade. The complexity of present-day models is directly limited by available computing resources, and can be expected to improve significantly as computer technology advances over the coming decade, if equipment upgrades are adequately funded.

Dynamics: Dynamical models have played an important role in our growing understanding of the Kuiper belt. Indeed, the first credible indication of the belt's existence came from efforts to model the origin of Jupiter-family comets in 1988. Models have also led to important conclusions that include: (*i*) orbits in the Kuiper belt are chaotic, (*ii*) the resonant structure of the Kuiper belt most likely has arisen from the outward migration of Neptune, (*iii*) there needed to be roughly 30 Earth-masses of material originally between 30 and 50 AU for large objects to grow, (*iv*) there was a scattered disk (before its discovery) and perhaps a fossilized scattered disk, and (*v*) the anomalously small eccentricity of Neptune may be the result of gravitational interactions with a primordial Kuiper belt. Understanding the origin of the dynamical structure of the Kuiper belt will rely on future dynamical models. These models will require not only simulations of the TNOs themselves, but also of formation and dynamical evolution of the giant planets.

Surfaces: Models of surface and interior evolution are needed to provide understanding of many processes which may have altered TNOs over the age of the solar system. As mentioned earlier, TNOs are relatively pristine remnants of the formation of the solar system. However, to understand what they can tell us about the chemistry of the protoplanetary nebula, we must first understand how they have evolved over the intervening time. Numerous processes are expected to have modified TNO surfaces, including accretion and erosion by impacting particles ranging from dust to many kilometer size, and loss of the most volatile species (such as CO) due to solar heating. Other effects of the radiation environment are also extremely important. In addition to visible solar photons, TNOs are bombarded by UV photons, solar wind particles, cosmic rays, etc., all of which drive structural and chemical evolution of TNO surfaces.

Interiors: The internal structure of TNOs is subject to modification by processes including the propagation of thermal waves, phase transitions such as the crystallization of amorphous H_2O ice, shock waves from large impacts, and effects of radiogenic heating and volatile transport. Efforts to model the influence of these types of processes on TNOs are essential. The possibility of differentiation critically depends on factors such as the amount of ^{26}Al available during and immediately following accretion, which is currently unknown. For C1 chondrite proportions of radioactive elements, coupled thermal and volatile-transport models suggest that the deep interiors of TNOs could be volatile-depleted by radiogenic heat (at least for the most volatile species, like CO) while surface layers could be devolatilized by solar heating. Mobilized species would partially escape and partially re-condense in cooler, intermediate strata, producing a complex layered structure including both volatile-depleted and volatile-enriched regions. These sorts of processes could result in strongly differentiated TNOs, but details are extremely sensitive to factors such as their rates of formation, shapes, sizes, thermal conductivities, porosities, and initial abundances of radioactive elements. Models are needed to explore these dependencies, and to work out probable interior/exterior structures of TNOs as functions of size, heliocentric distance, and rate of formation. Interior models will also be needed to interpret the transition from irregular to spherical shapes, when/if such a transition is discovered.

Radiative transfer: Finally, radiative transfer models are needed to interpret remote photometric and spectroscopic observations. Existing models will require extended capabilities to make them applicable to surfaces having very low albedos and small grain-sizes. Near-infrared reflectances and far-infrared emission have rich potential to provide compositional information about TNO surfaces, but until the necessary models are available, we will be severely handicapped in our ability to extract compositional information from spectral observations in these wavelength ranges.

3.5. Laboratory studies

Both observations and models of TNOs rely on material data from laboratory studies of outer solar system substances at appropriate cryogenic temperatures and thermodynamic conditions. Many critical data have never been measured, or have only been measured under conditions inapplicable to TNOs. Types of measurements urgently needed include measurements of optical, thermal, and mechanical properties, studies of radiolysis and photolysis, and investigations of gas-solid interactions. These measurements will lead to a better understanding of TNO compositions and of surface and internal processes affecting TNOs and can make significant contributions to the key science questions listed in Section 3.

Compositions: To extract compositional information from spectral observations of TNOs, data are needed for the optical properties of candidate species under the temperature and pressure conditions at TNO surfaces. Visible and near-infrared optical constants are needed for interpreting reflectance observations while far-infrared data are necessary for interpreting observations of thermal emission. Data for numerous ices, as well as photolytic and radiolytic products of common ices at TNO surface temperatures are particularly lacking at visible, near-infrared, and far-infrared wavelengths. Various mixtures of different species exhibit changes in the shapes and wavelengths of characteristic absorption bands of their constituent components, and data on these matrix effects are also in short supply.

Surfaces: To interpret color, albedo, or spectral distinctions between TNO classes we will need to understand the processes affecting their surfaces and how these processes modify observable reflectances. Processes expected to play important roles include photolysis, radiolysis, and impact resurfacing. Laboratory investigations of these processes are needed, if we are to interpret observations of the space-weathered surfaces of TNOs. These studies are difficult because time scales differ greatly between the laboratory and the outer solar system, so it is crucial to understand the limitations of extrapolations from laboratory time scales and doses. A better understanding of the chemical pathways whereby products of photolysis and radiolysis react to produce new species will lead to a clearer picture of surface compositions and how those compositions can change with time. Reaction rate constants for specific compounds are necessary for modeling their production over time.

Interiors: As with surfaces, interior compositions and structures result from formation conditions as well as the influence of various processes acting on TNOs subsequent to their formation. Models of these processes (see Section 4.4) depend critically on knowledge of material properties as inputs. Interior thermal and volatile transport models suffer from gaps in fundamental material properties data, with basic thermal parameters, rates of creep, grain growth, and deformation all being poorly constrained. Without appropriate equations of state, it is difficult to predict the results of large collisions in the trans-Neptunian region. Better understanding of the mechanical properties of ice will enable creation of more realistic models of accretion and fragmentation of TNOs. The phase state of H_2O ice is important for determining the amount of volatile gases (CO, CH_4) that may be trapped in the surface. Crystalline ice can effectively trap gases in clathrates in well measured ratios. However, if much of the water ice in TNOs is in the amorphous phase, it is unknown how much gas can be trapped. Since these gases are the source building blocks for complex organics, knowing the amount of gas available and the reaction pathways from photolysis and radiolysis can lead to predictions of abundances of numerous compounds, including many with potential astrobiological significance.

Facilities: Much of the existing laboratory work has been targeted at submicron interstellar dust grains, and thus has concentrated on temperatures too low for TNO surfaces and at mid-infrared wavelengths where reflectance observations of TNOs are nearly impossible. Also, the vast majority of existing studies have been performed on extremely thin (microns thick) films of ice. The physical characteristics of ices change drastically with the sample size, and processes like gas diffusion and radiolysis have length scales much longer than a micron. The H_2O amorphous–crystalline transition temperature appears to vary with the deposition environment and size of the sample. Working with larger samples can be difficult because of the temperature and pressure conditions needed, but these obstacles must be overcome to obtain the data needed for better predictions. Resources will need to be made available for upgrades in laboratory facilities so that the required sorts of studies can be done.

3.6. Environment

Past and present trans-Neptunian environments have yet to be well characterized. In-situ particles, fields, and dust measurements by existing and future spacecraft are essential for learning about this region, and need to be supported (see next section). We do not yet know enough about the environment beyond 30 AU to relate TNO surface properties to environmental factors, such as possible regions of preferential bombardment by dust or larger impactors, episodes of passage through the heliopause, external perturbations (radiative, dust, gas, or gravitational), etc. Observational and theoretical studies of these problems will enable us to understand TNOs in the context of the external conditions they are exposed to.

A potentially key factor in the chemical processing of TNO surfaces is the incidence of cosmic ray energetic particles arising from solar flares, the outwardly expanding solar wind and its heliospheric interaction with the local interstellar medium, as well as external, galactic sources. Expected increases in cosmic ray intensity near and beyond the heliospheric termination shock boundary somewhere near 90 to 100 AU are of special interest for their potential impact on SDOs.

Dust is likely ubiquitous throughout the Kuiper belt due to collisions among TNOs. In-situ dust measurements by spacecraft could conceivably push the solar system's known frontier outward dramatically, and would more firmly establish the suspected connection between Kuiper belts and the dusty disks observed around other stars.

3.7. Spacecraft exploration

The first spacecraft exploration of the Kuiper belt could occur soon after the close of the coming decade, if NASA's Pluto-Kuiper Belt (PKB) mission goes forward in a timely fashion. We endorse this mission vigorously, since it would substantially advance Kuiper belt science in ways completely impossible from Earth-based observations. For instance, spectra obtained by PKB would map regional surface compositions and could perhaps even detect tenuous atmospheres or exospheres. Close-up images revealing surface morphology would shed considerable light on the types and relative rates of processes shaping TNO surfaces. Material excavated in larger impacts could offer unique glimpses into TNO interiors, and comparisons between ejecta from older and younger craters could teach much about the evolution of interior materials following their exposure on TNO surfaces. Observations of the frequency and size distribution of impact craters could reveal the small end of the TNO size-frequency distribution.

As mentioned in the previous section, an on-board dust detector could be used to map the radial extent of the Kuiper belt, and such an instrument would be a valuable addition to the PKB payload. We note that no dust detector was included in the New Horizons proposal selected by NASA for PKB Phase B, and we recommend that one be incorporated into the flight instrument package if possible.

A potential next step after PKB might be the Interstellar Probe (IP) mission, currently under consideration by NASA. IP could completely traverse the Kuiper belt within its mission life span, and thus could offer a great opportunity to learn about radial distributions of dust, particles, and fields, especially in the more distant parts of the Kuiper belt. By extending measurements by the Pioneer, Voyager, and PKB space probes to $\sim$200 AU, Interstellar Probe could investigate the full heliocentric range of space environments affecting surfaces of most known TNOs.

It would constitute a major setback for subsequent decades of outer solar system studies if the PKB mission and its successors were not launched in a timely manner, or if Kuiper belt portions of upcoming missions were de-scoped. We express the strongest possible support for NASA's moving ahead promptly with adequate funding for spacecraft exploration of Pluto and the Kuiper belt beginning with PKB, ushering in a new era in which open competition and new technologies minimize costs while achieving spectacular scientific returns in the outer solar system.

The Future of Solar System Exploration, 2003-2013
ASP Conference Series, Vol. 272, 2002
Mark V. Sykes, ed.

Radio Science Investigations with the Deep Space Network

S.W. Asmar[1], K. Aksnes[2], J.D. Anderson[1], J.W. Armstrong[1], D.H. Atkinson[3], J.-P. Barriot[4], B. Bertotti[5], M.K. Bird[6], S.J. Bolton[1], R. Dutta-Roy[6], P. Edenhofer[7], W.M. Folkner[1], B. Häusler[8], L. Iess[9], E.A. Jensen[10], A.S. Konopliv[1], E.R. Kursinski[11], E.A. Marouf[12], M. Pätzold[13], D. Plettemeier[7], N.J. Rappaport[1], G.M. Resch[1,†], M.A. Slade[1], D.E. Smith[14], & G.L. Tyler[15]

[1] *Jet Propulsion Laboratory, California Institute of Technology, Pasadena, CA*
[2] *Institute for Theoretical Astrophysics, University of Oslo, Oslo, Norway*
[3] *Department of Electrical and Computer Engineering, University of Idaho, Moscow, ID*
[4] *Centre National D'Etudes Spatiales, Toulouse, France*
[5] *Dipartimento di Fisica Nucleare e Teorica, University of Pavia, Pavia, Italy*
[6] *Radioastronomisches Institut, University of Bonn, Bonn, Germany*
[7] *Institut für Hochfrequenztechnik, University of Bochum, Bochum, Germany*
[8] *Institut für Raumfahrttechnik, Universität der Bundeswehr München, Neubiberg, Germany*
[9] *Dipartimento Aerospaziale, Universita' "La Sapienza," Rome, Italy*
[10] *Institute of Geophysics and Planetary Physics, University of California, Los Angeles, CA*
[11] *Department of Atmospheric Sciences, University of Arizona, Tucson, AZ*
[12] *Department of Electrical Engineering, San Jose State University, San Jose, CA*
[13] *Institut für Geophysik und Meteorologie, University of Cologne, Cologne, Germany*
[14] *NASA Goddard Space Flight Center, Greenbelt, MD*
[15] *Center for Radar Astronomy, Stanford University, Stanford, CA*

Abstract. Scientists utilize the telecommunication links between spacecraft and Earth to examine changes in the phase/frequency, amplitude, line-width, or polarization, as well as round-trip light time, of radio signals to investigate: planetary atmospheres and ionospheres, planetary rings, planetary surface characteristics, shapes, gravitational fields, orbital motion and dynamics of solar system bodies, magnetic fields of the Sun and planets, the solar wind and corona, cometary atmospheres, grav-

---

[†] This report is dedicated to the memory of George M Resch, who passed away on November 22, 2001.

itational waves, gravitational redshift, relativistic time-delay, and other phenomena. Recommendations are made to maintain and expand the existing strong science capabilities in the network, carry them into the next generation DSN, continue advancing the technology to improve sensitivity, and to enable future experiments via an efficient multi-mission system that reduces cost via remote and automated operations.

REPORT

With strong international collaboration, Radio Science experiments have been conducted by almost every deep space mission with numerous published discoveries that address the NASA themes: Exploration of the Solar System, Structure and Evolution of the Universe, and the Sun and Earth Connection, as shown in these selected examples[§]:

- First estimate of Martian surface pressure
- Determination of ring structure and particle size distribution for Saturn and Uranus
- Measurement of Mars atmospheric density from spacecraft drag
- Derivation of the pressure and temperature profiles of the atmosphere of Mars over the annual cycle
- First measurement of surface pressure and detection of ionospheres of Titan and Triton
- Detection of ionospheres on the Galilean Satellites
- Measurement of the electron column density latitude profile of the Io Plasma Torus
- Discovery of lunar mascons
- Determination of masses and modeling of bulk compositions of Jupiter, Saturn, Uranus, Neptune, Titan, and Triton, Mars, Venus, and Moon
- Modeling the mass distribution in the interiors of the large moons of Jupiter
- High resolution gravitational fields of the Moon, Mars, Venus
- First detection of gravity field variations on a planet (Mars)
- First measurement determining the existence of a dense core on Mars from radio tracking of the Viking and Mars Pathfinder landers

---

[§]For a complete list of discoveries and publications, see http://radioscience.jpl.nasa.gov or contact sami.asmar@jpl.nasa.gov .

Figure 1. A 34-meter diameter station of the Deep Space Network with advanced Radio Science capabilities including simultaneous reception and transmission at X- and Ka-bands as well as advanced pointing techniques, frequency and timing standards. The inset shows the advanced water vapor radiometer adjacent to the station for the purpose of calibrating the earth troposphere.

- First gravity model of an asteroid
- Measurement of drag deceleration in the comae of comets Halley and Grigg-Skjellerup
- Description of large-scale solar coronal structure and electron densities in streamers and holes
- First evidence for the acceleration of coronal mass ejections far from the Sun
- Profile of deep winds on Jupiter via the Galileo probe Doppler Wind Experiment
- Surface characteristics of Venus, Moon, and Mars via bistatic radar experiments

Figure 2. Illustration of a Radio Science atmospheric occultation at Mars with the Mars Global Surveyor spacecraft: a radio beam travels through the atmosphere on the way to a receiving station on earth.

Radio Science has been a particularly productive field of investigation at the three complexes of NASA's Deep Space Network (DSN) comprised of 70-m and 34-m diameter stations with capabilities for transmission and reception at S- and X-bands. These capabilities have been augmented at one station at Ka-band frequencies for Radio Science experiments, along with advanced systems for pointing and media calibration. The limits of sensitivity of these experiments are based on: the frequency stability, the amplitude stability, the signal-to-noise ratio, navigation accuracy, and the ability to calibrate the intervening media.

Radio Science experiments utilize ultra-stable one-way or coherent two-way transmission with ground reception via open-loop or tracking receivers. The former provides for superior phase stability and more flexible handling of signal dynamics and multi-path as well as offering the capability for creative post-processing (also used for special spacecraft maneuvers). The tracking receiver provides Doppler and range data. DSN Radio Science also utilizes the frequency and timing system, which can include a hydrogen maser, super-conducting cavity maser oscillator or other devices, and a media calibration system that includes an advanced water vapor radiometer, a microwave profiler, meteorological and

GPS data (Figure 1, inset). Pointing techniques include blind pointing, conical scan, monopulse, and aberration correction.

The goals of the Radio Science user community are to maintain and enhance the existing strong science capabilities in the network and carry them into the next generation DSN, including arrays, optical links, and orbiting stations, to continue advancing the technology to improve the sensitivity, and to enable future experiments via an efficient multi-mission system that reduces cost via remote and automated operations. Towards these goals, we recommend:

- Expand the Ka-band system to the overseas DSN complexes, with accompanying pointing and media calibration systems.
- Implement an advanced ranging system that utilizes Ka-band.
- Increase occultation experiment sensitivity by expanding the capability for uplink Radio Science experiments using highly stable high power transmitters.
- Expand the capability to receive and transmit multi-frequency links on future missions and the DSN in order to calibrate or measure the charged particle effects for propagation experiments (e.g., Figure 2).
- Implement dual polarization receiving systems to 34-m stations.
- Support the program development of advanced ultra-stable oscillators, transponders, on-board open-loop receivers and signal processors, and the technology for quiet and precisely pointed spacecraft.
- Support the development of technology for spacecraft-to-spacecraft links, including multiple planetary orbiters, relay satellites from the backside of the Moon or Mars, as well as probes links to orbiters.
- Place spacecraft with Radio Science instrumentation in orbits that advance understanding of the solar system such as permanent solar occultation orbits for solar and relativity investigations.

The Future of Solar System Exploration, 2003-2013
ASP Conference Series, Vol. 272, 2002
Mark V. Sykes, ed.

Solar System Astrometry

D. Pascu[1], T.J. Johnson[1], J.R. Rohde[1], R.C. Stone[1], N. Zacharias[1], J.D. Giorgini[2], R.A. Jacobson[2], E.M. Standish[2], B.G. Marsden[3], & L.C. Ball[4]

[1] *U.S. Naval Observatory, Washington, DC*
[2] *Jet Propulsion Laboratory, California Institute of Technology, Pasadena, CA*
[3] *Smithsonian Astrophysical Observatory, Cambridge, MA*
[4] *Emerald Lane Observatory, Decatur, AL*

Abstract. The objective of solar system astrometry is the regular and accurate ground-based observation of positions and motions for all objects in the solar system – including spacecraft. We describe the status of the four principal observational techniques, especially the remarkable gains in accuracy of the last 10-20 years. Even though, this activity is indispensable to the support of several important applications, such as spacecraft navigation and collision hazard assessment, it will fall short of its main objective unless access to specialized and large telescopes, and to radar and VLBI facilities is increased to fulfil growing requirements. New facilities and continued support of astrometric observations are recommended.

EXECUTIVE SUMMARY

We describe the important but often overlooked ground support activity of astrometry in the solar system. This involves the regular, accurate observation of positions and motions for all major planets and their moons, minor planets (and their moons), Near Earth Objects (NEOs), Potentially Hazardous Asteroids (PHAs), Transneptunian Objects (TNOs) and comets. These observations support several key activities in solar system studies, such as: spacecraft navigation, impact hazard assessment, dynamical and structural properties of the solar system, and gravitational and relativity studies. Observations in the tangent plane, Very Long Baseline Interferometry (VLBI), radar and Lunar Laser Ranging (LLR) are the four principal techniques used for Solar System Astrometry.

The classical tangent-plane (focal-plane) methods are currently applied to all the above targets except the terrestrial planets and the Moon. Huge numbers of observations of primitive bodies are received by the Minor Planet Center each month. These observations come in large part from the large survey projects and from amateurs, and their accuracy ranges from 0.2 to 2 arcsec. In the U.S., the outer planets and their major satellites are regularly observed only at the U.S. Naval Observatory. The accuracy of these observations ranges from 0.05 to 0.15 arcsec. Contributing to the improved accuracy in recent years has been the introduction of high-precision and high-density star catalogs, and the introduction of CCD detectors.

The most accurate positional observations of the planets are obtained by differential VLBI (D-VLBI) measurements of spacecraft in the vicinity of a planet. Accuracies in RA and Dec range from 2 to 10 mas. These observations are taken directly with respect to the International Celestial Reference Frame (ICRF), which serves to orient the planetary ephemerides into the ICRF.

Radar delay and Doppler astrometry provide a powerful method of obtaining orbital knowledge of solar system bodies. Radar astrometry provides dynamically fundamental range and range-rate measurements that are orthogonal to spherical equatorial angles. Astrometry of Mercury, Venus and Mars is now exclusively radar and/or VLBI. Planetary radar range measurements show a scatter of a few hundred meters, due entirely to the planet's topography. For asteroids, shape reconstructions allow range determination to better than 10 meters and 0.1 mm/s in radial velocity. Radar commonly improves the accuracy of optical-only ephemerides by factors of 3 to 10.

McDonald Observatory, USA, and Grasse Observatory, France, have produced LLR observations in a routine and continuous way for more than three decades. These observatories are currently the only two in the world that still regularly produce LLR observations. Each routinely ranges with accuracies of 1-3 cm and each producing between two to three hundred normal points from their LLR observations. These observations are sufficient for the precise calculation of the Earth-Moon orbit.

Key science issues in astrometry are both direct and supportive:

- The status of the celestial coordinate system is critical to the accuracy of space navigation. While the accuracy of D-VLBI increases with the number and proximity of the ICRF sources, there are large gaps in the ICRF - especially along the southern ecliptic. Also, radio systems for planetary probes are moving to the Ka-Band (32 GHz) and a celestial reference frame will be required at sub-milliarcsecond accuracy to support deep space navigation with D-VLBI. In the optical, the need for accurate astrometry of fainter objects requires pushing the magnitude limit on star catalogs to fainter magnitudes.

- Radar observations can improve the accuracy of optical-only ephemerides of NEOs and PHAs by an order of magnitude. All NEOs and PHAs should be observed by radar, when possible, yet, only 4% of radar telescope time has been made available for such observations.

- Space reconnaissance now requires greater ephemeris precision to support more complex mission operations. Routine and continuous observations are needed to provide high-quality data from which these navigational ephemerides are produced. At present, such observational activity is only about a third of that which supported the Voyager missions.

- LLR observations have many scientific applications. Without continuous LLR data, the lunar ephemeris would be set back to pre-1970's accuracies with errors on the order of 10's to 100's of meters – insufficient for the present level of gravitational and relativistic studies.

- Interest has increased for a multitude of fainter, distant objects and those faint, rapidly moving NEOs. It is not now possible to keep up with demands for accurate and regular astrometric observations.

Recommendations for the next ten years are prioritized as follows:

- VLBI should be continued for densification of the ICRF and its extension to higher frequencies. In the optical, we recommend that an advanced ground-based telescope be built to produce a star catalog with a target accuracy of 20 mas at V=18 and 100 mas at V=21.
- A minimum of one radar facility should be built for the dedicated observation of solar system bodies, particularly PHAs and NEOs.
- Three domestic facilities should make regular astrometric observations of the planets and their major satellites.
- Support is needed for both current and planned LLR operations and station upgrades.
- One 4-meter class telescope in each hemisphere should be dedicated to the routine observation of very faint primitive bodies.

REPORT

1. Targets Under Panel's Purview

All major planets and their moons, minor planets (and their moons), Near Earth Objects (NEO), Potentially Hazardous Asteroids (PHA), Transneptunian Objects (TNO) and comets are included.

2. Objective

The objective is the continuous and accurate observation of positions and motions of these objects using ground-based observational techniques ranging from advanced classical tangent-plane methods to the most modern – radar, Very Long Baseline Interferometry (VLBI) and Lunar Laser Ranging (LLR). The observables include angular measures (RA, Dec, or ΔRA, ΔDec), time-delay (range) and Doppler (range rate).

Although the importance of this activity appears obvious, it is often neglected, resulting in inadequate orbital coverage for these objects. Few, if any, of the solar system bodies are observed astrometrically with any regularity or continuity or, for that matter, with state-of-the-art techniques. The primary reason for this shortcoming is simply the unavailability of the proper instrumentation for this application on a continuous basis.

The primary activities that astrometric observation supports in the solar system are:

- Spacecraft navigation,

- Impact hazard assessment,
- Fundamental dynamical and structural properties of the solar system,
- Gravitational, relativity and stability studies.

A fifth item critical to the accuracy of astrometric observations in the solar system is the status of the celestial reference frames (Johnston and de Vegt 1999).

Examples of these applications are given below in Section 3 and the current status of astrometric solar system observation for each of the observational techniques is found in Section 4. The consequences of the neglect of astrometric observations are addressed in Section 5, and our recommendations for their redress are given in Section 6.

3. Importance

Astrometric observations support the determination of fundamental solar system parameters, such as its scale and the mass of its constituents. Radar observations, for example, were used to determine the Astronomical Unit (AU) 30 years ago and the use of LLR data has lead to a four-orders- of-magnitude improvement in the accuracy of the Moon's distance from Earth (Dickey et al. 1994). And currently, astrometric observations of minor planets are being used to determine their masses from their mutual perturbations and from observations of their satellites.

This activity is indispensable to the construction of accurate ephemerides of the above targets. While these ephemerides support Earth-based and HST observations of the bodies for a variety of scientific studies, ephemerides of the highest accuracy, which are generated at JPL, are used for precision navigation of spacecraft to and around those targets. All U.S. space missions to the major planets and their moons, to minor planets and to comets have depended on these ephemerides for their success.

Astrometric observations support gravitational studies: radar refinement of planetary orbits provided the initial verification of second-order longitudinal terms in the special relativistic Doppler formulation and the non-Newtonian retardation of electromagnetic signals that pass by a massive body. And LLR data have been used for over 25 years to test and refine our understanding of the Strong Equivalence Principle (Shapiro et al. 1976).

Astrometric observation supports the detection and impact hazard assessment of NEOs (Giorgini et al. 2002, Ostro et al. 2002). While the detection and early astrometric observations are classical tangent-plane type, delay- Doppler data typically open up a useful close-approach prediction window (where the statistical uncertainty in time of close-approach is 10 days or less) of decades or centuries for single apparition asteroids, as compared to several years for an orbit solution based only on optical data (Ostro et al. 2002). For example, 2001 CP36 close approaches could meaningfully be predicted only between 1989-2004 until radar astrometry opened a prediction window from 1718-2225. This reduction of uncertainty allows more accurate assessment of impact hazards for radar-based orbits (Giorgini et al. 2002). Radar observations can also reduce the area of sky that needs to be searched for a next-apparition optical recovery by factors of 20 to 100 (Ostro et al. 2002). Thus, radar observations of all PHAs and NEOs are recommended when supported by adequate signal-to-noise ratios. Radar commonly improves the positional accuracy of optical-only ephemerides by reducing uncertainties by factors of between 3-10 (Ostro et al. 2002). Even for asteroids with long astrometric data arcs, radar measurements can shrink positional uncertainties significantly (Benner et al. 1997, Giorgini et al. 2002, Ostro et al. 1999, Ostro et al. 2002). Cumulative delay-Doppler astrometry can be used to determine an asteroid's three-dimensional shape and spin (Hudson 1993). This improved knowledge can greatly aid spacecraft navigation for future missions or potentially enable new missions that might not otherwise be considered.

In addition, astrometric observations have supported a wide spectrum of studies relevant to the dynamical evolution of the solar system, such as the delineation of asteroid families and structure of the asteroid belt.

4. Current Status of Astrometric Observing Programs

4.1. Tangent-plane astrometry

With this technique, spherical equatorial positions (RA, Dec) of solar system objects are derived from area detector (photographic plate, CCD, CMOS) images representing the plane tangent to the celestial sphere. Star images on the detector, having catalog positions, are used to perform a "plate solution" (König 1962). This is an interpolation scheme, using an appropriate mapping model often involving a power series in the measured coordinates. The accuracy of such observations has been improved in recent years through the introduction of high-precision star catalogs, such as Hipparcos (ESA 1997) and Tycho-2 (Høg et al. 2000), and high-density catalogs, such as USNO-A2.0 (Monet 1998) and GSC-II (Morrison et al. 2001). In addition, the new, highly accurate and dense USNO CCD Astrograph Catalog (UCAC) (Zacharias et al. 2000) now provides reference stars for observations at the 20 mas level. It is now possible, in principle, to obtain tangent- plane observations for most solar system bodies with an accuracy of 50 mas (Fienga et al. 1997). This was not possible a decade ago, when the accuracy was limited by the accuracy and density of the reference star catalogs. The accuracy of the observation is now limited by the measurement and representational errors of the object itself (Pascu et al. 1997). In large part, the improved accuracy is also due to the introduction of solid state detectors, especially for faint objects. Not only is it now possible to obtain high signal-to-noise images of the faintest objects, but most of the faint stellar images on the narrow field CCD now have reference catalog positions, permitting a plate solution.

While specialized astrometric techniques are applied to a select set of solar system bodies, usually with a substantial gain in precision, the tangent-plane observation is still the backbone of solar system astrometry: The great majority of bodies are observed in this way, and the technique is currently applied to all objects, except the Sun, Moon, and Terrestrial Planets.

Primitive bodies: By far, the largest astrometric program is for primitive bodies – main belt minor planets, Near Earth Objects (NEOs), Transneptunian Objects (TNOs), comets and faint outer satellites. In September 2001 alone, 720,000 positional observations of these objects were received and distributed by the Minor Planet Center. These observations resulted in 30,000 new and updated orbits (Marsden 2001). While the Minor Planet Center possesses the facilities to carry out this work, there does exist the problem of a shortage of manpower that needs urgently to be addressed by the international community. Although 60% of the observations came from the LINEAR project, as many as 150 facilities, including foreign and amateur, contributed to the total. The accuracy of these observations varies widely – from 0.2 to 2 arcsec. However when one of these bodies becomes of interest because of a planned spacecraft rendezvous or collision hazard assessment, advanced methods could be used to improve the astrometric precision to better than 80 mas, and radar techniques are used when possible.

Major planets and their satellites: Unfortunately, the astrometric programs for the major planets and their inner satellites are not as robust. Only the Naval Observatory has a reasonably comprehensive planet and satellite observing program, though somewhat more modest than in the past (Monet and Monet 1992, Pascu 1996, Pascu et al. 1998, Stone 2001). These observations are supplemented sporadically from abroad by a handful of sites. This is a substantial retreat from the heyday of the early 1970s when NASA funded such programs in support of the Voyager missions (Seidelmann 1979). At that time, astrometric observations of the planets were being made continuously with two transit instruments at USNO as well as by a number of such instruments abroad, and continuous observations of the satellites were made from at least three sites in the U.S.

4.2. Phenomena: "Photometric astrometry"

This refers to the application of photometry to observe astrometric phenomena. Included in this class are the planet/satellite eclipses and occultations, and the satellite/satellite (mutual) phenomena (mutual eclipses and occultations) of the Galilean satellites of Jupiter (and now Saturn also). It also includes stellar occultations and the Pluto/Charon mutual events. While the observations are photometric, they determine a specific geometric configuration (e.g., minimum separation) at a given instant. These data can be used to improve the orbits of the satellites (and the masses in the case of Pluto/Charon). The mutual events, in particular, are quite accurate but rare: every 6 years for the Galileans, every 15 years for the Saturnians, every 124 years for Pluto and Charon. The French IMCCE office has organized international campaigns to observe the Galilean mutual events since 1979 (Arlot et al. 1982).

4.3. Very Long Baseline Interferometry (VLBI)

The most accurate angular positional observations of the planets are now obtained by VLBI techniques. The differential VLBI technique, or D-VLBI, involves the measurement of a radio signal from a spacecraft, differentially with respect to one or more nearby International Celestial Reference Frame (ICRF) radio sources. Coupled with the planetocentric position of the spacecraft, this automatically orients the resulting planet's ephemeris onto the inertial ICRF reference frame (Standish 2000). Three antennas are needed to provide two non-parallel baselines in order to give two-dimensional positions. JPL's Deep Space Network (DSN) is usually used, with antennas at Goldstone (California), Madrid (Spain) and Canberra (Australia).

In addition to angular measure, a spacecraft with a transponder can also provide a highly accurate distance measurement between Earth and the spacecraft by measuring the round-trip light-time of a signal to and from the spacecraft. Again, this gives a tight constraint upon the planetary ephemeris, when used in conjunction with an accurate planetocentric spacecraft ephemeris.

The measurement accuracy of a single VLBI observing session is typically 8-10 mas (Folkner et al. 1996, Standish 2000); a number of such sessions establishes the planetary ephemeris at the 1-2 mas level. Such accuracy is possible due to the accuracy of the ICRF sources, which is 0.25 mas. To date, VLBI observations have been made for Venus, Mars (Standish 2000), Jupiter (Folkner et al. 1996) and Saturn (Border et al. 1982).

VLBI has two additional supportive roles: It is the fundamental technique used in defining and maintaining the ICRF catalog, and it is also fundamental for the determination of the Earth's orientation in space with respect to the ICRF.

4.4. Radar astrometry

Radar delay and Doppler astrometry provide a powerful method of obtaining orbital knowledge of solar system bodies (Butrica 1997, Ostro 2002). Radar astrometry provides dynamically fundamental range and range-rate measurements that are orthogonal to spherical equatorial angles (Ostro 2002). Astrometry of Mercury, Venus, and Mars is now exclusively radar and/or VLBI. Radar observation of these planets show a scatter of a few hundred meters, due entirely to the planet's topography. Though this seems like a large uncertainty, it is comparable to sub-milliarcsecond angular measurement.

For asteroids, range can be determined at Goldstone (DSS-14) and Arecibo to as fine as 7.5 meters and better than 0.1 Hertz in radial velocity (1.7 mm/s at the 8560 MHz X-band Goldstone/DSS-14 frequency) (Ostro 2002). As a result of such high fractional precision, where the measurement uncertainties are typically less than 10^{-8} the size of the measurement itself, a single radar measurement can secure knowledge of the orbit of a newly discovered asteroid for many decades (Ostro et al. 2002).

Radar interferometric observations of 6489 Golevka, with Goldstone/DSS-14 illuminating the asteroid and the Very Large Array observing the reflected signal, have produced angular astrometric plane-of-sky positions comparable to the best Hipparcos-reduced ground-based astrometry ($\sim$0.2 arcsec) (Ostro et al. 2002). Bistatic observations have been made in which Arecibo or Goldstone transmits and the echo is received at the other site or at third locations such as DSS-13 or Greenbank in the U.S. or, in one experiment, Kashima and Usuda in Japan, and Russia's Evpatoria site. Bistatic radar operations allow continuous observation such that a single site need not repeatedly switch between transmit and receive modes. Historically, about 3% and 5% of the time at DSS-14 and

Arecibo respectively has been used for asteroid radar. It has only been possible to obtain time on these heavily over-subscribed multi-purpose instruments to observe 3-4 targets per month, on average.

4.5. Lunar Laser Ranging (LLR)

The first successful Lunar Laser Ranging (LLR) observation was made in 1969 between the 3.1-m telescope at Lick Observatory and the Apollo 11 retro- reflector package located at Tranquillity Base. Over the past 30 years, LLR observations have been made at fewer than a dozen sites. However, only McDonald Observatory, USA, and Grasse Observatory, France, have produced LLR observations in a routine and continuous way for more than a decade. These two observatories are currently the only two in the world that still regularly produce LLR observations. Using a dedicated 0.76-m telescope with a 120 mJ per pulse, 200-ps pulse length, frequency-doubled neodymium-yttrium aluminum-garnet (Nd-YAG) laser, McDonald Observatory routinely ranges the Moon with accuracies of 1 cm. From January 1, 2001 to October 31, 2001, McDonald produced 153 normal points from their LLR observations. Grasse Observatory (CERGA) using a 1.5-m telescope and a 400 mJ per pulse, 300-ps pulse length, frequency-doubled Nd-YAG laser has produced ranges with accuracies approaching 0.5 cm and produced 225 normal points for the same time period.

About 15 years ago, LLR observations had the accuracy of 18 cm. Today, we consider a weighted RMS of 1.8 cm a reasonable value. However, considering the technological advances being made in computers, lasers, fiber optics and timing systems, tomorrow's stations could be making observations with accuracies of 1 mm or possibly less. These improvements would lead to an increase in the number and quality of the observations. Furthermore, an increase in the number and global distribution of stations would allow for an improved separation of lunar parameters and polar motion.

Currently, there are two new LLR stations in the planning stages. One in Matera, Italy and the other is part of a program called APOLLO (the Apache Point Observatory Lunar Laser-ranging Operation). The APOLLO program is being run by scientists at the University of Washington (personal communication, T.W. Murphy, Jr. 2002). The APOLLO program will utilize a 3.5-m astronomical telescope, advanced avalanche photodiode (APD) arrays as a detector, and a state-of-the-art timing system. APOLLO's photon detection rate will exceed that of current systems by more than two orders of magnitude, allowing sub-millimeter range precision. The APOLLO system will also be capable of performing laser ranging to Mars through the use of remote laser transponders.

The lunar ephemeris relies very heavily on LLR data. The Moon's elliptical orbit is strongly distorted by solar attraction. The instantaneous eccentricity of the Moon's orbit can vary between 0.03 and 0.07 (Newhall et al. 1983). Compared to the classical optical data, the use of LLR data has resulted in at least a two-orders-of-magnitude improvement in the accuracies of the orbit's orientation and a four-orders-of-magnitude improvement in the accuracies of the Moon's radial distance. However, the accuracies achieved using LLR data rapidly degrade when the orbit is extrapolated outside the span of the LLR observations (Dickey et al. 1994), motivating the need to continue a program of LLR observations.

LLR observations are sufficient for the precise calculation of the mass ratio of the Sun/(Earth + Moon), the relative orientation of the Earth-Moon system in its orbit about the Sun, the mutual orientation of the planes of the Earth's equator, the lunar orbit, and the ecliptic, the dynamical equinox, obliquity of the ecliptic, testing of the Strong Equivalence Principle, variations in the time variation of the gravitational constant (or strength of gravity), Earth's Rotation (both polar motion and length of day), the Earth's precession and nutation (Williams 1994), the Moon's tidal accelerations, relativistic geodetic precession of the lunar orbit, lunar rotational dissipation and the Moon's low-degree gravitational coefficients. LLR data also allow the orientation of the planetary ephemerides onto the fundamental astronomical reference frame at the milliarcsecond level (Dickey et al. 1994).

LLR research also has practical applications to both space and terrestrial navigation. For example, precision navigation requires a correction for the variations of the Earth's axial rotation rate (UT1). The correction for this variation can be determined from LLR observations.

5. Key Science Issues

Direct science issues have to do with the methods of obtaining and improving astrometric observations, such as detection, measurement and reduction. Supportive science has to do with how these observations are obtained and used to improve the orbital knowledge and ephemerides of solar system objects, and how such improvements can provide new knowledge of the structure, dynamics, origin and evolution of the solar system and its constituents.

A. For direct science, the principal need is the maintenance and improvement of the celestial coordinate system. The accuracy of D-VLBI for space navigation is substantially improved with number and proximity of the ICRF sources used. However, there are large gaps in the ICRF, especially along the southern ecliptic (Fomalont 2002). In addition, planetary probe radio systems will be moving to higher frequencies, which will be used for navigation in future space missions. The Ka-Band (32 GHz) will be used, beginning with the Mars '05 mission (Jacobs et al. 2002). This will require that the ICRF be extended to higher frequencies to support deep space navigation by D-VLBI.

   For the optical domain, this involves the increase in accuracy and in density of star catalogs used for solar system astrometry. While FAME, DIVA or GAIA will provide this in the decade following 2003-2013, ground-based work will have to suffice for the interim. Because of the small scale of CCD detectors, and the continued erosion of star catalog accuracy due to proper motion errors, star catalogs must be continuously reobserved, densified and pushed to fainter limiting magnitudes.

B. The most pressing issue for support science is obtaining enough observing time, especially on large telescopes, and on radar and VLBI facilities for adequate coverage of faint object motions. This is especially the case now and in the immediate future as our interest turns to fainter objects and the number of known objects grows geometrically.

C. For impact hazard assessment, routine and continuous observations of the NEOs and PHAs are crucial. Delay-Doppler measurements of these objects can provide decisive orbit information and, thus, are of great importance (Giorgini et al. 2002). Cumulative delay-Doppler astrometry also permits shape and spin solutions for asteroids (Hudson 1993, Ostro et al. 1999), simplifying spacecraft mission design and potentially reducing the cost of missions to such asteroids. As mentioned earlier, DSS-14 and Arecibo are heavily over-subscribed and only 3-4 targets per month have been observed on average. This is inadequate for the needs of impact hazard assessment.

D. There are two types of accuracy requirements for spacecraft missions: those which apply to the navigation of the spacecraft and those which apply to the pointing of instruments onboard the spacecraft. Cassini, for example, has repeated Titan flybys and will be observing other satellites with instruments that have small fields of view. For these maneuvers, the mission needs precise ephemerides. Moreover, because the mission would like to reduce the cost of continued optical navigation, ephemerides are required which will hold up for the duration of the mission. To get a navigational ephemeris that does not depend on optical navigation requires substantial high-quality ground-based observational support. While the most recent observations are the most important for this application, the most accurate ephemerides are for those bodies with extensive observational histories – demonstrating the value of continuous observation. At present, such observational activity is only about a third of that which successfully supported the Voyager mission.

E. LLR observations have many scientific applications. Without continuous LLR data, the lunar ephemeris would be set back to pre-1970's accuracies with errors on the order of 10's to 100's of meters. It would not be possible to test the Strong (and Weak) Equivalence Principle and achieve today's high accuracies, parameterized post-Newtonian parameter γ would not be constrained to 0.03%, and the time variation of Newton's gravitational constant would not be constrained to $< 10^{-12}$ fractional change per year. Upgrades to current stations and additional stations, like APOLLO, are needed to improve these accuracies. Finally, additional stations would allow for a better separation of lunar parameters and polar motion.

F. There is also a great deal of science to be learned in the outer solar system. We are still only scratching the surface of our knowledge of the types of orbits these objects pursue in the transneptunian region and how these objects interact with those in the Jupiter-Neptune region, including the presumably captured outer satellites of these planets. Very precise observations will also allow one to study better the effects of forces of a nongravitational nature, such as those due to outgassing for comets and the Yarkovsky effect on minor planets. Since these bodies are very faint, tangent-plane astrometric observations require the use of large telescopes.

6. Recommendations

In order of importance:

- Since all spherical equatorial observations require accurate reference catalogs, we recommend the continued support of VLBI observations for increasing the density of the ICRF, for its extension to higher frequencies, and for the determination of the Earth's orientation in space with respect to the ICRF. We also recommend that optical reference catalog's density and magnitude limit be increased to accommodate accurate astrometry of the faint NEOs, TNOs, comets and satellites. In particular, the UCAC needs to be completed soon, and a new astrometric instrument needs to be built which can produce a catalog with a target accuracy goal of 20 mas at V=18 and 100 mas at V=21 (addresses Science Issues A, C and F).

- A minimum of one radar facility should be built for the dedicated observation of solar system bodies. In particular, we recommend that radar observations be made for all PHAs and NEOs when supportable by adequate signal-to-noise ratios (addresses Science Issues B and C).

- Three domestic facilities should provide continuous and systematic tangent-plane astrometric observations of the planets and their major satellites, with an accuracy of 50 mas, using state-of-the-art methods. With the increased accuracy demands of spacecraft reconnaissance, and the threefold increase in the number of known satellites in the past 20 years, this would be the minimum needed to provide the same support level provided for the Voyager missions (addresses Science Issues B and D).

- Support of both current and planned LLR operations needs to be continued along with research into what upgrades can and should be made to the existing LLR stations. This funding should include monies for the production of value-added data products such as UT1 variations. The determination of the benefits and expenses of additional LLR stations should also be addressed (addresses Science Issue E).

- A 4-meter class telescope in each hemisphere should be dedicated to the routine observation of very faint NEOs, TNOs, comets and outer satellites. Such telescopes should be available not only for searches for new objects, but also for follow-up observations that may be necessary at critical times (addresses Science Issues B and F).

References

Arlot, J.-E., A. Bernard, P. Bouchet, J. Daguillon, G. Dourneau, A. Figer, G. Helmer, J. Lecacheux, Ph. Merlin, C. Meyer, P. Mianes, B. Morando, D. Naves, J. Rousseau, G. Soulie, A. Terzan, W. Thuillot, L. Vapillon, and G. Wlerick 1982. Les resultats de la campagne d'observation PHEMU79 des phenomenes mutuels des satellites galileens de Jupiter en 1979. *Astron. Astrophys.* **111**, 151-170.

Benner, L.A.M., S.J. Ostro, J.D. Giorgini, R.F. Jurgens, D.L. Mitchell, R. Rose, K.D. Rosema, M.A. Slade, R. Winkler, D.K. Yeomans, D.B. Campbell, J.F. Chandler, and I.I. Shapiro 1997. Radar detection of near-Earth asteroids 2062 Aten, 2101 Adonis, 3103 Eger, 4544 Xanthus, and 1992 QN. *Icarus* **130**, 296-312.

Border, J.S., F.F. Donivan, S.G. Finley, C.E. Hildebrand, B. Moultrie, and L.J. Skjerve 1982. AIAA paper 82-1471 presented at the 1982 AIAA/AAS Astrodynamics Conference, San Diego, California, August 1982.

Butrica, A.J. 1997. *To See the Unseen: A History of Planetary Radar Astronomy.* NASA History Series No. SP-4218, NASA, Houston.

Dickey, J.O., P.L. Bender, J.E. Faller, X.X. Newhall, R.L. Ricklefs, J.G. Ries, P.J. Shelus, C. Veillet, A.L. Whipple, J.R. Wiant, J.G. Williams, and C.F. Yoder 1994. Lunar laser ranging - A continuing legacy of the Apollo program. *Science* **265**, 482-490.

ESA 1997. *The Hipparcos and Tycho-2 Catalogues.* ESA Pub. SP-1200.

Fienga, A., J.-E. Arlot, and D. Pascu 1997. Impact of Hipparcos data on astrometric reduction of solar system bodies. In *Proceedings of the Hipparcos-Tycho Conference, Venice 1997* (B. Battrick, ed.), ESA SP-402, ESTEC, Noordwijk, The Netherlands, pp. 157-159.

Folkner, W.M., T.P. McElrath, and A.J. Mannucci 1996. Determination of Position of Jupiter from Very-Long Baseline Interferometry. *Astron. J.* **112**, 1294-1297.

Fomalont, E.B. 2002. Searching for high quality VLBI calibrators, paper 6-01P, Second IVS General Meeting, Tsukuba, Japan, 4-7 Feb. 2002.

Giorgini, J.D., S.J. Ostro, L.A.M. Benner, P.W. Chodas, S.R. Chesley, R.S. Hudson, M.C. Nolan, A.R. Klemola, E.M. Standish, R.F. Jurgens, R. Rose, A.B. Chamberlin, D.K. Yeomans, and J.-L. Margot 2002. Asteroid 1950 DA's encounter with Earth in 2880: Physical limits of collision probability prediction. *Science*, in press.

Høg, E., C. Fabricius, V.V. Makarov, U. Bastian, P. Schwekendiek, A. Wicenec, S. Urban, T. Corbin, and G. Wycoff 2000. Construction and verification of the Tycho-2 Catalogue. *Astron. Astrophys.* **357**, 367-386.

Hudson, R.S. 1993. Three-dimensional reconstruction of asteroids from radar observations. *Rem. Sensing Rev.* **8**, 195-203.

Jacobs, C.S., D.L. Jones, G.E. Lanyi, S.T. Lowe, C.J. Naudet, G.M. Resch, J.A. Steppe, L.D. Zhang, J.S. Ulvestad, G.B. Taylor, O.J. Sovers, C. Ma, D.G. Gordon, A.L. Fey, D.A. Boboltz, and P. Charlot 2002. Extending the ICRF to higher radio frequencies. In *IVS 2002 General Meeting Proceedings* (N.R. Vandenberg and K.D. Braver, Eds.). NASA/CP-2002-210002, in press.

Johnston, K.J. and C. de Vegt 1999. Reference frames in astronomy. *Annu. Rev. Astron. Astrophys.* **37**, 97-125.

König, A. 1962. Astrometry with astrographs. In *Astronomical Techniques* (W.A. Hiltner, Ed.), pp. 461-486. Univ. Chicago Press, Chicago.

Marsden, B.G. 2001. Editorial Notice. *Minor Planet Circ. No. 43423.*

Monet, D.G. 1998. The 526,280,881 objects in the USNO-A2.0 Catalog. *Bull. Am. Astron. Soc.* **30**, 1427.

Monet, D.G. and A.K.B. Monet 1992. Galilean Satellite Astrometry, U.S. Naval Observatory, Flagstaff Station, inter-office memo, 22 October 1992.

Morrison, J.E., B. McLean, and GSC-II Catalog Construction Team 2001. The Second Generation Guide Star Catalog. *Bull. Am. Astron. Soc.* **33**, 1194.

Newhall, X.X., E.M. Standish, and J.G. Williams 1983. DE 102 - A numerically integrated ephemeris of the moon and planets spanning forty-four centuries. *Astron. Astrophys.* **125**, 150-167.

Ostro, S. J. 2002. Planetary radar astronomy. In *Encyclopedia of Physical Science and Technology, 3rd edition* (R.A. Meyers, Ed.), Vol. 12, pp. 295-327. Academic Press, New York.

Ostro, S.J., R.S. Hudson, L.A.M. Benner, J.D. Giorgini, C. Magri, J.-L. Margot, and M.C. Nolan 2002. Asteroid radar astronomy. In *Asteroids III* (W. Bottke, A. Cellino, P. Paolicchi, and R.P. Binzel, Eds.), Univ. Arizona Press, Tucson, in press.

Ostro, S.J., R.S. Hudson, K.D. Rosema, J.D. Giorgini, R.F. Jurgens, D. Yeomans, P.W. Chodas, R. Winkler, R. Rose, D. Choate, R.A. Cormier, D. Kelley, R. Littlefair, L.A. M. Benner, M.L. Thomas, and M.A. Slade 1999. Asteroid 4179 Toutatis: 1996 radar observations. *Icarus* **137**, 122-139.

Pascu, D. 1996. Long-focus CCD astrometry of planetary satellites, in Dynamics, Ephemerides and Astrometry of the Solar System. In *Proceedings of IAU Symposium 172* (S. Ferraz-Mello, B. Morando, and J.-E. Arlot, Eds.), pp. 373-388. Kluwer, Dordrecht.

Pascu, D., A. Fienga and J.-E. Arlot, and J.-E. 1997. Archival high-precision astrometric observations of Jupiter. *Bull. Am. Astron. Soc.* **29**, 1103.

Pascu, D., J.R. Rohde, P.K. Seidelmann, E. Wells, C. Kowal, B. Zellner, A. Storrs, D.G. Currie, and D.M. Dowling 1998. HST astrometric observations and orbital mean motion corrections for the inner Uranian satellites. *Astron. J.* **115**, 1190-1194.

Seidelmann, P.K. 1979. Planetary satellites, a review of the past and assessment of the future. In *Natural and Artificial Satellite Motion,* (P.E. Nacozy and S. Ferraz-Mello, Eds.), pp. 3-12. Univ. Texas Press, Austin.

Shapiro, I.I., C.C. Counselman III, and R.W. King 1976. Verification of the Principle of Equivalence for massive bodies. *Phys. Rev. Lett.* **36**, 555-558.

Standish, E.M. 2000. Dynamical reference frame - Current relevance and future prospects. In *Towards Models and Constants for Sub-Microarcsecond Astrometry, Proc.IAU Coll. 180* (K.J. Johnston, D.D. McCarthy, B.J. Luzum, and G.H. Kaplan, Eds.), pp. 120-126. U.S. Naval Observatory, Washington, DC.

Stone, R.C. 2001. Positions of the outer planets and many of their satellites. V. *Astron. J.* **122**, 2723-2733.

Williams, J.G. 1994. Contributions to the Earth's obliquity rate, precession, and nutation. *Astron. J.* **108**, 711-724.

Zacharias, N., S.E. Urban, M.I. Zacharias, D.M. Hall, G.L. Wycoff, T.J. Rafferty, M.E. Germain, E.R. Holdenried, J.W. Pohlman, F.S. Gauss, D.G. Monet, and L. Winter 2000. The first U.S. Naval Observatory CCD Astrograph Catalog. *Astron. J.* **120**, 2131-2147.

The Future of Solar System Exploration, 2003-2013
ASP Conference Series, Vol. 272, 2002
Mark V. Sykes, ed.

The Role of NASA's Planetary Sub-Orbital Program in Our Exploration of the Solar System

W.M. Harris[1], S.A. Stern[2], J.T. Clarke[3], & D. Slater[2]

[1] *Space Astronomy Laboratory, University of Wisconsin, Madison, WI*
[2] *Southwest Research Institute, Boulder, CO*
[3] *Department of Astronomy, Boston University, Boston, MA*

Abstract. NASA's sub-orbital program dates back to the earliest days of the agency. It has been the source of fundamental discoveries in the solar system, interstellar space and galaxy, the Sun, and the Earth's atmosphere and magnetosphere, and it continues to provide a unique venue for the pursuit of targeted research that is not possible with existing or planned space instruments. In addition to their direct science benefit, sub-orbital research programs are also the foundation support for several space-flight instrument development laboratories and provide a crucial mechanism for low cost/risk technology development and technique validation in a space environment. Sub-orbital experimentation has also been a primary resource for the training of new instrument scientists and the next generation of investigators for space-based missions.

REPORT

1. The Planetary Sub-Orbital Program in the Next Decade

Sub-orbital research is cross-cutting, covering many scientific areas in addition to its role as a technology development resource. There are four sub-orbital research lines currently supported by NASA, astrophysics, space physics, solar, and planetary. In the next decade, solar system sub-orbital research will contribute significantly in each of the key areas of science, technology development, scientist training, and public outreach. We describe the history and future of solar system sub-orbital research, including the specific science, technical, and educational areas where it will play the most significant role. In addition we make specific recommendations about the scope of the program and its long-term viability as an asset to planetary science.

Sub-orbital research necessarily exploits those areas where existing instruments do not provide capability. The scientific justifications for sub-orbital research fall into many areas. Examples include:

- The requirement for observations that must be made above the Earth's atmospheric absorption in the UV and in the H_2O IR bands or to avoid bright airglow emissions. Experiments such as FUSE and HST provide

Figure 1. At left, the Wide-Field Imaging Survey Polarimeter (WISP) is mounted to a launch rail in preparation for a launch to observe comet Hale-Bopp in 1997. The WISP capabilities of narrow bandpass NUV and FUV imaging polarimetry over a $5^\circ \times 1.5^\circ$ are well matched to targets such as Hale-Bopp, but are not duplicated in any other space instrument. At right are the environmental test and engineering (T & E) facilities at the White Sands Missile Range (WSMR). The opportunity for student involvement in the T & E process adds to the instrumental development training opportunities of NASA's sub-orbital program.

coverage in these areas, but they are not suited to many types of observations that are of interest to solar system science.

- Observations that are not possible, or not permitted, with existing space-based instruments. For example, it is not possible to make observations inside of 0.7 AU or 45° from the Sun (e.g., of Mercury, Venus, comets, or solar wind/interplanetary medium interactions) using any existing space observatory. Rocket experiments can be designed, optimized, and/or targeted for space-based observations that the larger (and more expensive) missions cannot do.

- Campaigns of opportunity and coordinated observations. Sub-orbital experiments are frequently paired with other space based or ground based observing programs and are also operated in groups to study several aspects of a unique object. The recent case that most demonstrates this was the four mission campaign to study comet Hale-Bopp in 1997 (Figure 1). No other orbiting observatory was able to track this object when it was within 2.4 AU of the Sun. The sub-orbital initiative was the only NASA response to this event. Other opportunities for campaigns continue to occur, such as the Cassini flyby of Jupiter and the CONTOUR (Comets Encke in 2003 and Schwassmann-Wachmann 3 in 2006) and Deep Impact (Tempel 2 in 2005) comet encounters. Sub-orbital support science offers considerable value as collaborative efforts with much more expensive space missions. The compromises of deep space missions often require that instrumental capabilities be left out. In other cases space probes that operate in situ will lack a global system wide view. Supporting sub-orbital programs, both as part of the existing Planetary Sub-orbital line and from supplemental campaign opportunities, are able to fill many of these gaps.

- Observations of significant scientific interest, but not appropriate for satellite missions due to a limited target list, a small number of required observations, because the technique is not completely validated for satellite applications, or because there is not time to produce a satellite. The issue of technical targeting is particularly germane to Solar System exploration, because there is no NASA line for solar system space instruments under \$300M other than the sub-orbital program. Without the sub-orbital program opportunities for targeted science, timing, and technique validation would be frozen out.

The three existing planetary sub-orbital programs contain elements drawn from each above justifications. They provide a low cost means of filling the gaps in our observational capability and in driving new technological development. Virtually none of the research areas currently being pursued in this program could be conducted from any other existing or planned instrument. The opportunities presented by this program are primarily limited by the level of institutional support, which is less than 40% of the average in the other three sub-disciplines. Below we include several examples of observational techniques that have been, are, or could be potentially exploited in the Planetary Sub-orbital program.

- Spectral and imaging studies in the FUV-EUV below 115 nm. Several sub-orbital experiments in the Astrophysics and Planetary lines cover this range and have been used to some effect in solar system research (e.g., Stern et al. 2000, Harris et al. in preparation). There are hitchhiker experiments such as UVSTAR. Currently FUSE is providing capability in the FUV, however it lacks an imaging capability, is nearing the end of its design life, and has no planned replacement. Several Solar System targets are accessible from sub-orbital platforms in this bandpass. Some examples include:

- Exospheric He I emission from planetary coronae which are tracers of outflow and may be compared with the distribution of H Ly-α or Ly-β.

- Noble gas lines in planetary atmospheres and comets (Stern et al. 2000).

- Imaging emission line measurements from the Io torus (SII, OI, etc. - see above).

- Imaging of outer planet auroral emissions from the H_2 Werner band system, which may be measured by ratio to identify the energy spectrum incoming bombarding electrons as a function of local time and location along the oval.

- Observations of the inner solar system. Nearly all orbital observatories have programmatic restrictions against observing within 45° of the Sun, which closes off the entire region of the solar system within 0.7 AU. These areas similarly suffer in ground based observing, because the targets must be observed at high airmass. This area of the solar system includes Mercury, Venus, comets at their most active phase, and the interaction zone between the solar wind and IPH.

- Polarimetric observations. The scattering properties of dust and gas within the solar system offer a wealth of information on dynamical structure, opacity, radiative transfer, equilibrium state, and even on magnetic field properties. Sub-orbital research in polarimetry (including WUPPE) has proven to be very successful (e.g., Nordsieck et al. 1994, Cole et al. 2000, Harris et al. 1999) and several polarimetric sub-orbital instruments are available for use. Science targets in this area include:
  - Radiative transfer studies of planetary upper atmospheric, comet, and interplanetary medium emissions from resonance line polarimetry
  - Scattered solar Ly-α within the Jovian Ly-α bulge
  - Jovian H Ly-α auroral emission
  - Equilibrium state of cometary neutrals.
  - Radiative transfer of interplanetary H (IPH)
  - Scattering from comet dust:
  - Direct detection of magnetic fields from the Hanle Effect
- High spectral resolution studies of UV emission line profiles. A custom design for an echelle spectrograph has been developed to obtain high-resolution emission line profile measurements through a long, narrow aperture with spatial information along the aperture (Harris et al. 1993). This instrument can measure Doppler profiles of emission lines from planetary atmospheres and cometary comae, with a spectral resolution of 10^4-10^5 and a sensitivity of tens of Rayleighs. It may be used to detect superthermal distributions of hot atoms, and to measure isotopic ratios like D/H, which has been proposed for a Venus flight and may eventually be performed for comets.
- Interferometric observations of wide field diffuse emissions in the UV. Advances in UV interferometric technology is now making it possible to design simple, all-reflective instruments to study diffuse emissions as faint as $\sim$1 Rayleigh at spectral resolutions of $>10^5$. There are many science targets in the solar system, both from remote sensing platforms and as components of future space probes, where they are useful as atmospheric monitors. Sub-orbital interferometry targets include
  - The comet H coma brightness and velocity distribution
  - The geocorona and interplanetary hydrogen
  - The heliosphere
- Wide field UV and (H_2O blocked) IR observations: All existing and planned orbiting remote sensing observatories are narrow (arcminute or smaller) field of view (FOV). However, within the solar system there are many observational targets that are larger than this angular size and are better studied with instruments optimized for them. Examples of this include

- Comet comae and tails:
- The extended distribution of the Io plasma torus:
- Planetary coronal emissions

1.1. Student training

The training of students has always been a cornerstone of sub-orbital research, and is an ideal environment within which to develop new instrumentalists and/or researchers with a solid technical background. The culture of sub-orbital research makes it possible for a graduate student to carry a project from design, to construction, to calibration, to payload test and engineering, to flight, and finally post flight validation. There is no other mechanism existing within NASA space-flight programs where students, or even post-doctoral fellows, have such direct, definitive roles in the development and flight of experiments in space. A typical student in a sub-orbital research will gain experience with some or all of

- Optical, mechanical, thermal, electronic, and contamination design requirements for spaceflight.
- Flight operations, ground support, and remote uplink control of spacecraft.
- Mission support equipment such as telemetry formats and link margins, dedicated experiment processing, attitude control systems, power system distribution units, batteries, and recovery systems.
- Test and engineering (e.g., prototype testing, instrument vacuum calibration, vibration testing, spin, moment testing, sequence testing, integrated operation, coronal testing)

The value of the above skills for scientists with interest in space-based instrument design or use should not be underestimated. Students with this experience have the skills to design and implement the next generation of space-flight experiments and are in great demand once they graduate. They are able to critically evaluate the potential performance, cost, and support issues surrounding a mission, engage in useful negotiation with vendors on capabilities, reliability, and redundancy, and, most importantly, will be in a position to manage a mission program if they find themselves in charge of one. In the current NASA model of PI-class instruments, these skills are essential. The list of scientists with significant roles in space missions that received training or other experience in sub-orbital research includes (but is not limited to)

Robert Bless-HSP
Webster Cash - FUSE
Supriya Chakrabarti - SPIDR/SCARI
John Clarke - HST/WFPC 2/JMEX
Daniel Cotton - TERRIERS
Paul Feldman - HUT
James Green - COS/FUSE
Walter Harris - Planet Suborbital/Hale-Bopp Campaign
Roger Malina - EUVE

Christopher Martin - GALEX
Daniel McCammon - CHANDRA/ASTRO E
William McClintock - MESSENGER UVS/Cassini UVIS
Kenneth Nordsieck - WUPPE
John Nousek - SWIFT/CHANDRA/ASTRO E
Alan Stern - Pluto-Kuiper Express/Rosetta

In addition to the above list, many other scientists with suborbital training make up personnel of mission teams and instrument development groups for both ground and space based astronomy. These same researchers also form the core of the small group of scientists available to NASA for the evaluation of proposed spacecraft missions. In the absence of sub-orbital training, this experience must be gained on-the-job for a much larger project, where the risk to both the scientist and NASA is substantially greater, a risk that is higher in the planetary community because the missions are fewer in number and more expensive.

1.2. Technology development

A long-standing goal of all the sub-orbital research programs (astrophysics, solar, planetary, and space physics) is to use these missions to test, refine, and validate new instrumentation. Recent developments in UV detector technology, reflective spatial heterodyne spectroscopy, UV polarimetry, star trackers, deployment monitors, and aberration corrected holographic grating spectroscopy have all employed sub-orbital missions as test beds at some level. Many existing spacecraft instruments in these areas are logical extensions of sub-orbital experiments.

Where practical, planetary suborbital missions provide an important proving ground for new technologies, because the culture of such research is geared to facilitate the often rocky path from concept to space qualification. Sub-orbital research provides

- Laboratory based "proof of concept" instrument development in partnership with NASA engineers and support personnel.
- The opportunity to build instruments and prototypes without power and mass restrictions.
- Reduced exposure to the precision component, construction, and management requirements that drive up costs and inhibit innovation in larger higher profile missions.
- The flexibility to employ/validate low cost COTS technology for spaceflight environments.
- Flight operational experience and design refinement over successive missions.

Sub-orbital experience has been invaluable to the development of robust, flight ready support and scientific instrumentation. There is a substantial list of technology needs for the solar system community outlined in the decadal plan.

NASA's PIDDP program also provides an opportunity to develop new instruments, but not the support to test them in a space-flight environment. As a rule, it is imprudent to substitute engineering and models for space-flight experience. Indeed, mission support technology on sounding rockets must operate properly during three successive engineering flights before being used on other missions.

Sub-orbital missions in astrophysics, aeronomy and solar physics provide an opportunity to test several of the technologies described in the solar system decadal plan. However, without a robust planetary program we will not be in a position as a community to set the agenda on what is or is not developed. The existing and previous planetary sub-orbital programs have flight-tested reflective filter designs, CCD controller systems, star trackers, delay line anode microchannel plate detectors, and Wood's UV filters. Most recently, a novel star-tracker (the ST-5000) has been developed with rocket program funding, tested and flown on the latest planetary rocket flight. This promises to replace the old Ball Strap-V trackers, and may be used as well on low-cost Small Explorer and Discovery programs. The potential cost of not taking advantage of the sub-orbital validation opportunity is the flying of an unproven component that ultimately fails. By contrast, the cost of supporting the validation is very small. Sub-orbital missions cost ~5% of even the least expensive satellite missions ($20M-UNEX). There is an even larger gap in solar system exploration, where researchers do not have routine access to mission opportunities less costly than a Discovery mission, whose cost is the equivalent of 300 to 500 sub-orbital flights.

1.3. Education and public outreach

Sub-orbital research provides unique opportunities for bringing the experiences of space-based science to the general public in a complete form. Through these programs people can be exposed to all aspects of a project, including the difficult tasks of validation, field repair, rocket arming, and flight. Instrument construction, testing, flight, and recovery occur over a shorter period and in a smaller number of locations that make it easier to compartmentalize them for the public. The launch events themselves have a higher degree of drama because of the prototype nature of the experiments, the very short duration of a flight, and the opportunity for real-time interaction with an experiment. Several sub-orbital programs have active projects where student instruments are flown and/or classes built around specific missions or instruments. Other E/PO projects include the production of documentaries, teaching kits, and programs to provide experimental and other space-exposed equipment for school use.

2. Program Details

The Astrophysics, Solar Physics, and Aeronomy sub-orbital lines each have 4-8 individual programs that are funded at a level of 300-600 K/yr. The majority of these utilize short duration sounding rocket platforms for their research, although some employ long duration balloons or shuttle hitchhiker opportunities. An average of ~20 flights are supported each year from one of several launch sites including the White Sands Missile Range (New Mexico), Poker Flats Missile Range (Alaska), and remote launch sites in Puerto Rico, Brazil, and Australia.

Sub-orbital experiments are typically designed and constructed at the PI institution over a period of from 1-4 years. Following validation and prototype testing, these are integrated with rocket telemetry, power, guidance, and recovery systems, transported to field, and launched. The majority of the experiments are recoverable.

The characteristic total cost for a single instrument to go from development to first flight is ~\$1-2M, with the majority of these funds being spent on the launcher and its systems. Following refurbishment, repair, and upgrades, the average payload is returned several times for additional flights at substantially reduced cost. This re-use means that many experiments originally built as far back as the 1970s continue to be used for new scientific discovery or training. Long standing research programs will continue to develop new instruments that take advantage of emerging technologies and/or specialized observational techniques, thus providing an important research and development service to NASA and the astronomical community.

3. History and Current Status of the Planetary Sub-orbital Program

NASA's planetary sub-orbital program has existed since the mid 1980s, but has suffered in comparison with those in other areas. Until 1999, sub-orbital research was administered under the umbrella of the Planetary Atmospheres program, where it supported a small number of researchers at a level 20-30% that of similar projects in the Astrophysics sub-orbital program. In 1999, following a very successful campaign to study comet Hale-Bopp in 1997 (Harris et al. 1997, McPhate et al. 1999, Stern et al. 2000), a revised and expanded Planetary Sub-orbital line was added to the solar system exploration (SSE) theme and the number of funded researchers was to be increased to 4 and maintained at that number. However, before this could be formally implemented, the number of programs was reduced to 3 and the ramp up to a full complement of researchers was spread out over 2 years. More significantly, the funding levels proposed for the selected researchers, (150K/yr) are at a level between 25-40% that of programs in other areas and actually less in inflation adjusted dollars that the program provided in 1987.

Despite these limits, there has been a strong community interest in the program, with 14 proposals in the first year for 2 slots and a comparable number in second year for the remaining slot. With a proposal selection rate from 10-15%, the competition for slots in Planetary Sub-orbital is more competitive than any of the other SSE ROSS lines, and comparable to the selection rate for HST and Explorer/Discovery class missions.

The first flight of an instrument in the new program occurred on December 5, 2001 to observe comet C/2000 WM1 (LINEAR), and was a comprehensive success. Another flight is scheduled for Jupiter in 2002. Two new payloads are currently under construction with student involvement in the design, testing, and mission details. A subset of the published scientific results from the Planetary Suborbital program is included in the references (Stern et al. 2000, McPhate et al. 1999, Harris et al. 1999, Stern et al. 1996, Stern et al. 1995, Stern et al. 1999, Parker et al. 1998, Slater et al. 1995, Harris et al. 1993, Bush et al. 1991).

In Fall 2000 the Solar System Exploration Subcommittee (SSES) was asked to review issues related to sub-orbital research in all areas at NASA. The status and future of the planetary sub-orbital was reviewed and several recommendations made. The report of this panel (Appendix B) indicated that the planetary program received insufficient support, and was currently at an attrition level where researchers could only continue to launch experiments until something was damaged in-flight. They also concluded that the funding issue limits the current proposer pool to researchers at institutions with space flight support personnel that are supported through other programs and/or an existing instrument that can be reflown at low cost. Their recommendations were to raise the support levels for each program by at least 70% and to increase the number of programs to four.

The response of the NASA SSE administration has thus far run counter to the panel recommendations. Flat funding levels were approved for the FY2001 and 2002 years, which amount to a 7% cut in real support. In addition, there are no plans for an announcement of opportunity in the 2002 ROSS AO. With two of the three programs at the end of their first 3 year cycle, this decision removes their opportunity to propose a continuation of their research efforts. Such uncertainty also makes it nearly impossible to attract students or justify any new instrument development work at their host institutions. Another AO may be released in 2003, however the lack of continuity may be extremely damaging to the long-term prospects for this program.

4. Recommended Program Goals for the Coming Decade

Over the next decade we recommend that NASA implement a course of bringing the existing planetary sub-orbital program into conformity the SSES suggestions for the number of programs and level of funding. We also recommend increased use of targeted campaigns of opportunity and sub-orbital platforms as technology test beds.

- *Individual program support:* The funding level for each program should be increased from its current 150K/researcher-year to a minimum of between 200 and 300K. This is the minimum necessary to support an active program that can develop and maintain instruments at a non-attrition level, and is ~40-60% the average program size in Astrophysics and Solar sub-orbital programs. Funding levels below the current 150K/year are insufficient to justify the parallel investment in launch support.

- *Suborbital campaigns:* Parallel support for planetary sub-orbital research should be provided for targeted campaigns to support of NASA missions (e.g., Deep Impact, Cassini, Messenger, CONTOUR, etc.), and events-of-opportunity (e.g., C/1995O1 Hale-Bopp or C/2001 Q4 NEAT - in 2004). These should be open the investigators in the existing Planetary sub-orbital line and those in other areas of sub-orbital research.

- *Number of programs:* The total number of research programs supported should be a minimum of 3, and eventually expanded to 6 (the number supported in Astrophysics sub-orbital).

- *Technology:* Improved program links should be established between the sub-orbital program and other areas of instrument development to maximize the resource as a proving ground for new technologies and techniques.

The above recommendations would require the planetary sub-orbital program to be funded annually at 600K (minimum level for three programs) to 1800K (six programs at 300K each). A maximum investment by NASA of $18-20 M (approximately the cost of a single UNEX mission) over the next decade will enable researchers to pursue the above science agenda. Funding levels below the current 150K/researcher-year will not justify the parallel investment in launch support, and we recommend program termination or a reduction in researchers in the event that this is necessary.

Should the SSE management opt to discontinue administering the program as a separate line, the funding for it could be transferred into the Planetary Astronomy program within SSE (where a separate review panel would need to be convened to judge rocket proposals to ensure the expertise on space hardware missions), or possibly into the Astrophysics sub-orbital program (where a subgroup of planetary programs would be maintained within the guidelines outlined above).

References

Bush, B., D.M. Cotton, O.H. Siegmund, S. Chakrabarti, W.M. Harris, and J.T. Clarke 1991. High resolution, two-dimensional imaging, microchannel plate detector for use on a sounding rocket experiment. In *EUV, X-Ray, and Gamma-Ray Instrumentation for Astronomy II* (A93-34376 13-35), *Proc. SPIE* **1549**, pp. 290-301.

Cole, A.A., K.H. Nordsieck, S.J. Gibson, and W.M. Harris 1999. Ultraviolet imaging polarimetry of the Large Magellanic Cloud I: Observations. *Astron. J.* **118**, 2292-2305.

Harris, W., J.T. Clarke, J. Caldwell, P.D. Feldman, B. Bush, D. Cotton, and S. Chakrabarti 1993. High-resolution ultraviolet spectrograph for sounding rocket measurements of planetary emission line profiles. *Opt. Engin.* **32**, 3016-3027.

Harris, W., K.H. Nordsieck, F. Scherb, and E.J. Mierkiewicz 1999. Polarimetric observations of Comet C/1995 O1 Hale-Bopp. *Earth, Moon, and Planets* **74**, 161-167.

Parker, J., S.A. Stern, G.R. Gladstone, and J.M. Shull 1998. A re-analysis of the spectroscopic detection of Argon in the lunar atmosphere, *Astrophys. J.* **509**, L61-L64.

McPhate, J., P.D. Feldman, S.R. McCandliss, and E.B. Burgh 1999. Rocket-borne long-slit ultraviolet spectroscopy of Comet Hale-Bopp. *Astrophys. J.* **521**, 920-927.

Nordsieck, K.H. and W.M. Harris 1999. Ultraviolet astronomical polarimetry: some results and prospects. In *Ultraviolet and X-Ray Detection, Spectroscopy, and Polarimetry II* (S. Fineschi, B.E. Woodgate, and R.A. Kimble, Eds.) *Proc. SPIE* **3764**, pp. 124-133.

Nordsieck, K.H., P. Marcum, K.A. Jaehnig, and D.E. Michalski 1994. New techniques in ultraviolet astronomical polarimetry: Wide-field imaging and far-ultraviolet spectropolarimetry. *X-Ray and Ultraviolet Polarimetry* (S. Fineschi, Ed.)*Proc. SPIE* **2010**, pp.28-36.

Slater, D., S.A. Stern, J. Scherrer, W.C. Cash, J.C. Green, and E. Wilkinson 1995. The Extreme Ultraviolet Spectrograph Sounding Rocket payload: Recent modifications for planetary observations. In *EUV/FUV, EUV, X-Ray, and Gamma Ray Instrumentation for Astronomy VI* (O.H. Siegmund and J.V. Vallerga, Eds.), *Proc. SPIE* **2518**, pp. 211-222.

Stern, S.A., D.C. Slater, M.C. Festou, J.W. Parker, G.R. Gladstone, M.F. A'Hearn, and E. Wilkinson 2000. The discovery of Argon in Comet Hale-Bopp. *Astrophys. J.* **544**, L169-L172.

Stern, S.A., W.B. Colwell, M.C. Festou, P.M. Tamblyn, J.W. Parker, D.C. Slater, P.R. Weissman, and L.J. Paxton 1999. Near 2.3 post-perihelion: SWUIS measurements of the H_2O and dust production. *Astron. J.* **118**, 1120-1125.

Stern, S.A., D.C. Slater, G.R. Gladstone, E. Wilkenson, W.C. Cash, J.C. Green, D.M. Hunten, T.C. Owen, and L. Paxton 1996. The 825-1110ÅEUV spectrum of Venus. *Icarus* **122**, 200-204.

Stern, S.A., D. Slater, W. Cash, E. Wilkinson, J. Green, and G.R. Gladstone 1995. Rocket FUV observations of the Io plasma torus during the Shoemaker-Levy 9 impacts. *Geophys. Res. Lett.* **22**, 1837.

Appendix A: SSES Recommendations for the Planetary Sounding Rocket Program

November 27, 2000

Dr. Jay Bergstralh
Code S
NASA Headquarters
Washington, D.C. 20546

Dear Jay:

The Solar System Exploration Subcommittee (SSES) of the Space Science Advisory Committee (SScAC) met on October 30 and 31, 2000 in Pasadena, California. The purpose of this letter is to summarize the findings and recommendations of that meeting.

Attached Payloads

The SSES considered the problem of the suborbital/attached payloads program that has found itself fitted within the Planetary Atmospheres program as an unfunded mandate. For FY00 that amount is $300K out of the $8M Planetary Atmospheres budget. It was widely agreed that level of funding is far below any reasonable support level for the program and that the mandate must necessarily grow to 450K in FY01 and 600K in FY02. The consensus of the SSES is that if that program support remains within Planetary Atmospheres, it should continue to be separated by a firewall from the rest of Planetary Atmospheres. The two programs are distinctly different in terms of their average grant size (about a factor of two) and in their objectives (basic investigator science vs. flight missions, technology development, space hardware training). The SSES sees as the most important objective to find a natural home for the planetary-suborbital/attached payloads program, perhaps as a new, free standing Research and Analysis Program for suborbital/attached payloads that cuts across themes. This would allow SSE proposals to compete more directly with their naturally complementary counterparts from other themes. The financial support at the current or projected level from the SSE suborbital grants would be contributed to this program. This kind of cross-theme program is not unprecedented. Such a program currently exists in the Flight Information Systems area. The SSES recommends that the SScAC review the suborbital/attached payloads issue to advise on the possibility of creating a cross-theme program. The availability of the Space Shuttle and the International Space Station should not be overlooked by the Solar System Exploration Program.

With kindest regards.

Sincerely,

Michael J. Drake, Chair
Solar System Exploration Subcommittee

The Future of Solar System Exploration, 2003-2013
ASP Conference Series, Vol. 272, 2002
Mark V. Sykes, ed.

Instrument Technology Development

J.P. Hoffman[1], J.R. Piepmeier[2], T.L. Akins[1], A.C. Berkun[1], F. Carsey[1], D.R. DeBoer[3], W.N. Edelstein[1], H.B. Hammel[4], R.D. Lorenz[5], B. Macintosh[6], & R.A. Tucker[7]

[1] *Jet Propulsion Laboratory, California Institute of Technology, Pasadena, CA*
[2] *NASA Goddard Space Flight Center, Greenbelt, MD*
[3] *SETI Institute, Mountain View, CA*
[4] *Space Science Institute, Boulder, CO*
[5] *Lunar and Planetary Laboratory, University of Arizona, Tucson, AZ*
[6] *Lawrence Livermore National Laboratory, Livermore, CA*
[7] *National Optical Astronomy Observatory, Tucson, AZ*

Abstract. The Instrumentation Technology Development Panel has attempted to identify the areas of recent innovations and continuing deficiencies in instrument technology as it pertains to planetary science goals and missions. Collaboration between science and technology aspects of all planetary science disciplines is crucial to the success of this task. We conclude that the technology advances most necessary for the success of future missions (Mars Sounders, Europa Orbiter, Venus Landers, Titan Landers) are increased downlink capacity (i.e., Deep Space Network), improvements in radiation hardened and extreme temperature electronics, increased aperture sizes for radio science, on-board processing power, and increased collaboration with principle investigators.

REPORT

1. Current State of Knowledge

While it is obvious that mission success, even mission plausibility, is strongly dependent on technical capability, it may be less obvious that improvements in technology can also lead to a greater number and/or new types of missions. Therefore, we are interested not only in identifying new types of instrumentation, but also in improving existing (even mature) technologies by reducing size, complexity, and cost. Improvements of this type not only lower mission costs, but also enable more complex instrument suites to be considered for advanced data fusion measurement concepts.

We sought to identify the short-term and long-term technological needs (new instruments) and bottlenecks (better instruments) for each of the planetary science disciplines.

2. Advances in Instrument Development

2.1. Basic research

Advances in instrument development may benefit from work in basic research, device research, instrument sub-system development, and instrument system development. Almost by definition, the returns on investments into basic research are difficult to predict. The planetary science community has neither the resources nor the impetus to invest substantial resources in basic research that may benefit instrument development. However, investments in device development may yield substantial returns to planetary science. One reason for this is the divergence of interests between commercial and science communities. With profit margins intimately linked to the cost of large quantities of inexpensive devices, which may sell for pennies to dollars per unit, commercial interests seek ever less expensive low power devices with statistically acceptable yields. Obviously, this is not a proper business model for space-based instrumentation development. For this reason, the success of the commercial electronics market might be hindering the advance of technologies suitable to extreme environments.

While it is true that the commercial markets emphasis on cost, power, and size is increasingly in line with current space mission philosophy, the very technologies that enable cost and size reductions work to preclude their use in extreme environments. For example, CMOS (Complementary Metal Oxide Semiconductor) technology is favored by the commercial markets. It is reliable, inexpensive, and widely available. Space qualifying individual chips, however, is time consuming and costly. Serious effort needs to be expended on identifying nascent semiconductor technologies that will meet the needs of the next decade of space exploration. It is possible, given current trends, that the commercial markets will have little impact on improving the quality or capability of electronics for space-borne instruments over the next decade. Indeed, the market for space-qualified electronics is a tiny fraction of the global semiconductor market. Therefore, whether success is found in CMOS, MEMS (MicroElectroMechanical Systems), Silicon Carbide (SiC), Gallium Nitride (GaN), or some other semiconductor technology, the burden of advancing these technologies for space applications may fall on the science community. While direct financial support for research into low-level device technology may not be practical for planetary science interests and budgets, formal and informal lobbying for support into this research would be a way to ensure future needs of planetary science are met.

2.2. Device development

By its very nature it is difficult to predict the capabilities of an entirely new device, however, a number of new technologies are currently showing great promise. Microelectromechanical Systems or MEMS technology shows great promise as a building block for extremely small, highly rugged instrumentation. Some recent examples of MEMS applications pertinent to the planetary science community are listed:

- Adaptive Lenses for Microwave Antennas
- Micropropulsion

- Thermal Systems
- RF Switches

MEMS may be able to address some of the difficulties associated with instruments for extreme environments, such as the high-temperatures of Venus, the low temperatures of the outer planets, or the radiation belts of Jupiter. Missions currently under serious consideration by various researchers include the environments described above.

3. System Development

Since it is far easier to envision the utility of systems-level instrument development, as opposed to basic research or even device development, members of the planetary science community, but outside of specific instrument developments, were able to aid in identifying improvements that would lead to the success or enrichment of missions in their own areas of expertise. Therefore, this section contains both general input into the direction of systems-level instrument development and suggestions for improvements with specific science goals in mind.

Some obstacles to improved instruments are fairly obvious and highly dependent on device technologies and applications, such as lower power, higher resolution, increased sensitivity. Many of these traits involve trade-offs, with limitations imposed by optimization around more pressing traits. Such trade-offs are the purview of the particular instrument designer and principle investigator. However, from our survey, we have identified some of the main bottlenecks that, if improved, would yield specific increases in science volume or quality.

3.1. Improvements to ground-based systems

Unlike space-borne instrumentation, ground-based instruments are freely capitalizing on the consumer electronics market. Distributed Computing and consumer electronics may enable order-of-magnitude increases in capabilities for ground-based systems, such as the Deep Space Network (DSN) and radio telescopes.

The DSN is not a new piece of technology, but is currently being upgraded. Even so, with its limitations in capability and availability, data rates will become an increasingly important technology driver for designers of space-based instrumentation. It is incumbent upon the space science community to push for increased capacity of the DSN if large numbers of data intensive instruments are to be viable. Imaging instruments of any type (e.g., Synthetic Aperture Radar, optical, IR) tend to be popular with scientists and the public alike, but are notorious for generating copious amounts of data return. Serious consideration should be given to increasing the capabilities of the DSN further and to developing new communication technologies such as optical communication systems.

An example of new instrumentation for ground-based technology is the SETI Institute's Allan Telescope Array (ATA). The designers of this array are leveraging consumer electronics technology to construct an advanced instrument. The ATA will be capable of covering the spectrum from 0.5 to 11.2 GHz, with up to four concurrent users. This will enable its use as both a science instrument and as a possible tracking station for multiple Mars missions. The base technology for this instrument is derived from commercial satellite broadcasting. Of course the necessary modifications and integration issues are nontrivial, but by leveraging off the commercial product, the designers take advantage of market driven technology innovation, which no science mission, regardless of funding level, could hope to match.

3.2. Improvements to space-based systems

The limits placed on the resolution of radio frequency, microwave, and millimeter wave instruments by fundamental physics demand larger aperture sizes. To meet the demand for space-based instruments of this class, serious efforts need to be expended on developing technologies such as large (10's of meters) deployable antennas and antenna constellations. More development is also required for efficient far-infrared detectors and antennas. Research into this area would make possible or improve missions to detect sub-surface water on Mars, sub-ice oceans on Europa, high-resolution mapping of the surface of Io, Triton, and other interesting moons.

Along with antenna technology, these same missions/instruments will require increased DSN time unless on-board processing technology can be improved and approved. This means that so-called smart systems must not only be developed, they must be pressed into service in actual science gathering missions. High-resolution passive or active imagers produce large data sets that are cumbersome to down link. Efficient compression algorithms can reduce data rates, but only so far. More efficient lossy compression algorithms are available, but some within the science community deem their use unacceptable. On-board processing must be employed if large numbers of high-resolution imagery missions, such as Mars and Europa missions, are to be successful. Such on-board processing requires more than simply research into computer models or so-called artificial intelligence agents; to fly it requires hardware. Further investment into radiation-hardened solid state RAM and programmable logic, such as field programmable gate arrays (FPGAs), is required before serious consideration can be given to missions that envision long-term investigation of high-value targets such as Europa. Space qualified programmable logic devices are useful because one device can be targeted to many applications. For example, specific microprocessors can be programmed into an FPGA, or software algorithms conventionally executed on a CPU can be implemented in hardware using a programmable logic device. Qualifying one line of FPGAs can make many different processors and algorithms functionally available to space flight systems designers.

Another option for relieving the strain on the DSN, enabling order of magnitude increases in data rates, is optical communications. Many technological challenges currently make this option difficult to implement, but success in this area would bring real returns in data collection capacity.

Missions to extreme environments require devices and systems capable of withstanding those environments. Research into alternatives to standard silicon technology, such as GaN, SiC, and MEMS, are necessary to mounting serious missions to extremely hostile locations. Ultra-low temperature electronics are required for in-situ missions to Titan ($\sim$95 K) or Triton ($\sim$50 K). Ultra-high temperature ($\sim$500 K) electronics are required for in-situ missions to Venus and perhaps to Io's volcanoes. For mission concepts, such as the Titan Aerobot (and various Mars Scout mission concepts), these devices must be light and low power, as well as rugged. As previously stated, current technology may allow one to optimize one of these parameters, but meeting all requirements is possible only with advances in technology.

A popular concept for planetary research is the use of multiple sensors/craft (anywhere from a handful to many dozens) to blanket large areas with detailed in-situ sensor coverage. This requires advances in almost every technology area previously mentioned. Lightweight, low power probes have been envisioned to cover sections of Mars, Europa, and Venus, and to swarm through the atmospheres of the Gas Giants, Mars, and Venus. These missions, while extremely interesting, require advances in autonomy (even if the probes are 'dumb,' the 'mothership' must have some decision making capability), lightweight electronics, low power electronics, and down link capacity. Every year, interesting missions of this type are proposed, at least on posters, but are limited in their plausibility and utility by lack of technology and by risk aversion.

4. Maintaining Capabilities for the Future (and Outreach)

Even highly motivated students may find difficulty pursuing careers in technology development in the specific market of space exploration. University research laboratories routinely receive large donations in money and equipment in exchange for research into high-risk commercial ventures. This practice benefits the university, the commercial interest, and usually the student, but funnels future researchers into highly specific commercial fields. Increasing expenditures to university researchers is an obvious remedy, but difficult for the decreasing budgets in planetary science. A less expensive approach is to develop and nurture programs to expose interested students to high-profile missions.

Another approach to expose students to technology development for planetary exploration is through leveraging existing programs, such as E-Week (National Engineers Week), which occurs around February each year. Since engineering students are the most likely to get involved in technology development, this type of exposure would bring high visibility. The involvement of the planetary community could be as great as sending representatives to individual E-Week activities are as little as sending information regarding technology development to E-week organizers.

Similar programs for younger students could also benefit from increased support from the science community. Technology competitions, such as NASA's FIRST Robotics competition, are excellent opportunities for high-profile scientists to show their support for advancing technology for space exploration

A valuable (and inexpensive) way to increase support for technology development is to increase communication between the planetary science community and the technology development community. A few versatile people make serious contributions to both fields and should be sought as conduits for information flow.

5. Summary of Technology Needs

Currently identified areas (non-inclusive) requiring technology development:

- Deployable large (10's of meters) microwave antennas
- FIR (sub/mm wave) detectors and antennas
- Extreme-temperature semiconductors for Venus (high-temp), Titan (low-temp)
- Low power radiation-hard electronics
- Radiation-hard on-board processing power and memory
- Increased DSN capacity
- Optical interplanetary communications
- Lightweight deployable optics
- Lightweight, inexpensive, in-situ atmospheric probes

The Future of Solar System Exploration, 2003-2013
ASP Conference Series, Vol. 272, 2002
Mark V. Sykes, ed.

Defining Long Term Goals and Setting Priorities for Education and Public Outreach

J.A. Grier[1], D.H. Atkinson[2], N.G. Barlow[3], I.P. Griffin[4], J.P. Hoffman[5], B.A. Kelly-Serrato[6], L.P. Keszthelyi[7], M.J. Klein[5], S.L. Klug[8], R.A. Kolvoord[9], P.D. Lanagan[7], L.A. Lebofsky[7], D.J. Lien[10], M.M. Lindstrom[6], R.M.C. Lopes[5], L.L. Lowes[5], J.D. Manifold[11], R.M.E. Mastrapa[7], M.P. Milazzo[7], E.D. Miner[5], P.A. Morris-Smith[12], A.S. Rivkin[13], C.J. Runyon[14], A.M. Sohus[5], M.L. Urquhart[15], A.R. Vasavada[16], R.L. Warasila[17], P.G. Withers[7], & C.A. Wood[18]

[1] *Planetary Science Institute, Tucson, AZ*
[2] *Department of Electrical and Computer Engineering, University of Idaho, Moscow, ID*
[3] *Department of Physics, University of Central Florida, Orlando, FL*
[4] *Space Telescope Science Institute, Baltimore, MD*
[5] *Jet Propulsion Laboratory, California Institute of Technology, Pasadena, CA*
[6] *NASA Johnson Space Center, Houston, TX*
[7] *Lunar and Planetary Laboratory, University of Arizona, Tucson, AZ*
[8] *Department of Geological Sciences, Arizona State University, Tempe, AZ*
[9] *Department of Integrated Science and Technology, James Madison University, Harrisonburg, VA*
[10] *Department of Physics, Oklahoma State University, Stillwater, OK*
[11] *Applied Space, Hicksville, NY*
[12] *Department of Natural Science, University of Houston Downtown, Houston, TX*
[13] *Department of Earth, Atmospheric, and Planetary Sciences, Massachusetts Institute of Technology, Cambridge, MA*
[14] *Department of Geology and Environmental Sciences, College of Charleston, Charleston, SC*
[15] *NASA Ames Research Center, Moffett Field, CA*
[16] *Amercian Association for the Advancement of Science, Washington, DC*
[17] *Department of Physical Science, Suffolk County Community College, Selden, NY*
[18] *Columbia University Biosphere 2 Center, Oracle, AZ*

Abstract. The Education and Public Outreach Panel for the 2003-2013 Decadal Survey has generated this report to assist NASA and the planetary sciences community with planning and prioritization of E/PO activities. We recognize the significant achievements of the NASA Office of Space Science (OSS) E/PO efforts in recent years which address many of the issues noted in this report. The key issues identified by the panel are: Communication, Scientist Involvement, Public Relations, Access to

Resources, Underrepresented Groups, Leveraging, and Feedback to the Decadal Survey. Recommendations have been made in each of these key areas to enhance E/PO activities through the years 2003-2013. Also included are suggestions for maintaining future capabilities in several areas including Workforce Development and Teacher Training.

EXECUTIVE SUMMARY

The Education and Public Outreach Panel for the Decadal Survey was assembled under the auspices of the Division of Planetary Sciences of the American Astronomical Society. The members of the panel represent a variety of scientific interests and involvement in E/PO activities. Members cover the range from full time E/PO professionals to full time scientists, engineers and students. Panel members volunteered from large universities, NASA centers, small colleges, private industries, and other science and education institutions. This report is a compilation of the ideas generated by this diverse panel suggesting what key issues, priorities, and recommendations should be highlighted in continuted E/PO activities, by NASA and the greater planetary sciences community, over the 2003-2013 decade.

Education and Public Outreach (E/PO) activities are an integral part of NASA's mandated mission and detailed in its Strategic Plan. The Office of Space Science Solar System Exploration (OSS SSE) E/PO program has made great strides in defining priorities and achieving its goals in the last five years. The issues and recommendations identified by this panel have in many cases already been targeted and acted upon. They appear here so that the panel may present a complete list and underscore their importance in planning for the 2003-2013 decade. Key issues identified by this panel are: Communication, Scientist Involvement, Public Relations, Access to Resources, Underrepresented Groups, Leveraging, and Feedback to the Survey. The panel has developed recommendations to augment OSS SSE E/PO activities in the next decade in each of these key areas.

The communication between planetary scientists and the OSS E/PO Support Network needs significant strengthening. Such communication is a two way street, with both educators and scientists needing to take an active role. OSS should consider a promotional campaign to underscore what the OSS E/PO is doing, why scientists should want to be involved, and how scientists can get into E/PO activities most effectively.

Because of tight funding in all phases of their work, planetary scientists are spending most of their time just maintaining status quo doing research science. Old paradigms of scientific elitism and lack of encouragement for E/PO activities stop many scientists from getting involved. Having limited venues for involvement, and problems with funding cycles and mission-oriented funding create problems for scientists interested in E/PO activities. The crowning barrier to deeper involvement is the seeming lack of rewards or incentives for scientists from their home institutions and professional societies. Change must be initiated by management of those places where scientists are employed so that E/PO is truly rewarded with funds and tangible career progression.

Planetary scientists, in general, do not understand and are confused by the distinction between E/PO and Public Affairs at NASA. NASA's codes should fully integrate their needs and plans and present a consistent, unified strategy and requirements to the community it funds. OSS (Code S) and Public Affairs (Code P) should together define the goals for public affairs and E/PO. Synergy between E/PO and Public Relations (Public Affairs) will allow each to work more effectively, and eliminate confusion in the public and scientific community.

The internet is the most high profile method to access resources at the present time. NASA should present one very easy to find and navigate site (or modify an existing site) for use as a clearinghouse by educators without the organizational divisions that have no significance outside of NASA. Such a site should have its content evaluated for accuracy by appointed scientists and educators. Those who maintain the site should also develop a dynamic plan for updating it so that it continues to provide value.

Further work needs to be done to assure that women, minorities, the financially disadvantaged and the disabled have equal access and are being effectively reached at all levels by science education endeavors. It may be helpful to have more materials translated into languages such as Spanish or presented in Braille. There is a large subset of the population that cannot be reached by internet programs because they have little or no access to computers. NASA TV should be examined to see if there are other, more productive means of utilizing it as a E/PO resource, especially for those who do not have computer access.

Creative and innovative relationships and methods have evolved from OSS activities. Efforts need to continue to make good programs and products more universally available. The best means of generating more at lower cost is to better utilize existing programs within NASA as well as state-based organizations. The OSS should emphasize partnering with community colleges as a cost effective means of reaching large numbers of college students. In addition to the OSS Education Conference, NASA should sponsor a national meeting bringing scientists and educators together to develop a strategic plan for how NASA can achieve the greatest impact for its investment.

The Decadal Survey as a whole will generate a series of science recommendations for the OSS to follow over the next ten years. These will have direct impact on E/PO. It is important for the OSS to know how the planetary science community views the relevance of these recommendations to the public, and all audiences served by OSS E/PO programs. Once the science priorities have been determined the OSS should return to the consideration of E/PO issues, and determine how scientists view their public engagement potential.

In addition to the key issues and recommendations, this panel has also made several suggestions to maintain future capabilities. A comprehensive plan should be initiated to ensure workforce development in the areas of science, engineering and technology. Funding sources should remain stable to ensure that the investment made in developing new resources is not lost when the time comes to maintain those resources. Emphasis on the professional development of teachers will continue to be very important for science education in the 2003-2013 decade.

REPORT

1. Scope

This report is intended for use by the National Aeronautics and Space Administration (NASA) to help identify, define, and implement improvements and new programs related to its education and public outreach (E/PO) efforts. Specifically, this report is targeted at the needs of the Office of Space Science (OSS) division of Solar System Exploration, and the planetary sciences community at large, for the decade 2003-2013. However, since education and outreach are a part of every division of the OSS, overlap with other programs is to be both expected and desired. Therefore many of the issues and recommendations discussed in this report are relevant to all areas of science education at NASA. The reverse is also true; how NASA conducts E/PO in other divisions and departments is relevant to the functioning of OSS Solar System E/PO endeavors.

2. Current State of Knowledge - Background and Perspective

Before defining goals and priorities for the next decade, we first present a general look at NASA's overall motivations for involvement in E/PO and what large scale structure has been established. This provides a context within the current NASA E/PO framework for the interpretation of the issues and recommendations that we present in this report.

2.1. NASA motivations and objectives

Why is NASA involved in Education and Public Outreach, and what are its stated motivations and objectives? Both the National Space Act of 1958, and NASA's Strategic Plan define the foundation on which NASA has built its programs for E/PO.

The National Space Act of 1958 mandates that NASA Administration "provide for the widest practicable and appropriate dissemination of information concerning its activities and the results thereof." A key element of NASA's strategic plan is to "Communicate Knowledge". The strategic plan further states "The knowledge generated by NASA's activities is without purpose if it is not shared with those who can use it. This includes ... science and technology communicators such as educators, publishers, museums, and the media, and every citizen of the United States and the world." The Plan outlines specific goals in E/PO with respect to its major areas of operation in: Space Science (Share the excitement and knowledge generated by scientific discovery and improve science education), Biological and Physical Research (Use space research opportunities to improve academic achievement and the quality of life), and Human Exploration and Development of Space (Share the Experience and Benefits of Discovery).

2.2. NASA programs to achieve education objectives

NASA has developed programs both in and outside of the OSS to define and meet E/PO goals and objectives. Further information on NASA's present position, current direction and "state of knowledge" in education can be therefore found by looking, for example, at the NASA Education Implementation Plan, the Space Grant Strategic Plan, and OSS E/PO Strategic Plan, amongst others.

The NASA Education Implementation Plan notes NASA's desire to strive for "Educational Excellence", involving "the educational community in our endeavors to inspire America's students, create learning opportunities, and enlighten inquisitive minds." The Mission statement in this Plan is for NASA to "use its resources to support educational excellence for all." The customers of this department are the formal and informal education community. ("The formal community is defined as K-12, Community College, Undergraduate, Graduate, and Post doctoral. The informal community includes both K-12 and postsecondary levels, as well as science and technology centers, museums, planetariums and other nonprofit education organizations.") This department is targeting certain improvement initiatives: Focus and Coordinate State-Based Efforts, Enhance Instructional Products and Dissemination, Improve Education Program Integration and Coordination, Facilitate NASA Research in the Education Community, Support Pre-service Education, and Target Informal Education.

The Strategic Plan of the NASA Space Grant College and Fellowship Program lists a number of priority items which show a grasp of current trends and needs for E/PO in general, although naturally specifically targeted for undergraduate education including: Teacher preparation, classroom teaching, research opportunities for undergraduates, and interdisciplinary curricula; Formation of partnerships between engineering schools and education institutions, industry, and government; Stress development of interdisciplinary courses and curricula; Enhance pre-college teacher education (Pre-service) programs through collaborations among education, science, and engineering disciplines; Promote development of community college initiatives, high technology and distance learning projects, career retraining, and the development of new technical courses; Aeronautics, science, and technology-based programs or activities that serve industry, state or local government, nonprofit organizations, children and adults such as non-technical courses, lectures, science fairs, and assorted media format, museum and planetarium programs, and other activities that are offered outside of a formal academic setting and; Focus on involving women, underrepresented groups, and persons with disabilities in all aspects of education.

The Earth Science Enterprise has established a set of basic principles which for the framework that supports the development of the ESE Education Program. Earth science is strongly connected to Solar System science, providing the best in-depth look at a planet. The ESE principals are: to demonstrate relevance to the society, to operate and work within NASA's strategy for education, to focus the implementation of a sustainable Earth system science education program that is consistent with externally imposed education standards, to increase the involvement of ESE scientists in education, to involve teachers in the development and decision-making aspects of education activities, to coordinate (and perhaps integrate) strategy and programs with other agencies and organizations, to ensure equity and diversity in all ESE education activities, and finally to leverage the resources of external groups.

2.3. Office of Space Science E/PO

Along with E/PO conducted by these programs, the Office of Space Science (OSS) has developed its own program for E/PO. OSS has stated its vision and mission for education in its Strategic Plan as follows "The Office of Space Science will use its knowledge and discoveries about the Sun, the solar system, the galaxy, and the universe to develop education and public outreach opportunities and activities that enhance science, mathematics, and technology education and the scientific and technological literacy of all Americans. The education mission of OSS is to engage its community of research scientists, managers, engineers, and support staff across the nation in education and public outreach activities to fulfill its vision. To this end, OSS will use all its resources, specifically drawing on the inspirational nature of space science; the remarkable results from its missions; and the special talents of its research community in universities, institutes, and laboratories throughout the country."

To meet its goals the OSS has developed a "Support Network" of Forums and Brokers/Facilitators. Each Forum is associated with a scientific theme, and each Broker/Facilitator associated with a geographical region. The Forums work both independently and in partnership on certain programs. E/PO is supported by the OSS in several core ways in addition to the Broker/Facilitator and Forum programs. E/PO is an integral part of all space science missions. E/PO is optionally available to all OSS supported research activities, and also is funded independently of these activities through the IDEAS program - which supports scientist-educator partnership activities. The OSS specifically has targeted areas for continued development as listed in The Annual Report of the OSS E/PO.

The specific goals of the OSS E/PO are:

- Use space science missions and research programs and the talents and resources of its research and development communities to make significant and measurable contributions to meeting national goals for the reform of science, mathematics, and technology education, particularly at the K-14 level, and the general elevation of scientific and technological literacy throughout the country.
- Continue to support the education and training of graduate students and postdoctoral fellows in space science in order to create the talented scientific workforce needed for the 21st century. OSS will expand its efforts in support of the education and training of undergraduate students.
- Promote the involvement of women, underrepresented minorities, and students with disabilities in its educational programs and their participation in space science research and development activities.
- Facilitate and cultivate strong and lasting partnerships on local, regional, and national scales between the space science research and development communities and the professional communities in science, mathematics, and technology education.

2.4. OSS Solar System Exploration E/PO

As listed above, NASA has identified and outlined important objectives that are common to all of its E/PO programs, including those of the OSS. These programs tend to work independently of both one another and of public relations activities. The specific Solar System Exploration E/PO program of the OSS (that which is most relevant to the scope of this report) places emphasis on the following objectives:

- Engaging its community of planetary scientists in education and public outreach activities
- Inclusion of minorities, traditionally underrepresented groups, women and the disabled
- Development of partnerships with groups and organizations outside of NASA
- Improving access of scientists and educators to resources for E/PO

The SSE E/PO community has already put in motion several programs to address these objectives, the details of many of which follow as each key issue is discussed. This panel hopes that this report will provide useful recommendations to the OSS SSE E/PO effort, and the general planetary sciences community, for further addressing these stated objectives and the key issues detailed below.

3. Key Education and Outreach Issues

Key issues which this panel believes need to be addressed in the next decade include both items that have been previously identified (as noted in the Background and Perspective section of this report) as well as others that have been seen to require new emphasis based on recent changes in both science and education. Some of these issues are very general, and apply to all E/PO endeavors, and some are tightly focussed to the programs and needs of the OSS.

It is important to note that the OSS has already independently identified and begun to work on addressing some of these issues. The presentation of those issues here reinforces their importance as seen by this panel and the communities they represent. We recognize efforts of the OSS on some of these issues where the panel is aware of them. In many cases this work has not yet had time to generate all the results that are expected, therefore further ventures in these areas that may result from recommendations in this report must be considered in the light of any existing programs.

The major issues identified by this panel are:

- Communication: The need for more communication with NASA OSS regarding E/PO, and assistance in understanding the needs of the education community that NASA OSS is being asked to serve.
- Scientist Involvement: The need for more effective and widespread involvement of planetary science professionals in E/PO activities. This requires addressing several topics:
    - Eliminating Elitism
    - Validating of E/PO for Scientists

  - Expanding Roles for Scientist Involvement
  - Improving Funding Cycles/Mission Centered Funding

- Public Relations: The perceived confusion between NASA E/PO and public relations
- Access to Resources: Access to resources for EPO for scientists, educators and partner organizations.
- Underrepresented Groups: Reaching traditionally underrepresented groups, women, minorities, and the disabled.
- Leveraging: Making consistent use of existing programs and resources.
- Feedback to Survey: Putting in place a system to gain feedback from scientists and others on the public engagement potential of the final science recommendations of the Decadal Survey.

3.1. Communication

At the crux of a successful implementation of E/PO as part of NASA OSS activities is the essential link of communications between the communities involved. Open and effective methods of dialog and information exchange must be provided for between the E/PO infrastructure of NASA OSS/SSE and the planetary science community. The responsibility of this communication rests equally on both groups.

Communication is also important because it allows for the free flow of information from one group to another so time is not wasted developing resources that exist elsewhere. It also allows for fruitful collaborations to be created which can stretch budget dollars further for all organizations involved.

In addition, the planetary science community needs to understand the needs of the communities it is being asked to serve through its E/PO efforts. Creating useful programs for educators requires an understanding of their needs based on their audience and their general experiences. We anticipate the first OSS-sponsored education conference June 12-14, 2002, and the associated proceedings document, as an opportunity for developing communication with and knowledge about the education community.

3.2. Scientist involvement

Getting scientists involved *effectively* in E/PO activities is one of the most important issues that must be addressed in the next ten years. This is a highly complex topic as current paradigms, programs, and approaches to scientist involvement are all in need of change in order to improve the level of involvement of planetary scientists in E/PO activities. We attempt here to mention some of the key points that we have identified as critical to improving scientist involvement.

Eliminating elitism: Scientific Elitism, in its many forms, is the basis of the current paradigm that separates scientists from more personal and effective involvement in E/PO activities. What follows is a stereotypical description of the current paradigm, but the overall impression it provides is accurate. From the scientist perspective, research science is the pinnacle of learning. The mindset has been one of tenured faculty taking on apprentice graduate students, and then in due time producing a "clone" of the mentor who continues working in the same area and who, of course, obtains the same tenured faculty position at a university. While this was an accurate model of the development of the mentor/student relationship 30 years ago, it is no longer a viable model. This mindset includes the idea that other pursuits such as education and public outreach are "lesser" and a waste of time, and those involved in these endeavors are also somehow lesser because they could not "cut it" to go into, or remain in, science. The mindset propagates into the perception of how the scientist relates to the public.

The public is viewed as simple, uneducated and lacking in critical thinking skills. It is not generally viewed as the client or end consumer of what the scientist produces. This elitism also makes any synergistic communication with the education community problematic at best. Eliminating this ineffective, unrealistic, and isolationist paradigm is a critical issue which needs to be addressed. Of course not all scientists feel this way, but some aspects or ripples of the elitism paradigm effect almost every program or institution that includes research scientists.

Validating E/PO for scientists: One of the contributing factors to Elitism is the fact that E/PO activities for scientists are not validated and rewarded. The scientific culture of "publish or perish" leaves little room for time to be spent on E/PO. Any amount of time spent on an area that is not perceived to further one's career or that does not directly contribute to one's livelihood is, not surprisingly, considered a poor use of time. Although many institutions say they support E/PO activities, when the time comes to reward or advance a scientist who has given time to these activities (say the award of tenure for a scientist in a university environment) some institutions use the involvement as evidence against the serious career research intentions of the scientist. Culturally, scientists did not go into their field as a public service. In order to change the view of scientists, the organizations that employ and fund them must also change their paradigm to view E/PO as necessary and valuable, and reward and fund scientists for their involvement.

Expanding roles for scientist involvement: At present there is a perception that there are few acceptable ways for a scientist to be involved in E/PO, such as teaching a formal class or giving a public talk. Many scientists do not want to be involved in E/PO because they are not good at teaching or public speaking. Some believe that E/PO must include interfacing with young people, and they are unsure of their abilities with children. They therefore avoid E/PO activities. However, education and outreach can be targeted for each scientist according to their strengths. Those who have abilities with the web could be called upon to help evaluate and post content to education websites. For those who do well mentoring students, opportunities (such as Space Grant) should be provided for matching students with prospective scientist mentors. For those who are good writers, education pamphlets or even books for public consumption could be an approved E/PO activity. NASA OSS is currently addressing this issue by raising awareness within, as well as identifying opportunities for, the scientific community, a variety of types of participation. The "Explanatory Guide to NASA Office of Space Science Education and Public Outreach Evaluation Criteria" discusses general types of activities that involve scientists and which are supported in mission and NRA E/PO programs. We recognize the recent restructuring of the NRA E/PO proposal process. This is accompanied by the Internet-based "Menu of Opportunities for Scientists in Education" (MOSIE), which lists specific opportunities for scientist participation in support of E/PO for NRAs awarded by OSS or volunteer efforts, and we anticipate its further population. Further access to the informal education community is provided by the "Space Science Access" website for museums and planetariums.

Improving funding cycles: The means whereby E/PO is funded is not amenable to easy scientist involvement. Most scientists are funded in a "soft money" situation. In order for funding to be useful it must be of a significant amount and time span. The attachment of E/PO proposals to science proposals through the OSS is a frustrating enterprise. The E/PO proposal is an afterthought because it is rarely funded, even when scientists receive excellent reviews for their E/PO proposals. It is no longer viewed as a good use of time to submit such proposals at all.

The mission funding cycles are also not realistically thought out. E/PO for the missions are initiated as much as ten years in advance of a mission returning any data at all. The requirement that some funds be allocated for E/PO for each funded mission is not well thought out, and presents impediments for more than a small number of scientists to be constructively involved.

The idea of attaching education to major programs has many problems that are not related directly to encouraging scientist involvement, one being the quality of programs and products emerging from them varies greatly. In some instances, the focus still is not on what the students (and educators) need or want. The focus in these cases is often on the major program. The emphasis on educational needs and standards can become an afterthought to the promotion of the mission or research itself.

Further, attempting to promote the relevancy of each mission to members of the general public may not be the most effective way to communicate why NASA's exploration of the solar system matters to them.

3.3. Public affairs/Public relations

There is perceived confusion between how NASA views E/PO and how it views public affairs (also referred to as public relations). These are not the same thing, although there can be overlap in areas such as press releases, press conferences, etc. NASA further divides the two by operating each out of different codes at HQ, who are apparently not synergistic in their needs and even communications about what is important or how scientists might address their needs.

3.4. Access to resources

Particularly as related to the Internet, the concept of "data" storage appears to be very important. "Data" in this context could mean actual science data, but more to the point would be lesson plans, laboratory write-ups, reports, etc. that could be of use to educators. These data can be highlighted by a single Internet site designed to serve as one point which all educators could turn to in order to find space science related resources and materials that are in some way related to NASA and its mission.

We recognize the recent unveiling of the on-line Office of Space Science Education Resource Directory as a significant movement in this direction. This searchable directory is focused on the needs of educators, and provides access to OSS educational and supplementary materials that are available in electronic format. In addition, this site is structured in a standard way that allows the materials to be fed into the Eisenhower National Clearinghouse website that provides access to all educational material produced by government agencies. The OSS site provides ready links for scientists so they have access to existing, tested, and approved materials. The site was developed with considerable beta-testing in the education community.

It is also critical to provide materials that are scientifically validated. We recognize the recent implementation by OSS of an independent product review process involving the evaluation of OSS materials by educators and scientists.

K-12 teachers in particular need to be involved early and at all stages of planning new programs. Any activity that does not respect their knowledge, ability and experience is unlikely to succeed in the long term. The projects that bring these teachers in as partners from the outset have the best chance of being successful. NASA OSS should therefore acquire K-12 teacher input on new plans and activities, as well as this white paper.

3.5. Underrepresented groups

We recognize that OSS has made some valiant forages into serving the needs of underrepresented groups with its Minority University Initiative, its work with Minority Professional Societies, and a large number of other collaborations (see, for example, the OSS Support Network efforts documented at http://serch.cofc.edu/serch/special.htm). Nevertheless, further work needs to be done to assure that women, minorities, and the disabled have equal access and are being effectively reached at all levels by science education endeavors. Also of note are the financially disadvantaged. With the advent of home computers and the Internet in the last few decades, science education has a new and powerful arena in which to operate. However there is a large subset of the population that cannot be reached in this fashion because they have little or no access to computers.

3.6. Leveraging

In an attempt to develop E/PO programs from a grass-roots level, many creative and innovative relationships and methods have evolved from OSS activities. This takes advantage of different external expertise and experience and avoids favoritism or focus on single organizations. However, much of the lessons learned and knowledge gained in these relationships and programs remains with those directly involved with the individual program. Efforts need to continue to make good programs and products more universally available. Individual partnerships between scientists and educators should not be overlooked, and can eventually lead to broader involvement.

3.7. Feedback to survey

It is extremely important for those involved with OSS SSE E/PO to know how the planetary community views the public engagement potential of the final science recommendations that come from this Decadal Study. The study will generate a large number of recommendations and possibly initiate restructuring of science priorities in the NASA OSS. These will have direct bearing on E/PO programs and activities. It is important that the SSE E/PO community be able to ascertain the view of the planetary science community on the relevance of any new priorities and recommendations to the public, and all audiences served by SSE E/PO activities.

4. Recommendations

4.1. Communications

Communication is the critical link for improving E/PO over the next decade. Clearly, this is a very complex endeavor and requires the expertise of many people and organizations with diverse skill sets (scientists, teachers, science educators, teacher educators, informal science educators, educational technology specialists, etc.). OSS E/PO should continue to be inclusive, and find new ways to communicate with this large and varied population. While scientist involvement is critical, the input of people from all these areas will be required to meet the goals of the next ten years and beyond.

NASA OSS must seek the most effective ways of disseminating information for easy access to the planetary science community. Equally, the planetary science community must take responsibility upon itself to recognize the importance of reading and absorbing this information and seeking assistance and guidance from NASA OSS in its implementation. One suggestion to improve the dissemination of information is to have a publicity campaign about why scientists should do E/PO, what types of E/PO are available, and where to go for more information (and help when necessary). This campaign would have several key events, as well as smaller events at conferences, meetings, and perhaps a traveling "exhibit" to participating institutions.

The "Space Science Access" website for museums and planetariums should be expanded and advertised fully in the planetary science community, who should also be encouraged to contribute collaborative ideas and projects.

4.2. Scientist involvement

Scientists must view involvement in E/PO as a means for furthering their career. They need to recognize that this is an integral part of every scientist's job, particularly those that are being funded from public money. This may require a redefinition or rethinking of what E/PO is for scientists and how they are rewarded for this activity. Innovative ways must be identified and implemented to reward and encourage scientists for active involvement in E/PO. Ideas must be collected about how the culture can be encouraged to change and evolve. E/PO activities must carry as much prestige for scientists as research does. The home institutions of scientists like universities, institutes, centers, etc., must recognize, require, and reward E/PO involvement.

This panel recommends that NASA OSS convene a group of scientists, administrators of institutions that house the scientists, and experts and advocates for scientist involvement in the communication of science (National Science Foundation and the American Association for the Advancement of Science) to specifically address these issues. NASA should allocate a specific amount of funds to both fund the group and to carry out their imperatives. Questions they could ask are what would it take to change the way scientists feel about education, where should the culture be going, and what is NASA's overall vision for continued communication between the education and scientist communities. Such a panel should not primarily consist of people who have vested interests in existing E/PO programs and architectures at NASA.

We have generated some suggestions which the recommended group of experts could consider. Some E/PO funds cannot be used for salary, only materials. Institutions should be encouraged or required to provide matching funds, which can be used for salary. That would encourage scientists to apply for E/PO money. NASA has recently been giving very visible awards for E/PO, such as NASA's Exceptional Service Medal being awarded for solar system outreach. This should be encouraged. Major societies (like AGU, Meteoritical Society, etc.) should offer prestigious awards to those who excel at outreach activities. The DPS offers such an award. In addition to this, award of medals which are not directly assigned on the basis of E/PO activities should not be hindered by the scientist having been involved in E/PO. Many scientists will not "jeopardize" their long-term chances for a prestigious society metal by spending time in E/PO, so such activities should be viewed favorably by committees which choose awardees. In addition to these suggestions, programs targeting the education and involvement of graduate students could support the needed culture change.

Some scientists are not involved in E/PO because they do not understand the system in place to help them get connected. Part of this problem goes back to Communications, already discussed. NASA OSS should carefully define the nature of the Broker/Facilitator and Forum representatives. NASA should consider working to eliminate any possible conflicts of interest between scientists and educators/brokers proposing for E/PO money.

We await the availability of the electronic Journal of Astronomy in Education developed by the National Optical Astronomy Observatories and the Astronomical Society of the Pacific as an avenue for peer-reviewed publication of work done in this area by planetary scientists. Such a journal or journals dedicated to publication of peer-reviewed work will allow scientists to continue to publish scholarly work at the same rate (or better) while involved in high level E/PO activities.

We recommend that further study into more flexible avenues for scientist involvement be done (or better promotion for those studies in this area which have been done already) which may result in more scientists being interested in these opportunities rather than turned off or intimidated by them. PI's do not want to have the only option for involvement be "signing on" to prepackaged, large programs. The system should allow the flexibility to encourage an investigator involved in small, local programs to succeed alone. If PI's can only subcontract parts of their proposals to established institutes to succeed then opportunities for innovative scientist involvement will be missed.

NASA should carefully examine how funding cycles impact the desire of scientists to participate in education and outreach. Understanding that the vast majority of scientists are NOT hard money, but instead live grant-to-grant will promote a more friendly and workable environment for scientists to be involved and active in E/PO.

NASA OSS must understand its major project/mission goals versus education and outreach goals. The needs of students and educators for a useful end product should be the top priority of E/PO. Having each E/PO activity directly based to specific missions may not best suit this purpose. NASA OSS should emphasize the funding of a portion of E/PO activities independent of missions and proposals to lessen the possibility for conflict. Integration activities such as the "National Science Education Standards Quilt" (from JPL) for access to standards-based solar system exploration products and the space science concepts matrix (from GSFC) should be supported and integrated NASA-wide. Programs such as the IDEAS program should be re-examined and enhanced if necessary to continue to provide a venue for non-mission related activities.

In order to make our missions and research more relevant to the public, NASA should emphasize an integrated picture of the solar system and its exploration. The PI of any large mission which has E/PO funds attached must give high priority to E/PO. The rest of the mission team will take their lead from the PI - if the PI regards E/PO as important and necessary, then the rest of the team will be willing to assist those involved with E/PO in their efforts.

4.3. Public affairs/Public relations

In order to provide a consistent story to its constituents, NASA's codes should fully integrate their needs and plans and present consistent, unified strategy and requirements to the community it funds. In particular, OSS (Code S) and Public Affairs (Code P) should together define the goals for public affairs and E/PO. Particularly when initiating a new mission activity, the public affairs goals and budget, and the E/PO goals and budget should be set out clearly and separately. The same holds true for NASA's Education Division (Code FE). The E/PO goals must be centered on what educators and students need, not from a scientist or mission oriented perspective. These messages should be consistently and coherently communicated and practices in all these offices. Experience with previous mission E/PO activities (such as NEAR Shoemaker) have pointed out that a clear division between E/PO activities and PR activities needs to be made early in the mission process. Having E/PO and PR activities defined well and separately, but working in tandem will allow for the most cost effective outreach. PR and E/PO should work synergistically, with PR getting the public "turned on" to science, and then E/PO activities providing a wide net to effectively turn this interest into understanding.

4.4. Access to resources

NASA should present one very easy to find and navigate site (or modify an existing site) for use as a clearinghouse by educators. There appear to be such central sites for each NASA code, as well as the education division, but "one-stop-shopping" with a seamless access to all NASA material should be available, and presented without the organizational divisions that have no significance outside of NASA. The content of the unified site should be synthesized, evaluated, and actively moderated by a special group of educators and scientists. Only approved materials should be included in the repository. The information that resides in the repository could include duplicate content available elsewhere (with permission). NASA OSS should consider coordinating this effort with other nascent Digital Library for Science, Math, Engineering and Technology projects going on around the country. NASA should consider allowing non-NASA materials (or materials produced without NASA funding) to be included provide they pass the review process. This will allow for better leveraging and cost effective presentation of material.

The moderator of the web site should collect information and comments from users of the site as to how the site could be improved and how it should change and evolve to meet the needs of students and educators. Evaluation of any resource is a critical component to its development and upkeep.

While the OSS site provides "one-stop-shopping" access to material that is available and/or downloadable electronically, educators still need access to print and multimedia material. The relationship with the NASA Central Operation for Resources for Educators (CORE) should evolve to include access to these OSS materials, at low-cost to educators and at no reproduction cost to the providing mission or scientist.

Websites that are related to outreach for current, high profile missions need special attention. Such sites (as we know from Pathfinder) can take massive numbers of hits in a short time period.

4.5. Underrepresented groups

All NASA programs have mentioned the need to reach out to underrepresented groups. This should persist as a priority. Special programs should be continued and augmented to help reach women and minorities. Programs for the disabled have been the most lacking, and a great deal more can be done here to reach this population with science education. For both of these groups it may be helpful to have more materials translated into languages such as Spanish or presented in Braille. For the financially disadvantaged, computer access is a major issue. For this group it is likely that television is still one of the best ways to reach and teach. This should be examined more closely, and television should not be dismissed as "old hat" as compared to Internet-based teaching, but instead each outreach method should be evaluated for its particular strengths in education and employed on that basis. NASA TV should be examined to see if there are other, more productive means of utilizing it as a E/PO resource, especially for those who do not have computer access. In addition to this, for those who do not have cable access, we recommend the OSS continue study on the potential of broadcast TV.

4.6. Leveraging

Further communication of the elements of successful programs and relationships should be communicated and shared among those OSS-funded E/PO efforts. An emphasis should be placed on modeling or expanding existing approaches to working with the education community and to bringing the results of our programs to the public, and on growth in new directions. NASA must be flexible to reach its E/PO goals. The OSS E/PO should continue to look outward (as well as inward) to establish partnerships with other agencies/ organizations.

The strongest and most long-lasting leverage of existing groups may come from modeling and expansion of programs which work with the pre-service teachers (teachers in training) and finding further ways to understand what their needs are, especially for K-6 teachers whose content domain spans the entire universe of topics. This requires both work with content courses and the methods courses in our Schools/Colleges of Education. It is important to note that such educators may be limited in their interactions by rigid state rules for teacher accreditation that make change a challenge. This should include a priority to expand partnerships with colleges, universities, and K-12.

Of note are community colleges, which educate vast numbers of first and second year college students, but have not acquired the same amount of perceived importance as traditional four year universities. Partnering with community colleges may be a very cost effective means to reach large numbers of college students.

The best means of generating more at lower cost is to better utilize existing programs within NASA (NASA Space Grant program, Solar System Ambassadors, Solar System Educators, OSS Broker and Facilitator, NASA's Educator Resource Center network), between NASA and other agencies (EPSCoR), as well as state-based organizations including the NSF/ED Systemic Initiative, the National Alliance of State Science and Mathematics Coalitions, the Aerospace States Association, the National Aerospace Education Alliance, the Space Grant Directors Council, the Association of State Supervisors of Mathematics, the Council of State Science Supervisors, and relevant education associations (NSTA, NCTM, ITEA, ASEE, USRA, and GENIP) and the NASA OSS. This will require continued emphasis on communication and partnership building. Further communication should be developed to promote and utilize NASA on-line educational resources such as NASA Spacelink, NASA Television, and the NASA Learning Technologies Project. The OSS should also continue to leverage partnerships with AAPT, AAS - Education Office, ASP and NGTA.

In addition to the OSS Education Conference, NASA should sponsor a national meeting bringing scientists, educators, and folks that work with both populations (science educators, educational technologists, Planetary Society, ASP) to brainstorm a strategic plan for how NASA can achieve the greatest impact for its investment. DPS members could be called upon to provide a leadership role in such a meeting. NASA OSS should be sure that representatives of "informal education" (science centers, museums, planetariums, etc.) are well-represented at such a meeting, and in offering input on new initiatives and programs that influence them.

NASA OSS should be certain to recognize the achievements of smaller programs which may be involved in large (or small) missions or other E/PO endeavors. Being willing to properly and publicly promote and share proper credit (particularly with organizations NOT related to NASA) will encourage partnership with all organizations, and eliminate some of the frustration of being a small (but important) part of a large E/PO project.

4.7. Feedback to survey

Obtaining feedback from the planetary science community on the E/PO engagement potential of the science priorities from this survey will need to be an iterative process. After the science recommendations and new priorities have been established, we recommend that the science community be tapped to discover their viewpoint on the relevance of the new priorities to the public and other outreach audiences. This could be worked out through the NRC and/or the DPS. This information will be of critical use to those conducting E/PO programs. This issue should therefore be kept open, to be addressed after the science survey recommendations have been finalized.

5. Maintaining Capabilities for the Future

It is clear that a consistent, reliable source of funding is the most important factor in maintaining the programs, partnerships and resources developed during the next ten years. Excellent web sites have been developed using NASA funding, but in spite of overwhelmingly positive feedback, funding was eventually dropped (see Volcano World). Web sites need constant upkeep and maintenance, or the investment originally placed to create them can become lost. When NASA agrees that the creation of a resource, such as a web site, is in order, then NASA should commit the funds for both the development and maintenance of that resource.

There is interest and demand for a variety of products from straight public affairs shots to lesson plans to original data, however there is little market for it. Any steps suggested or taken by NASA OSS need to have some method of support that does not involve schools buying products or subscriptions, because the funds for such transactions are simply not there. NASA will be required to support such initiatives on its own.

Any plans for change, new programs, activities and initiatives taken by the NASA OSS should be emphasized and targeted for the long term. Reform trends which are not carefully implemented with a consideration for their long term effectiveness waste time and money. Any new programs that are not practical and useful will eventually disappear. Teachers and other educators will just wait out changes they think are ill-advised, or have had no input on until they pass (the half-life seems to be about 5 years).

NASA OSS should consider involvement in a comprehensive plan of workforce development. This will help assure that over the next ten years that young people continue to consider technical careers. This plan might also branch out to include investigating how to encourage those with the necessary interest and education to continue their involvement or employment in technical/science careers and endeavors. Such a plan should be developed in tandem with NASA's overall program to enhance the number of young people seeking employment in technical fields.

An effective approach to long term development is to emphasize teacher education. An expert teacher who is always present at a school or museum is an important resource. NASA OSS should continue to support and investigate ways of promoting the development of Expert or "master" science teachers who can then educate both students and other teachers.

Obtaining feedback from the planetary science community on the E/PO engagement potential of the science priorities from this survey will need to be an iterative process. After the science recommendations and new priorities have been established, we recommend that the science community be tapped to discover their viewpoint on the relevance of the new priorities to the public and other outreach audiences. This could be worked out through the NRC and/or the DPS. This information will be of critical use to those conducting E/PO programs. This issue should therefore be kept open, to be addressed after the science survey recommendations have been finalized.

References, Sources, and Additional Reading

Sakimoto, P., Ed. 2001. *OSS E/PO Annual Report FY 2001*, NASA
http://ossim.hq.nasa.gov/ossepo/

Partners in Education: A Strategy for Integrating Education and Public Outreach into NASA's Space Science Programs, OSS E/PO Strategic Plan
http://spacescience.nasa.gov/admin/pubs/edu/educov.htm

NASA Strategic Plan and other Plans
http://www.hq.nasa.gov/office/codez/plans.html

Jesse, J.K. 2002. *Education and Human Resources in the FY 2002 Budget*, American Association for the Advancement of Science.
http://www.aaas.org/spp/dspp/rd/xxvi/chap5.htm

Chapter 8 of the National Science Foundation's *Science and Engineering Indicators 2000,*
http://www.nsf.gov/sbe/srs/seind00/frames.htm

National education standards
http://education.nasa.gov/k12.html

NASA Education Website
http://education.nasa.gov/

http://ssibroker.colorado.edu/Broker/Eval_criteria/Guide/

NASA Space Science Education Resource Directory
http://teachspacescience.stsci.edu

NASA Central Operation of Resources for Educators
http://core.nasa.gov/

NASA Space Science Education page
http://spacescience.nasa.gov/education/index.htm

http://spacescience.nasa.gov/education/resources/ecosystem/index.htm

OSS E/PO 2001 Evaluation Report
http://spacescience.nasa.gov/education/resources/strategy/2001_Final_Report.PDF

Menu of Opportunities for Scientists in Education
http://ssibroker.colorado.edu/broker/mosie/

Space Science Access for museums and planetariums
http://cfa-www.harvard.edu/seuforum/wateringHole/

VolcanoWorld
http://volcano.und.edu

http://www.spacescience.org/Education/ResourcesForScientists/Workshops
/Four-Day/Resources/Articles/venn_paper.pdf

Earth Science Enterprise
http://www.earth.nasa.gov/education/edstratplan/index.html

IDEAS program
http://ideas.stsci.edu/

Subject Index

Author Index

A LIST OF THE VOLUMES

Published
by

THE ASTRONOMICAL SOCIETY OF THE PACIFIC
(ASP)

An international, nonprofit, scientific and educational organization founded in 1889

All book orders or inquiries concerning

THE ASTRONOMICAL SOCIETY OF THE PACIFIC
CONFERENCE SERIES
(ASP - CS)

and

INTERNATIONAL ASTRONOMICAL UNION VOLUMES
(IAU)

should be directed to the:

The Astronomical Society of the Pacific Conference Series
390 Ashton Avenue
San Francisco CA 94112-1722 USA

Phone: 800-335-2624 (Within USA)
Phone: 415-337-2126
Fax: 415-337-5205

E-mail: service@astrosociety.org
Web Site: http://www.astrosociety.org

Complete lists of proceedings of past IAU Meetings are maintained at the IAU Web site at the URL: http://www.iau.org/publicat.html

Volumes 32 - 189 in the IAU Symposia Series may be ordered from:

Kluwer Academic Publishers
P. O. Box 117
NL 3300 AA Dordrecht
The Netherlands

Kluwer@wKap.com

ASP CONFERENCE SERIES VOLUMES

Published by the Astronomical Society of the Pacific

PUBLISHED: 1988 (* asterisk means OUT OF STOCK)

Vol. CS -1 PROGRESS AND OPPORTUNITIES IN SOUTHERN HEMISPHERE OPTICAL ASTRONOMY: CTIO 25TH Anniversary Symposium
eds. V. M. Blanco and M. M. Phillips
ISBN 0-937707-18-X

Vol. CS-2 PROCEEDINGS OF A WORKSHOP ON OPTICAL SURVEYS FOR QUASARS
eds. Patrick S. Osmer, Alain C. Porter, Richard F. Green, and Craig B. Foltz
ISBN 0-937707-19-8

Vol. CS-3 FIBER OPTICS IN ASTRONOMY
ed. Samuel C. Barden
ISBN 0-937707-20-1

Vol. CS-4 THE EXTRAGALACTIC DISTANCE SCALE: Proceedings of the ASP 100th Anniversary Symposium
eds. Sidney van den Bergh and Christopher J. Pritchet
ISBN 0-937707-21-X

Vol. CS-5 THE MINNESOTA LECTURES ON CLUSTERS OF GALAXIES AND LARGE-SCALE STRUCTURE
ed. John M. Dickey
ISBN 0-937707-22-8

PUBLISHED: 1989

Vol. CS-6 SYNTHESIS IMAGING IN RADIO ASTRONOMY: A Collection of Lectures from the Third NRAO Synthesis Imaging Summer School
eds. Richard A. Perley, Frederic R. Schwab, and Alan H. Bridle
ISBN 0-937707-23-6

PUBLISHED: 1990

Vol. CS-7 PROPERTIES OF HOT LUMINOUS STARS: Boulder-Munich Workshop
ed. Catharine D. Garmany
ISBN 0-937707-24-4

Vol. CS-8* CCDs IN ASTRONOMY
ed. George H. Jacoby
ISBN 0-937707-25-2

Vol. CS-9 COOL STARS, STELLAR SYSTEMS, AND THE SUN: Sixth Cambridge Workshop
ed. George Wallerstein
ISBN 0-937707-27-9

Vol. CS-10* EVOLUTION OF THE UNIVERSE OF GALAXIES: Edwin Hubble Centennial Symposium
ed. Richard G. Kron
ISBN 0-937707-28-7

Vol. CS-11 CONFRONTATION BETWEEN STELLAR PULSATION AND EVOLUTION
eds. Carla Cacciari and Gisella Clementini
ISBN 0-937707-30-9

Vol. CS-12 THE EVOLUTION OF THE INTERSTELLAR MEDIUM
ed. Leo Blitz
ISBN 0-937707-31-7

PUBLISHED: 1991

Vol. CS-13 THE FORMATION AND EVOLUTION OF STAR CLUSTERS
ed. Kenneth Janes
ISBN 0-937707-32-5

ASP CONFERENCE SERIES VOLUMES

Published by the Astronomical Society of the Pacific

PUBLISHED: 1991 (* asterisk means OUT OF STOCK)

Vol. CS-14 ASTROPHYSICS WITH INFRARED ARRAYS
ed. Richard Elston
ISBN 0-937707-33-3

Vol. CS-15 LARGE-SCALE STRUCTURES AND PECULIAR MOTIONS IN THE UNIVERSE
eds. David W. Latham and L. A. Nicolaci da Costa
ISBN 0-937707-34-1

Vol. CS-16 Proceedings of the 3rd Haystack Observatory Conference on ATOMS, IONS, AND MOLECULES: NEW RESULTS IN SPECTRAL LINE ASTROPHYSICS
eds. Aubrey D. Haschick and Paul T. P. Ho
ISBN 0-937707-35-X

Vol. CS-17 LIGHT POLLUTION, RADIO INTERFERENCE, AND SPACE DEBRIS
ed. David L. Crawford
ISBN 0-937707-36-8

Vol. CS-18 THE INTERPRETATION OF MODERN SYNTHESIS OBSERVATIONS OF SPIRAL GALAXIES
eds. Nebojsa Duric and Patrick C. Crane
ISBN 0-937707-37-6

Vol. CS-19 RADIO INTERFEROMETRY: THEORY, TECHNIQUES, AND APPLICATIONS, IAU Colloquium 131
eds. T. J. Cornwell and R. A. Perley
ISBN 0-937707-38-4

Vol. CS-20 FRONTIERS OF STELLAR EVOLUTION:
50th Anniversary McDonald Observatory (1939-1989)
ed. David L. Lambert
ISBN 0-937707-39-2

Vol. CS-21 THE SPACE DISTRIBUTION OF QUASARS
ed . David Crampton
ISBN 0-937707-40-6

PUBLISHED: 1992

Vol. CS-22 NONISOTROPIC AND VARIABLE OUTFLOWS FROM STARS
eds. Laurent Drissen, Claus Leitherer, and Antonella Nota
ISBN 0-937707-41-4

Vol CS-23 ASTRONOMICAL CCD OBSERVING AND REDUCTION TECHNIQUES
ed. Steve B. Howell
ISBN 0-937707-42-4

Vol. CS-24 COSMOLOGY AND LARGE-SCALE STRUCTURE IN THE UNIVERSE
ed. Reinaldo R. de Carvalho
ISBN 0-937707-43-0

Vol. CS-25 ASTRONOMICAL DATA ANALYSIS, SOFTWARE AND SYSTEMS I - (ADASS I)
eds. Diana M. Worrall, Chris Biemesderfer, and Jeannette Barnes
ISBN 0-937707-44-9

Vol. CS-26 COOL STARS, STELLAR SYSTEMS, AND THE SUN:
Seventh Cambridge Workshop
eds. Mark S. Giampapa and Jay A. Bookbinder
ISBN 0-937707-45-7

Vol. CS-27 THE SOLAR CYCLE: Proceedings of the
National Solar Observatory/Sacramento Peak 12th Summer Workshop
ed. Karen L. Harvey
ISBN 0-937707-46-5

ASP CONFERENCE SERIES VOLUMES

Published by the Astronomical Society of the Pacific

PUBLISHED: 1992 (asterisk means OUT OF STOCK)

Vol. CS-28	AUTOMATED TELESCOPES FOR PHOTOMETRY AND IMAGING eds. Saul J. Adelman, Robert J. Dukes, Jr., and Carol J. Adelman ISBN 0-937707-47-3
Vol. CS-29	Viña del Mar Workshop on CATACLYSMIC VARIABLE STARS ed. Nikolaus Vogt ISBN 0-937707-48-1
Vol. CS-30	VARIABLE STARS AND GALAXIES ed. Brian Warner ISBN 0-937707-49-X
Vol. CS-31	RELATIONSHIPS BETWEEN ACTIVE GALACTIC NUCLEI AND STARBURST GALAXIES ed. Alexei V. Filippenko ISBN 0-937707-50-3
Vol. CS-32	COMPLEMENTARY APPROACHES TO DOUBLE AND MULTIPLE STAR RESEARCH, IAU Colloquium 135 eds. Harold A. McAlister and William I. Hartkopf ISBN 0-937707-51-1
Vol. CS-33	RESEARCH AMATEUR ASTRONOMY ed. Stephen J. Edberg ISBN 0-937707-52-X
Vol. CS-34	ROBOTIC TELESCOPES IN THE 1990's ed. Alexei V. Filippenko ISBN 0-937707-53-8

PUBLISHED: 1993

Vol. CS-35*	MASSIVE STARS: THEIR LIVES IN THE INTERSTELLAR MEDIUM eds. Joseph P. Cassinelli and Edward B. Churchwell ISBN 0-937707-54-6
Vol. CS-36	PLANETS AROUND PULSARS ed. J. A. Phillips, S. E. Thorsett, and S. R. Kulkarni ISBN 0-937707-55-4
Vol. CS-37	FIBER OPTICS IN ASTRONOMY II ed. Peter M. Gray ISBN 0-937707-56-2
Vol. CS-38	NEW FRONTIERS IN BINARY STAR RESEARCH: Pacific Rim Colloquium eds. K. C. Leung and I.-S. Nha ISBN 0-937707-57-0
Vol. CS-39	THE MINNESOTA LECTURES ON THE STRUCTURE AND DYNAMICS OF THE MILKY WAY ed. Roberta M. Humphreys ISBN 0-937707-58-9
Vol. CS-40	INSIDE THE STARS, IAU Colloquium 137 eds. Werner W. Weiss and Annie Baglin ISBN 0-937707-59-7
Vol. CS-41	ASTRONOMICAL INFRARED SPECTROSCOPY: FUTURE OBSERVATIONAL DIRECTIONS ed. Sun Kwok ISBN 0-937707-60-0

ASP CONFERENCE SERIES VOLUMES

Published by the Astronomical Society of the Pacific

PUBLISHED: 1993 (* asterisk means OUT OF STOCK)

Vol. CS-42 GONG 1992: SEISMIC INVESTIGATION OF THE SUN AND STARS
ed. Timothy M. Brown
ISBN 0-937707-61-9

Vol. CS-43 SKY SURVEYS: PROTOSTARS TO PROTOGALAXIES
ed. B. T. Soifer
ISBN 0-937707-62-7

Vol. CS-44 PECULIAR VERSUS NORMAL PHENOMENA IN A-TYPE AND RELATED STARS, IAU Colloquium 138
eds. M. M. Dworetsky, F. Castelli, and R. Faraggiana
ISBN 0-937707-63-5

Vol. CS-45 LUMINOUS HIGH-LATITUDE STARS
ed. Dimitar D. Sasselov
ISBN 0-937707-64-3

Vol. CS-46 THE MAGNETIC AND VELOCITY FIELDS OF SOLAR ACTIVE REGIONS, IAU Colloquium 141
eds. Harold Zirin, Guoxiang Ai, and Haimin Wang
ISBN 0-937707-65-1

Vol. CS-47 THIRD DECENNIAL US-USSR CONFERENCE ON SETI -- Santa Cruz, California, USA
ed. G. Seth Shostak
ISBN 0-937707-66-X

Vol. CS-48 THE GLOBULAR CLUSTER-GALAXY CONNECTION
eds. Graeme H. Smith and Jean P. Brodie
ISBN 0-937707-67-8

Vol. CS-49 GALAXY EVOLUTION: THE MILKY WAY PERSPECTIVE
ed. Steven R. Majewski
ISBN 0-937707-68-6

Vol. CS-50 STRUCTURE AND DYNAMICS OF GLOBULAR CLUSTERS
eds. S. G. Djorgovski and G. Meylan
ISBN 0-937707-69-4

Vol. CS-51 OBSERVATIONAL COSMOLOGY
eds. Guido Chincarini, Angela Iovino, Tommaso Maccacaro, and Dario Maccagni
ISBN 0-937707-70-8

Vol. CS-52 ASTRONOMICAL DATA ANALYSIS SOFTWARE AND SYSTEMS II - (ADASS II)
eds. R. J. Hanisch, R. J. V. Brissenden, and Jeannette Barnes
ISBN 0-937707-71-6

Vol. CS-53 BLUE STRAGGLERS
ed. Rex A. Saffer
ISBN 0-937707-72-4

PUBLISHED: 1994

Vol. CS-54* THE FIRST STROMLO SYMPOSIUM: THE PHYSICS OF ACTIVE GALAXIES
eds. Geoffrey V. Bicknell, Michael A. Dopita, and Peter J. Quinn
ISBN 0-937707-73-2

Vol. CS-55 OPTICAL ASTRONOMY FROM THE EARTH AND MOON
eds. Diane M. Pyper and Ronald J. Angione
ISBN 0-937707-74-0

Vol. CS-56 INTERACTING BINARY STARS
ed. Allen W. Shafter
ISBN 0-937707-75-9

ASP CONFERENCE SERIES VOLUMES

Published by the Astronomical Society of the Pacific

PUBLISHED: 1994 (* asterisk means OUT OF STOCK)

Vol. CS-57 STELLAR AND CIRCUMSTELLAR ASTROPHYSICS
eds. George Wallerstein and Alberto Noriega-Crespo
ISBN 0-937707-76-7

Vol. CS-58* THE FIRST SYMPOSIUM ON THE INFRARED CIRRUS
AND DIFFUSE INTERSTELLAR CLOUDS
eds. Roc M. Cutri and William B. Latter
ISBN 0-937707-77-5

Vol. CS-59 ASTRONOMY WITH MILLIMETER AND SUBMILLIMETER WAVE INTERFEROMETRY,
IAU Colloquium 140
eds. M. Ishiguro and Wm. J. Welch
ISBN 0-937707-78-3

Vol. CS-60 THE MK PROCESS AT 50 YEARS: A POWERFUL TOOL FOR ASTROPHYSICAL INSIGHT, A Workshop of the Vatican Observatory --Tucson, Arizona, USA
eds. C. J. Corbally, R. O. Gray, and R. F. Garrison
ISBN 0-937707-79-1

Vol. CS-61 ASTRONOMICAL DATA ANALYSIS SOFTWARE AND SYSTEMS III - (ADASS III)
eds. Dennis R. Crabtree, R. J. Hanisch, and Jeannette Barnes
ISBN 0-937707-80-5

Vol. CS-62 THE NATURE AND EVOLUTIONARY STATUS OF HERBIG Ae/Be STARS
eds. Pik Sin Thé, Mario R. Pérez, and Ed P. J. van den Heuvel
ISBN 0-9837707-81-3

Vol. CS-63 SEVENTY-FIVE YEARS OF HIRAYAMA ASTEROID FAMILIES:
THE ROLE OF COLLISIONS IN THE SOLAR SYSTEM HISTORY
eds. Yoshihide Kozai, Richard P. Binzel, and Tomohiro Hirayama
ISBN 0-937707-82-1

Vol. CS-64* COOL STARS, STELLAR SYSTEMS, AND THE SUN:
Eighth Cambridge Workshop
ed. Jean-Pierre Caillault
ISBN 0-937707-83-X

Vol. CS-65* CLOUDS, CORES, AND LOW MASS STARS:
The Fourth Haystack Observatory Conference
eds. Dan P. Clemens and Richard Barvainis
ISBN 0-937707-84-8

Vol. CS-66* PHYSICS OF THE GASEOUS AND STELLAR DISKS OF THE GALAXY
ed. Ivan R. King
ISBN 0-937707-85-6

Vol. CS-67 UNVEILING LARGE-SCALE STRUCTURES BEHIND THE MILKY WAY
eds. C. Balkowski and R. C. Kraan-Korteweg
ISBN 0-937707-86-4

Vol. CS-68* SOLAR ACTIVE REGION EVOLUTION:
COMPARING MODELS WITH OBSERVATIONS
eds. K. S. Balasubramaniam and George W. Simon
ISBN 0-937707-87-2

Vol. CS-69 REVERBERATION MAPPING OF THE BROAD-LINE REGION
IN ACTIVE GALACTIC NUCLEI
eds. P. M. Gondhalekar, K. Horne, and B. M. Peterson
ISBN 0-937707-88-0

Vol. CS-70* GROUPS OF GALAXIES
eds. Otto-G. Richter and Kirk Borne
ISBN 0-937707-89-9

ASP CONFERENCE SERIES VOLUMES

Published by the Astronomical Society of the Pacific

PUBLISHED: 1995 (* asterisk means OUT OF STOCK)

Vol. CS-71	TRIDIMENSIONAL OPTICAL SPECTROSCOPIC METHODS IN ASTROPHYSICS, IAU Colloquium 149 eds. Georges Comte and Michel Marcelin ISBN 0-937707-90-2
Vol. CS-72	MILLISECOND PULSARS: A DECADE OF SURPRISE eds. A. S Fruchter, M. Tavani, and D. C. Backer ISBN 0-937707-91-0
Vol. CS-73	AIRBORNE ASTRONOMY SYMPOSIUM ON THE GALACTIC ECOSYSTEM: FROM GAS TO STARS TO DUST eds. Michael R. Haas, Jacqueline A. Davidson, and Edwin F. Erickson ISBN 0-937707-92-9
Vol. CS-74	PROGRESS IN THE SEARCH FOR EXTRATERRESTRIAL LIFE: 1993 Bioastronomy Symposium ed. G. Seth Shostak ISBN 0-937707-93-7
Vol. CS-75	MULTI-FEED SYSTEMS FOR RADIO TELESCOPES eds. Darrel T. Emerson and John M. Payne ISBN 0-937707-94-5
Vol. CS-76	GONG '94: HELIO- AND ASTERO-SEISMOLOGY FROM THE EARTH AND SPACE eds. Roger K. Ulrich, Edward J. Rhodes, Jr., and Werner Däppen ISBN 0-937707-95-3
Vol. CS-77	ASTRONOMICAL DATA ANALYSIS SOFTWARE AND SYSTEMS IV - (ADASS IV) eds. R. A. Shaw, H. E. Payne, and J. J. E. Hayes ISBN 0-937707-96-1
Vol. CS-78	ASTROPHYSICAL APPLICATIONS OF POWERFUL NEW DATABASES: Joint Discussion No. 16 of the 22nd General Assembly of the IAU eds. S. J. Adelman and W. L. Wiese ISBN 0-937707-97-X
Vol. CS-79*	ROBOTIC TELESCOPES: CURRENT CAPABILITIES, PRESENT DEVELOPMENTS, AND FUTURE PROSPECTS FOR AUTOMATED ASTRONOMY eds. Gregory W. Henry and Joel A. Eaton ISBN 0-937707-98-8
Vol. CS-80*	THE PHYSICS OF THE INTERSTELLAR MEDIUM AND INTERGALACTIC MEDIUM eds. A. Ferrara, C. F. McKee, C. Heiles, and P. R. Shapiro ISBN 0-937707-99-6
Vol. CS-81	LABORATORY AND ASTRONOMICAL HIGH RESOLUTION SPECTRA eds. A. J. Sauval, R. Blomme, and N. Grevesse ISBN 1-886733-01-5
Vol. CS-82*	VERY LONG BASELINE INTERFEROMETRY AND THE VLBA eds. J. A. Zensus, P. J. Diamond, and P. J. Napier ISBN 1-886733-02-3
Vol. CS-83*	ASTROPHYSICAL APPLICATIONS OF STELLAR PULSATION, IAU Colloquium 155 eds. R. S. Stobie and P. A. Whitelock ISBN 1-886733-03-1
ATLAS	INFRARED ATLAS OF THE ARCTURUS SPECTRUM, 0.9 - 5.3 μm eds. Kenneth Hinkle, Lloyd Wallace, and William Livingston ISBN: 1-886733-04-X

ASP CONFERENCE SERIES VOLUMES

Published by the Astronomical Society of the Pacific

PUBLISHED: 1995 (* asterisk means OUT OF STOCK)

Vol. CS-84 THE FUTURE UTILIZATION OF SCHMIDT TELESCOPES, IAU Colloquium 148
eds. Jessica Chapman, Russell Cannon, Sandra Harrison, and Bambang Hidayat
ISBN 1-886733-05-8

Vol. CS-85* CAPE WORKSHOP ON MAGNETIC CATACLYSMIC VARIABLES
eds. D. A. H. Buckley and B. Warner
ISBN 1-886733-06-6

Vol. CS-86 FRESH VIEWS OF ELLIPTICAL GALAXIES
eds. Alberto Buzzoni, Alvio Renzini, and Alfonso Serrano
ISBN 1-886733-07-4

PUBLISHED: 1996

Vol. CS-87 NEW OBSERVING MODES FOR THE NEXT CENTURY
eds. Todd Boroson, John Davies, and Ian Robson
ISBN 1-886733-08-2

Vol. CS-88* CLUSTERS, LENSING, AND THE FUTURE OF THE UNIVERSE
eds. Virginia Trimble and Andreas Reisenegger
ISBN 1-886733-09-0

Vol. CS-89 ASTRONOMY EDUCATION: CURRENT DEVELOPMENTS, FUTURE COORDINATION
ed. John R. Percy
ISBN 1-886733-10-4

Vol. CS-90 THE ORIGINS, EVOLUTION, AND DESTINIES OF BINARY STARS IN CLUSTERS
eds. E. F. Milone and J. -C. Mermilliod
ISBN 1-886733-11-2

Vol. CS-91 BARRED GALAXIES, IAU Colloquium 157
eds. R. Buta, D. A. Crocker, and B. G. Elmegreen
ISBN 1-886733-12-0

Vol. CS-92* FORMATION OF THE GALACTIC HALO .... INSIDE AND OUT
eds. Heather L. Morrison and Ata Sarajedini
ISBN 1-886733-13-9

Vol. CS-93 RADIO EMISSION FROM THE STARS AND THE SUN
eds. A. R. Taylor and J. M. Paredes
ISBN 1-886733-14-7

Vol. CS-94 MAPPING, MEASURING, AND MODELING THE UNIVERSE
eds. Peter Coles, Vicent J. Martinez, and Maria-Jesus Pons-Borderia
ISBN 1-886733-15-5

Vol. CS-95 SOLAR DRIVERS OF INTERPLANETARY AND TERRESTRIAL DISTURBANCES: Proceedings of 16th International Workshop National Solar Observatory/Sacramento Peak
eds. K. S. Balasubramaniam, Stephen L. Keil, and Raymond N. Smartt
ISBN 1-886733-16-3

Vol. CS-96 HYDROGEN-DEFICIENT STARS
eds. C. S. Jeffery and U. Heber
ISBN 1-886733-17-1

Vol. CS-97 POLARIMETRY OF THE INTERSTELLAR MEDIUM
eds. W. G. Roberge and D. C. B. Whittet
ISBN 1-886733-18-X

ASP CONFERENCE SERIES VOLUMES

Published by the Astronomical Society of the Pacific

PUBLISHED: 1996 (* asterisk means OUT OF STOCK)

Vol. CS-98 FROM STARS TO GALAXIES: THE IMPACT OF STELLAR PHYSICS ON GALAXY EVOLUTION
eds. Claus Leitherer, Uta Fritze-von Alvensleben, and John Huchra
ISBN 1-886733-19-8

Vol. CS-99 COSMIC ABUNDANCES:
Proceedings of the 6th Annual October Astrophysics Conference
eds. Stephen S. Holt and George Sonneborn
ISBN 1-886733-20-1

Vol. CS-100 ENERGY TRANSPORT IN RADIO GALAXIES AND QUASARS
eds. P. E. Hardee, A. H. Bridle, and J. A. Zensus
ISBN 1-886733-21-X

Vol. CS-101 ASTRONOMICAL DATA ANALYSIS SOFTWARE AND SYSTEMS V – (ADASS V)
eds. George H. Jacoby and Jeannette Barnes
ISBN 1080-7926

Vol. CS-102 THE GALACTIC CENTER, 4th ESO/CTIO Workshop
ed. Roland Gredel
ISBN 1-886733-22-8

Vol. CS-103 THE PHYSICS OF LINERS IN VIEW OF RECENT OBSERVATIONS
eds. M. Eracleous, A. Koratkar, C. Leitherer, and L. Ho
ISBN 1-886733-23-6

Vol. CS-104 PHYSICS, CHEMISTRY, AND DYNAMICS OF INTERPLANETARY DUST, IAU Colloquium 150
eds. Bo Å. S. Gustafson and Martha S. Hanner
ISBN 1-886733-24-4

Vol. CS-105 PULSARS: PROBLEMS AND PROGRESS, IAU Colloquium 160
ed. S. Johnston, M. A. Walker, and M. Bailes
ISBN 1-886733-25-2

Vol. CS-106 THE MINNESOTA LECTURES ON EXTRAGALACTIC NEUTRAL HYDROGEN
ed. Evan D. Skillman
ISBN 1-886733-26-0

Vol. CS-107 COMPLETING THE INVENTORY OF THE SOLAR SYSTEM:
A Symposium held in conjunction with the 106th Annual Meeting of the ASP
eds. Terrence W. Rettig and Joseph M. Hahn
ISBN 1-886733-27-9

Vol. CS-108 M.A.S.S. -- MODEL ATMOSPHERES AND SPECTRUM SYNTHESIS:
5th Vienna - Workshop
eds. Saul J. Adelman, Friedrich Kupka, and Werner W. Weiss
ISBN 1-886733-28-7

Vol. CS-109 COOL STARS, STELLAR SYSTEMS, AND THE SUN: Ninth Cambridge Workshop
eds. Roberto Pallavicini and Andrea K. Dupree
ISBN 1-886733-29-5

Vol. CS-110 BLAZAR CONTINUUM VARIABILITY
eds. H. R. Miller, J. R. Webb, and J. C. Noble
ISBN 1-886733-30-9

Vol. CS-111 MAGNETIC RECONNECTION IN THE SOLAR ATMOSPHERE:
Proceedings of a Yohkoh Conference
eds. R. D. Bentley and J. T. Mariska
ISBN 1-886733-31-7

ASP CONFERENCE SERIES VOLUMES

Published by the Astronomical Society of the Pacific

PUBLISHED: 1996 (* asterisk means OUT OF STOCK)

Vol. CS-112 THE HISTORY OF THE MILKY WAY AND ITS SATELLITE SYSTEM
eds. Andreas Burkert, Dieter H. Hartmann, and Steven R. Majewski
ISBN 1-886733-32-5

PUBLISHED: 1997

Vol. CS-113 EMISSION LINES IN ACTIVE GALAXIES: NEW METHODS AND TECHNIQUES, IAU Colloquium 159
eds. B. M. Peterson, F.-Z. Cheng, and A. S. Wilson
ISBN 1-886733-33-3

Vol. CS-114 YOUNG GALAXIES AND QSO ABSORPTION-LINE SYSTEMS
eds. Sueli M. Viegas, Ruth Gruenwald, and Reinaldo R. de Carvalho
ISBN 1-886733-34-1

Vol. CS-115 GALACTIC CLUSTER COOLING FLOWS
ed. Noam Soker
ISBN 1-886733-35-X

Vol. CS-116 THE SECOND STROMLO SYMPOSIUM: THE NATURE OF ELLIPTICAL GALAXIES
eds. M. Arnaboldi, G. S. Da Costa, and P. Saha
ISBN 1-886733-36-8

Vol. CS-117 DARK AND VISIBLE MATTER IN GALAXIES
eds. Massimo Persic and Paolo Salucci
ISBN-1-886733-37-6

Vol. CS-118 FIRST ADVANCES IN SOLAR PHYSICS EUROCONFERENCE: ADVANCES IN THE PHYSICS OF SUNSPOTS
eds. B. Schmieder. J. C. del Toro Iniesta, and M. Vázquez
ISBN 1-886733-38-4

Vol. CS-119 PLANETS BEYOND THE SOLAR SYSTEM AND THE NEXT GENERATION OF SPACE MISSIONS
ed. David R. Soderblom
ISBN 1-886733-39-2

Vol. CS-120 LUMINOUS BLUE VARIABLES: MASSIVE STARS IN TRANSITION
eds. Antonella Nota and Henny J. G. L. M. Lamers
ISBN 1-886733-40-6

Vol. CS-121 ACCRETION PHENOMENA AND RELATED OUTFLOWS, IAU Colloquium 163
eds. D. T. Wickramasinghe, G. V. Bicknell, and L. Ferrario
ISBN 1-886733-41-4

Vol. CS-122 FROM STARDUST TO PLANETESIMALS:
Symposium held as part of the 108th Annual Meeting of the ASP
eds. Yvonne J. Pendleton and A. G. G. M. Tielens
ISBN 1-886733-42-2

Vol. CS-123 THE 12th 'KINGSTON MEETING': COMPUTATIONAL ASTROPHYSICS
eds. David A. Clarke and Michael J. West
ISBN 1-886733-43-0

Vol. CS-124 DIFFUSE INFRARED RADIATION AND THE IRTS
eds. Haruyuki Okuda, Toshio Matsumoto, and Thomas Roellig
ISBN 1-886733-44-9

Vol. CS-125 ASTRONOMICAL DATA ANALYSIS SOFTWARE AND SYSTEMS VI
eds. Gareth Hunt and H. E. Payne
ISBN 1-886733-45-7

ASP CONFERENCE SERIES VOLUMES

Published by the Astronomical Society of the Pacific

PUBLISHED: 1997 (* asterisk means OUT OF STOCK)

Vol. CS-126 FROM QUANTUM FLUCTUATIONS TO COSMOLOGICAL STRUCTURES
eds. David Valls-Gabaud, Martin A. Hendry, Paolo Molaro, and Khalil Chamcham
ISBN 1-886733-46-5

Vol. CS-127 PROPER MOTIONS AND GALACTIC ASTRONOMY
ed. Roberta M. Humphreys
ISBN 1-886733-47-3

Vol. CS-128 MASS EJECTION FROM AGN (Active Galactic Nuclei)
eds. N. Arav, I. Shlosman, and R. J. Weymann
ISBN 1-886733-48-1

Vol. CS-129 THE GEORGE GAMOW SYMPOSIUM
eds. E. Harper, W. C. Parke, and G. D. Anderson
ISBN 1-886733-49-X

Vol. CS-130 THE THIRD PACIFIC RIM CONFERENCE ON RECENT DEVELOPMENT ON BINARY STAR RESEARCH
eds. Kam-Ching Leung
ISBN 1-886733-50-3

PUBLISHED: 1998

Vol. CS-131 BOULDER-MUNICH II: PROPERTIES OF HOT, LUMINOUS STARS
ed. Ian D. Howarth
ISBN 1-886733-51-1

Vol. CS-132 STAR FORMATION WITH THE INFRARED SPACE OBSERVATORY (ISO)
eds. João L. Yun and René Liseau
ISBN 1-886733-52-X

Vol. CS-133 SCIENCE WITH THE NGST (Next Generation Space Telescope)
eds. Eric P. Smith and Anuradha Koratkar
ISBN 1-886733-53-8

Vol. CS-134 BROWN DWARFS AND EXTRASOLAR PLANETS
eds. Rafael Rebolo, Eduardo L. Martin, and Maria Rosa Zapatero Osorio
ISBN 1-886733-54-6

Vol. CS-135 A HALF CENTURY OF STELLAR PULSATION INTERPRETATIONS: A TRIBUTE TO ARTHUR N. COX
eds. P. A. Bradley and J. A. Guzik
ISBN 1-886733-55-4

Vol. CS-136 GALACTIC HALOS: A UC SANTA CRUZ WORKSHOP
ed. Dennis Zaritsky
ISBN 1-886733-56-2

Vol. CS-137 WILD STARS IN THE OLD WEST: PROCEEDINGS OF THE 13th NORTH AMERICAN WORKSHOP ON CATACLYSMIC VARIABLES AND RELATED OBJECTS
eds. S. Howell, E. Kuulkers, and C. Woodward
ISBN 1-886733-57-0

Vol. CS-138 1997 PACIFIC RIM CONFERENCE ON STELLAR ASTROPHYSICS
eds. Kwing Lam Chan, K. S. Cheng, and H. P. Singh
ISBN 1-886733-58-9

Vol. CS-139 PRESERVING THE ASTRONOMICAL WINDOWS: Proceedings of Joint Discussion No. 5 of the 23rd General Assembly of the IAU
eds. Syuzo Isobe and Tomohiro Hirayama
ISBN 1-886733-59-7

ASP CONFERENCE SERIES VOLUMES

Published by the Astronomical Society of the Pacific

PUBLISHED: 1998 (* asterisk means OUT OF STOCK)

Vol. CS-140 SYNOPTIC SOLAR PHYSICS --18th NSO/Sacramento Peak Summer Workshop
eds. K. S. Balasubramaniam, J. W. Harvey, and D. M. Rabin
ISBN 1-886733-60-0

Vol. CS-141 ASTROPHYSICS FROM ANTARCTICA:
A Symposium held as a part of the 109th Annual Meeting of the ASP
eds. Giles Novak and Randall H. Landsberg
ISBN 1-886733-61-9

Vol. CS-142 THE STELLAR INITIAL MASS FUNCTION: 38th Herstmonceux Conference
eds. Gerry Gilmore and Debbie Howell
ISBN 1-886733-62-7

Vol. CS-143* THE SCIENTIFIC IMPACT OF THE GODDARD HIGH RESOLUTION SPECTROGRAPH (GHRS)
eds. John C. Brandt, Thomas B. Ake III, and Carolyn Collins Petersen
ISBN 1-886733-63-5

Vol. CS-144 RADIO EMISSION FROM GALACTIC AND EXTRAGALACTIC COMPACT SOURCES, IAU Colloquium 164
eds. J. Anton Zensus, G. B. Taylor, and J. M. Wrobel
ISBN 1-886733-64-3

Vol. CS-145 ASTRONOMICAL DATA ANALYSIS SOFTWARE AND SYSTEMS VII – (ADASS VII)
eds. Rudolf Albrecht, Richard N. Hook, and Howard A. Bushouse
ISBN 1-886733-65-1

Vol. CS-146 THE YOUNG UNIVERSE GALAXY FORMATION
AND EVOLUTION AT INTERMEDIATE AND HIGH REDSHIFT
eds. S. D'Odorico, A. Fontana, and E. Giallongo
ISBN 1-886733-66-X

Vol. CS-147 ABUNDANCE PROFILES: DIAGNOSTIC TOOLS FOR GALAXY HISTORY
eds. Daniel Friedli, Mike Edmunds, Carmelle Robert, and Laurent Drissen
ISBN 1-886733-67-8

Vol. CS-148 ORIGINS
eds. Charles E. Woodward, J. Michael Shull, and Harley A. Thronson, Jr.
ISBN 1-886733-68-6

Vol. CS-149 SOLAR SYSTEM FORMATION AND EVOLUTION
eds. D. Lazzaro, R. Vieira Martins, S. Ferraz-Mello, J. Fernández, and C. Beaugé
ISBN 1-886733-69-4

Vol. CS-150 NEW PERSPECTIVES ON SOLAR PROMINENCES, IAU Colloquium 167
eds. David Webb, David Rust, and Brigitte Schmieder
ISBN 1-886733-70-8

Vol. CS-151 COSMIC MICROWAVE BACKGROUND
AND LARGE SCALE STRUCTURES OF THE UNIVERSE
eds. Yong-Ik Byun and Kin-Wang Ng
ISBN 1-886733-71-6

Vol. CS-152 FIBER OPTICS IN ASTRONOMY III
eds. S. Arribas, E. Mediavilla, and F. Watson
ISBN 1-886733-72-4

Vol. CS-153 LIBRARY AND INFORMATION SERVICES IN ASTRONOMY III -- (LISA III)
eds. Uta Grothkopf, Heinz Andernach, Sarah Stevens-Rayburn,
and Monique Gomez
ISBN 1-886733-73-2

ASP CONFERENCE SERIES VOLUMES

Published by the Astronomical Society of the Pacific

PUBLISHED: 1998 (* asterisk means OUT OF STOCK)

Vol. CS-154 COOL STARS, STELLAR SYSTEMS AND THE SUN: Tenth Cambridge Workshop
eds. Robert A. Donahue and Jay A. Bookbinder
ISBN 1-886733-74-0

Vol. CS-155 SECOND ADVANCES IN SOLAR PHYSICS EUROCONFERENCE: THREE-DIMENSIONAL STRUCTURE OF SOLAR ACTIVE REGIONS
eds. Costas E. Alissandrakis and Brigitte Schmieder
ISBN 1-886733-75-9

PUBLISHED: 1999

Vol. CS-156 HIGHLY REDSHIFTED RADIO LINES
eds. C. L. Carilli, S. J. E. Radford, K. M. Menten, and G. I. Langston
ISBN 1-886733-76-7

Vol. CS-157 ANNAPOLIS WORKSHOP ON MAGNETIC CATACLYSMIC VARIABLES
eds. Coel Hellier and Koji Mukai
ISBN 1-886733-77-5

Vol. CS-158 SOLAR AND STELLAR ACTIVITY: SIMILARITIES AND DIFFERENCES
eds. C. J. Butler and J. G. Doyle
ISBN 1-886733-78-3

Vol. CS-159 BL LAC PHENOMENON
eds. Leo O. Takalo and Aimo Sillanpää
ISBN 1-886733-79-1

Vol. CS-160 ASTROPHYSICAL DISCS: An EC Summer School
eds. J. A. Sellwood and Jeremy Goodman
ISBN 1-886733-80-5

Vol. CS-161 HIGH ENERGY PROCESSES IN ACCRETING BLACK HOLES
eds. Juri Poutanen and Roland Svensson
ISBN 1-886733-81-3

Vol. CS-162 QUASARS AND COSMOLOGY
eds. Gary Ferland and Jack Baldwin
ISBN 1-886733-83-X

Vol. CS-163 STAR FORMATION IN EARLY-TYPE GALAXIES
eds. Jordi Cepa and Patricia Carral
ISBN 1-886733-84-8

Vol. CS-164 ULTRAVIOLET–OPTICAL SPACE ASTRONOMY BEYOND HST
eds. Jon A. Morse, J. Michael Shull, and Anne L. Kinney
ISBN 1-886733-85-6

Vol. CS-165 THE THIRD STROMLO SYMPOSIUM: THE GALACTIC HALO
eds. Brad K. Gibson, Tim S. Axelrod, and Mary E. Putman
ISBN 1-886733-86-4

Vol. CS-166 STROMLO WORKSHOP ON HIGH-VELOCITY CLOUDS
eds. Brad K. Gibson and Mary E. Putman
ISBN 1-886733-87-2

Vol. CS-167 HARMONIZING COSMIC DISTANCE SCALES IN A POST-HIPPARCOS ERA
eds. Daniel Egret and André Heck
ISBN 1-886733-88-0

Vol. CS-168 NEW PERSPECTIVES ON THE INTERSTELLAR MEDIUM
eds. A. R. Taylor, T. L. Landecker, and G. Joncas
ISBN 1-886733-89-9

ASP CONFERENCE SERIES VOLUMES

Published by the Astronomical Society of the Pacific

PUBLISHED: 1999 (* asterisk means OUT OF STOCK)

Vol. CS-169 — 11th EUROPEAN WORKSHOP ON WHITE DWARFS
eds. J.-E. Solheim and E. G. Meištas
ISBN 1-886733-91-0

Vol. CS-170 — THE LOW SURFACE BRIGHTNESS UNIVERSE, IAU Colloquium 171
eds. J. I. Davies, C. Impey, and S. Phillipps
ISBN 1-886733-92-9

Vol. CS-171 — LiBeB, COSMIC RAYS, AND RELATED X- AND GAMMA-RAYS
eds. Reuven Ramaty, Elisabeth Vangioni-Flam, Michel Cassé, and Keith Olive
ISBN 1-886733-93-7

Vol. CS-172 — ASTRONOMICAL DATA ANALYSIS SOFTWARE AND SYSTEMS VIII
eds. David M. Mehringer, Raymond L. Plante, and Douglas A. Roberts
ISBN 1-886733-94-5

Vol. CS-173 — THEORY AND TESTS OF CONVECTION IN STELLAR STRUCTURE: First Granada Workshop
ed. Álvaro Giménez, Edward F. Guinan, and Benjamín Montesinos
ISBN 1-886733-95-3

Vol. CS-174 — CATCHING THE PERFECT WAVE: ADAPTIVE OPTICS AND INTERFEROMETRY IN THE 21st CENTURY,
A Symposium held as a part of the 110th Annual Meeting of the ASP
eds. Sergio R. Restaino, William Junor, and Nebojsa Duric
ISBN 1-886733-96-1

Vol. CS-175 — STRUCTURE AND KINEMATICS OF QUASAR BROAD LINE REGIONS
eds. C. M. Gaskell, W. N. Brandt, M. Dietrich, D. Dultzin-Hacyan, and M. Eracleous
ISBN 1-886733-97-X

Vol. CS-176 — OBSERVATIONAL COSMOLOGY: THE DEVELOPMENT OF GALAXY SYSTEMS
eds. Giuliano Giuricin, Marino Mezzetti, and Paolo Salucci
ISBN 1-58381-000-5

Vol. CS-177 — ASTROPHYSICS WITH INFRARED SURVEYS: A Prelude to SIRTF
eds. Michael D. Bicay, Chas A. Beichman, Roc M. Cutri, and Barry F. Madore
ISBN 1-58381-001-3

Vol. CS-178 — STELLAR DYNAMOS: NONLINEARITY AND CHAOTIC FLOWS
eds. Manuel Núñez and Antonio Ferriz-Mas
ISBN 1-58381-002-1

Vol. CS-179 — ETA CARINAE AT THE MILLENNIUM
eds. Jon A. Morse, Roberta M. Humphreys, and Augusto Damineli
ISBN 1-58381-003-X

Vol. CS-180 — SYNTHESIS IMAGING IN RADIO ASTRONOMY II
eds. G. B. Taylor, C. L. Carilli, and R. A. Perley
ISBN 1-58381-005-6

Vol. CS-181 — MICROWAVE FOREGROUNDS
eds. Angelica de Oliveira-Costa and Max Tegmark
ISBN 1-58381-006-4

Vol. CS-182 — GALAXY DYNAMICS: A Rutgers Symposium
eds. David Merritt, J. A. Sellwood, and Monica Valluri
ISBN 1-58381-007-2

Vol. CS-183 — HIGH RESOLUTION SOLAR PHYSICS: THEORY, OBSERVATIONS, AND TECHNIQUES
eds. T. R. Rimmele, K. S. Balasubramaniam, and R. R. Radick
ISBN 1-58381-009-9

ASP CONFERENCE SERIES VOLUMES

Published by the Astronomical Society of the Pacific

PUBLISHED: 1999 (* asterisk means OUT OF STOCK)

Vol. CS-184 THIRD ADVANCES IN SOLAR PHYSICS EUROCONFERENCE: MAGNETIC FIELDS AND OSCILLATIONS
eds. B. Schmieder, A. Hofmann, and J. Staude
ISBN 1-58381-010-2

Vol. CS-185 PRECISE STELLAR RADIAL VELOCITIES, IAU Colloquium 170
eds. J. B. Hearnshaw and C. D. Scarfe
ISBN 1-58381-011-0

Vol. CS-186 THE CENTRAL PARSECS OF THE GALAXY
eds. Heino Falcke, Angela Cotera, Wolfgang J. Duschl, Fulvio Melia, and Marcia J. Rieke
ISBN 1-58381-012-9

Vol. CS-187 THE EVOLUTION OF GALAXIES ON COSMOLOGICAL TIMESCALES
eds. J. E. Beckman and T. J. Mahoney
ISBN 1-58381-013-7

Vol. CS-188 OPTICAL AND INFRARED SPECTROSCOPY OF CIRCUMSTELLAR MATTER
eds. Eike W. Guenther, Bringfried Stecklum, and Sylvio Klose
ISBN 1-58381-014-5

Vol. CS-189 CCD PRECISION PHOTOMETRY WORKSHOP
eds. Eric R. Craine, Roy A. Tucker, and Jeannette Barnes
ISBN 1-58381-015-3

Vol. CS-190 GAMMA-RAY BURSTS: THE FIRST THREE MINUTES
eds. Juri Poutanen and Roland Svensson
ISBN 1-58381-016-1

Vol. CS-191 PHOTOMETRIC REDSHIFTS AND HIGH REDSHIFT GALAXIES
eds. Ray J. Weymann, Lisa J. Storrie-Lombardi, Marcin Sawicki, and Robert J. Brunner
ISBN 1-58381-017-X

Vol. CS-192 SPECTROPHOTOMETRIC DATING OF STARS AND GALAXIES
ed. I. Hubeny, S. R. Heap, and R. H. Cornett
ISBN 1-58381-018-8

Vol. CS-193 THE HY-REDSHIFT UNIVERSE: GALAXY FORMATION AND EVOLUTION AT HIGH REDSHIFT
eds. Andrew J. Bunker and Wil J. M. van Breugel
ISBN 1-58381-019-6

Vol. CS-194 WORKING ON THE FRINGE: OPTICAL AND IR INTERFEROMETRY FROM GROUND AND SPACE
eds. Stephen Unwin and Robert Stachnik
ISBN 1-58381-020-X

PUBLISHED: 2000

Vol. CS-195 IMAGING THE UNIVERSE IN THREE DIMENSIONS: Astrophysics with Advanced Multi-Wavelength Imaging Devices
eds. W. van Breugel and J. Bland-Hawthorn
ISBN 1-58381-022-6

Vol. CS-196 THERMAL EMISSION SPECTROSCOPY AND ANALYSIS OF DUST, DISKS, AND REGOLITHS
eds. Michael L. Sitko, Ann L. Sprague, and David K. Lynch
ISBN: 1-58381-023-4

Vol. CS-197 XVth IAP MEETING DYNAMICS OF GALAXIES: FROM THE EARLY UNIVERSE TO THE PRESENT
eds. F. Combes, G. A. Mamon, and V. Charmandaris
ISBN: 1-58381-24-2

ASP CONFERENCE SERIES VOLUMES

Published by the Astronomical Society of the Pacific

PUBLISHED: 2000 (* asterisk means OUT OF STOCK)

Vol. CS-198	EUROCONFERENCE ON "STELLAR CLUSTERS AND ASSOCIATIONS: CONVECTION, ROTATION, AND DYNAMOS" eds. R. Pallavicini, G. Micela, and S. Sciortino ISBN: 1-58381-25-0
Vol. CS-199	ASYMMETRICAL PLANETARY NEBULAE II: FROM ORIGINS TO MICROSTRUCTURES eds. J. H. Kastner, N. Soker, and S. Rappaport ISBN: 1-58381-026-9
Vol. CS-200	CLUSTERING AT HIGH REDSHIFT eds. A. Mazure, O. Le Fèvre, and V. Le Brun ISBN: 1-58381-027-7
Vol. CS-201	COSMIC FLOWS 1999: TOWARDS AN UNDERSTANDING OF LARGE-SCALE STRUCTURES eds. Stéphane Courteau, Michael A. Strauss, and Jeffrey A. Willick ISBN: 1-58381-028-5
Vol. CS-202*	PULSAR ASTRONOMY – 2000 AND BEYOND, IAU Colloquium 177 eds. M. Kramer, N. Wex, and R. Wielebinski ISBN: 1-58381-029-3
Vol. CS-203	THE IMPACT OF LARGE-SCALE SURVEYS ON PULSATING STAR RESEARCH, IAU Colloquium 176 eds. L. Szabados and D. W. Kurtz ISBN: 1-58381-030-7
Vol. CS-204	THERMAL AND IONIZATION ASPECTS OF FLOWS FROM HOT STARS: OBSERVATIONS AND THEORY eds. Henny J. G. L. M. Lamers and Arved Sapar ISBN: 1-58381-031-5
Vol. CS-205	THE LAST TOTAL SOLAR ECLIPSE OF THE MILLENNIUM IN TURKEY eds. W. C. Livingston and A. Özgüç ISBN: 1-58381-032-3
Vol. CS-206	HIGH ENERGY SOLAR PHYSICS – *ANTICIPATING HESSI* eds. Reuven Ramaty and Natalie Mandzhavidze ISBN: 1-58381-033-1
Vol. CS-207	NGST SCIENCE AND TECHNOLOGY EXPOSITION eds. Eric P. Smith and Knox S. Long ISBN: 1-58381-036-6
ATLAS	VISIBLE AND NEAR INFRARED ATLAS OF THE ARCTURUS SPECTRUM 3727-9300 Å eds. Kenneth Hinkle, Lloyd Wallace, Jeff Valenti, and Dianne Harmer ISBN: 1-58381-037-4
Vol. CS-208	POLAR MOTION: HISTORICAL AND SCIENTIFIC PROBLEMS, IAU Colloquium 178 eds. Steven Dick, Dennis McCarthy, and Brian Luzum ISBN: 1-58381-039-0
Vol. CS-209	SMALL GALAXY GROUPS, IAU Colloquium 174 eds. Mauri J. Valtonen and Chris Flynn ISBN: 1-58381-040-4
Vol. CS-210	DELTA SCUTI AND RELATED STARS: Reference Handbook and Proceedings of the 6th Vienna Workshop in Astrophysics eds. Michel Breger and Michael Houston Montgomery ISBN: 1-58381-043-9

ASP CONFERENCE SERIES VOLUMES

Published by the Astronomical Society of the Pacific

PUBLISHED: 2000 (* asterisk means OUT OF STOCK)

Vol. CS-211 MASSIVE STELLAR CLUSTERS
eds. Ariane Lançon and Christian M. Boily
ISBN: 1-58381-042-0

Vol. CS-212 FROM GIANT PLANETS TO COOL STARS
eds. Caitlin A. Griffith and Mark S. Marley
ISBN: 1-58381-041-2

Vol. CS-213 BIOASTRONOMY '99: A NEW ERA IN BIOASTRONOMY
eds. Guillermo A. Lemarchand and Karen J. Meech
ISBN: 1-58381-044-7

Vol. CS-214 THE Be PHENOMENON IN EARLY-TYPE STARS, IAU Colloquium 175
eds. Myron A. Smith, Huib F. Henrichs and Juan Fabregat
ISBN: 1-58381-045-5

Vol. CS-215 COSMIC EVOLUTION AND GALAXY FORMATION:
STRUCTURE, INTERACTIONS AND FEEDBACK
The 3rd Guillermo Haro Astrophysics Conference
eds. José Franco, Elena Terlevich, Omar López-Cruz, and Itziar Aretxaga
ISBN: 1-58381-046-3

Vol. CS-216 ASTRONOMICAL DATA ANALYSIS SOFTWARE AND SYSTEMS IX
eds. Nadine Manset, Christian Veillet, and Dennis Crabtree
ISBN: 1-58381-047-1 ISSN: 1080-7926

Vol. CS-217 IMAGING AT RADIO THROUGH SUBMILLIMETER WAVELENGTHS
eds. Jeffrey G. Mangum and Simon J. E. Radford
ISBN: 1-58381-049-8

Vol. CS-218 MAPPING THE HIDDEN UNIVERSE: THE UNIVERSE BEHIND THE MILKY WAY
THE UNIVERSE IN HI
eds. Renée C. Kraan-Korteweg, Patricia A. Henning, and Heinz Andernach
ISBN: 1-58381-050-1

Vol. CS-219 DISKS, PLANETESIMALS, AND PLANETS
eds. F. Garzón, C. Eiroa, D. de Winter, and T. J. Mahoney
ISBN: 1-58381-051-X

Vol. CS-220 AMATEUR - PROFESSIONAL PARTNERSHIPS IN ASTRONOMY:
The 111th Annual Meeting of the ASP
eds. John R. Percy and Joseph B. Wilson
ISBN: 1-58381-052-8

Vol. CS-221 STARS, GAS AND DUST IN GALAXIES: EXPLORING THE LINKS
eds. Danielle Alloin, Knut Olsen, and Gaspar Galaz
ISBN: 1-58381-053-6

PUBLISHED: 2001

Vol. CS-222 THE PHYSICS OF GALAXY FORMATION
eds. M. Umemura and H. Susa
ISBN: 1-58381-054-4

Vol. CS-223 COOL STARS, STELLAR SYSTEMS AND THE SUN:
Eleventh Cambridge Workshop
eds. Ramón J. García López, Rafael Rebolo, and María Zapatero Osorio
ISBN: 1-58381-056-0

Vol. CS-224 PROBING THE PHYSICS OF ACTIVE GALACTIC NUCLEI
BY MULTIWAVELENGTH MONITORING
eds. Bradley M. Peterson, Ronald S. Polidan, and Richard W. Pogge
ISBN: 1-58381-055-2

ASP CONFERENCE SERIES VOLUMES

Published by the Astronomical Society of the Pacific

PUBLISHED: 2001 (* asterisk means OUT OF STOCK)

Vol. CS-225 VIRTUAL OBSERVATORIES OF THE FUTURE
eds. Robert J. Brunner, S. George Djorgovski, and Alex S. Szalay
ISBN: 1-58381-057-9

Vol. CS-226 12th EUROPEAN CONFERENCE ON WHITE DWARFS
eds. J. L. Provencal, H. L. Shipman, J. MacDonald, and S. Goodchild
ISBN: 1-58381-058-7

Vol. CS-227 BLAZAR DEMOGRAPHICS AND PHYSICS
eds. Paolo Padovani and C. Megan Urry
ISBN: 1-58381-059-5

Vol. CS-228 DYNAMICS OF STAR CLUSTERS AND THE MILKY WAY
eds. S. Deiters, B. Fuchs, A. Just, R. Spurzem, and R. Wielen
ISBN: 1-58381-060-9

Vol. CS-229 EVOLUTION OF BINARY AND MULTIPLE STAR SYSTEMS
A Meeting in Celebration of Peter Eggleton's 60th Birthday
eds. Ph. Podsiadlowski, S. Rappaport, A. R. King, F. D'Antona, and L. Burderi
IBSN: 1-58381-061-7

Vol. CS-230 GALAXY DISKS AND DISK GALAXIES
eds. Jose G. Funes, S. J. and Enrico Maria Corsini
ISBN: 1-58381-063-3

Vol. CS-231 TETONS 4: GALACTIC STRUCTURE, STARS, AND THE INTERSTELLAR MEDIUM
eds. Charles E. Woodward, Michael D. Bicay, and J. Michael Shull
ISBN: 1-58381-064-1

Vol. CS-232 THE NEW ERA OF WIDE FIELD ASTRONOMY
eds. Roger Clowes, Andrew Adamson, and Gordon Bromage
ISBN: 1-58381-065-X

Vol. CS-233 P CYGNI 2000: 400 YEARS OF PROGRESS
eds. Mart de Groot and Christiaan Sterken
ISBN: 1-58381-070-6

Vol. CS-234 X-RAY ASTRONOMY 2000
eds. R. Giacconi, S. Serio, and L. Stella
ISBN: 1-58381-071-4

Vol. CS-235 SCIENCE WITH THE ATACAMA LARGE MILLIMETER ARRAY (ALMA)
ed. Alwyn Wootten
ISBN: 1-58381-072-2

Vol. CS-236 ADVANCED SOLAR POLARIMETRY: THEORY, OBSERVATION, AND INSTRUMENTATION, The 20th Sacramento Peak Summer Workshop
ed. M. Sigwarth
ISBN: 1-58381-073-0

Vol. CS-237 GRAVITATIONAL LENSING: RECENT PROGRESS AND FUTURE GOALS
eds. Tereasa G. Brainerd and Christopher S. Kochanek
ISBN: 1-58381-074-9

Vol. CS-238 ASTRONOMICAL DATA ANALYSIS SOFTWARE AND SYSTEMS X
eds. F. R. Harnden, Jr., Francis A. Primini, and Harry E. Payne
ISBN: 1-58381-075-7

Vol. CS-239 MICROLENSING 2000: A NEW ERA OF MICROLENSING ASTROPHYSICS
ed. John Menzies and Penny D. Sackett
ISBN: 1-58381-076-5

ASP CONFERENCE SERIES VOLUMES

Published by the Astronomical Society of the Pacific

PUBLISHED: 2001 (* asterisk means OUT OF STOCK)

Vol. CS-240 GAS AND GALAXY EVOLUTION,
A Conference in Honor of the 20th Anniversary of the VLA
eds. J. E. Hibbard, M. P. Rupen, and J. H. van Gorkom
ISBN: 1-58381-077-3

Vol. CS-241 CS-241 THE 7TH TAIPEI ASTROPHYSICS WORKSHOP ON COSMIC RAYS IN THE UNIVERSE
ed. Chung-Ming Ko
ISBN: 1-58381-079-X

Vol. CS-242 ETA CARINAE AND OTHER MYSTERIOUS STARS: THE HIDDEN OPPORTUNITIES OF EMISSION SPECTROSCOPY
eds. Theodore R. Gull, Sveneric Johannson, and Kris Davidson
ISBN: 1-58381-080-3

Vol. CS-243 FROM DARKNESS TO LIGHT: ORIGIN AND EVOLUTION OF YOUNG STELLAR CLUSTERS
eds. Thierry Montmerle and Philippe André
ISBN: 1-58381-081-1

Vol. CS-244 YOUNG STARS NEAR EARTH: PROGRESS AND PROSPECTS
eds. Ray Jayawardhana and Thomas P. Greene
ISBN: 1-58381-082-X

Vol. CS-245 ASTROPHYSICAL AGES AND TIME SCALES
eds. Ted von Hippel, Chris Simpson, and Nadine Manset
ISBN: 1-58381-083-8

Vol. CS-246 SMALL TELESCOPE ASTRONOMY ON GLOBAL SCALES, IAU Colloquium 183
eds. Wen-Ping Chen, Claudia Lemme, and Bohdan Paczyński
ISBN: 1-58381-084-6

Vol. CS-247 SPECTROSCOPIC CHALLENGES OF PHOTOIONIZED PLASMAS
eds. Gary Ferland and Daniel Wolf Savin
ISBN: 1-58381-085-4

Vol. CS-248 MAGNETIC FIELDS ACROSS THE HERTZSPRUNG-RUSSELL DIAGRAM
eds. G. Mathys, S. K. Solanki, and D. T. Wickramasinghe
ISBN: 1-58381-088-9

Vol. CS-249 THE CENTRAL KILOPARSEC OF STARBURSTS AND AGN: THE LA PALMA CONNECTION
eds. J. H. Knapen, J. E. Beckman, I. Shlosman, and T. J. Mahoney
ISBN: 1-58381-089-7

Vol. CS-250 PARTICLES AND FIELDS IN RADIO GALAXIES CONFERENCE
eds. Robert A. Laing and Katherine M. Blundell
ISBN: 1-58381-090-0

Vol. CS-251 NEW CENTURY OF X-RAY ASTRONOMY
eds. H. Inoue and H. Kunieda
ISBN: 1-58381-091-9

Vol. CS-252 HISTORICAL DEVELOPMENT OF MODERN COSMOLOGY
eds. Vicent J. Martínez, Virginia Trimble, and María Jesús Pons-Bordería
ISBN: 1-58381-092-7

PUBLISHED: 2002

Vol. CS-253 CHEMICAL ENRICHMENT OF INTRACLUSTER AND INTERGALACTIC MEDIUM
eds. Roberto Fusco-Femiano and Francesca Matteucci
ISBN: 1-58381-093-5

ASP CONFERENCE SERIES VOLUMES

Published by the Astronomical Society of the Pacific

PUBLISHED: 2002 (* asterisk means OUT OF STOCK)

Vol. CS-254 EXTRAGALACTIC GAS AT LOW REDSHIFT
eds. John S. Mulchaey and John T. Stocke
ISBN: 1-58381-094-3

Vol. CS-255 MASS OUTFLOW IN ACTIVE GALACTIC NUCLEI: NEW PERSPECTIVES
eds. D. M. Crenshaw, S. B. Kraemer, and I. M. George
ISBN: 1-58381-095-1

Vol. CS-256 OBSERVATIONAL ASPECTS OF PULSATING B AND A STARS
eds. Christiaan Sterken and Donald W. Kurtz
ISBN: 1-58381-096-X

Vol. CS-257 AMiBA 2001: HIGH-Z CLUSTERS, MISSING BARYONS, AND CMB POLARIZATION
eds. Lin-Wen Chen, Chung-Pei Ma, Kin-Wang Ng, and Ue-Li Pen
ISBN: 1-58381-097-8

Vol. CS-258 ISSUES IN UNIFICATION OF ACTIVE GALACTIC NUCLEI
eds. Roberto Maiolino, Alessandro Marconi, and Neil Nagar
ISBN: 1-58381-098-6

Vol. CS-259 RADIAL AND NONRADIAL PULSATIONS AS PROBES OF STELLAR PHYSICS
eds. Conny Aerts, Timothy R. Bedding, and Jørgen Christensen-Dalsgaard
ISBN: 1-58381-099-4

Vol. CS-260 INTERACTING WINDS FROM MASSIVE STARS
eds. Anthony F. J. Moffat and Nicole St-Louis
ISBN: 1-58381-100-1

Vol. CS-261 THE PHYSICS OF CATACLYSMIC VARIABLES AND RELATED OBJECTS
eds. B. T. Gänsicke, K. Beuermann, and K. Reinsch
ISBN: 1-58381-101-X

Vol. CS-262 THE HIGH ENERGY UNIVERSE AT SHARP FOCUS: CHANDRA SCIENCE
The 113th Annual Meeting of the ASP
eds. Eric M. Schlegel and Saeqa Dil Vrtilek
ISBN: 1-58381-102-8

Vol. CS-263 STELLAR COLLISIONS, MERGERS AND THEIR CONSEQUENCES
ed. Michael M. Shara
ISBN: 1-58381-103-6

Vol. CS-264 CONTINUING THE CHALLENGE OF EUV ASTRONOMY: CURRENT ANALYSIS AND PROSPECTS FOR THE FUTURE
eds. Steve B. Howell, Jean Dupuis, Daniel Golombek, Frederick M. Walter, and Jennifer Cullison
ISBN: 1-58381-104-4

Vol. CS-265 ω CENTAURI, A UNIQUE WINDOW INTO ASTROPHYSICS
eds. Floor van Leeuwen, Joanne D. Hughes, and Giampaolo Piotto
ISBN: 1-58381-105-2

Vol. CS-266 ASTRONOMICAL SITE EVALUATION IN THE VISIBLE AND RADIO RANGE
eds. J. Vernin, Z. Benkhaldoun, and C. Muñoz-Tuñón
ISBN: 1-58381-106-0

Vol. CS-267 HOT STAR WORKSHOP III: THE EARLIEST STAGES OF MASSIVE STAR BIRTH
ed. Paul A. Crowther
ISBN: 1-58381-107-9

Vol. CS-268 TRACING COSMIC EVOLUTION WITH GALAXY CLUSTERS
eds. Stefano Borgani, Marino Mezzetti, and Riccardo Valdarnini
ISBN: 1-58381-108-7

ASP CONFERENCE SERIES VOLUMES

Published by the Astronomical Society of the Pacific

PUBLISHED: 2002 (* asterisk means OUT OF STOCK)

Vol. CS-269 THE EVOLVING SUN AND ITS INFLUENCE ON PLANETARY ENVIRONMENTS
eds. Benjamín Montesinos, Álvaro Giménez, and Edward F. Guinan
ISBN: 1-58381-109-5

Vol. CS-270 ASTRONOMICAL INSTRUMENTATION AND THE BIRTH AND GROWTH OF ASTROPHYSICS: A Symposium held in honor of Robert G. Tull
eds. Frank N. Bash and Christopher Sneden
ISBN: 1-58381-110-9

Vol. CS-271 NEUTRON STARS IN SUPERNOVA REMNANTS
eds. Patrick O. Slane and Bryan M. Gaensler
ISBN: 1-58381-111-7

Vol. CS-272 THE FUTURE OF SOLAR SYSTEM EXPLORATION, 2003-2013
Community Contributions to the NRC Solar System Exploration Decadal Survey
ed. Mark V. Sykes
ISBN: 1-58381-113-3

A LISTING OF IAU VOLUMES MAY BE FOUND ON THE NEXT PAGE

INTERNATIONAL ASTRONOMICAL UNION (IAU) VOLUMES

Published by the Astronomical Society of the Pacific

PUBLISHED: 1999 (* asterisk means OUT OF STOCK)

Vol. No. 190 NEW VIEWS OF THE MAGELLANIC CLOUDS
eds. You-Hua Chu, Nicholas B. Suntzeff, James E. Hesser, and David A. Bohlender
ISBN: 1-58381-021-8

Vol. No. 191 ASYMPTOTIC GIANT BRANCH STARS
eds. T. Le Bertre, A. Lèbre, and C. Waelkens
ISBN: 1-886733-90-2

Vol. No. 192 THE STELLAR CONTENT OF LOCAL GROUP GALAXIES
eds. Patricia Whitelock and Russell Cannon
ISBN: 1-886733-82-1

Vol. No. 193 WOLF-RAYET PHENOMENA IN MASSIVE STARS AND STARBURST GALAXIES
eds. Karel A. van der Hucht, Gloria Koenigsberger, and Philippe R. J. Eenens
ISBN: 1-58381-004-8

Vol. No. 194 ACTIVE GALACTIC NUCLEI AND RELATED PHENOMENA
eds. Yervant Terzian, Daniel Weedman, and Edward Khachikian
ISBN: 1-58381-008-0

PUBLISHED: 2000

Vol. XXIVA TRANSACTIONS OF THE INTERNATIONAL ASTRONOMICAL UNION REPORTS ON ASTRONOMY 1996-1999
ed. Johannes Andersen
ISBN: 1-58381-035-8

Vol. No. 195 HIGHLY ENERGETIC PHYSICAL PROCESSES AND MECHANISMS FOR EMISSION FROM ASTROPHYSICAL PLASMAS
eds. P. C. H. Martens, S. Tsuruta, and M. A. Weber
ISBN: 1-58381-038-2

Vol. No. 197 ASTROCHEMISTRY: FROM MOLECULAR CLOUDS TO PLANETARY SYSTEMS
eds. Y. C. Minh and E. F. van Dishoeck
ISBN: 1-58381-034-X

Vol. No. 198 THE LIGHT ELEMENTS AND THEIR EVOLUTION
eds. L. da Silva, M. Spite, and J. R. de Medeiros
ISBN: 1-58381-048-X

PUBLISHED: 2001

IAU SPS ASTRONOMY FOR DEVELOPING COUNTRIES
Special Session of the XXIV General Assembly of the IAU
ed. Alan H. Batten
ISBN: 1-58381-067-6

Vol. No. 196 PRESERVING THE ASTRONOMICAL SKY
eds. R. J. Cohen and W. T. Sullivan, III
ISBN: 1-58381-078-1

Vol. No. 200 THE FORMATION OF BINARY STARS
eds. Hans Zinnecker and Robert D. Mathieu
ISBN: 1-58381-068-4

Vol. No. 203 RECENT INSIGHTS INTO THE PHYSICS OF THE SUN AND HELIOSPHERE: HIGHLIGHTS FROM SOHO AND OTHER SPACE MISSIONS
eds. Pål Brekke, Bernhard Fleck, and Joseph B. Gurman
ISBN: 1-58381-069-2

Vol. No. 204 THE EXTRAGALACTIC INFRARED BACKGROUND AND ITS COSMOLOGICAL IMPLICATIONS
eds. Martin Harwit and Michael G. Hauser
ISBN: 1-58381-062-5

INTERNATIONAL ASTRONOMICAL UNION (IAU) VOLUMES

Published by the Astronomical Society of the Pacific

PUBLISHED: 2001 (* asterisk means OUT OF STOCK)

Vol. No. 205 — GALAXIES AND THEIR CONSTITUENTS AT THE HIGHEST ANGULAR RESOLUTIONS
eds. Richard T. Schilizzi, Stuart N. Vogel, Francesco Paresce, and Martin S. Elvis
ISBN: 1-58381-066-8

Vol. XXIVB — TRANSACTIONS OF THE INTERNATIONAL ASTRONOMICAL UNION REPORTS ON ASTRONOMY
ed. Hans Rickman
ISBN: 1-58381-087-0

PUBLISHED: 2002

Vol. No. 12 — HIGHLIGHTS OF ASTRONOMY
ed. Hans Rickman
ISBN: 1-58381-086-2

Vol. No. 206 — COSMIC MASERS: FROM PROTOSTARS TO BLACKHOLES
eds. Victor Migenes, Mark J. Reid, and Everton Ludke
ISBN: 1-58381-112-5

Ordering information is available at the beginning of the listing